Series in Optics and Photonics — Vol. 7

AN INTRODUCTION TO OPTOELECTRONIC SENSORS

Series in Optics and Photonics

Series Editor: S L Chin *(Laval University, Canada)*

Published

Vol. 1 Fundamentals of Laser Optoelectronics
by S. L. Chin

Vol. 2 Photonic Networks, Components and Applications
edited by J. Chrostowski and J. Terry

Vol. 3 Intense Laser Phenomena and Related Subjects
edited by I. Yu. Kiyan and M. Yu. Ivanov

Vol. 4 Radiation of Atoms in a Resonant Environment
by V. P. Bykov

Vol. 5 Optical Fiber Theory:
A Supplement to Applied Electromagnetism
by Pierre-A. Bélanger

Vol. 6 Multiphoton Processes
edtied by D. K. Evans and S. L. Chin

Series in Optics and Photonics — Vol. 7

AN INTRODUCTION TO OPTOELECTRONIC SENSORS

Editors

Giancarlo C. Righini
CNR, Italy

Antonella Tajani
CNR, Italy

Antonello Cutolo
University of Sannio, Italy

World Scientific

NEW JERSEY • LONDON • SINGAPORE • BEIJING • SHANGHAI • HONG KONG • TAIPEI • CHENNAI

Published by

World Scientific Publishing Co. Pte. Ltd.
5 Toh Tuck Link, Singapore 596224
USA office: 27 Warren Street, Suite 401-402, Hackensack, NJ 07601
UK office: 57 Shelton Street, Covent Garden, London WC2H 9HE

British Library Cataloguing-in-Publication Data
A catalogue record for this book is available from the British Library.

AN INTRODUCTION TO OPTOELECTRONIC SENSORS
Series in Optics and Photonics — Vol. 7

ISBN-13 978-981-283-412-6
ISBN-10 981-283-412-5

Printed by FuIsland Offset Printing (S) Pte Ltd, Singapore

To Marta and Nicoletta
(GCR)

To Maria Emilia, Maria Teresa, Maria Alessandra and my Parents
(AC)

To my Family
(AT)

According to Michael Lebby, President and CEO of OIDA, the Optoelectronics Industry Development Association (USA), the potential market for photonics sensors alone for 2009 exceeds 5 billion dollars, with fiber optic sensors taking the lead (Fig. 1).[a]

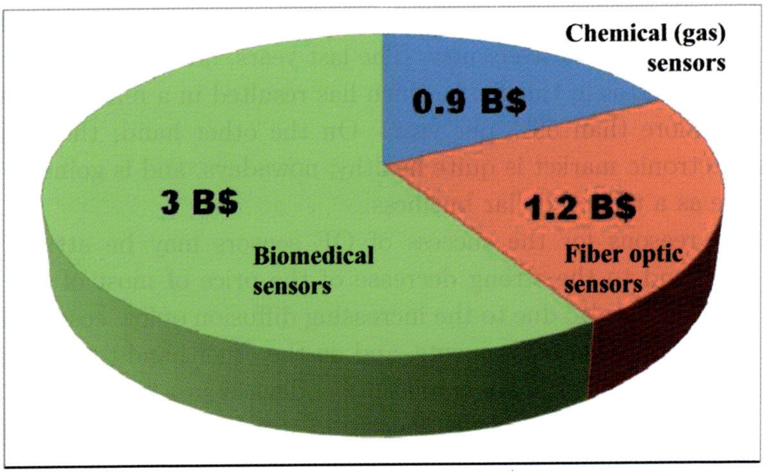

Fig. 1 Photonic Sensor Market Potential in 2009.

Far from being exhaustive, and according to its title, this book aims at providing a basic background in the field of optoelectronic sensors for graduate students or for people approaching this field. We hope, however, that the information provided will be of valuable interest to physicists, engineers, material scientists and systems designer who wish to obtain a broad review on the subject. Additionally, this book provides an excellent overview of the state of the art of the R&D in this field in Italy, boasting contributions from renowned academic and industrial experts.

For their relevance in a great variety of practical applications, particular attention has been paid here to the field of optical fiber sensors (OFS). Taking advantage of the integration with different materials and appropriate micro- and nano-structuring, OFSs have revealed an enormous potential for the design and production of innovative instrumentation.

[a]http://www.optofluidics.caltech.edu/publications/industry.html

In the first part of the book, attention is paid to the basic principles and technologies of the most relevant OE sensor classes, from the "classical" infrared detectors to the most innovative photonic crystal structures, without neglecting fashion THz sensing technologies. Examples of relevant applications are also provided.

The in-depth analysis of some application areas is the subject of the second part of the book, where OE sensors for structural health monitoring, environmental monitoring, medicine, materials and process control, are described and discussed.

We would like to thank all the authors for their excellent and timely contributions to this volume; particular thanks are due to the IFAC-CNR authors (Gualtiero, Ilaria, Massimo, Silvia, Simone, Stefano) who also helped with the revision of the text. We are also grateful to Lakshmi Narayanan and the WSPC staff for their patience and support.

Giancarlo C. Righini
Antonella Tajani
Antonello Cutolo

The support of CNR — Deparment of Materials and Devices is gratefully acknowledged.

CONTENTS

FIBER AND INTEGRATED OPTICS SENSORS: FUNDAMENTALS AND APPLICATIONS

Giancarlo C. Righini,[a,b,*] Anna Grazia Mignani,[b]
Ilaria Cacciari[b] and Massimo Brenci[b]

[a]*Consiglio Nazionale delle Ricerche, Dipartimento Materiali e Dispositivi*
Via dei Taurini, 19, 00185 Rome, Italy
[b]*Istituto di Fisica Applicata 'Nello Carrara', CNR*
Via Madonna del Piano, 10, 50019 Sesto Fiorentino (FI), Italy
[*]*E-mail: giancarlo.righini@cnr.it*

The chapter summarizes the fundamentals of light propagation in fiber and integrated optics and explains the basic working principles of optical sensors making use of these waveguides. Outstanding applications where these sensors have been used are also presented.

1. Introduction

Optical techniques have always been used for a large number of metrological and sensing applications. The conventional methods based on free-space interferometry and spectroscopy, for example, are outstanding examples of optics capabilities. This kind of free-space monitoring, however, is effective only for line of sight and suffers from undesired misalignments and external perturbations. Guided-wave sensing adds to intrinsic advantages of optical techniques the possibility of guiding the light beam in a confined and inaccessible medium, thus allowing more versatile and less perturbed measurements.

Fiber- and integrated- optics technologies were primarily developed for telecommunication applications. However, the advances in the development of high quality and competitive price optoelectronic components and fibers have largely contributed to the expansion of guided wave technology for sensing as well. The main reasons which make guided wave optics attractive for sensing can be summarized as follows:

- Non-electrical method of operation, which is explosion-proof and offers intrinsic immunity to radio frequency and, more generally, to any kind of electromagnetic interference;
- Small size/weight and great flexibility, that allow access to otherwise restricted areas;
- Capability of resisting to chemically aggressive and ionizing environments;
- Easy interface with optical data communication systems and secure data transmission.

The guided wave sensors that have been proposed to solve problems in industrial, automotive, avionic, military, geophysical, environmental and biomedical applications are countless. This chapter aims at providing some fundamentals in this field. Sensors are presented in a relatively simple and straightforward way to give a tour through the subject by minimizing theoretical explanations and showing outstanding examples of what guided wave technology is able to offer for sensing. References to the extensive literature in this area are provided, where the interested reader can find more details. An increasing number of textbooks is also available.[1-4]

2. Fiber and Integrated Optics: Fundamentals of Waveguiding

In accordance with the ray theory of light propagation, when light impinges at the interface between two transparent media, it is partially reflected and partially refracted. The Snell's law describes the refraction phenomena as (Fig. 1):

$$n_1 \sin \theta_1 = n_2 \sin \theta_2 \tag{1}$$

When $n_2 < n_1$, any ray impinging at the interface with an incident angle greater than θc is totally reflected inside the first medium.

$$\theta_c = \sin^{-1} \frac{n_2}{n_1} \tag{2}$$

An optical fiber consists of layered cylinders of glass or plastic, as shown in Fig. 2. Inner and outer cylinders, namely 'core' and 'cladding', have refractive indices n_1 and n_2, respectively. Any ray impinging at the core-cladding interface with an incident angle greater than θ_c is undergoing multiple reflections within the core, in which it results trapped and propagates.

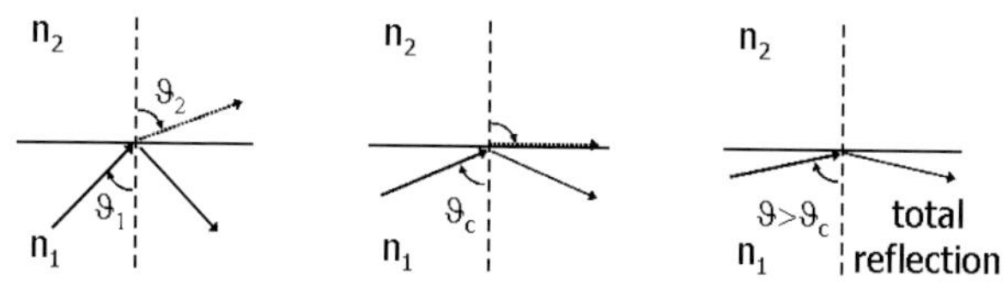

Figure 1. Reflection and refraction at the interface between two transparent media: the Snell law.

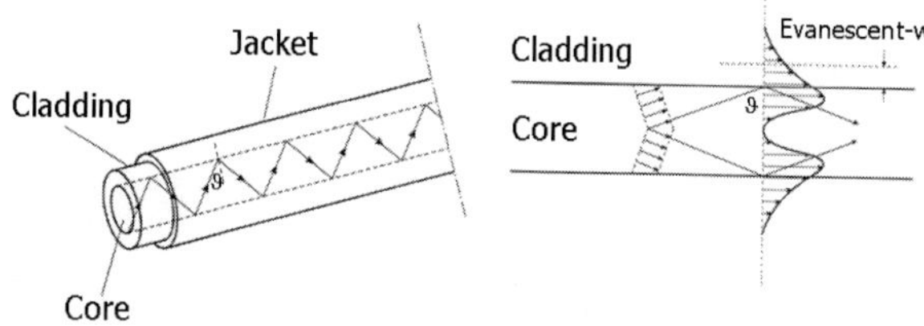

Figure 2. Ray propagation in the optical fiber.

An integrated optical waveguide, on the other hand, consists of a thin-film structure supported by a substrate. The simplest structure is shown in Fig. 3a, where the guiding layer (the core, with refractive index n_1) is deposited on a transparent substrate (having refractive index n_0) and is covered by another layer (the cladding, with refractive index n_2). If $n_2=n_0$ we have a symmetrical structure, analogous to an optical fiber; in fact, while the x-y cross sections of the fiber and the slab waveguide are different from each other, their x-z cross sections are identical and one can expect that their waveguiding properties are fundamentally the same ones. In most cases, however, the cladding is air ($n_2=1$), and we speak about a planar asymmetric waveguide. In this case light is confined only along the x direction, while the light energy can diffract in the y-z plane.

The confinement of light also along the y direction is obtained by a strip waveguide, as shown in Fig. 3b, where total reflections of light rays occur also at side walls. For both fiber and slab waveguide the dependence of the refractive index on the x coordinate *n(x)* is called the refractive index profile. In the simplest case, i.e. $n(x) = n_1 =$ constant, we refer to step-index waveguides; otherwise, we speak about gradient-index waveguides, and sophisticated profiles may be produced as well, either by a multi-layer deposition technique or by a diffusion process.

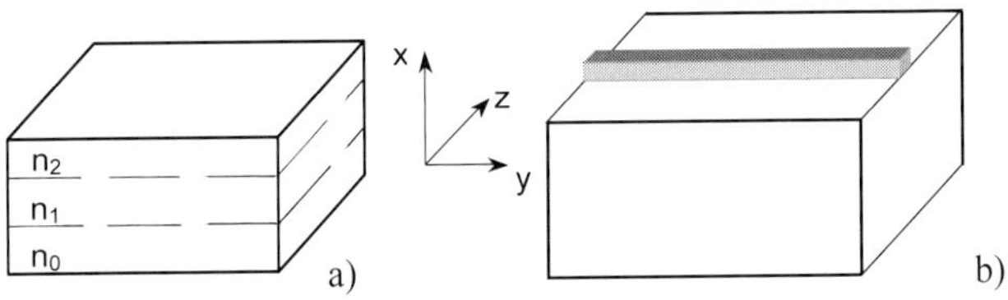

Figure 3. Slab waveguide (a), and strip waveguide (b).

Taking into account that light is an electromagnetic wave phenomenon, a more accurate description of light propagation within a waveguide is obtained by means of Maxwell's equations. When the geometric boundary conditions at media interfaces are introduced, only discrete solutions of the wave equations are permitted. This means that only discrete waves can propagate, namely 'modes', characterized by discrete amplitudes and discrete velocities.[5,6] Waveguides can be single-mode or multimode according to whether a single or a multiplicity of modes can propagate. Once the materials constituting the waveguide are set for a given wavelength, the number of supported modes depends on waveguide dimension, namely on the fiber core radius or the planar waveguide thickness.

A characteristic of a guided mode which is particularly important for sensing devices is its spatial amplitude distribution. Often, in fact, the interaction between the propagating mode and the quantity to be measured (the measurand) occurs through the evanescent field of the mode itself, namely its exponentially-decreasing tail.

3. Waveguide Sensors: Basic Working Principle

Although trapped within the dielectric medium of the optical waveguide, the radiation that propagates inside the waveguide can be perturbed by the external environment, and this perturbation can be used to draw useful information for sensing purposes. In fact, the interaction of the parameter of interest, that is the measurand, with the waveguide produces a modulation in the propagation constants of the guided light beam. That modulation represents the sensitive function of the measurand of interest.

As shown in Fig. 4, the basic elements constituting a guided wave sensor are: an optical source, an optical interface for source-to-waveguide light coupling, the waveguide itself where the measurand-induced light modulation occurs, a photodetector and the electronics for amplification, signal processing and data display.

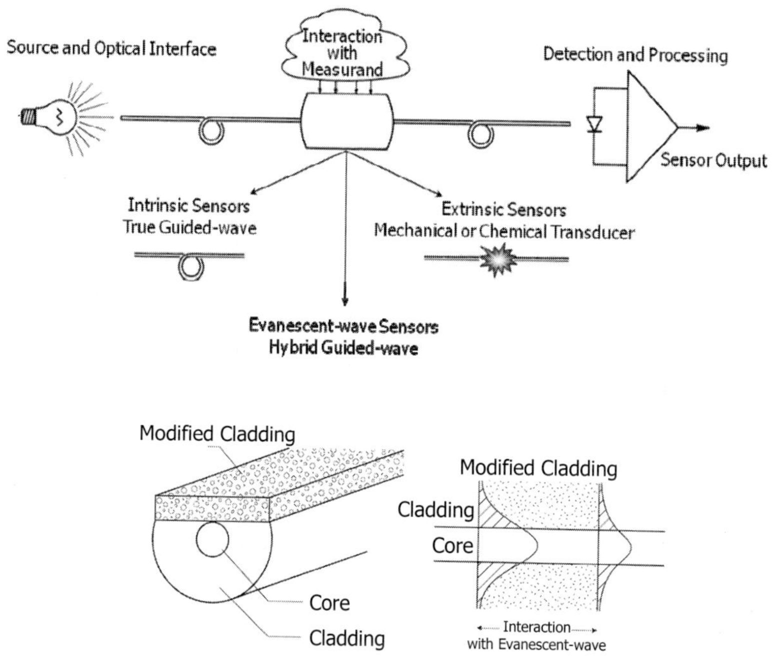

Figure 4. The waveguide sensor: general working principle.

In accordance to the optical parameter, which is modulated by the measurand, waveguide sensors can be divided into four basic categories:

- phase-modulated,

- polarisation-modulated,
- wavelength-modulated,
- intensity-modulated.

Waveguide sensors are further subdivided as intrinsic, extrinsic, or evanescent-wave sensors. Intrinsic sensors are true waveguide sensors in which the sensing element is the waveguide itself. Extrinsic sensors make use of an optical transducer coupled to waveguide, the optical constants of which are modulated by the measurand. Evanescent-wave sensors are hybrid intrinsic/extrinsic sensors, since measurand-induced modulation occurs in the waveguide itself, in most cases because of the presence of a measurand-sensitive cladding section.

The following section refers in particular to fiber optic sensors, but most considerations on the operation principle apply to integrated optical sensors as well.

4. Fiber Optic Sensors

Here, a brief overview of fiber optic sensors (FOSs) is given, according to their operational classification. An indication of their commercial availability is also provided.

4.1. Phase-Modulated Sensors

The action of measurand producing a variation of the waveguide length, ΔL, causes a phase shift of the guided lightwave, $\Delta\phi$, which is expressed as follows:

$$\Delta\phi = \frac{2\pi}{\lambda}\Delta L \tag{3}$$

where λ is the wavelength of light propagating in the waveguide. Being able to detect phase shifts as small as 10-7 rad, and assuming $\lambda \approx 1$ µm, a perturbation causing length differences as small as ΔL=10-8 µm can be detected. Consequently, phase-modulated sensors are capable to offer extremely high sensitivity.

The previous expression given for $\Delta\phi$ takes into account length variations only. However, it should be noted that the wavelength is dependent on the waveguide refractive index, n_w, which, in turn, can be perturbed by a measurand. A more exact expression for the phase shift is the following:

$$\Delta\phi = \frac{2\pi}{\lambda_{vacuum}}\left(n_w \Delta L + L \Delta n_w\right) \tag{4}$$

Phase shifts are usually measured by means of interferometric schemes.[7] Actually, phase-modulated waveguide sensors are classical interferometers, the legs of which are single-mode optical fibers or waveguide channels. Because many different measurands can perturb waveguide length and refractive index, cross sensitivity can occur. This is why most phase-modulated waveguide sensors have the sensing leg covered by means of an additional jacketing. The material of the jacketing is suitable to provide specific sensitivity to a certain measurand and also to amplify the length variation while desensitizing to refractive index, or vice versa.

The interferometric configurations most widely used by waveguide sensors are: the Mach-Zehnder, the Sagnac, and the Fabry-Perot, as shown in Fig. 5.

The Mach-Zehnder configuration is a two-beam interferometer. The light from a highly coherent laser is split by means of a beam splitter and injected into two optical fibers or waveguide channels which follow separate paths, one of which is exposed to the action of the measurand.

When they are recombined by means of another beam splitter, interference fringes appear. The phase of these fringes is proportional to measurand-induced optical path difference between co-propagating beams within the legs of the interferometer. The Mach-Zehnder scheme is the basic principle of most fiber optic hydrophones, also arranged in very dense arrays,[8-13] as well as current and magnetic field sensors.[14-16]

The Sagnac configuration is a two-beam interferometer the primary application of which is in rotation sensing.[17-19] The light beam is split by means of a beam splitter and injected as two counter-propagating beams into the same optical fiber arranged in a coil. When the coil is held stationary, clockwise and counter-clockwise beams return on the detector in phase after having travelled along the same path in opposite direction. If a rotation rate is applied to the fiber coil, the co-rotating beam reaches the starting point having travelled a longer path with respect to the counter-rotating beam, and the path length difference results in a phase difference. Several gyroscopes for military and civil applications are now commercially available.[20-23]

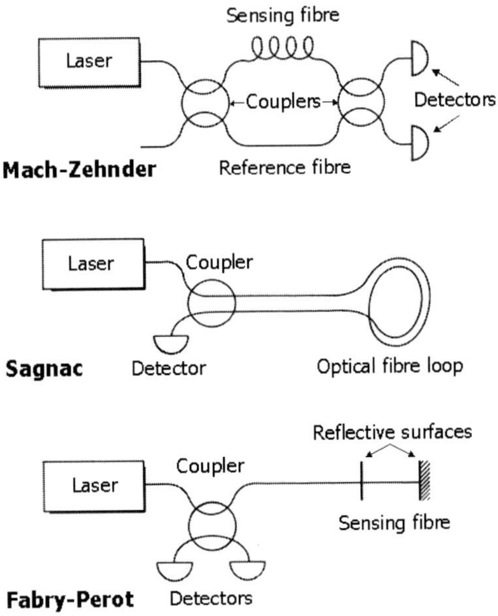

Figure 5. Interferometric arrangements for waveguide sensors.

The Fabry-Perot interferometer is a multiple-beam interferometer that does not make use of a reference fiber, since the interference results from multiple reflections of the light beam inside a single optical fiber.[24] The laser light is coupled to a single-mode optical fiber by means of a beam splitter. A resonant cavity is created by splicing a mirrored fiber section to the fiber end. The light beam is partially reflected and partially transmitted inside the cavity, which is exposed to the measurand action. Because of the reflectivity of the distal fiber section, the light beam impinging the cavity undergoes multiple reflections, the measurand acting on the light at every pass, thus magnifying the phase difference.[25,26] A measurand can act on the Fabry-Perot cavity at fiber end by changing its length and/or refractive index. This type of device is particularly suitable for temperature and pressure sensing, and these sensors are also commercially available.[27]

In addition to intrinsic Fabry-Perot sensors, also extrinsic sensors have been implemented, which make use of an etalon or a MEMS as resonant cavity at fiber end. Also these sensors are commercially available for pressure, strain, acceleration, temperature and vibration monitoring.[28-30]

The previously discussed interferometers do not produce absolute data, unless extra complexity is added to the sensor system. The generation of fringes is dependent on the two beams being able to interfere at the detector. This requires that the beams have the same polarization, the same wavelength and a path length difference less than the coherence length of the source.

Unfortunately, all real sources possess finite bandwidth and size. In addition, single-mode optical fibers are actually two-mode fibers, having the two modes different polarization states. All these factors affect fringe visibility thus impairing interferometer performance. To overcome the mentioned problems, a smart interrogation technique was setup, namely white-light interferometry.[31,32] A low-coherence optical source is used to illuminate two cascaded interferometers, one of which responds to the measurand, while the other is a reference interferometer. The reference interferometer is used as processing interferometer, having a known optical path difference which can be scanned by means of a piezoelectric system over a known range. The maximum of the intensity on the detector will appear when the optical path differences of the two interferometers are equal (zero-order fringe). Consequently, the optical path difference of the sensing interferometer can be measured by the known optical path difference of the processing interferometer, thus achieving absolute measurements.

The most modern fiber optic interferometers are based on white-light interferometry, which is particularly suitable for processing Fabry-Perot and Michelson interferometers.[33,34]

4.2. Polarization-Modulated Sensors

Because of slightly noncircular core and asymmetric thermal stress distributions, single-mode fibers are in reality dual-mode fibers, with the fundamental mode split into two orthogonally polarized states.[35] The modes propagate with slightly different propagation constants, and the fiber is said to have a modal birefringence. Highly birefringent fibers are particularly suitable for current and magnetic field measurements.[36-39]

The monitoring of electromagnetic phenomena is critical for power utilities, and optical sensors are particularly attractive, being able to offer high electrical insulation and total immunity to electromagnetic interferences. For this reason, many devices are now commercially available.[29,40-42]

4.3. Wavelength-Modulated Sensors

Truly wavelength-modulated sensors are those making use of gratings inscribed inside the optical fiber. The following paragraph 5 describes the operating principle and applications of sensors making use of optical fiber long-period gratings, while we refer the reader to Chapter 2 for details on the sensors that use optical fiber Bragg gratings.

Other wavelength-modulated sensors are of the extrinsic type, and make use of optical or chemical transducers joint at fiber end.

A typical example of an optical transducer is a section of sapphire fiber joint to a conventional silica fiber. Sapphire acts as a black-body cavity emitting a broad band radiation which is wavelength modulated by temperature conditions. This radiation is remotely transmitted to the detector unit by means of the silica fiber, in order to perform a remote pyrometry.[43] Sensors of this type are commercially available since many years; their wide sensing range (up to 2000 °C) and good sensitivity (0.1 °C) are particularly attractive for many industrial applications.[44,45]

Wavelength-modulated guided wave sensors making use of a chemical transducer are also called *optrodes*, by the combination of the two terms 'optical' and 'electrode'. Interaction of the measurand with the chemistry changes the spectral properties of the chemistry itself, the measurement of which makes it possible to monitor measurand status. A variety of guided wave sensors have been implemented, based on this type of indirect chemically-mediated spectroscopy.[46,47] Absorption- or fluorescence-based optrodes have been experimented for the monitoring of physical, chemical (pH, for instance, is one of the most considered ones) and environmental parameters.[48-49] The measurement of pH is frequently carried out by means of optrodes based on chromophores or fluorophores usually bond on polymeric or sol-gel supports.[50] The sensitive chemistry can be butt-coupled to the fiber-end for single-point measurements, or can constitute the fiber cladding for distributed monitoring by means of evanescent-wave sensing.[51] Oxygen is another parameter widely measured by means of optrodes, usually making use of a ruthenium complex as sensitive transducer.[52,53] A good selection of commercial products is available since many years, thanks to the reliability and good sensitivity of the probes. These products are offered for a wide range of sectors, including environmental, medical and food applications.[54-56]

4.4. Intensity-Modulated Sensors

Since it is relatively easy to perturb the intensity of the light guided by an optical fiber, intensity-modulated sensors represent the most experimented fiber sensors. They can use multimode fibers and simple optoelectronic devices, making thus possible the implementation of low-cost sensing devices.

Intensity-modulated sensors can be sub-divided into two main classed: extrinsic sensors making use of mechanical transducers positioned in front or in-between an optical fiber strand (Fig. 6), and intrinsic sensors, which measure the loss produced by the measurand on the fiber itself (Fig. 7).

Extrinsic-type sensors are photocells and intrusion detectors, which are widely commercially available,[57-61] and position, pressure or vibration sensors implemented for medical or industrial applications.[62-66]

Intrinsic-type sensors make use of an optical fiber squeezed between a periodic structure, or a plastic spiral wrapped around the optical fiber. Impact, edge, anti-squeeze and weight-in-motion sensors are based on this simple concept and are commercially available. They are often realized by embedding the fiber in a mat or ribbon.[67,68]

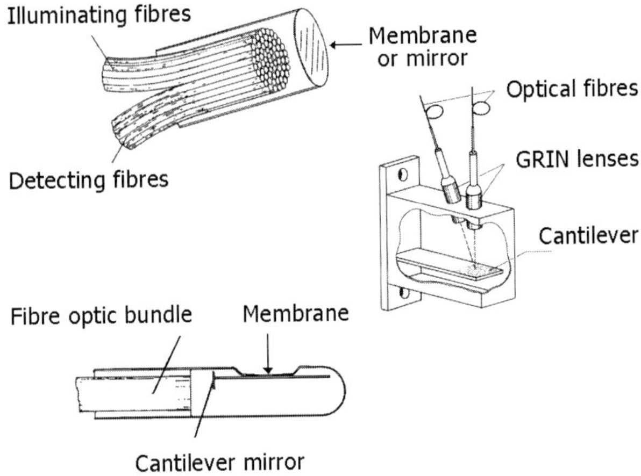

Figure 6. Examples of fiber optic pressure or vibration sensors based on a mechanical transducer.

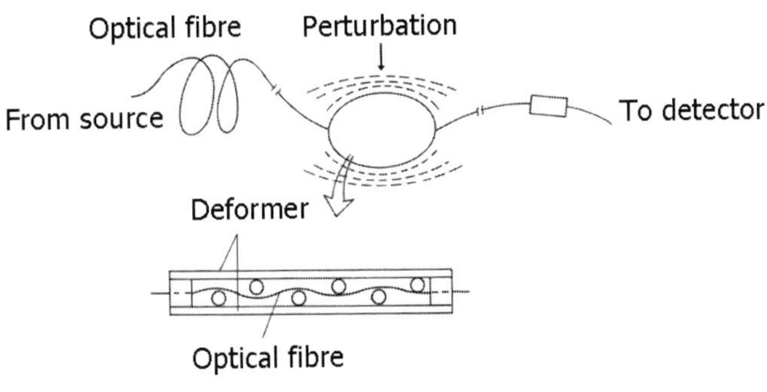

Figure 7. Fiber optic pressure or vibration sensor based on a microbending effect.

4.5. Fiberized Sensors

The geometrical versatility of optical fibers, together with their capability of guiding light with very low attenuation, make them ideal tools to replace free space architectures in conventional optical instrumentation. Indeed, many optical sensors have been *fiberized* by using fiber optic strands for illuminating and detecting means. It is worth mentioning that vibrometer and other Laser Doppler-based instruments have been *fiberized* to easily achieve localized measurements without complex bulk-optics architectures. A number of them is now commercially available.[69,70]

Optical fibers have been advantageously exploited also for dynamic and static light scattering measurements. Especially for dynamic light scattering measurements, single-mode fiber-based instrumentation not only offers measurement flexibility, but is also able to provide improved performance than that achieved by conventional bulk-optics systems.[71]

The early warning of cataract on-set is an outstanding example of what optical fibers are able to offer to dynamic light scattering measurements.[72,73] As far as static light scattering measurements are concerned, countless are the applications in which optical fibers have played an essential role, ranging from the monitoring of smokes, steams and aerosols, to the characterization of water-suspended sediments.[74-76]

Most commercially available spectrophotometers and spectro-fluorimeters are now equipped with fiber optic probes for localized measurements without sample drawing. This is particularly useful in many industrial process control in which avoiding sample handling represents a cost effective approach. Also, the availability of miniaturized spectrometers and bright LEDs makes it possible the implementation of compact spectrophotometers which can be used for monitoring parameters in a wide arrange of industrial and biomedical applications. Custom probes are now commercially available to face both the most common or simple measurement requirements.[77-79]

5. Long-Period Optical Fiber Grating Sensors

An optical fiber grating consists on a periodic modulation of the properties of an optical fiber (usually the refraction index of the core).

These structures have been actively studied since several years,[80-81] but now have a considerable impact on the development of fiber optic communication systems, laser sources, instrumentation for the detection and the measurement of various physical, chemical, biological and environmental quantities.

Depending on the period of the grating, fiber gratings are categorized into two types: Short Period Fiber gratings (or Fiber Bragg Grating-FBG), which have a sub-micron period, and Long Period Fiber Gratings (LPFG), which have typically a period in the range 100-1000 micron. The FBGs act as narrow-band reflection filters (or narrow-band rejection filters if used in transmission). Sensors making use of FBGs are examined in detail in Chapter 2. In the following, a short review of LPFG sensors is presented.

An LPFG is an optical fiber structure in which the energy typically couples from the fundamental core propagation mode to forward propagating cladding modes. As the cladding modes undergo a rapid attenuation due to scattering, bends of the fiber and absorption by the fiber jacketing, the transmission spectrum of the LPFG is characterized by a number of attenuation bands centered at discrete wavelengths.[82-84] Each of these attenuation bands corresponds to a coupling of the energy of the core mode to a distinct cladding mode (Fig. 8).

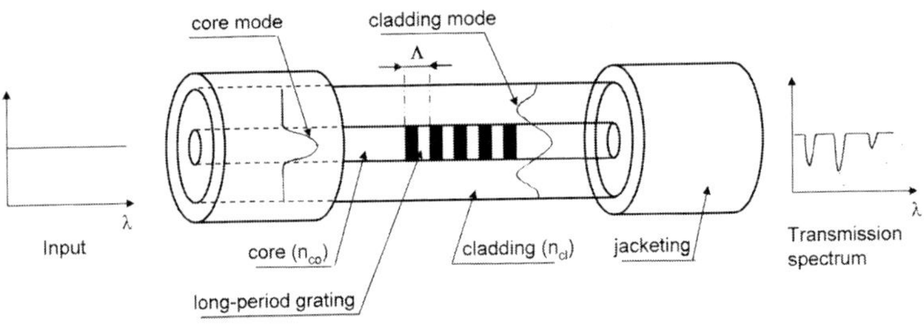

Figure 8. Schematic of a long-period fiber grating.

With the help of the coupled-modes theory[85] the central wavelength λm at which this coupling occurs can be expressed with:[86] $\lambda_m = \left(n_{co} - n_{cl}^{m}\right)\Lambda$, where λ_m = mth resonance wavelength, n_{co} = effective refractive index of the core mode, n^{m}_{cl} = effective refractive index of the mth cladding mode, Λ is the grating period.

Any modulation in the grating period Λ or in the effective refractive index of the core (n_{co}) and cladding (n_{cl}) modes induces changes in the distribution of the light between the core and the cladding modes and, as a consequence, gives rise to changes in the spectral response of a long-period fiber grating. As these changes in the spectral response can be measured, this behavior of the LPFGs can be utilized for sensing purpose.[87] Bend, strain, temperature, refractive index of the surrounding medium are some of the typical parameters that a LPFG can measure.

As to the fabrication method, LPFGs can be produced in various types of fibers, from standard telecommunication fibers to microstructured ones. So far, many techniques have been developed, such as point-by-point exposure to UV radiation,[84] CO_2 laser pulses,[88] infrared femtosecond laser pulses[89] or electric arc discharges.[90]

5.1. Bending Measurements

Two main effects occur in LPFGs subject to bend, which can be utilized to detect the bend itself: a) the attenuation bands in the spectral response, which are present when the LPFG is straight, shift in wavelength and change in depth as LPFG is bent (*wavelength shift detection*); b) each attenuation band can split into two peaks when the LPFG is curved and the resulting two peaks tend to separate as the bend increases (*resonance splitting detection*).

An LPFG bend sensor using the *wavelength shift detection* method has been proposed for shape sensing in smart structures.[91] The sensing of curvatures up to 4.4 m^{-1} has been demonstrated, and detection of curvature changes as small as 2×10^{-3} m^{-1} seems to be possible. A significantly higher sensitivity has been obtained by other authors using bend sensors based on the *resonance splitting detection.*[92,93] Over 80 nm mode splitting was measured under a bend curvature of 5.6 m^{-1}, giving a bend sensitivity of 14.5 nm/m^{-1}, which is nearly four times higher than the value demonstrated by the wavelength shift detection method.

The exact physical interpretation of the resonance mode splitting in an LPFG under bending is quite complex, and several works related to this matter have been published.[6,7,94,95]

Using two LPFG bonded to either side of a bent structure, it is possible to determine magnitude and sign of curvature. One grating is utilized for negative, and the other for positive curvature measurement.[96]

the birefringence increases with increasing loading, the related spectral peaks separation provides a measurement of the transverse loading.

Corrugated long-period gratings can be used to form tensile stress sensors and torsion sensors.[105,106] A corrugated LPFG consists of a periodicity of etched and non etched regions along the fiber. When a conventional LPGF is twisted, the induced refractive index perturbation is small because the fiber structure is uniform. On the contrary the application of axial load, torsion and bending to a corrugated LPFG, owing to the photoelastic effect, causes a periodic modulation of the refractive index of the fiber and results in mode coupling between the fundamental core mode and the forward-propagating cladding modes with the effect of changing the central wavelength of the LPFG attenuation bands. Therefore, corrugated LPFG are sensitive to the external stresses and can act as strain and torsion sensor.

5.4. Sensors Based on the Response to External Refractive Index

The attenuation spectrum of an LPFG is highly sensitive to the ambient refractive index. This sensitivity results from the dependence of the attenuation bands wavelength on the effective refractive index of the cladding modes, which are dependent upon the difference between the refractive index of the cladding and that of the medium surrounding it.

Several chemical sensors based on the response of LPFGs to the changes on the refractive index of the external medium have been proposed. For instance, LPFGs have been used to determine the concentration of antifreeze in water,[107,108] or for online concentration measurements of aqueous solutions with sodium chloride, calcium chloride and ethylene glycol.[109] As optical fiber sensors can be safely used in inflammable environments, LPFG sensors can be used to monitor organic aromatic compounds in the petrochemical industry. For such applications they offer the possibility of continuous in situ control measurements and can therefore be an attractive alternative to the current monitoring techniques, such as high performance liquid chromatograph (HPLC) and UV spectroscopy.[110,111]

The sensitivity to the ambient refractive index of an LPFG can be improved by coating the fiber grating with a thin film of material with higher refractive index than that of the fiber cladding.[112,113] As an example, an opto-chemical sensor employing LPFGs coated with polymeric sensitive overlays (syndiotactic polystyrene (sPS) in the nanoporous crystalline δ form) has been proposed.[114] A monolayer of colloidal gold nano-particles has also been proposed for improving the spectral sensitivity and detection limit of long-period gratings.[115,116]

This kind of sensors have been demonstrated to be able to measure refractive indices in the range of 1.34 to 1.39 with resolution of 10^{-3} to 10^{-4}, suggesting that these devices may be suitable for use with aqueous solutions in applications such as medical diagnostics, biochemical sensing, and environmental monitoring.[117]

6. Micro-structured Fiber Sensors

Photonic Crystal Fibers (PCFs) constitute a class of optical fibers that has a large potential for sensing applications. Their novel structure, with a lattice of air holes running along the length of the fiber, offers extraordinary control over the waveguiding properties in a way that is not possible with conventional fibers.

PCFs are commonly classified by the light-guidance mechanism in two categories, namely index guiding and photonic band gap (PBG) fibers (Fig. 9). In the two types, the microstructured cladding surrounds a solid and a hollow core, respectively.

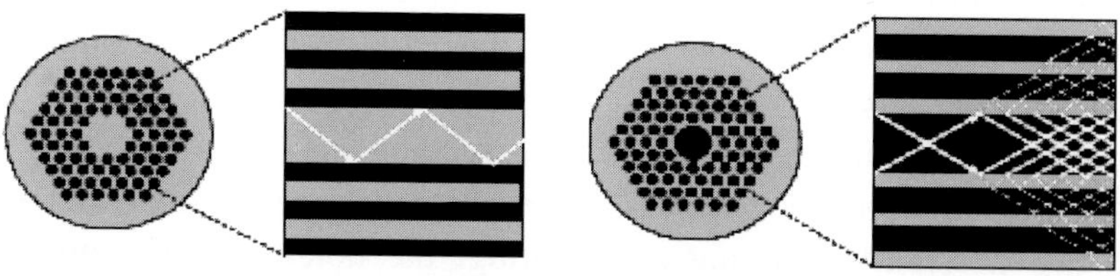

Figure 9. Structure of index guiding (*left*) and photonic band-gap (*right*) fibers.

In the index guiding fibers, the refractive index of the core is higher than the effective refractive index of the cladding and a modified form of total internal reflection guides the light; in the second type photonic band gap effect provides guidance, allowing for novel features such as light confinement to a low-index core.[118]

The PCFs design suggests a variety of strategies for optical sensing of different physical parameters (temperature, hydrostatic pressure, elongation, force, bending, etc.). The most studied approach involves the interaction with an evanescent field of PCF modes for the detection and analysis of liquid and gas phases species infiltrated in air holes of the cladding in index-guiding PCF[119,120] or with a guided field in hollow core of the PBG fibers.[121] Even if the majority of sensors reported in literature is based on index guiding fibers, because first introduced into the market, for sensing applications there are considerable advantages also in band-gap fibers: one of them is the possibility of guiding light in hollow cores filled with liquid or gas solutions of molecules.

In a well designed photonic band gap fiber, the largest part of the mode field (< 90%) is guided in the sample volume, thus providing a strong interaction between molecules and light over several tens of centimetres using few microliters of sample.[122] For acetylene detection with high sensitivity, Ritari et al.[123] investigated the feasibility of using PBG fibers. A significant interaction between light and molecules in the air holes of the cladding can take place also in index guiding PCFs, but the effect is smaller because it concerns only the evanescent field. For evanescent-wave sensing of biomolecules, such as DNA or proteins,[124] this effect can be enhanced using index guiding PCFs based on polymers.

An improvement of this sensing technique is represented by new geometries of cladding holes and, very recently, by the development of defected solid core.[125]

When the evanescent field sensing method may be impractical or inconvenient, an improvement is achieved by tapering the fiber.[126] There are two possibilities during the tapering: the holes structure may be preserved or may collapse. In both cases the guided mode of the PBG fiber spreads out, and the tapered PBG fiber results highly sensitive to external environment. The mechanism is very similar to what happens in tapered conventional fibers.[127] The collapse of the holes makes the core mode to couple to multiple modes of the solid taper waist, which is a solid multimode fiber. Several interference peaks appear from the beating of the multiple modes of the collapsed region, and they shift as external index changes.[128] The tapering technique has been successfully employed with fibers formed by a Germanium doped core surrounded by large air holes in the cladding,[129] demonstrating to be of particular interest also for biophotonic sensing.

The optical properties of PCFs are strongly controlled by the geometry of the holey region, and in sensing application this tunability is widely employed. One of the most promising advantages of PCF is the possibility of fabrication of multi-core fibers. A two-core index guiding fiber[130] bends in the plane containing the two cores, each of them supports a single guided mode. Because of the bending, the outer and the inner cores undergo an increase and a reduction in length, respectively. The PCF with these two cores acts as a two arms Mach-Zehnder interferometer in which the phase difference is a function of curvature in the plane containing the two cores fiber, demonstrating a resolution of about 170 μrad/rad.[132]

A particular advantage of PCF based sensors is the possibility of writing additional periodic structures on fibers such as Bragg and Long Period Gratings (LPGs). Standard grating fabrication techniques applied to PCFs have enabled the fabrication of gratings with original properties, mainly due to the complex index profile and dispersion properties of PCFs. From the point of view of fabrication, LPGs are generally easily fabricated, and can also generate well-isolated resonance, by proper selection of cladding mode for coupling, that can be highly sensitive to different measurands such as temperature, bending, strain and external refractive index. In particular for DNA sensing, this kind of gratings can be employed to detect the average thickness of a biomolecules layer within a few nm with sensitivity of approximately 1.4 nm / 1 nm in terms of shift in resonance wavelength per thickness of DNA layer.[131]

The control of the dispersion properties of core and cladding can be used, in principle, to increase the sensitivity to one measurand and to make the device insensitive to another. Recently it has been reported that LPGs inscribed in a dopant free endlessly single mode (ESM) PCF and in a large mode area PCF by electric arc discharges eliminate the cross-sensitivity[132,133] to temperature perturbations.

Another sensing technique makes use of birefringence in PCFs, which can indeed be made highly birefringent: the large index contrast facilitates high form birefringence, allowing the development of a new generation of polarimetric fiber sensors which use polarization (phase) modulation induced by external perturbations. Different methods have been developed to introduce birefringence into PCFs, such as using elliptical air holes[134,135] and/or asymmetric core or asymmetric distribution of holes.[136,137]

Important engineering areas can be influenced by future advances of polarimetric PCF based sensors, in particular thanks to their direct sensitivity to strain. The best example is the measurement of axial strain for structural monitoring. The essential mechanisms for strain and pressure sensing are almost the same: physical changes in fiber dimensions and the elasto-optic effect. Taking advantage of these two effects, one can implement distributed sensing elements to assess length changes, internal stresses or pressure in civil engineering structures. Based on elasto-optical measurements of the polarization state of the fiber output, it is possible to determine the fiber birefringence (beat length) for different wavelengths and compare it with numerical simulations. A new and quite important application of highly birifringent PCF is in dynamic pressure sensing for tsunami detection,[138] making use of standard polarimetric technique.

7. Integrated Optic Sensors

While the basic principles on which integrated optic sensors (IOSs) are based are the same as for fiber optic sensors, the two fields have developed at different paces and with slight different targets.

Fibers have the unique capability of operating over extended gauge lengths (even km!) in either point sensing or distributed sensing format. In the former case, the FOS is configured in such a way that monitoring of the measurand occurs at a specified location along the fiber (generally at its distal end); in the latter case, the values of the measurand (e.g. temperature or strain) are probed as a function of the position along the fiber. Remote measurements are made possible by the low attenuation characteristic of an optical fiber. Integrated optics (IO), on the other hand, has been developed with the aim of implementing multi-functional miniaturized circuits, possibly of size of a few cm, if not mm.

High-quality fibers, for both telecommunications and sensing, are mostly made of a silica core (even if, of course, there are alternative materials, including polymers). IO waveguides can be fabricated in a variety of materials, from dielectrics to polymers, from liquid crystals to semiconductors, and none of them has so far emerged as the key material. The lack of a unique solution for IO in terms of material and fabrication technology, however, is at the same time its major limit and its greater advantage: it permits, in fact, very great flexibility both in design and manufacture. Thus, an IOS may fully exploit the combination of thin films technology with other planar technologies, such as surface acousto-optic interaction, laser writing, silicon micromachining, micro-electro-mechanical systems (MEMS), optoelectronics integration on a semiconductor substrate, etc. Since two papers on IOSs, a temperature and a displacement sensor, respectively, were first published in 1982,[139,140] many other integrated optical devices for sensing have been proposed and demonstrated.[141-147] In the following, some examples of IOSs will briefly presented and discussed.

7.1. Integrated Optical Interferometers

Mach-Zehnder Interferometers (MZI) are easily fabricated in integrated optics, by means of standard photolithographic processes, and are one of the most common structures exploited for the detection of the phase shift induced by a measurand. While the free-space configuration requires several optical components and a tight alignment, a single IO circuit a few mm long represents a very stable and efficient solution. The schematic structure of an integrated optical MZI is shown in Fig. 10a, while the field distribution in the waveguide (and the interacting evanescent field) is sketched in Fig. 10b. MZI IO sensors have been fabricated in various materials, from glass to lithium niobate, from silicon-oxynitride on silicon to silicon-on-insulator. Several sensing devices have been demonstrated, e.g. for the detection of displacement, for refractometry and for bio-sensing.[148-156] Some sensors of this type, especially for biomedical applications, are also commercially available.[157]

The technique which is becoming a key tool for characterizing biomolecular interaction is that based on surface plasmon resonance (SPR).[164] The optical excitation of surface plasmons by the method of attenuated total reflection (ATR) was demonstrated in the late Sixties, and very soon it was applied for characterization of metal thin films.[165] In early Eighties the use of SPR for gas sensing and biosensing was demonstrated[166,167] and since then SPR sensor technology has continued to grow up[168-172] and it is now commercialized.[173]

An example of low-cost SPR sensor is represented by polymer-based chips, exploiting replication fabrication processes, which include a prism, microchannels and a chamber at microscale dimensions.[174]

The reader is also referred to Chapter 5 for a more detailed discussion on SPR-based sensors and biosensors.

8. Conclusions

Optical waveguide sensors have certain advantages that include immunity to electromagnetic interference, lightweight, small size, high sensitivity, large bandwidth, and ease in implementing multiplexed or distributed sensors.

Strain, temperature and pressure are the most widely studied measurands for optical fiber sensors, but biomedical applications are becoming the most interesting area for both fiber and integrated optic sensors. Nowadays, some success has been gained in the commercialization of optical waveguide sensors, even if in various fields they still suffer from competition with other mature sensor technologies.

New ideas, materials and structures, however, are being continuously developed and tested not only for the traditional measurands but also for new applications. As an example, we can conceive that further advances in the fabrication and understanding of microstructured fibers and photonic crystal structures will provide a platform for new sensors, aiming at being alternatives for standard sensing technologies.

Brilliant perspectives also exist for new "smart" optical sensors which mix nanoelectronics and micro/nano optical devices on the same silicon chip. These fully integrated optosensors would have the same, or better, characteristics of current sensors, while being much smaller, lighter and lower power than the existing systems.

References

1. B. Culshaw and J. P. Dakin Eds., Optical Fiber Sensors: Vol. 1, Principles and Components – 1988; Vol. 2 Systems and Applications – 1989; Vol. 3 Components and Subsystems – 1996; Vol. 4 Applications, Analysis and Future Trends – 1997 (Artech House, Nordwood MA).
2. K. T. V. Grattan and B. T. Meggit Eds., *Optical Fiber Sensor Technology* (Kluwer Academic Pbl., Dordrecht, 1999).
3. D. A. Krohn, *Fiber Optic Sensors: Fundamentals and Applications* (Instrumentation Society of America Pbl., Research Triangle Park, NC, 2000).
4. M. Lòpez-Higuera Ed., *Handbook of Optical Fiber Sensing Technology* (John Wiley & Sons Ltd., Chichester UK, 2002).
5. A. W. Snyder and J. D. Love, *Optical Waveguide Theory* (Chapman & Hall, London and New York, 1983).
6. R. Marz, *Integrated Optics: Design and Modeling* (Artech House, Norwood, 1995).
7. W. H. Steele, *Interferometry* (Cambridge University Press, 1983).
8. T. G. Giallorenzi, J. A. Bucaro, A. Dandridge, G. H. Sigel, J. H. Cole, S. C. Rashleigh and R. G. Priest, *IEEE J. Quant. Electr.* **QE18**, 626 (1983).
9. T. G. Giallorenzi, in *Optical Fiber Sensors*, NATO ASI Series E, Vol. 132 (Martinus Nijhoff, Dordrecht, 1987) p. 35.
10. G. B. Hocker, *Appl. Opt.* **18**, 3679 (1979).
11. N. Lagakos and J. A. Bucaro, *Appl. Opt.* **20**, 2716 (1981).
12. A. Dandridge, A. Tveten, A. D. Kersey, A. Yurek, *IEEE J. Light. Technol.* LT5, 947 (1987).
13. A. R. Davis, C. K. Kirkendall, A. Dandridge and A. D. Kersey, in *12th Intl. Conference on Optical Fiber Sensors* (OSA Technical Digest Series Vol. 16, 1997) p. 616.
14. A. Dandridge, A. B. Tveten and T. G. Giallorenzi, *Electr. Lett.* **17**, 523 (1981).
15. A. Yariv and H. W. Winsor, *Opt. Lett.* **5**, 87 (1980).
16. K. P. Koo and G. H. Sigel Jr., *Opt. Lett. Opt. Lett*, 334 (1982).
17. E. J. Post, *Rev. Mod. Phys.* **39**, 475 (1967).
18. B. Kim and H. Shaw, *IEEE Spectr.* **23**, 54 (1986).
19. E. Udd, H. C. Lefevre and K. Hotate, Eds., *Fiber Optic Gyros: 20th Anniversary Conference, Proc. SPIE* Vol. 2837 (1996).
20. Crossbow Technology Inc., USA, http://www.xbow.com.
21. Japan Aviation Electronics Industry Ltd., http://www.jae.co.jp/e-top/index.html
22. IXSEA (formerly Photonetics), France, http://www.ixsea.com.
23. KVH Industries Inc., USA, http://www.kvh.com/FiberOpt/.

24. M. Born and E. Wolf, *Principle of Optics* 6th Ed. (Pergamon Press, Oxford, 1986).
25. C. E. Lee and H. F. Taylor, *Electr. Lett.* **24**, 193 (1988).
26. C. E. Lee, H. F. Taylor, A. M. Markus and E. Udd, *Opt. Lett.* **14**, 1225 (1989).
27. R. A. Atkins, J. H. Gardner, W. H. Gibler, C. E. Lee, M. D. Oakland, M. O. Spears, V. P. Swenson, H. F. Taylor, J. J. McCoy and G. Beshouri, *Appl. Opt.* **33**, 1315 (1994).
28. RJC Enterprises, USA, http://rjcentreprises.net
29. Samba Sensors AB, Sweden, http://www.samba.se
30. Davidson Instruments Inc., USA, http://www.davidson-instruments.com
31. A. S. Gerge, F. Farahi, T. P Newson, J. D. C. Jones and D. A. Jackson, *Electr. Lett.* **23**, 1110 (1987).
32. C. E. Lee and H. F. Taylor, *IEEE J. Light. Technol.* **9**, 129 (1991).
33. SMARTEC SA, Switzerland, http://www.smartec.ch
34. FISO Technologies Inc., Canada, http://www.fiso.com
35. S. C. Rashleigh, *IEEE J. Light. Technol.* **1**, 312 (1983).
36. I. P. Kaminow, *IEEE J. Quant. Electr.* **17**, 15 (1981).
37. Y. N. Ning, Z. P. Wang, A. W. Palmer, K. T. V. Grattan and D. A. Jackson, *Rev. Sci. Instrum.* **66**, 3097 (1995).
38. S. Ishizuka, N. Itoh and H. Minemoto, *Opt. Rev.* 4, 45 (1997).
39. K. B. Rochford, A. H. Rose and G. W. Day, *IEEE Trans. Magn.* **32**, 4113 (1996).
40. N. Itoh, H. Minemoto, D. Ishiko and S. Ishizuka, in *12th International Conference on Optical Fiber Sensors* (OSA Technical Digest Series Vol. 16, 1997) p 92.
41. Nxtar Technologies Inc., Taiwan, http://www.nxtar.com
42. ABB Group, http://www.abb.com
43. R. R. Diles, *J. Appl. Phys.* **54**, 1198 (1983).
44. Williamson Corp., USA, http://www.williamsonir.com
45. Conax Buffalo Technologies, USA, http://www.conaxbuffalo.com/
46. O. S. Wolfbeis Ed., *Fiber Optic Chemical Sensors and Biosensors*, Vols. I and II (CRC Press, Boca Raton FL, 1991).
47. G. Boisdé and A. Harmer, Chemical and Biochemical Sensing with Optical Fibers and Waveguides (Artech House Inc., Norwood MA, 1996).
48. P. T. Sotomayor, I. M. Raimundo, A. J. G. Zarbin, J. J. R. Rohwedder, G. O. Neto, O. L. Alves, *Sensors & Actuators* **B**, **74**, 157-162 (2001).
49. P. Roche, R. Al-Jowder, R. Narayanaswamy, J. Young and P. Scully, *Anal. Bioanal. Chem.* **386**, 1245 (2006).
50. A. Lobnik, in Optical Chemical Sensors, F. Baldini, A. N. Chester, J. Homola, S. Martellucci, Eds., NATO *Science Series* Vol. 224 (Springer, Dordrecht, 2006), p. 77.
51. F. Baldini, *Trends in Appl. Spectr.* **2**, 119 (1998).
52. D. B. Papkowski, in Optical Chemical Sensors, F. Baldini, A. N. Chester, J. Homola, S. Martellucci, Eds., NATO *Science Series* Vol. 224 (Springer, Dordrecht, 2006), p. 501.

53. G. Orellana, in Optical Chemical Sensors, F. Baldini, A. N. Chester, J. Homola, S. Martellucci, Eds., NATO *Science Series* Vol. 224 (Springer, Dordrecht, 2006), p. 99.
54. Grupo Interlab, Spain, http://www.interlab.es/
55. Presens GmbH, Germany, http://www.presens.de
56. Ocean Optics Inc., USA, http://www.oceanoptics.com/
57. Banner Engineering Corp., USA, http://www.bannerengineering.com/
58. Sunx Ltd., Japan, http://www.sunx.jp/en/
59. Dinel, France, http://www.dinel.com
60. ECSI International Inc., USA, http://www.anti-terrorism.com/
61. Optellios Inc., USA, http://www.fiberpatrol.com
62. T. E. Hansen, *Sensors & Actuators* **4**, 545 (1984).
63. A. G. Mignani, A. Mencaglia, M. Brenci, A. M. Scheggi, in *Diffractive Optics and Optical Microsystems*, S. Martellucci, A. N. Chester, Eds. (Plenum Press, NY, 1997), 311.
64. M. Brenci, A. Mencaglia and A. G. Mignani, *Appl. Opt.* **30**, 2947 (1991).
65. Optrand Inc., USA, http://www.optrand.com.
66. Integra Lifesciences Corp., USA, http://www.integra-ls.com/.
67. Abacus Optical Mechanics Inc., USA, http://www.abacusa.com.
68. Herga Electric Ltd., UK, http://www.herga.com.
69. Polytec GmbH, Germany, http://www.polytec.de/polytec-com.
70. Perimed AB, Sweden, http://www.perimed.se.
71. M. Brenci, A. Mencaglia, A.G. Mignani, M. Pieraccini, *Appl. Opt*. **35**, 6775 (1996).
72. H. S. Dhadwal, R. R. Ansari and M. A. Dalla Vecchia, *Opt. Eng*. **32**, 233 (1993).
73. F. Könz, J. Rička, M. Frenz and F. Fankhauser, *Opt. Eng*. **34**, 2390 (1995).
74. K. Tatsuno and S. Nagao, *J. Heat Transfer* **108**, 939 (1986).
75. M. Brenci, D. Guzzi, A. Mencaglia, A. G. Mignani and M. Pieraccini, *Sensors & Actuators A* **48**, 23 (1995).
76. L. Ciaccheri, P. R. Smith and A. G. Mignani, in *15th International Conference on Optical Fiber Sensors* (IEEE Technical Digest Vol. 02EX533, 2002), p. 253.
77. Avantes Inc., USA, http://www.avantes.com
78. Control Development Inc., USA, http://www.controldevelopment.com
79. Stellarnet Inc., USA, http://www.stellarnet-inc.com
80. G. Meltz, W. Morey, and W. H. Glenn, *Opt. Lett*., 14, 823 (1989).
81. A. Othonos, *Rev. Sci. Instrum.,* 68, 4309, (1997).
82. A. M. Vengsarkar, P. J. Lemaire, J. B. Judkins, V. Bhatia, T. Erdogan and J. E. Sipe, J. *Light. Technol*., **14**, 58 (1996).
83. S. A. Vasiliev, E. M. Dianov, A. S. Kurkov, O. I. Medvedkov and V. N. Protopopov, *Quantum. Electron.* **27**, 146 (1997).
84. T. Erdogan, *J. Opt. Soc. Am. A* **14**, 1760 (1997).
85. A. Yariv, *IEEE J. Quant. Electr*. **9**, 919 (1973).
86. T. Erdogan, *J. Ligh. Technol*. **46**, 1277 (1997).

87. V. Bhatia and A. M. Vengsarkar, *Opt. Lett.* **21**, 692 (1996).
88. C. D. Poole, H. M Presby, J. P. Meester, *Electron. Lett.* **30**, 1437 (1994).
89. Y. Kondo, K. Nouchi, T. Mitsuyu, M. Watanabe, P. G. Kazansky, K. Hirao, *Opt. Lett.* **24**, 646 (1999).
90. G. Rego, O. Okhotnikov, E. Dianov, V. Sulimov, *J. Lightwave Technol.* **19**, 1574 (2001).
91. H. J. Patick, C. C. Chang and S. T. Vohra, *Electr. Lett.* **34**, 1773 (1998).
92. Y. Liu, L. Zhang, J.A. Williams, I. Bennion, *IEEE Photon. Technol. Lett.* **12**, 531 (2000).
93. Y. Liu, L. Zhang, J. A. Williams and I. Bennion, *Opt. Comm.* **193**, 69 (2001).
94. V. V. Steblina, J. D. Love, R. H. Stolen, J. S. Wang, *Opt. Comm.* **156**, 271 (1998).
95. U. L. Block, V. Dangui, M. J. F. Digonnet and M. M. Fejer, *J. Light. Technol.* **24**, 1027 (2006).
96. H. J. Patrick and S. T. Vohra, in 13th International Conference on Optical Fiber Sensors, Proc.of SPIE Vol. **3746**, p. 561.
97. B. H. Lee, Y. Chung, W. T. Han and U. C. Paek, *EIECE Trans. Electr.* **E83-C**, 287 (special issue on Optical Fiber Sensor) (2000).
98. S. Khaliq, S. W. James and R. P. Tatam, *Meas. Sci. Technol.* **13**, 792 (2002).
99. V. Bathia, D. K. Campbell, D. Sherr, T. G. D'Alberto, N. A. Zabaronick, G. A. Ten Eyck, K. A. Murphy and R. O. Claus, *Opt. Eng.* **36**, 1872 (1997).
100. V. Bhatia, D. Campbell, R. O. Claus and A. M. Vengsarkar, *Opt. Lett.* **22**, 648 (1997).
101. Y.G. Han, S. H. Kim and S. B. Lee, *16th International Conference on Optical Fiber Sensors Technical Digest*, (IEICE, Japan, Tokyo, 2003), paper Tu2-6, p. 54.
102. Y. G. Han, S. B. Lee, C. S. Kim, J. U. Kang, U. C. Paek and Y. Chung, *Opt. Expr.* **11**, 476 (2003).
103. Y. Liu, L. Zhang and I. Bennion, *Electr. Lett.* **35**, 661 (1999).
104. L. Zhang, Y. Liu, L. Everall, J. A. R. Williams and I. Bennion, *IEEE J. Select. Topics Quant. Electr.* **5**, 1373 (1999).
105. C. Y Lin and L. A. Wang, *J. Light. Technol.* **19**, 1159 (2001).
106. L. A. Wang, C. Y. Lin and G. W. Chern, *Meas. Sci. Technol.* **12**, 793 (2001).
107. H. J. Patrick, A. D. Kersey, F. Bucholtz, K. J. Ewing, J. B. Judkins and A. M. Vengsarkar, in *Proc. Conf. on Lasers and Electro-Optics*, CLEO'97, paper CThQ5, 11, 420 (1997).
108. H. J. Patrick, A. D. Kersey and F. Bucholtz , *J. Light. Technol.* **16**, 1606 (1998).
109. R. Falciai, A. G. Mignani and A. Vannini, *Sens Act.* **B74**, 74 (2001).
110. T. Allsop, L. Zhang and I. Bennion, *Opt. Comm.* **191**, 181 (2001).
111. R. Falate, R. C. Kamikawachi, J. L. Fabris, M. Muller, H. J. Kalinowski, F. A. S. Ferri and L. K. Czelusniak, in *Proc. Internat. Microwave and Optoelectronics Conference, IMOC-2003*, 2, 907 (2003).

112. I. Isahq, A. Quintela, S. W. James, G. J. Ashwell, J. M. Lopez-Higuera and R. P. Tatam, *16th International Conference on Optical Fiber Sensors Technical Digest*, (IEICE, Japan, Tokyo, 2003), paper ThP-3, p. 578.
113. I. Del Villar, I. R. Matias and F. J. Arregui, *Opt. Expr.* **13**, 56 (2005).
114. A. Cusano, A. Iadicicco, P. Pilla, L. Contessa, S. Campopiano, A. Cutolo, M. Giordano and G. Guerra, *IEEE J. Light. Technol.* **24**, 1776 (2006).
115. T. Okamoto, I. Yamaguchi and T. Kobayashi, *Opt. Lett.* **25**, 372 (2000).
116. J. L. Tang, S. F. Cheng, W. T. Hsu, T. Y. Chiang and L. K. Chau, *Sensors & Actuators* **B 119**, 105 (2006).
117. J. L. Tang and J. N. Wang, *Sensors* **8**, 171 (2008).
118. J. Broeng, S. E. Barkou, T. Søndergaard and A. Bjarklev, *Opt. Lett.* **25**, 96 (2000).
119. Y. L. Hoo, W. Jin, C. Z. Shi, H. Lo, D. N. Wang and S. C. Ruan, *Appl. Opt.* **42**, 3509 (2003).
120. T. M. Monro, W. Belardi, K. Furusawa, J. C. Baggett, N. G. R. Broderick and D. J Richardson, *Meas. Sci, Technol.*, **12**, 854 (2001).
121. P. J. Roberts, F. County, H. Sabert, B. J. Mangan, D. P. Williams, L. Farr, M. W. Mason, A. Tomlinson, T. A. Birks, J. C. Knight and P. S. J. Russell, *Opt. Expr.* **13**, 236 (2005).
122. J. B. Jensen, L. H. Pedersen, P. E. Hoiby, L. B. Nielsen, T. P. Hansen, J. R. Folkenberg, J. Riishede, D. Noordengraaf, K. Nielsen, A. Carlsen and A. Bjarklev, *Opt. Lett.* **29**, 1974 (2004).
123. T. Ritari, J. Tuominem, H. Ludvigsen, J. C. Petersen, T. Sorensen, T. P. Hansen and H. R. Simonsen, *Opt. Expr.* **12**, 4080 (2004).
124. J. B Jensen, P. E. Hoiby, G. Emiliyanov, O. Bang, L. H. Pedersen and A. Bjarklev, *Opt. Expr.* **13**, 5883 (2005).
125. X. Yu, G. B. Ren, P. Shum, N. Q. Ngo and Y. C. Kwok, *IEEE Photon. Technol. Lett.* **20**, 336 (2008).
126. H. C. Nguyen, B. T. Kuhlmey, E. C. Magi, M. J. Steel, P. Domashuck, C. L. Smith and B. J. Eggleton, *Appl. Phys.* **B 81**, 377 (2005).
127. S. Lacroix, F. Gonthier, R. J. Black and J. Bures, *Opt. Lett.* **13**, 395 (1988).
128. V. P. Minkovich, J. Villatoro, D. Mozon-Hernandex, S. Calixto, A. B. Stotsky and L. I. Sotskava, *Opt. Expr.* **13**, 7609 (2005).
129. E. C. Mägi, H. C. Nguyen and B. J. Eggleton, *Opt. Expr.* **13**, 453 (2005).
130. W. N. MacPherson, M. J. Gander, R. Mc Bride, J. D. C. Jones, P. M. Blanchard, J. G. Burnett, A. H. Greenway, B. Mangan, T. A. Birks, J. C. Knight and P. St. J. Russell, *Opt. Comm.* **193**, 97 (2001).
131. L. Rindorf, J. B. Jensen, M. Dufva, L. Hagsholm Pederson, P. E. Hoiby and O. Bang, *Opt. Expr.* **14**, 8224 (2006).
132. H. Dobb and K. Kalli, *Electr. Lett.* **40**, 657 (2004).
133. C. L. Zhao, L. Xiao, J. Ju, M. S. Demokan and W. Jin, *J. Light. Technol.* **26**, 220 (2008).
134. M. J. Steel and J. R. M. Osgood, *J. Light. Technol.* **19**, 495 (2001).

FIBER BRAGG GRATING SENSORS: INDUSTRIAL APPLICATIONS

Carmen Ambrosino,[a] Agostino Iadicicco,[b] Stefania Campopiano,[b] Antonello Cutolo,[a] Michele Giordano[c] and Andrea Cusano[a, *]

[a] *Dipartimento di Ingegneria, Divisione di Optoelettronica, Università del Sannio, Corso Garibaldi 107, 82100 Benevento, Italy*

[b] *Università degli Studi di Napoli "Parthenope", Facoltà di Ingegneria, Centro Direzionale Napoli, Isola C4, 80143 Napoli, Italy*

[c] *Istituto per i Materiali Compositi e Biomedici, CNR Piazzale Enrico Fermi 1, 80055, Portici (Napoli), Italy*

[*] *E-mail: a.cusano@unisannio.it*

Over the last few years, optical fiber sensors have seen increased acceptance and widespread use for a variety of applications. Among the large number of fiber optic sensing configurations, Fiber Bragg Grating (FBG) based sensors, more than any other particular sensor type, have become widely known and popular within and out the photonics community and seen a rise in their utilization and commercial growth. The capability of FBGs to measure a multitude of parameters such as strain, temperature and pressure and many others coupled with their flexibility of design to be used as single point or multi-point sensing arrays and their relative low cost, make them ideal devices to be adopted for a multitude of different sensing applications and implemented in different fields and industries. This work, involving the present and next chapter, reports on recent FBG sensing applications in several industrial fields. In particular, we first summarize the FBG major milestones of their technological evolution in thirty years from the discovery of Kenneth Hill in 1978 and then focus the attention on FBG recent application in civil engineering. We also report on FBG applications in aerospace, energy, oil and gas, transportation and underwater industrial fields. In particular relevant works ranging from structural sensing and health monitoring of composites and structures in aeronautic areas, to pressure and temperature sensors for oil and gas reservoir monitoring, to acoustic sensors for underwater applications, to high voltage and high current sensing systems for the power industry to name just a few, proposed by research groups and industries in last years are discussed.

1. Introduction

The fiber optics field has undergone a tremendous growth and advancement over the past 40 years. Initially conceived as a medium to carry light and images for medical endoscopic applications, optical fibers were later proposed in the mid 1960's as an adequate information-carrying medium for telecommunication applications. This has all been documented with awe over the past several decades. Among the reasons why optical fibers are such an attractive are their low loss, high bandwidth, EMI immunity, small size, lightweight, safety, relatively low cost, low maintenance, etc. As optical fibers cemented their position in the telecommunications industry and its technology and commercial markets matured, parallel efforts were carried out by a number of different groups around the world to exploit some of the key fiber features and utilize them in sensing applications.[1,2] Initially, fiber sensors were lab curiosities and simple proof-of-concept demonstrations. Today optical fiber sensing mechanism is involved in bio-medical laser delivery systems, military gyro sensors, as well as automotive lighting and control - to name just a few. This transition has taken the better part of 20 years and reached the point where fiber sensors enjoy increased acceptance as well as a widespread use for structural sensing and monitoring applications in civil engineering, aerospace, marine, oil & gas, composites, smart structures, electric power industry and many others.[3,4]

Optical fiber sensor operation and instrumentation have become well understood and developed. And a variety of commercial discrete sensors based on Fabry-Perot (FP) cavities and Fiber Bragg Gratings (FBGs), as well as distributed sensors based on Raman and Brillouin scattering methods, are readily available along with pertinent interrogation instruments. Among all of these, FBG based sensors have become widely known, researched and popular within and out the photonics community.

This work, involving the present and next chapter, reviews the major milestones of their technological evolution during the thirty years from the discovery of Kenneth Hill in 1978. Moreover, since the current FBGs technological assessment makes them widely involved in several industrial applications, in the following most relevant FBG applications published in last years are discussed.

2. Fiber Bragg Gratings History

Figure 1 illustrates the significant milestones and timeline evolution of the FBG industry over the past 30 years.

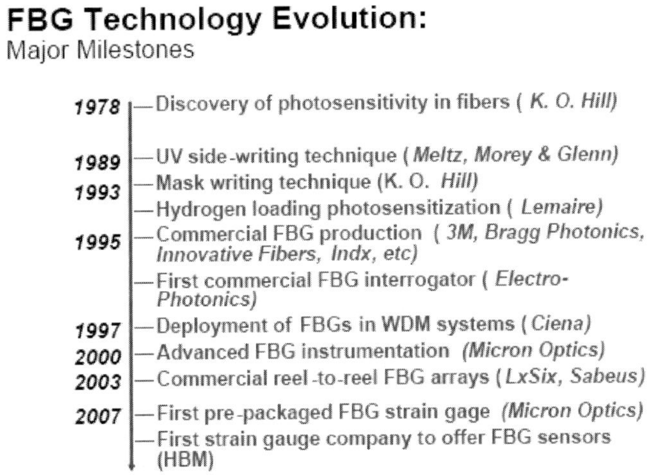

Figure 1. FBG Technology Evolution (Source: A. Mendez).

The formation of permanent grating was first demonstrated by Hill et al. in 1978.[5] They excited a germania-doped optical fiber with intense argon-ion laser radiation at 488 nm and observed that after several minutes the intensity of reflected light increased until eventually almost all the light was reflected from the fiber. The growth in back reflected light was explained in terms of non linear effect called photosensitivity, which permits the index of refraction in the core of the fiber to be increased by exposure to intense laser radiation. In this early experiment, a fiber Bragg grating was formed when a small amount of the laser light reflected back from the end of the optical fiber interferes with the exciting laser light to establish a standing wave pattern.

"Photosensitivity" causes the index of refraction to be increased to a much greater extent at position where constructive interference results in a maximum of laser intensity. As the strength of the grating (proportional to the depth of its index modulations) increases the intensity of the back-reflected light increases until it saturates near 100%.

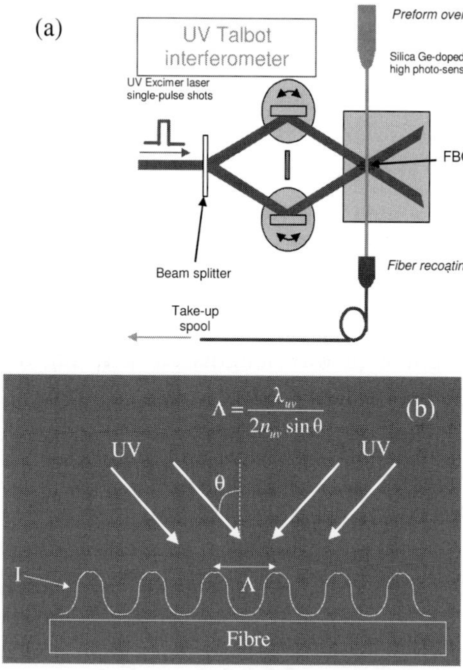

Figure 2. (a) Schematic of interferometric configuration used by Meltz in 1989 (Source: IPHT Jena); and (b) Grating pitch depends on intersecting angle between UV beams.

Although photosensitivity appeared to be an ideal means for fabricating these early "Hill gratings" in optical fibers, their usefulness was extremely limited because they only reflected at wavelengths in the visible close to the wavelength of the writing light, were spread along the optical fiber with varying strength and took a long time to produce. These limitations were overcome 10 years later by Meltz et al. in 1989[6] who recognized from the work of Lam and Garside[7] that photosensitivity was a two photon-process that could be made more efficient if it were a one-photon process corresponding to the germania oxygen vacancy defect band, at a wavelength of 245 nm (i.e. 5 eV)[8]. In the experiment of Meltz (1989) the fiber was irradiated form the side with two intersecting coherent ultraviolet laser beams of wavelength 244 nm, (see Fig.2(a)), which corresponds to one half of the 488 nm, the wavelength of the blue argon laser line.

The transverse holographic method worked since fiber cladding is transparent to UV light, whereas fiber core is highly absorbing of this radiation. The principal advantage with regard grating fabrication is related to the fact that spatial period of photo-induced perturbation depends on intersecting angle between the two interfering beams. This permits a versatile and efficient fabrication of custom Bragg gratings operating at much longer wavelengths than the writing wavelength as shown in Fig. 2(b). The periodic perturbation of the core index of refraction gives rise to successive coherent scattering for a narrowband band of the incident light. The grating thus effectively acts as a stop-band filter, reflecting light with wavelengths close to the Bragg wavelength, and transmitting wavelengths sufficiently different from resonance condition. Each reflection from a peak in the index perturbation is in phase with the reflection from the next peak when the wavelength of the light corresponds to the Bragg wavelength as shown in Fig. 3.

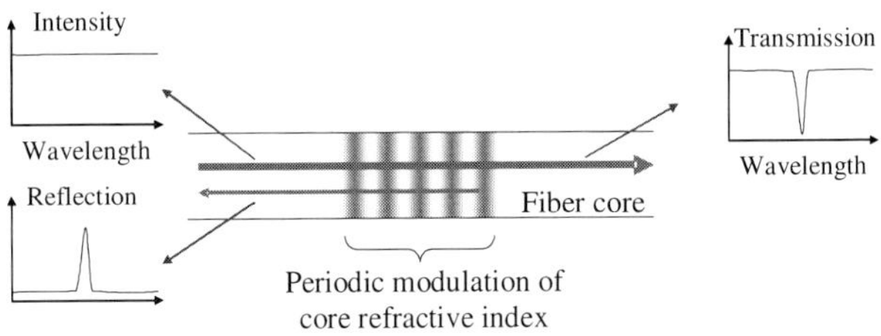

Figure 3. Principle of operation of FBGs.

Theoretical formulations based on coupled mode theory[9] have been developed to analyze fiber grating spectra by Erdogan et al.[10] Successively, a variety of different continuous wave and pulsed lasers with wavelengths ranging from the visible to the vacuum UV have been used to write gratings in optical fiber. In practice, krypton-fluoride (KrF) and Argon fluoride (ArF) excimer lasers that generate (10ns) pulses at wavelengths of 148 and 193 nm, respectively, are used most frequently to produce FBGs. The exposure required to produce a FBG is typically a few minutes with laser intensities of 100 to 1000 mJ/cm^2 and pulse rates of 50 to 75 s^{-1}. Under these conditions, the change in the core index of refraction is between 10^{-5} and 10^{-3} in germanium doped single-mode optical fiber. Techniques such as hydrogen loading proposed by Lemaire in 1993 can be used to enhance the optical fiber photosensitivity prior to laser irradiation.[11] Hydrogen diffusion makes the core more susceptible to UV laser radiation. Changes in refractive index of the order of 10^{-2} have been achieved by this means.

Successively, the transverse holographic method of writing fiber Bragg gratings has largely been superseded by the phase mask technique in 1993.[12] Phase mask is a thin slab of silica glass into which is etched (using photolitographic techniques) a one-dimensional square wave periodic surface relief structure as shown in Fig. 4. Since this material is transparent to UV laser radiation the primary effect of the phase mask is to diffract the light into the 0, +1 and –1 diffraction orders. Careful control of the depth of the corrugations in the phase mask suppresses zero-order diffraction, leaving the +/- 1 diffracted beams to interfere and produce the periodic pattern of intense laser radiation needed to photoimprint a Bragg grating in the core of an optical fiber. If Λ_{mask} is the phase mask period, the photoimprinted index grating is $\Lambda_{mask}/2$. Note that grating period is independent of the writing radiation wavelength.

Although, the usual practice brings the optical fiber almost into contact with phase mask, Othonos in 1995 demonstrated the improvements in the spatial coherence of the laser writing, relaxing the need for such close contact.[13] The phase mask technique greatly simplifies the manufacture of FBGs through easier alignment, reduced stability requirements on the photoimprinting apparatus, and lower coherence demands on the laser beam. It also permits the use of cheaper UV excimer laser source and tends to consistently yield high performance gratings. The prospect of manufacturing high performance gratings at low cost is critical to the large scale implementation of this technology for sensing applications. The main drawback associated to this approach relies to the need of separate phase mask for each grating with a different operating wavelength. On the other hand, it results very flexible since it can be used to fabricate gratings with controlled spectral responses characteristics. As a consequence of technological assessment, in the mid 1990's many research groups have been engaged in the study and realization of new grating devices through more complex refractive index modulation profiles. Examples include apodized FBGs, chirped FBGs, tilted FBGs, phase shift FBGs and long period fiber gratings.[14-19]

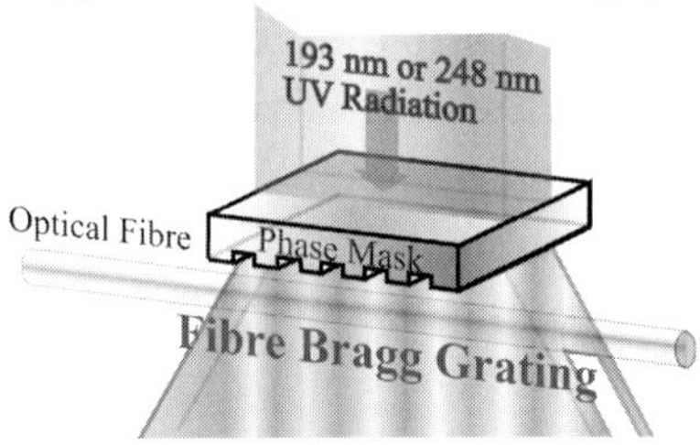

Figure 4. Diffracted UV beams from phase mask.

The commercial transition did not happen until the mid-1990s and was subsequently strongly driven by communications needs and the ramping up of the telecommunications "bubble", which saw a tremendous explosion of the number of companies and research groups engaged with the design, fabrication, packaging and use of gratings. First companies to produce commercial FBGs were 3M, Photonetics and Bragg Photonics in 1995. At the same time, Innovative Fibers was founded by Benoit Lavigne and Bernard Malo in 1995 and was a leader in the design and manufacture of FBG based components for the fiber optics industry including gain flattening filters, 50 GHz and 100 GHz Dense Wavelength Division Multiplexing (DWDM) filters and 980 nm and 1480 nm pump laser stabilizers. Successively, in 1997, Ciena Corp a manufacturer of Wavelength Division Multiplexing (WDM) devices became the largest public start-up company in corporate history and with first year earnings of ~U$ 200 million had the fastest revenue track ever.

Soon after the telecommunications bubble collapse, there was a significant shift by many players in the industry from communications to sensing applications. At the time, this was a prudent and strategic move on the part of FBG manufacturers to keep exploiting the technical and manufacturing infrastructure. As FBGs made the transition from optical communications devices to sensing elements in the 1990s, the bulk of the sensing applications centered on discrete, single-point sensing of specific parameters-such as strain and temperature-using sensors based on embedded or packaged gratings. These early gratings were typically written using phase masks or side exposure interferometric techniques.

These fabrication methods initially relied heavily on manual skills and labor, severely limiting many of the features and performance of the gratings in terms of production capacity, repeatability, mechanical strength, as well as number and quantity of FBGs written on a continuous fiber. Due to this increasing interest in FBG sensing technology, many research studies were devoted to the conception of optoelectronic unit able to demodulate FBGs based sensors. As matter of fact, the first optoelectronic unit able to interrogate FBGs sensors was developed 1996 by ElectroPhotonics corporate solutions and was based on the edge filtering concept.[4,20] However, the sensor industry is much more cost sensitive, demanding multiple sensing points and greater mechanical strength. Such requirements also call for the capability to fabricate an array of multiple FBGs at different locations along a same length of optical fiber. Such needs are being addressed by more sophisticated, on-line, reel-to-reel fabrication processes and systems that allow the writing of complex FBG arrays along a single fiber spool.

3. Fiber Bragg Gratings as Sensors

As described in the previous section and with reference to Fig. 3, the fiber optic intracore grating relies on the narrowband reflection from a region of periodic variation in the core index of refraction of a single mode optical fiber.[21] The central wavelength of the reflected Bragg signal is generally called Bragg wavelength and is linearly dependent upon the product of the effective index of refraction of the fundamental mode and the grating pitch: $\lambda_B=2n_{eff}\Lambda$. This means that changes in strain or temperature to which the optical fiber is subjected linearly shift the Bragg wavelength leading to a wavelength encoded measurements that is self referencing.[22-24] Furthermore, intrinsic wavelength encoding also provides a simple method for serial sensor multiplexing.[4]

The present FBG sensor market is primarily composed of 3 key segments: 1) sensing devices, 2) instrumentation, and 3) system integration & installation services.[25] The sensing devices segment is composed of bare FBGs for sensing applications, packaged FBG sensors and FBG arrays. The instrumentation market segment is composed of FBG interrogating instruments and related ancillary components such as multiplexors, switches, data acquisition systems, software and graphical user interfaces. Finally, the third segment is mostly covering services-rather than products-and entails all project management and engineering aspects related to implementing sensing solutions and system installations such as design, planning, system integration, customer training, service and on-site installation.

Today several companies are active in the development of efficient FBG demodulator systems. These could be classified into three main groups: 1) passive detection schemes based on pass band edge detection systems using fixed filters; 2) active detection schemes including tunable filters and interferometric systems; and 3) other schemes such as wavelength tunable sources and laser frequency modulation.[4] With regard to multiplexing capability, commercial interrogators fall into two main categories: time division multiplexing (TDM) and wavelength division multiplexing (WDM).[26] TDM discriminates between many sensors on a single optical fiber by gauging the time required for a pulse of light to return to the detection system. Blue Road Inc. has successfully developed FBG interrogators based on such idea. However, the most popular approach is WDM. WDM systems discriminate individual sensors by wavelength. Most WDM read-out systems are designed using one of two basic configurations: broadband source and swept detector and laser source and broadband detector. In the former approach, usually, few tens of sensors on a single fiber can be investigated whereas laser-based interrogators can illuminate more than 100 sensors per channel.

For instance, Micron Optics Inc. has developed such kind of interrogators.[27] In fact they propose wavelength scanning systems with sub-picometer peak wavelength resolution, broad-spectrum (80 nm) capability, but with a relatively slow-scan, data acquisition rate typically from 1 Hz to 250 Hz up to 512 sensors on four fibers. Also, Micron Optics Inc. offers the si920 high-speed optical sensing interrogator capable of monitoring FBG sensors up to four simultaneous channels with acquisition rates as fast as 500 kHz on a single channel or 100 kHz on four parallel channels. It is built with a patented architecture, using parallel fiber Fabry-Perot tunable filters.

Alternative wavelength scanning systems are available such as the FiberPro2 from Luna Innovations (Roanoke, Va.), operating at data sampling rates of 1 kHz; the HS-FOIS produced by AEDP (Lanham, Md.) with data rates of up to 3.5 kHz; the I*Sense systems produced by IFOS (Santa Clara, Calif.) with data rates of up to 5 kHz; and the FBG read-out systems from Blue Road Research (Gresham, Oreg.) with data rates of up to 2 MHz. Further commercial systems offering from 1 to 16 input channels are available. In 2008, Micron Optics, Inc. has announced an enhancement in available dynamic Optical Sensing Interrogator, in terms of increase in scanning range to over 160 nm which means more sensors per channel; up to 640 sensors on four fibers.[27] Finally it's important to note that, in May 2007 HBM[28]- the world's largest supplier of strain sensing systems- began offering optical strain gages and interrogators based on FBG technology.

Also, to favorite a wide spread out of FBG sensors, the development of appropriate packages was demanded. In particular, there was a need to develop appropriate protective coatings and housings for fiber sensors; to investigate the fundamental transfer of strains, stresses, pressure and temperature from the host specimen or matrix to the sensing fiber and the associated materials inter-play; as well as the development of field installation processes and deployment techniques suitable for different applications and environments.[29]

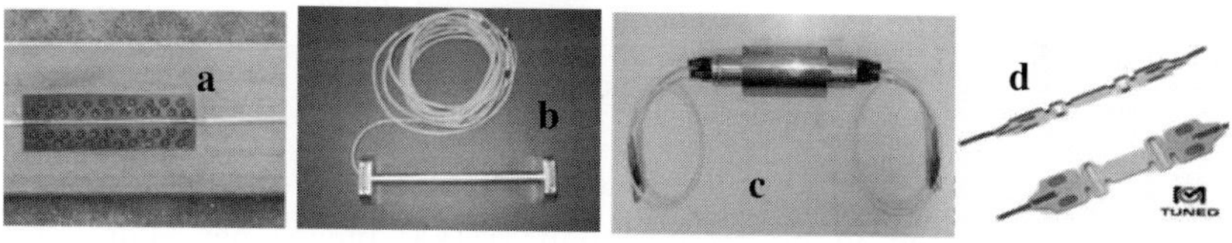

Figure 5. (a) Strain sensor welded to a stainless steel bar (Source: Ref. 30); (b) Temperature compensated strain sensor. (Source: Ref. 31); (c) FBG bending gauge (Source: Ref. 32); (d) Micron Optic os310 Strain Sensor (Source: Ref. 27).

According to this line of argument, FiberSensing[30] has developed a weldable FBG strain gauge for the monitoring of large steel structures (see Fig. 5(a)). Moyo et al. [31] investigated an FBG package consisting of an FBG sandwiched between layers of carbon composite material, for application in concrete structures (see Fig. 5(b)). Wen Wu et al. [32] with Prime Optical Fiber Corporation have presented the applications of the

FBG bending gauge (see Fig. 5(c)). Micron Optics Inc. presented in 2007 an opto-mechanical strain sensor (FlexPatch) based on a FBG mounted into miniature metallic flexure[27] (see Fig. 5 (d)).

On the bases of the FBG principle, a large number of solutions based on it have been proposed in the last decades, for strain, temperature, acoustic waves, ultrasound measurements as well as pressure and magnetic fields.[22-24] FBG-based sensors have been proposed, designed and developed for a wide variety of mechanical sensing applications including monitoring of civil structures, smart manufacturing and non-destructive testing, remote sensing, underwater applications and transportation (see Fig. 6).

Figure 6. FBGs applications.

The present worldwide volume demand for bare and packaged FBG sensors is estimated to be greater than 10,000 pieces per year. The worldwide volume demand for FBG arrays is estimated at several 100s to 1,000s arrays per year. The combined present global market size of this segment is estimated to be in the range of \$ 15 M to \$ 35 MUSD a year, with an annual growth rate of 15% to 25%. The instrumentation market has been growing steadily over the past three years, in part due to a variety of new fiber sensing projects and installations throughout Asia. Furthermore, the global volume for FBG interrogating instruments is estimated at several hundred units a year, with an annual growth rate of 20% to 30%. The total market size is estimated to be in excess of \$ 50

MUSD. In following sections the most relevant FBG industrial applications achieved in last years are reported.

4. Civil Applications

Since the first field application of FBG sensors for bridge health monitoring demonstrated in 1995, in Calgary, Alberta, Canada[33] there was a large interest in research and industry community during these years. Recently, Fiber Sensing[34] developed a strengthening technique, based on the introduction of carbon fiber laminates with embedded FBG into thin slits opened on the concrete cover of the elements to be strengthened. The high performance of this technique was already assessed on the flexural strengthening of concrete structures but preliminary tests have indicated that this performance can be still more significant for the shear strengthening of reinforced concrete beams. Besides, Fos&S[35] has been interested in the Structural Health Monitoring (SHM) of the steel roof structure of the Velodrome and the Olympic Stadium in Athens, by means of a FBG sensing network. An important application has been developed by Grattan et al in 2007.[36] They implemented a sensor protection system involving 16 FBGs for two concrete foundation piles to enable real-time and in situ acquisition of strain and temperature data during the whole construction phase of a 13-storey building at Bankside 123, London UK. The cages were approximately 46 m long and 1.5 m in diameter.

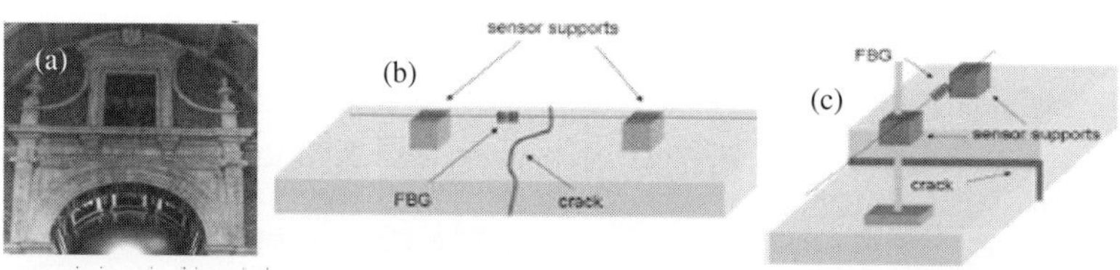

Figure 7. (a) Photo of the studied arch; Sensor designs used in: (b) the planar regions; and (c) the non-planar regions. (Source: Ref. 37).

Another important aspect in civil structure monitoring concerns the heritage structures and historical monuments. In 2007 a SHM system based on FBGs and interrogator unit from Micron-Optics Inc. (model sm125), was installed in the church of Santa Casa da Misericordia of Aveiro (see Fig. 7) by Kalinowski et al. [37] This system comprises 19 displacement sensors and 5 temperature sensors and was successfully tested in period April-December 2006.

4.1. Monitoring of High Performance Bridges

The advantages of FBG based SHM sensor systems have widely attract attention in Bridge monitoring.[38-39] Existent bridges, particularly those made of reinforced concrete, are deteriorating at a rapid rate. In this context, Ou with Micron Optics Inc. in these years has developed several SHM systems based on FBG sensors to be applied in several large-span bridges, like Songhua River Bridge in Heilongjiang and using up to 1800 FBGs as in the Dongying Yellow River Bridge in Shandong[26] (see Fig. 8(a) and (b). In 2005, Grattan et al. [40] have tested a network of 32 FBGs with a measurement bandwidth of up to 200 Hz over an 18-month period on a 346 m road bridge in Norway (see Fig. 8 (c)), for structural integrity monitoring purposes.

During the period 2005-2007 SMARTEC (http://www.smartec.ch) have monitored the Manhattan cable stayed bridge that crosses the East River in New York City by using 4 FBGs. Purpose of these sensors was to measure strain on main cable and one hanger as a function of: temperature variations, time of day (sunshine), time of year (seasons) and traffic conditions (day / night).

Figure 8. (a) Songhua River Bridge in Heilongjiang (2003, over 50 FBGs uses) (Source: Ref. 26); (b) Dongying Yellow River Bridge in Shandong (2003, over1800 FBGs used) (Source: Ref. 26); (c) Mjosund bridge during field trials (Source: Ref. 40).

In general, FBG installation on bridge cable can be achieved in different methods: adhering FBG directly on steel wire of the cable and digging a slot along the wire and embedding bare FBG in it. Alternative procedures based on Fiber Reinforced Polymers and Optical Fiber Bragg Grating (FRP-OFBG) are also proposed for bridge SHM and cable corrosion monitoring.[26,41]

4.2. Tunnel

The monitoring of existing road tunnel also results important for human safety. Recently Smart Fibres[42] has been involved in a project for the installation of Optical Fiber Sensor (OFS) system to monitor the movements of a road tunnel in Spain during remedial grouting works. A new sensor type, SmartRod (consisting of a composite pultrusion into which one or more arrays of FBG strain sensors are installed) has been developed for long-term tunnel deformation monitoring. The rods were fixed to the tunnel wall using rigid clamping plates and the sensor responses were recorded during the whole grouting works (see Fig. 9 (a)). In Ref. 43, optical FBG-based strain sensor modules embedded in a three dimensional geo-mechanical model of a forked tunnel model is presented. A number of strain blocks using the same material of the tunnel model were made and three different FBGs into different directions of the strain block were glued (see Fig 9 (b)). Ninety-nine strain blocks with a total of 297 FBGs strain blocks in the entire tunnel model were installed.

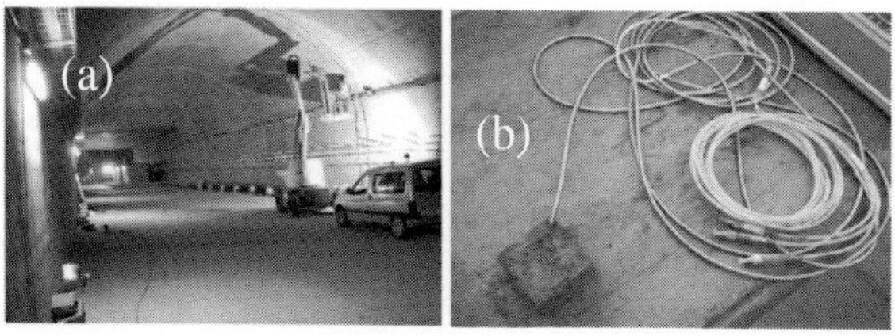

Figure 9. (a) Road Tunnel (Source: Ref. 42); (b) FBG strain sensor (Source: Ref. 43).

4.3. Geotechnical Investigations

In this section several applications of FBG sensors for geotechnical investigations, exploited in recent years, are presented.

4.3.1. Soil Pressure Sensors

Soil pressure sensors aim to monitor underground, rock and soil. Unfortunately, measurement of stress within a granular material has always been problematic because the three-phase of the material complicate the identification of what is being measured. In recent years FBG based soil pressure sensors have been successfully employed.[44,45] In Ref. 44 the feasibility of a FBG stress cell where sensor was encapsulated in a silicone rubber (in order to enhance FBG transverse sensitivity), for use in geotechnical environment has been reported (see Fig. 10 (a)). Zhou el al.[45] demonstrated a new kind of FBG soil pressure sensor with temperature compensation as shown in Fig. 10(b).

The soil pressure sensor has been calibrated under oil using an FBG interrogator from Micron Optics Inc. showing a good accuracy and high precision characteristics.[45] A new attempt in the use of FBG in soil monitoring relies on an innovative geotextile-based monitoring system, developed for the measurement of strain and deformation of earthworks structures reinforced with geosynthetics[46-48] by several companies (ID FOS Research, Bidim Geosynthetics SAS, FOS&S).[46-48]

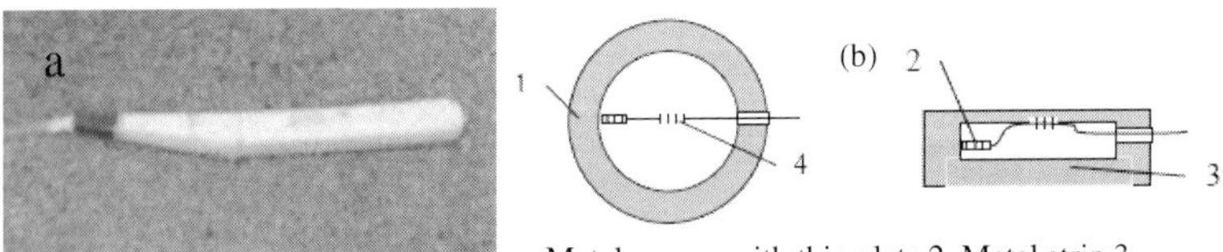

Metal groove with thin plate 2. Metal strip 3 Capping cover 4 FBG

Figure 10. (a) Stress cell (Source: Ref. 44); (b) A schematic illustration of FBG-based soil pressure sensor with temperature sensing described in (Source: Ref. 45).

4.3.2. Seismic Wave Detection

Seismic and volcanic events prevention can be considered as one of the most important tasks. The feasibility to cover all the frequencies of interest (0.001–0.01Hz) with a single sensor as it can be done with FBG based one would be a great advantage. In 2007, an innovative FBG seismic sensor was proposed by Optosmart.[49-50] The prototype has a cylindrical structure with a stiff plexiglass frame on which an empty plexiglass tube with a 1kg steel mass at its top is mounted. The symmetrical structure has been chosen in order to have the same bending response in all directions. The sensing system is composed of three FBG sensors within the same optical fiber and bonded on the inner surface of the empty tube forming an angle of 120°, as shown in Fig. 11 (a).

Preliminary dynamic characterization has been carried out by using an accelerometer as reference sensor and an instrumented hammer to excite the structure. Dynamic interrogation unit based on broadband interrogation and optical filtering[51] has been used, employing a WDM demultiplexer for simultaneous interrogation of four sensors.[52]

For a given direction of the impact, a different strain field is induced at the different sensing location enabling the system capability to detect amplitude and direction of the seismic wave. In order to characterize the frequency response of the designed structure, Fourier transforms of the FBGs and accelerometer time responses have been evaluated and compared as shown in Fig. 11(b).

The most widely used technique in the detection of ground movements is related to inclinometers.[53] Recently, a technique to be referred as the FBG Pipe Strain Gage was experimented by Ho Yen-Te et al. [54] A series of FBG strain sensors are attached to the outside of a flexible PVC pipe and then grouted in ground and used to monitor the deformation of a laterally loaded pre-cast concrete pile driven in a reclaimed silty sand deposit. In 2007, Fujihashi et al. [55] for NTT InfraNet Corporation have been working on the development of FBG-based accelerometers and tsunami sensors, which will provide high reliability while greatly reducing costs. Verification tests were performed at sea in the ocean to the west of the Koshikijima islands in Kagoshima prefecture (Japan).

5. Aerospace Applications

In last decades, large efforts have been provided to retrieve efficient SHM methods useful for aerospace crafts. In aerospace industry undetected damages or damages growth can have catastrophic results. FBG devices well meet the aerospace stringent demand in sensing capabilities. The examples included in this section report the recent progress of the FBG technologies in different sub-area of aerospace engineering.

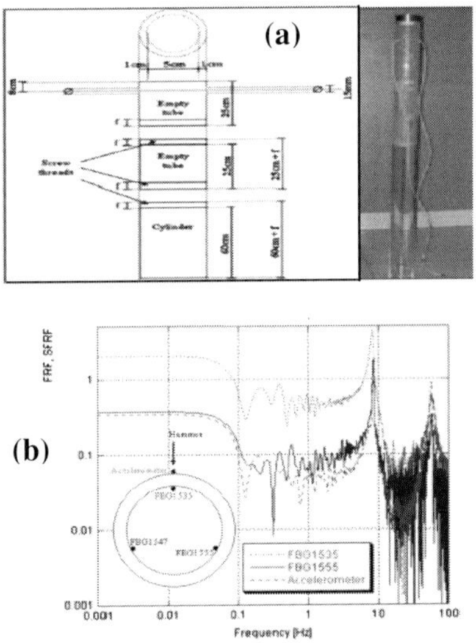

Figure 11. (a) FBG-based seismic sensor, construction scheme and photograph; (b) Frequency response function of the FBGs and the reference accelerometer as a result of the hammer impact on the 1535 sensor.

5.1. Aeronautic Applications

The use of carbon fiber reinforced plastic (CFRP) has been increasing especially in civil aviation aircraft. Due to their small size and low weight, FBG sensors can easily be integrated into CFRP as suitable solution for SHM. Accordingly, FBG devices have been applied to detect damages that causes such strains in CFRP laminates.[56] Recently Takeda and Hitachi Cable, Ltd have succeeded in developing a small-diameter

optical fiber involving FBG sensors for embedment inside a lamina of composite laminates without strength reduction to be used in SHM.[57]
Also a new damage detection system for quantitative evaluation of delamination length in CFRP laminates, which is the most important damage for structural design of composite laminates, was successfully proposed.[58-60] Moreover, University of Tokyo and Kawasaki Heavy Industries demonstrated real-time detection of impact damage by embedding these new optical fiber sensors in a CFRP fuselage structure with a diameter of 1.5 m and a length of 3 m (see Fig. 12 (a)).[58,61,62]

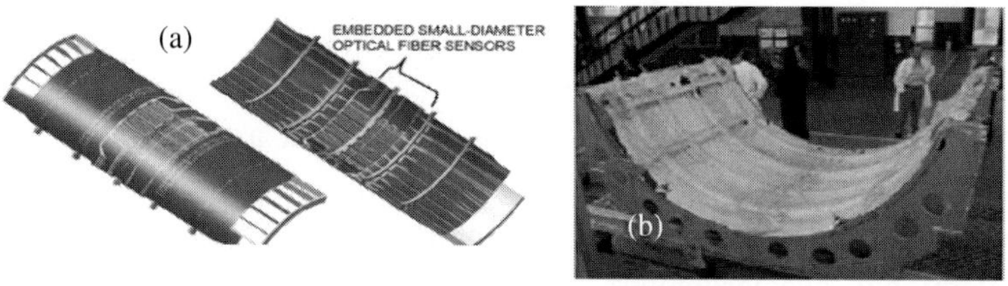

Figure 12. (a) Arrangement of embedded small diameter FBG sensors in upper panel of composite fuselage demonstrator (source: Ref. 58); (b) A340-600 fan cowl inside the vacuum bag before the demolding (source: Ref .62).

FBGs have also been incorporated in monolithic structures composed by a light CFRP skin with stringers. According to this, recently, Airbus Espana with Guemes et al. [63] have both embedded and attached FBGs and piezoeletric devices (only bonded) over the surface of a sample monolithic specimen: a CFRP skin with a co-bonded stiffner of a test panel extracted from an Airbus A340-600 fan cowl (See Fig. 12 (b)) and tested in damage induced experiments. In these years sandwich structures with advanced composite face sheets are attracting much attention as a solution to maximize the potential of composite materials but these structures are prone to damage. According to this, several researchers (Kuang et al.,[64] Dawood et al.,[65] Takeda et al. [66]) have attempted to utilize optical fiber sensors for monitoring manufacturing process and damage development.
Recently in aerospace engineering the concept of using bonded composite repairs for the maintenance of aging metallic aircraft has been

demonstrated. With regards to this issue, Kressel et al. [67] showed how FBG sensors can track the initiation of structural bonding and measure the residual strains during bonded composite patch curing.

Also EADS with Weis[68] applied FBGs on the CFRP surface of an aircraft fuselage stringer stiffened panel and detected impact damage by measuring the change in strain during and after the impact. Davis with Air Vehicles Division of DSTO (Australia)[69] presented a comparative analysis of strain measurements on arm fillets of an F/A-18 stabilator spindle between FBGs sensors and electrical resistance foil gauges. A good agreement by both technologies has been demonstrated whilst cabling weight and complexities were significantly reduced by using FBGs.

In 2006, Cusano et al.,[70] demonstrated the feasibility analysis to perform experimental modal analysis by using FBGs on a composite wing of aircraft model (see Fig. 13 (a)) by FRF approach. Excitation has been provided by an instrumented impact hammer while embedded FBGs and conventional accelerometers bonded to the structure were used as reference sensing elements. Experimental results demonstrated the good agreement between the displacement modes provided by both sensing technologies. Successively, the same group reported results of damage detection tests on an ad hoc steel structure with FBGs bonded on it.[71] The structure used in the dynamic tests was obtained soldering two beams as shown in Fig. 13 (b). Two identical FBGs were bonded on the "A" and "B" beams respectively. Damage detection tests were performed on the "B" beam. As excitation a piezoelectric element bonded on the horizontal beam was chosen. On the rear side of the "B" beam, in correspondence of the FBG, an accelerometer was placed. After the first acquisition on the undamaged test sample, clay masses were added in order to simulate the presence of damages. FRFs and (SFRFs) resonant frequencies shifts induced by structural alterations were the same for both reference accelerometers and FBGs for the most part of the retrieved modes. A comparison of sensors response at 1060 Hz is shown in Fig. 14. Both sensing technologies exhibit variations at each state.

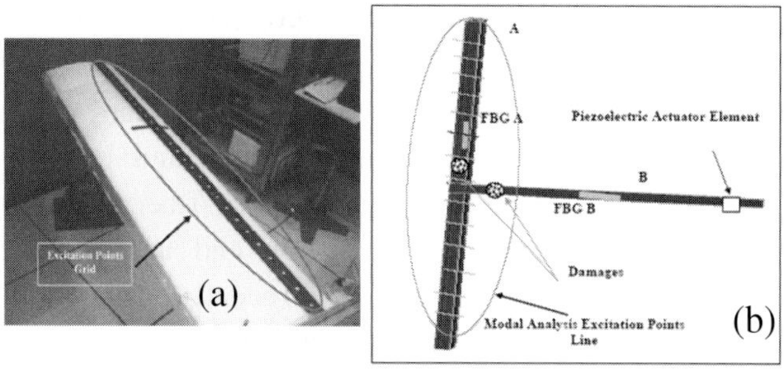

Figure 13. (a) Photograph of the composite wing with excitation point grid; (b) A scheme of the test structure: the beam "A" is an AISI 4340 steel hollow sample while the "B" is a thin AISI 4340 steel beam and is soldered at its midspan.

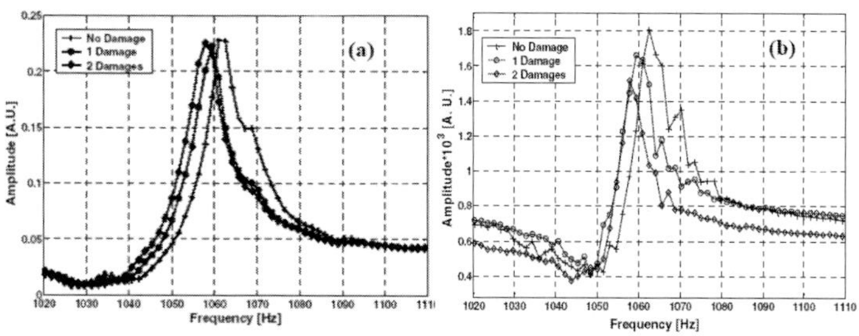

Figure 14. Comparison between the FRF's (a) and SFRF's (b) amplitudes for the 1060 Hz resonant frequency.

5.1.1. Damage Detection using Lamb Waves

A very attractive method for detecting and monitoring damages in composite or metal structures employs ultrasonic Lamb waves.[72] Lamb waves can be generated with small piezoelectric disc shaped actuators by means of a pulse with a known Fourier transform. In the low frequency range it is possible to generate only flexural waves. The interaction of these flexural waves with a defect induces an echo signal, which can then be detected, normally via the same (PZT) discs. The energy produced by this echo signal has a strong correlation with the size of the damage and may be used to follow its evolution.[72] Quite a few investigators have described successful attempts to measure Lamb waves using FBGs,[73-77] either in surface-attached or embedded forms. Betz et al. [73] have glued FBGs to the surface of Perspex and Aluminum plates whereas in Ref. 74

impact damage was detected by Lamb waves, using FBG bonded on composite ply. In both cases, FBG responses to Lamb wave propagated through the damaged area were comparable to standard piezoceramic sensors ones. The combination of piezoelectrics and FBGs was also proposed by Qing et al. as hybrid piezoelectric/fiber optic diagnostic system for quick non-destructive evaluation and long term SHM of aerospace vehicles and structures.[75] Quite recently, Takeda et al., [58,76] published detailed reports describing their extensive experience with glued and embedded small-diameter FBG sensors in composites. They measured the directional sensitivity of their surface-attached FBGs, and studied Lamb wave propagation in composite laminates, using a novel interrogation system. Takeda with Fuji Heavy Industries Ltd and Hitachi Cable, Ltd, also applied this system to a skin/stringer structural element of airplanes made of CFRP laminates.[76] An ultrasonic wave at 300 kHz was propagated through the debonded region, and the wavelet transform was applied to the received waveform.

5.1.2. Active Vibration Control

In the last two decades, Active Vibration Control (AVC) methods, focused on reducing the sound radiation of light structures by voluntary addition of controlled signals, have attracted increasing interest due to the numerous applications in which they could be successfully adopted. Chau et al. [78] have experimentally demonstrated the use of FBG Strain Sensors for structural vibration control. A cantilevered flexible aluminum beam is used as the object for vibration control. A piezoceramic patch surface-bonded to the cantilevered end of the beam is used as an actuator to suppress the beam vibration.

Kim et al. also investigated a hybrid system (PZT/FBG) for the flutter suppression of a composite plate structure.[79] The effectiveness of the flutter suppression system has been evaluated via wind-tunnel testing.

Cheng et al. presented an experimental study on the closed-loop control of the vortex-induced vibration of a flexible square cylinder, fixed at both ends, in a cross-flow.[80] Curved piezoceramic actuators were embedded underneath one cylinder surface to generate a controllable motion to perturb the interaction between flow and structure.

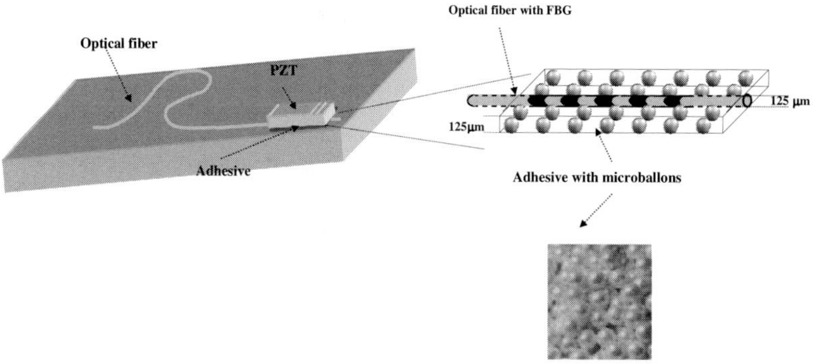

Figure 15. Schematic bonding configuration for one of the four FBG /PZT pairs: in the inset: a photograph of microballoons.

Recently, Cusano et al. investigated the feasibility of an AVC system using FBG sensors and PZT actuators for vibration suppression in co-located configuration.[81] To this aim, a test aluminum proof in fixed-fixed beam configuration has been equipped with four FBG sensors/PZT couples A numerical analysis of the structure have been carried out. According to this, in order to plan couples' position and control the higher number of modes with the minimum number of sensor/actuator cells, a modal superposition technique was used: sensors/actuators locations were set by considering the condition of higher strain field along the optical fiber axis for the greater number of vibration modes simultaneously. FBGs have been bonded to the upper part of the aluminum plate by using a 2-component fast curing adhesive. Glass micro-balloons with a diameter of approximately 125 µm have been added to the adhesive, providing a well supported adhesion between structure, sensing and actuator elements. Finally, the piezoelectric elements have been bonded upon the fiber surface. A schematic bonding configuration for one of the four FBG and PZT pair is reported in Fig.15. Preliminary tests in closed loop configuration have been carried out. The structure has been excited with single frequency signals simulating the unwanted vibrations from external environment. The couples PZT/FBG have been used in order to provide the control loop. For each excitation frequency, the FBGs time responses have been continuously used as input signals for the proper designed Proportional-Derivative (PD) controller determining the actual driving signals for the active actuators in the same locations. From these preliminary results, for an excitation frequency of 80 Hz, a maximum vibration reduction of slightly less than 17 dB was observed.

5.2. Astronautic Applications

Spacecraft monitoring is critical for the successful operation of any space mission. Space is a very challenging environment for any sensing system as it is characterized by microgravity, vacuum, presence of radiation, large thermal variations, mechanical vibrations and shock resulting from launch. In recent years, the European Space Agency (ESA) has been investigated embedded and surface mounted FBG sensors for space structures. As an example,[82] a tripod demonstrator (typical of a telescope structure) with embedded FBG sensors and actuators in one of the 3 legs was developed. Although the overall sensor/actuator design and embedding technique has to be optimized, the demonstrator showed that this type of structure can be successfully operated adaptively to counteract environmentally induced deformations. Another example always in Ref. 82 by ESA is a flywheel support that was actively damped to reduce the coupled vibration. Also Blue Road Research in these years has investigated in FBG spacecraft applications and conducted tests in collaboration with NASA Marshall and NASA White Sands on composite pressure vessels[83-84] used to support the Space Shuttle. These vessels were being qualified for continued usage beyond their design lifetime and were tested to failure. The surface strain field was mapped by using an array of single axis FBGs applied to the surface. During the tests the surface mapping technique was able to localize the point of burst to less than 2 cm in all cases.[83-84]

In 2006, Cusano et al. [85] have been instrumented an aluminum prototype of the AMICA (Astro Mapper for Instrument Check of Attitude) Star Tracker Support (ASTS) of the AMS_02 (Alpha Magnetic Spectrometer) (see Fig. 16(a)) space experiment developed by the Center for Advanced research in Space Optics (CARSO) in Trieste, Italy, with FBGs. In order to verify whether this structure was able to survive to launch stress and to very harsh operative environment like open space, its modal dynamic features were experimentally evaluated by using the classic modal analysis approach. Also, to evaluate dynamic features of the structure, an excitation point grid is traced in its lower side (see Fig. 16 (b)). The 1-st and 2-nd bending modes are both simulated with NASTRAN™ software and retrieved from experimental data. Comparisons between experimental and numerical data show good agreement and demonstrate the capability of FBG sensors to be efficiently used for the dynamic characterization of complex structures.

Reusable Launch Vehicles (RLVs) seem to offer the potential for major cost reduction of access to space. ESA has looked at two distinct applications of FOS for RLV health monitoring: in the structure of large reusable cryogenic tanks and in the inter-tank structure. In the case of cryogenic tanks a suit of embedded FBG sensors has been investigated for the combined monitoring of strain, temperature and H_2 leakage.[86] The preliminary conclusions are that the FBGs will function as strain gauges (- 1000 $\mu\varepsilon$ to + 3000 $\mu\varepsilon$) over a wide range of temperatures down to cryogenic temperatures of 20 K. The temperature sensor also operates down to these temperatures where the fiber needs to be encapsulated in a special glass capillary. However, at cryogenic temperatures the palladium coated FBG H_2 sensors are not practical as they exhibit an inadequate response time below - 30° C. One possible solution is to locally heat the sensors.

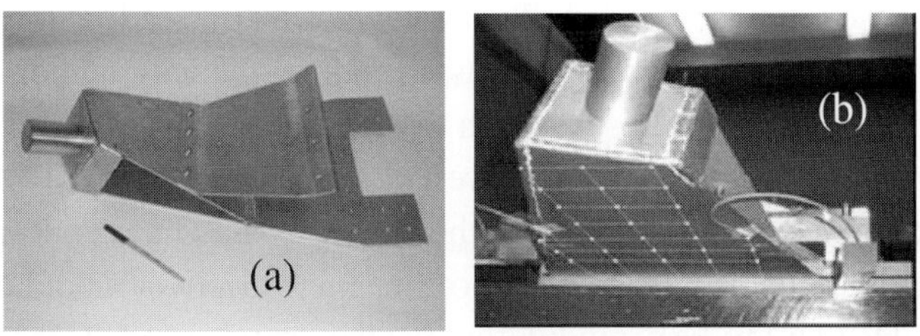

Figure 16. ASTS structure: (a) a lateral view; (b) the lower face with the excitation grid.

Also, ESA has monitored inter-tank CFRP structure with embedded and surface mounted FBGs on a reduced scale demonstrator of an RLV to measure both static and dynamic strain.[87]

Takeda with Mitsubishi Electric Co. presented in 2006[88] a real-time strain measurement of a composite liquid hydrogen (LH_2) tank using FBG sensors. The tank was composed of CFRP and an aluminum liner was fabricated by the filament winding method and mounted on a reusable rocket. This rocket (vertical takeoff and landing) was developed by the Institute of Space and Astronautical Science of the Japan Aerospace Exploration Agency (ISAS/JAXA). A real-time strain measurement of the composite LH_2 tank using FBG sensors during rocket operations has been attempted.

6. Energy Applications

Globalization has caused high demand for electric power; fuel resources are generally not unlimited causing a continuous process of fuel price increase during the recent years.[89]

In the following the various sections of the energy sector will be described in the way of their potential for the application for optical fiber sensors and particularly for FBG based sensing system.

6.1. Power Transmission and Distribution

Over the decades the power demand has continually raised leading to the fact that today some energy transmission lines are reaching their constructed load limit under peak load condition. On this topic, efficient current sensors are demanded. Today several distributed or multiplexed fiber optic systems have been discussed, but so far only demonstrated.[89]

In most of FBG based sensors a calibrated Gaint Magnetostrictive Materials (GMM) is used to convert an electromagnetic field due to electric current to strain applied to the FBG.[90-95] The base concept uses a FBG rigidly attached to a piece of magnetostrictive modulator. The Terfenol-D (Tb-Dy-Fe) modulator is the most used GMM.[90] The modulator responds to the magnetic field by producing a bulk strain that is proportional to the square of the magnetic field strength[96] with consequent shift in Bragg wavelength. Satpathi et al. proposed a Terfenol-D and FBG based sensor scheme for accurate current measurements up to 1000 A.[90] Also enhanced configuration to simultaneous measurement of current and temperature have been demonstrated.[93-95] First, Chiang proposed a temperature-compensated FBG-based magnetostrictive sensor for dc and ac current.[93] The sensor consists of a FBG bonded on two joined pieces of metal alloys, one being Terfenol-D and the other MONEL 400. In order to improve and fit the performance of Terfenol-D based fiber Bragg grating magnetic sensor, the dependence of the magnetostrictive response on the pre-stress has been used by Cusano et al. [97] The possibility to tune sensitivity with a suitable mechanical load allows to work at different operative conditions and to develop advanced sensors with reconfigurable sensitivity. Performance improvements in terms of magnetic resolution up to 0.0116 A/m have been demonstrated. Moreover the non negligible rate-independent memory effects (i.e. hysteresis) can be taken into account with adequate and optimized techniques for hysteresis compensation as shown in Ref. 98.

6.2. Power Generation

Condition monitoring systems are more in use in power generation plants. SmartFibres, Insensys, AOS GmbH, FiberSensing are only a few of the companies actively involved in the different power generation fields. For instance Siemens AG with Bosselmann successfully demonstrated the use of FBG for temperature and dynamic strain measurements in power generators[99] compatible with high voltage nature (15KV). Figure 17(a) shows 4 FBGs attached to the edges of stator winding inside a power generator during a shop test.

6.2.1. Gas and Stream Turbine

Gas turbines for power generation are being operated on the physical limits of materials and structures to meet the increasing demand. Fiber optic systems have chances here, especially for high temperature applications. Willsch et al. (Siemens AG) successfully installed an array of 6 FBG temperature sensors in the cooling air area of a 200 MVA gas turbine.[100] With regard to this issue, also special design grating have been developed: chemical composition gratings (CCGs) (where the grating is

vehicles.[125] In Refs. 126-128, instead, the sensors system was characterized by a proper package involving a piece of steel plate which supports the weight of the traveling vehicle. Compared with other designs of fiber-optic WIM systems this design is easy and reliable. Especially it's suitable for heavy vehicles because of its large capacity, such as military vehicles, trucks and trailers. Over 40-ton load was applied on the system exhibiting a resolution of about 10 kg.

9. Underwater Applications

The development of the efficient hydrophone has remained at the forefront of FBG sensing technology. The operating principle of a FBG-based hydrophone is typically based on the intensity modulation of the laser light due to the shift of transmission power spectrum curve of the sensing element under the influence of the acoustic field.[129-131]

Unfortunately, this class of sensor exhibits low sensitivity to acoustic pressure due to Young module of optical fiber while the sensitivity of FBG response decrease when reducing the ultrasound wavelength below the grating length.[132] In order to increase the sensitivity, FBGs have been coated with proper materials characterized by elastic modulus much lower than the fiber one, as demonstrated by Cusano et al. [129,130] and Yang et al. [131]

In order to analyze the behavior of the FBG hydrophone in terms of sensitivity and bandwidth, FBGs have been coated with different materials and dimensions.[129-130] For a given acoustic pressure, the basic effect of the FBG coating, if thick enough, is to enhance the dynamic strain experienced by the sensor of a factor given by the ratio between the fiber and the coating elastic modulus. This effect can be efficiently adopted to enhance the acoustic sensitivity, if materials with low acoustic damping and acoustic impedance approaching that of the water are used.

The experimental set-up used in Refs. 129-130 is reported in Fig. 21(a).

The acoustic field is generated by a PZT acoustic transducer immersed in a very large water tank (11 x 5 x 7 m) together with the reference PZT hydrophone and the hydrophone under test. The utilized FBGs were embedded in a polymer of cylindrical geometry with diameter of 4mm and length of 25 mm (exhibiting an elastic modulus of ~ 100 MPa) and in a polymeric material of spherical geometry with diameter of 4.4 cm (exhibiting elastic modulus lower than 100 Mpa), respectively. The tested hydrophones are shown in Fig. 21(b). Figure 22(a) shows a comparison between the typical temporal response of the cylindrical-coated FBG hydrophone under test and the reference PZT hydrophone to a sound pressure pulse of 2 kPa at the frequency of 10 kHz. The phase difference between the two responses is due to the different distance from the acoustic source whereas the fluctuations at the end of the traces are due to the eco signals from the wall of the tank.

Using the measured signal-to-noise ratio from the FFT of sensor response, the minimum detectable pressure level was estimated to be about 10 Pa. Finally, also the sensitivities of the both FBG hydrophone configurations have been retrieved and reported in Fig. 22(b).

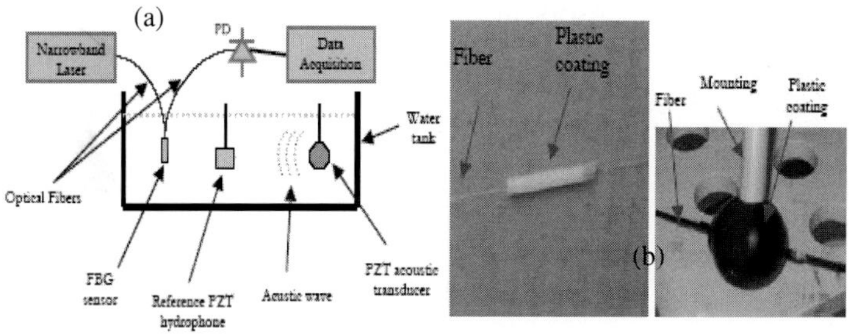

Figure 21. Lateral view of the experimental set-up (a) - Photographs of the tested hydrophones (b).

It can be seen a decreasing response from low frequencies up to about 27 kHz and 16 kHz for the cylindrical and spherical coating, respectively. For upper frequencies the signal to noise ratio approaches the unity around the value of -235 dB reV/μPa. Furthermore, with respect to the cylindrical coating, the spherical-coated hydrophone exhibits a higher sensitivity (up to 30 dB re V/microPa) due to a better acoustic impedance matching with the water, to a slightly smaller acoustic modulus and to a lower damping.

Also FBG-based laser hydrophone has been successfully investigated by different researcher groups.[133,134] Besides, recently, Tam et al. [135] presented a cladding-etched Distributed Bragg Reflector (DBR) fiber laser hydrophone for high-frequency ultrasound sensing applications. A wet etching technique is utilized to reduce the fiber diameter of the DBR laser. They demonstrated that with decreasing diameter of the fiber cladding, the frequency response of the sensor becomes flatter and the peak response frequency increases so that the sensitivity in the high-frequency region is improved. In fact the peak response frequency moves from 21 to 40 MHz when the fiber diameter was reduced from 125 to 68 μm.

For a practical use of such sensors, multiplexed sensing and thermally stabilized operation in the sound detection is also desirable. Takahashi et al recently[136-138] proposed a complex detection configuration involving a feedback control circuit of a tunable laser to enable simultaneous measurement of underwater acoustic field and temperature using FBG sensor array. TDM is demonstrated by constructing a FBG sensor array in which two FBG sensors are parallely arranged by using an optical switch. They achieved a resolution of the temperature measurement of 0.038 °C.

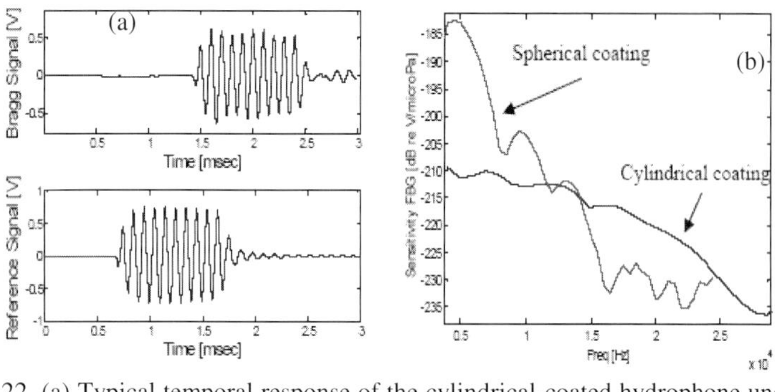

Figure 22. (a) Typical temporal response of the cylindrical-coated hydrophone under test (upper) and the reference hydrophone PZT (lower) to a sound pressure pulse of 2 kPa at the frequency of 10 kHz; (b) The calculated sensitivity curves of the FBG hydrophones.

10. Perspective and Challenges

Numberless examples exhibited in this and previous chapter allow considering FBG as ideal devices to be adopted for a multitude of different sensing applications in light of their intrinsic capability to measure a multitude of parameters such as strain, temperature, pressure —and many others—coupled with their flexibility of design to be used as single point or multi-point sensing arrays. Although FBG-based sensors have attracted commercial interest and developed some lucrative niche markets, there are a number of significant technical hurdles and market barriers to overcome. The most significant barriers that have prevented a more widespread use and commercial diffusion of FBG sensors are inadequate reliability of some existing products and excessive cost.[139]

Reliability is a key feature that needs to be taken very seriously and incorporated in every aspect of the fiber sensing design and production facets. Another significant barrier is the fact that most of the sensor developers and manufacturers only provide one piece of the complete sensing solution puzzle. Customers and end users require, in most cases, complete turn-key solutions that encompass all the necessary sensing components as well as all the necessary software and data processing algorithms and, most importantly, the actual sensing system design and installation.[139] Also precise and accurate standards could favorite FBG spread-out. At the present time, there is no FBG sensor standard in place.[139] This has lead to a broad variability in available grating designs and specifications offered by commercial vendors, as well as a variation in the performance of FBG-based sensors when used in conjunction with instruments from different vendors. In general, custom products are always more expensive and difficult to manufacture than standardized ones. Hence, sensor interrogation systems need to be standardized as well. Several groups in North America, Europe and Asia are active in standards for fiber optic sensors[139] including OIDA (Optoelectronic Industry Development Association, in Washington, D.C.), [140] ISIS Canada,[141] the European Union COST 270/299 Committee[142] and RILEM.[143] Future applications of FBG sensors will rely heavily on cost reduction and development of specialized and application specific packaging. It is expected that more conventional and popular applications such as discrete strain and temperature sensing will continue to evolve and grow and acquire greater market shares. Similarly, applications calling for multi-grating arrays will become more popular as prices come down, allowing to compete more directly with truly alternative optical or electrical sensing technologies. High temperature resistant FBGs such as chemical composition gratings or those written on n-doped, pure Si fibers will open up opportunities in harsh environment sectors such as power plants, turbines, combustion, and aerospace. Similarly, the prospects of using polymer optical fibers (POF) in sensing applications is expected also to open up the door to the development of POF FBGs to be used as inexpensive, simple and low-cost disposable sensors. Notwithstanding these caveats, growth for the FBG sensor market is forecasted to be strong for the commercial markets, with domestic and international government investments likely providing an additional boost. FBGs have reached an inflection point where technology, pricing, and needs have converged.

References

1. B. Culshaw and J. Dakin, *Optical Fiber Sensors: Principle and Components*, (Artech House inc., Norwood, 1988).
2. E. Udd, in *Fiber Optic Sensors: An Introduction for Engineering and Scientists,* Ed. John Wiley and Sons, New York (1991).
3. B. Culshaw and J. Dakin in *Optical Fiber Sensors: Applications, analysis, and future Trends*, Artech House inc., Norwood (1997).
4. R. M. Measures, in *Structural monitoring with fiber optic Technology*, Academic Press, London (2001).
5. K. O. Hill, Y. Frujii, D. C. Johnson and B. S. Kawasaky, *Appl. Phys. Lett.*, **32**, 647 (1978).
6. G. Meltz, W. W. Morey and W. H. Glam, *Optics Letters*, **14**, 823 (1989).
7. D. Lam and B. Garside, *Applied Optics*, **20**, 440 (1981).
8. K. O. Hill and G. Meltz, *J. of Lightwave Technology*, **15**, 1263 (1997).
9. A. Yariv, *IEEE J. of Quantum Electronics* **QE-9**, 919 (1973).
10. T. Erdogan, *J. of Lightwave Technology*, **15**, 1277 (1997).
11. P. J. Lemaire, R. M. Atkins, V. Mizrahi, K. L. Walker, K. S. Kranz and W. A. Reed, *Electronics Letters*, **29**, 1191 (1993).
12. K. O. Hill, B. Malo, F. Bilodeau, D.C. Johnson and J. Albert, *Applied Physical Letters.*, **62**, 1035 (1993).
13. A. Othonos and X. Lee, *IEEE Photonics Technology. Letters*, **7**, 1183 (1995).
14. B. Malo, S. Theriault, D. C. Johnson, F. Bilodeau, J. Albert and K. O.Hill, *Electronics Letters*, **31**, 223 (1995).
15. J. E. Sipe, L. Poladian and C. M. de Sterke, *J. Optic Society America* **A 11**, 1307 (1994).
16. T. Erdogan and J. E. Sipe, *J. of Optic Society of America* **A 13**, 296 (1996).

17. W. H. Loh and R. I. Laming, *Electronics Letters*, **31**, 1440 (1995).
18. B. J. Eggleton, P. A. Krug, L. Poladian and F. Ouellette, *Electr. Lett.*, **30**, 1620 (1994).
19. T. Erdogan, *J. of Optic Society of America* **A 14,** 1760 (1997).
20. S. M. Melle, K. Liu and R. Measures, *IEEE Photon Techn. Letters*, **4**, 516 (1992).
21. P. St. J. Russel, and J. L. Archambault, *Fiber Gratings in Optical Fiber Sensors,* Culshaw and J. Dakin Eds., Artech House, 9-67 (1997).
22. A. D. Kersey, M. A. Davis, H. J. Patrick, M. LeBlanc, K. P. Koo, C. G. Askins, M. A. Putnam and E. J. Friebele, *J. of Lightwave Technology*, **15**, 1442 (1997),
23. A.Othonos and K. Kalli, in *Fiber Bragg Gratings Fundamentals and Applications in Telecommunications and Sensing* Boston: Artech House (1999).
24. R.Kashyap, in *Fiber Bragg Gratings* San Diego: Academic Press (1999).
25. A. Mendez, Proceedings of SPIE **6619**, 661905 (2007).
26. Zhi Zhou and Jinping Ou, Proceeding of North American Euro-Pacific Workshop, USA 2004, http://www.micronoptics.com.
27. http://www.micronoptics.com/sensing.htm.
28. http://www.hbm.com/.
29. A. Csipkes, S. Ferguson, T.W. Graver, T. C. Haber, A. Méndez and J. W. Miller in *The maturing of optical sensing technology for commercial applications*, Micron Optics (USA), www.micronoptics.com/.
30. C. Barbosa, N. Costa, L. A. Ferreira,, F. M. Araújo, H. Varum and A. Costa, 18th International Optical Fiber Sensors Conference Technical Digest Mf4, (2006).
31. P. Moyo, J. M. W. Brownjoh, R. Suresh, S. C. Tjin, *Engine Struct.*, **27**, 1828 (2005).
32. I. W. Wu, C. Y. Wang, M. H. Chen, H. L. Wang, A. Cheng, P. Tsai, D. Wu, H. Chien, Shang, in *FBG Bending Gauge on bridges: An effort towards standardization of bridge structural health monitoring,* http://www.micronoptics.com.
33. R. Measures, A. T. Alavie, R. Maaskant, M. Ohn, S. Karr and S. Huang, *Smart Materials and Structures*, **4,** 20 (1995).
34. http://www. fibersensing.com/.
35. http://www.fos-s.be/.
36. G. Kister, D. Winter, Y. M. Gebremichael, J. Leighton, R. A. Badcock, P. D. Tester, S. Krishnamurthy, W. J. O. Boyle, K. T. V. Grattan and G. F. Fernando, *Engineering Structures* **29,** 2048 (2007).
37. H. F. Lima, R. Vicente , R. N. Nogueira, I. Abe, P. Andréa, C. Fernandes, H. Rodrigues, H. Varum, H. J. Kalinowski, A. Costa and J. L. Pinto, Proceedings of SPIE **6619**, 661941 (2007).
38. Y. Bin Lin, C. L. Pan, Y. H. Kuo, K. C. Chang and J. C. Chern, *Smart Materials and Structures* **14**, 1075 (2005).
39. T. H. T. Chan, L. Yu, H. Y. Tam, Y. Q. Ni, S. Y. Liu, W. H. Chung and L. K. Cheng, *Engineering Structures*, **28**, 648 (2006).

40. Y. M. Gebremichael, W. Li, B. T. Meggitt, W. J. O. Boyle, K. T. V. Grattan, B. McKinley, L. F. Boswell, K. A. Aarnes, S. E. Aasen, B. Tynes, Y. Fonjallaz and T. Triantafillou, *IEEE Sensors J.*, **5,** 510 (2005).
41. S. K. T. Grattan, P. A. M. Basheer, S. E. Taylor, W. Zhao, T. Sun and K. V. T. Grattan, *J. of Physics: Conference* **76**, 012018 (2007).
42. http://www.smartfibres.com/.
43. T. Chang, D. Li, Q. Sui, L. Jia, Z. Wei and H. Cui, Proceedings of the IEEE International Conference on Automation and Logistics August 18 - 21, Jinan, China, 1652 (2007).
44. T Francis, H Legge, P. L Swart, G. van Zyl and A. A Chtcherbakov, *Measurement Science and Technology*, **17,** 1173 (2006).
45. Z. Zhou, H. Wang, J. Ou, 18th International Optical Fiber Sensors Conference Technical Digest (Optical Society of America, Washington, DC, ThE90 (2006).
46. L. Briançon, A. Nancey, F. Caquel, P. Villard, Proceedings of EUROGEO 3, March 1–3, Munich, Germany, 471 (2004).
47. M. Voet, A. Nancey, J. Vlekken, Proceedings of SPIE, **5855**, 214 (2005).
48. L. Briançon A. Nancey P. Villard, Studia Geotech Mechanica, **XXVII**, 21 (2005).
49. A. Laudati, F. Mennella, M. Giordano, G. D'Altrui, C. Calisti Tassini, and A. Cusano, *IEEE Photonics Technology Letters*, **19**, 1991 (2007).
50. www.optosmart.com.
51. A. Cusano, A. Cutolo, J. Nasser, M. Giordano and A. Calabrò, *Sensors and Actuators A, Physical*, **110**, 276 (2004).
52. A. Cusano, A. Cutolo, G. Breglio, M. Giordano and L. Nicolais, *Optical Engineering*, **5706**, 1 (2005).
53. G. E. Green and P. E. Mikkelsen, in *Deformation Measurements with Inclinometers*, Transportation Research Record 1169, TRB, National Research Council, Washington, D.C., 1 (1988).
54. H. Y. Te, H. A. Bin, M. Jiming, Z. Baishan and C. Jingang, Proceedings of SPIE **5855**, 1020 (2005).
55. K. Fujihashi, T. Aoki, M. Okutsu, K. Arai, T. Komori, H. Fujita, Y. Kurosawa, Y. Fujinawa, and K. Sasaki, in *Development of Seafloor Seismic and Tsunami Observation System*, Symposium on Underwater Technology and Workshop on Scientific Use of Submarine Cables and Related Technologies, 349 (2007).
56. Y. Okabe, R. Tsuji and N. Takeda, in *Composites. Part A: Applied Science and manufacturing*, **35,** 59 (2004).
57. H. Tsutsui, A. Kawamata, J Kimoto, A Isoe, Y Hirose, T Sanda and N Takeda, Proceeding of SPIE **5054**, 184 (2003).
58. N. Takeda, Y. Okabe, J. Kuwahara, S. Kojima and T. Ogisu, *Composite Science and Technologies*, **65**, 2575 (2005).
59. S. Takeda, S. Minakuchi, Y. Okabe and N. Takeda, *Composites. Part A: Applied Science and manufacturing*, **36,** 971 (2005).

60. T. Yamamoto, S. Takeda, Y. Okabe and N. Takeda, Proceedings of the 8th Japan Inter SAMPE Symposium, Tokyo, SAMPE-Japan, 159 (2003)
61. H. Tsutsui, A. Kawamata, T. Sanda and N. Takeda, *Smart Materials and Structures*, **13**, 1284 (2004)
62. H. Tsutsui, A. Kawamata, J. Kimoto, A. Isoe, Y. Hirose, T. Sanda and N. Takeda, *Advanced Composites Materials* **13,** 43 (2004)
63. J. M. Menendez Martin and A. G. Gordo, 18th International Optical Fiber Sensors Conference Technical Digest (2006), Mf1.
64. K. S. C. Kuang, L. Zhang, W. J. Cantwell and I. Bennion, *Composites Science and Technology* **65**, 669 (2005).
65. T. A. Dawood, R. A. Shenoi, and M. Sahin, *Composites Part A: Applied Science and manufacturing*, **38**, 217 (2007).
66. N. Takeda S. Minakuchi and Y. Okabe, *J. of Solid Mechanics and Materials Engineering*, **1**, 3 (2007).
67. U. Ben-Simon, I. Kressel, Y. Botsev, A.K. Green, G. Ghilai, N. Gorbatov, M. Tur and S. Gali, Proceedings of SPIE **6619**, 661944 (2007).
68. M. Weis, J. Hoflin, P. Deimel, R. Bilgram and K. Drechsler, *Damage Detection using Fiber Bragg Grating Sensors* 9th European Conference NDT ECNDT (2006).
69. C. Davis, in *Strain Survey of an F/A-18 Stabilator Spindle Using High Density Bragg Grating Arrays*, Published by DSTO Platforms Sciences Laboratory 506 Lorimer St. Fishermans Bend, Victoria 3207, Australia (2005).
70. A. Cusano, P. Capoluongo, S. Campopiano, A. Cutolo, M. Giordano, F. Felli, A. Paolozzi, and M. Caponero, *IEEE Sensors J.*, **6**, 67 (2006).
71. P. Capoluongo, C. Ambrosino, S. Campopiano, M. Giordano, I. Bovio, L. Lecce, A. Cutolo and A. Cusano, *Sensors and Actuators A: Physical*, **133**, 415 (2007).
72. R. D. Reed, D. Osmont and M. Dupont, 2nd European Workshop on Structural Health Monitoring SHM 2004, 1051 (2004).
73. D. Betz, G. Thursby, B. Culshaw and W. Staszewski, *Smart Materials and Structures* **12**, 122 (2003).
74. H. Tsuda, N. Toyoma, J. Takatsubo, *J. of Material Science*, **39**, 2211 (2004).
75. X. Qing, A. Kumar, C. Zhang, I. Gonzales, G. Guo and F. Chang, *Smart Materials and Structures*, **14**, 98 (2005).
76. Y. Okabe, J. Kuwahara, K. Natori, N. Takeda, T. Ogisu, S. Kojima and S. Komatsuzaki, *Smart Materials and Structures*, **16,** 1370 (2007).
77. Y. Botsev, M. Tur, E. Dery, I. Kressel, U. Ben-Simon, S. Gali and D. Osmont, 18th International Optical Fiber Sensors Conference Technical Digest (Optical Society of America, Washington, DC, ThD3 (2006).
78. K. K. Chau, B. Moslehi, G. Song and V. Sethi, Proceedings SPIE **5391**, 753 (2004).
79. D. H. Kim, J. H. Han, and I. Lee, *AIAA J. of Aircraft*, **41**, 409 (2004).
80. L. Cheng, Y. Zhou, M. M. Zhang, *J. of Sound and Vibration*, **292**, 279 (2006).

81. C. Ambrosino, G. Diodati, A. Laudati, A. Gianvito A. Concilio, R. Sorrentino, G. Breglio, A. Cutolo and A. Cusano, Proceedings of SPIE **6619**, 661940 (2007).
82. I. Mckenzie and N. Karafolas, Proceedings of SPIE **5855,** 262 (2005).
83. M. Kunzler, E. Udd, S. Kreger, M. Johnson and V. Henrie, Proceedings of SPIE **5758**, 168 (2005).
84. E. Udd, 18th International Optical Fiber Sensors Conference Technical Digest, Optical Society of America, Washington, DC, Mf2 (2006).
85. A. Cusano, P. Capoluongo, S. Campopiano, A. Cutolo, M. Giordano, M. Caponero, F. Felli and A. Paolozzi, *Smart Materials and Structures,* **15,** 441 (2006).
86. I. Latka, W. Ecke, B. Hofer, C. Chojetzki A. Reutlinger, Proceedings of SPIE **5579**, 195 (2004)
87. V. Diaz, N. Eaton, S. Merillat, A. Wurth, G. Ramusat, 54th Internat Astronautical Congress of the International Astronautical Federation, Germany (2003).
88. T. Mizutani, N. Takeda and H. Takaya, *Structural Health Monitoring*, **5**, 205 (2006).
89. T. Bosselmann, Proceedings of SPIE **6619**, 661903, (2007).
90. D. Satpathi, J.A. Moore and M.G. Ennis, *IEEE Sensors Journal*, **5**, 1057 (2005).
91. Z. Hong, X. Yanling, Z. Jian and L. Yuelan, Conference on Lasers and Electro-Optics - Pacific Rim, 1 (2007).
92. Z. Hong, X. Yanling, Z. Jian, W. Shurong, in *Measurement of Power Frequency AC Current Using FBG and GMM*, Proceedings of 2005 International Symposium on Electrical Insulating, 741 (2005).
93. K.S. Chiang, R. Kancheti and V. Rastogi, *Optical Engineering*, **42,** 1906 (2003).
94. D. Reilly, A. J. Willshire, G. Fusiek, P. Niewczas and J. R. McDonald, Proceedings of IEEE Sensors, 1426 (2004).
95. D. Reilly, A. J. Willshire, G. Fusiek, P. Niewczas and J. R. McDonald, *IEEE Sensors Journal*, **6**, 1539 (2006).
96. G. Engdahl, Ed., in *Handbook of Giant Magnetostrictive Materials* (New York: Academic, 2000).
97. C. Ambrosino, S. Campopiano, A. Cutolo, A. Cusano, "Sensitivity Tuning in Terfenol-D Based Fiber Bragg Grating Magnetic Sensors", *IEEE Sens. Lett.* - in press
98. D. Davino, C. Visone, C. Ambrosino; S. Campopiano, A. Cusano and A. Cutolo, " Compensation of hysteresis in magnetic field sensors employing Fiber Bragg Grating and magnetoelastic materials", *Sensors and Actuators* **A** –in press.
99. T. Bosselmann, Proceedings of the SPIE **5855**, 188 (2005).
100. M. Willsch, T. Bosselmann, N. M. Theune, Sens Conf. Proceedings IEEE **1**, 20 (2004).
101. A.F. Fernandez, B. Brichard, F. Berghmans, H.E. Rabii, M. Fokine and M. Popov, *IEEE Transactions on Nuclear Science*, **53**, 1607 (2006).

102. K. Y. Lee, S. S. Yin, and A. Boehman, "Intelligent Monitoring System With High Temperature Distributed Fiberoptic Sensor For Power Plant Combustion Processes", Final Technical Report Submitted to U. S. Department of Energy, Period: 09/27/2002 to 09/26/2006. http://www.micronoptics.com (2006).
103. K. Krebber, W.Habel, T.Gutmann, C.Schram, Proceedings of SPIE **5855**, 1036 (2005).
104. K. Schroeder,W. Ecke, J Apitz, E. Lembke and G. Lenschow, *Measurement Science Technology* **17**, 1167 (2006).
105. K. S. Kim, Y. C. Song, G. S. Pang and D. J. Yoon, Proceedings of SPIE **5765**, 584 (2005)
106. R. Damon, *Oil Gas Magazine*, **35**, (2007).
107. P. J. Wright and W. Womack, Offshore Technology Conference, Houston, Texas, U.S.A., OTC 18121, (2006).
108. F.AI-Bani , H.AI-Sarrani, I. Arnaout , A. Anderson, Y. Aubed and E. S. Johansen, Intelligent Well Completion (www.worldoil.com) **228** (2007).
109. O.V.Butov, K.M.Golant, V.I. Grifer, Ya.V.Gusev, A.V.Kholodkov, A.V.Lanin, R.A. Maksutov and G.I Orlov, in 18th International Optical Fiber Sensors Conference Technical Digest, TuB6 (2006) .
110. G.D. Lloyd, L.A. Everall, K. Sugden, I. Bennion, Proceedings of SPIE **5855**, 218 (2005).
111. G. Lloyd, L. Everall, K. Sugden and I. Bennion, *Optics Commun.,* **244**, 193 (2005).
112. D. Roberts and T. Moros, *Deepwater Technology Journal* **227**, (2006).
113. Y. Zhang, S. Li, Z. Yin, B. Chen, H. L. Cui, J. Ning, *Optic. Eng.,* **45**, 084404 (2006).
114. L. Ren, H.N. Li, J. Zhou, D.S. Li and L. Sun, *Optic. Eng.* **45**, 084401 (2006).
115. www.fbgs-technologies.com.
116. D. Inaudi and B. Glisic, in 18th International Optical Fiber Sensors ConferenceTechnical Digest, FB3 (2006).
117. S. Grosswig, E. Hurtig, S. Luebbecke, B. Vogel, Proceedings of SPIE **5855**, 226 (2005).
118. N. Singh, S.C. Jain, V. Mishra, G. C. Poddar, P. Kaur, H. Singla, A. K. Aggarwal and P. Kapur, *Current Science*, **90**, (2006).
119. M.G. Shlyagin, S.V. Miridonov, V.V. Spirin, R. Martinez Manuel, I. Márquez Borbón, S.A. Kukushkin, V.V. Kulikov and V. I. Belotitskii, in 18th International Optical Fiber Sensors Conference Technical Digest, ThE50 (2006).
120. T.L. Yeo, T. Sun, K.T.V. Grattan, D. Parry, R. Lade and B.D. Powell, *Sensors and Actuators B: Chemical,* **110**, 148 (2005).
121. N. .M. Theune, T. Bosselmann, M. Willsch, J. Kaiser and H. Hertsch, Proceedings of SPIE **5502**, 536 (2004).
122. F. Mennella, A. Laudati, M. Esposito, A. Cusano, A. Cutolo, M. Giordano, S. Campopiano and G. Breglio, Proceedings of SPIE **6619**, 66193H, (2007).

123. Z. Zhou, J. Liu, H. Li and J. Ou, Proceeding of SPIE **5855**, 735 (2005).
124. M. A. Caponero, D. Colonna, M. Gruppi, M. Pallotta and R. Salvatori, Proceedings SPIE **5502**, 480 (2005).
125. S. Berardis, M. A. Caponero, F. Felli, F. Rocco, Proceedings of SPIE **5855**, 695 (2005).
126. K. Wang, Z. Wie, B. Chen, C. H. Liang, Proceed SPIE **6004**, 60040S.1 (2005).
127. W. Li, B. Jiang, Q. Zhang and F. Zhu, Proceedings of SPIE **6830**, 683023, (2007).
128. K. Wang, Z. Wei, B. Chen and H. L. Cui, Proceedings of SPIE **5778**, 220 (2005).
129. A. Cusano, S. D'Addio, S. Campopiano, M. Balbi, S. Balzarini, M. Giordano and A. Cutolo, *Sensors & Transducers J.,* **82**, 1450 (2007).
130. A. Cusano, S. Campopiano, S. D'Addio, M. Balbi, S. Balzarini, M. Giordano, A. Cutolo, 18th International Optical Fiber Sensors Conference, ThE85 (2006).
131. X. Ni, Y. Zhao and J. Yang , *Sensors and Actuators A* **138,** 76 (2007).
132. A. Minardo, A. Cusano, R. Bernini, L. Zeni, M. Giordano, *Ultrasonics, Ferroelectrics and Frequency Control* **52,** 304 (2005).
133. B. O. Guan, H. Y. Tam, S. T. Lau and H. L. Chan, Proceedings of SPIE **5502**, 116 (2004).
134. S. Foster, A. Tikhomirov, M. Milnes, J. Velzen, G. Hardy, Proceedings of SPIE **5855**,627 (2005).
135. L. Y. Shao, S. T. Lau, X. Dong, A. Ping Zhang, H. L. W. Chan, H.Y. Tam, and Sailing He, *IEEE Photonics Technology Letters* **20**, 548 (2008).
136. S. Tanaka, T. Ogawa, H. Yokosuka, and N. Takahashi, *Japanese J. Applied Physics Part 1*, **43**, 2969 (2004).
137. S. Tanaka, H. Yokosuka, N. Takahashi and J. Marine, *Acoustic Society Jpn.*, **33**, (2006).
138. H. Yokosuka, S. Tanaka, K. Inamoto and N. Takahashi, in 18th International Optical Fiber Sensors Conference Technical Digest, TuE83 (2006).
139. A. Mendez, Proceedings of SPIE **6619**, 661905 (2007).
140. www.oida.org.
141. www.isiscanada.com.
142. www.cost299.org.
143. www.rilem.net.

DISTRIBUTED OPTICAL FIBER SENSORS

Romeo Bernini,[a] Aldo Minardo[b] and Luigi Zeni[b,*]

[a]Istituto per il Rilevamento Magnetico dell'Ambiente, CNR
Via Diocleziano 128, 80124 Napoli, Italy
[b]Seconda Università di Napoli
Dipartimento di Ingegneria dell'Informazione
Via Roma 29, 81031 Aversa, Italy
[*]E-mail: luigi.zeni@unina2.it

Optical fibers offer the unique advantage of allowing spatially distributed sensing of several quantities. This is especially important for the monitoring of large, critical structures. In this chapter we review the main techniques for distributed sensing using optical fibers.

1. Introduction

Optical fibers are made from fused silica, are about the diameter of a human hair, and transmit light over large distances with very little loss. They can also be made to be sensitive to their state and environment and are therefore well suited as sensors. Optical fibers sensors have been the subject of a remarkable interest in the last 20 years, since they present some distinct advantages over other technologies. The principal single attractive feature of optical-fiber sensors is undoubtedly their ability to function without any interaction with electromagnetic fields. This opens applications in the electrical power industry (where nothing else can do the job) and assists very significantly where long transmission distances of relatively weak signals are an essential part of the sensing process. The lack of electrical connections has other, broader implications. Optical sensors have major advantages when conductive fluids, such as blood or sea water, are involved. Also, the need for intrinsic safety (for example, in monitoring the presence of explosives gases or in assessing petrochemical plants) is often paramount. The optical fiber is also remarkably strong, elastic, and durable, and has found its place as an instrumentation medium for addressing smart structures, where the sensors must tolerate the environment to which the structure is subjected and therefore to be immune to large physical strain excursions, substantial temperature excursions, and often a chemically corrosive operating environment.

A very important and unique feature of fiber-optics technology consists in its capability for long-range distributed sensing. These measurements allow the values of the measurand of interest to be extracted, as a function of position, along the length of the sensing fiber. Distributed sensors are particularly attractive for use in applications where monitoring of the measurand is required at a large number of points or continuously over the path of the fiber. Typical examples of applications areas include:

- stress monitoring of large structures such as buildings, bridges, dams, storage tanks, pipelines, ships;
- temperature profiling in electrical power transformers, generators, reactor systems, process control systems and fire detection systems;
- leakage detection in pipelines, fault diagnostics and detection of magnetic/electrical field anomalies in power distribution systems and intrusion alarm systems;
- embedded sensors in composite materials for use in real-time evaluation of stress, vibration and temperature in structures.

Truly distributed sensing techniques are commonly based on some kind of light scattering mechanism occurring inside the fiber. Spatial resolution is typically achieved by using the optical time domain reflectometry (OTDR),[1] in which optical pulses are launched into an optical fiber and the variations in backscattering intensity caused by measurand is detected as a function of time. Alternative detection techniques, such as frequency-domain approaches, have been also demonstrated.

Distributed optical fiber systems can be classified into three primary sub-classes:[2]

Linear backscattering: in this class the optical pulse propagation lies within the linear regime and light backscattered from the pulse is time-resolved and analyzed to provide the spatial distribution of the measurand field (see Fig. 1a). Backscattered light keeps in this case the same wavelength of the incident pulse.

Non-linear backscattering: the difference here is that the optical pulse has sufficient peak power to enter the nonlinear regime and the backscattered power has to be analyzed differently (Fig. 1b). The backscattered light will have, in this case, a different wavelength than that of the incident light. The advantages of entering the non-linear regime are that there is a diverse range of non-linear optical effects offering specific responses to external measurands and ready discrimination at the detector. The main disadvantage is that the magnitude of the effect is strongly dependent upon the optical power and thus can vary significantly along the fiber as a result of attenuation.

Non-linear forward-scattering: another advantage of the non-linear regime is that it allows independent optical signals to interact. Thus it is possible for counter-propagating radiations (e.g. a pulse and a continuous wave (CW), or two pulses) to interact (see Fig.1c). When the interaction is influenced by the external field, the field can be mapped by the action of a forward-scattered (as opposed to a backscattered) propagation of light. However, the same disadvantage of strong power dependence also applies, of course, to this mode of operation. Linear systems are less complex; in particular they are less demanding with respect to source requirements and fiber properties. Non-linear backscattering systems require, generally, high-power pulse sources and, sometimes, fibers appropriate for the nonlinear effect in question, but they do provide a broader range of measurand interactions and a ready discrimination at the detector.

Non-linear forward-scattering systems possess the same advantages and disadvantages as non-linear backscattering systems but have the added advantage of a much higher signal level, and thus a larger signal-to-noise ratio, and the added disadvantage of requiring in most cases two high-performance optical sources and access to both ends of the fiber.

A better appreciation of all of these features will be acquired as we now illustrate how each of these methods operates in practice, by describing examples of specific arrangements which have been demonstrated.

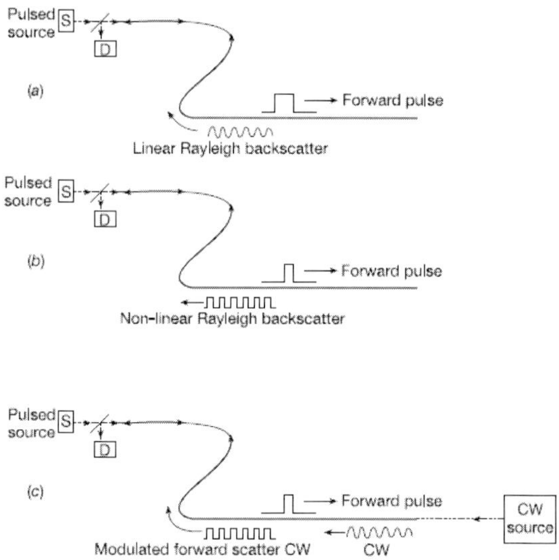

Figure 1. Schemes for fully distributed sensing: (a) linear backscattering, (b) non-linear backscattering (c) non-linear forwards scattering (After A. Rogers, Meas. Sci. Technol., 10, R75 (1999)).

2. Linear Backscattering Systems

Distributed sensing in the linear regime is commonly based on the use of OTDR systems. The latter were first developed, in order to locate fiber breaks or bad splices along a fiber link. Within the linear regime, backscattered light is due to Rayleigh scattering, and exhibits the same wavelength than that of the incident pulse. Measuring the intensity of Rayleigh backscatter as function of time, optical attenuation can be measured all along the fiber. Spatial resolution is directly related to the pulse temporal width: narrower pulses give rise to higher spatial resolution (but also weaker backscatter signals). When using standard OTDR equipment with a spatial resolution of 1 meter or higher, Rayleigh backscatter in standard fibers gives information only about optical attenuation, and it can not be related to other measurands such as temperature or strain. In order to achieve sensitivity to these measurands, specialty fibers must be employed, such as liquid-core fibers[3] or doped fibers,[4] by which the measurand of interest modulates the optical attenuation. These approaches have been demonstrated with a relatively poor spatial resolution (in the order of 10 meters) and temperature accuracy (in order of 2 °C). Consequently, whilst such systems may well have uses in particular applications in which high accuracy and resolution are not required (e.g. fire alarms in buildings), they are unlikely to find general application as temperature monitors in industrial plant.

A recent, very interesting approach makes use of the very high spatial resolution allowed by swept-wavelength interferometry, in order to correlate temperature and strain of the fiber with the spectrum of the Rayleigh backscatter spatial fluctuations.[5] The most appreciable features of this approach are: a) standard telecommunication fibers can be used, instead of specialty fibers; b) very high spatial resolution (a few millimeters) has been demonstrated, allowing, for example, monitoring of aerospace structures where very large strain gradients may exist. On the other hand, the main disadvantages are the equipment cost (a tunable laser source is needed for the measurements), and the limited number of sensing points (≈ 100).

Another approach, in which standard fibers can be used for temperature and strain monitoring, is the Polarization-Optical Time-Domain Reflectometry (POTDR). It represents a polarimetric extension of OTDR.[6,7] Whereas, in OTDR, the power level of the Rayleigh-backscattered radiation, from a propagating optical pulse, is time resolved to provide the distribution of attenuation along the length of the fiber, in POTDR it is the polarization state of the backscattered light which is time resolved; this provides the spatial distribution of the fiber's polarization properties. With the determination of the spatial distribution of the polarization properties of the fiber comes the capability of measuring the distribution of any external field which modifies those properties, such as strain, pressure, temperature, electrical field and magnetic field. While the reported performances for POTDR systems compare very well with other techniques (e.g. Ref. 8 reports an inaccuracy of 1% for measurement of 3 μstrain over 0.1 m of spatial resolution), the technique possesses several disadvantages. Firstly, it cannot discriminate amongst the various effects (e.g. simultaneous temperature and strain) all of which are capable of modifying the polarization properties. Secondly, polarization information is partly lost in backscattering. Any rotation of the polarization state which occurs on the forward passage of light through the fiber is cancelled on back-reflection through the fiber: hence all knowledge of a pure rotation is lost in backscatter. Consequently, the loss of information prevents one from having full knowledge of the distribution of the fiber's polarization properties.[2]

3. Non-Linear Backscattering Systems

As discussed before, nonlinear effects provide the opportunity to employ the fiber for measuring some quantities which have some influence on the nonlinear process. The nonlinear processes having higher efficiencies in standard silica optical fibers are the Raman and the Brillouin effects. Both processes result in some backscattered light within the fiber, wavelength-shifted with respect to the incident light.

Spontaneous Raman scattering arises from molecular vibrations and rotations within a medium. The Raman spectrum for silica is a result of the spread of bond energies in an amorphous solid. The higher energy (shorter wavelength) radiation is known as anti-Stokes light whilst the lower energy component is known as Stokes light. As the ratio between anti-Stokes radiation and Stokes level is temperature-dependent, measurements of Raman scattering can be employed to estimate fiber temperature.[9] Still, the principle of time-domain reflectometry can be employed in order to spatially resolve the measurand of interest. Intensive development of this system has led to various important improvements in design and performance. Amongst these are the uses of two wavelengths so that Stokes and anti-Stokes wavelengths suffer no differential loss,[10] the use of photon counting to improve the spatial resolution down to a few centimeters,[11] or carrying out the measurements in the frequency-domain for synchronous detection.[12] Currently available commercial systems provide impressive performances of ± 0.5 °C temperature resolution with a spatial resolution of 1 m for distances up to 10 km, and of 5 m for distances up to 30 km. Measurement times are of the order of tens of seconds.

The Brillouin effect is similar to the Raman effect in that an optical pump causes excitation of molecules either from the ground state or from an excited state and decay of these states leads to Stokes (a longer wavelength than that of the pump) or anti-Stokes (a shorter wavelength than that of the pump) components (Fig. 2), just as in the Raman case.

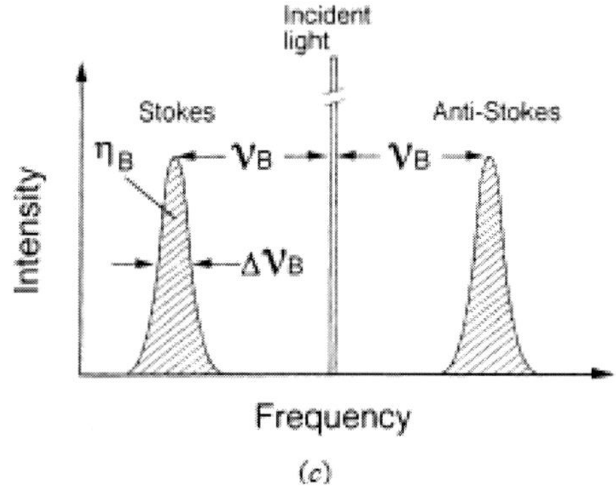

Figure 2. The Brillouin spectrum.

The important difference is that, for Brillouin scattering, the real excited states are due to bulk movement of the molecules rather than to the rotations and vibrations of the individual molecules.[13] Essentially, scattering occurs as a result of a Bragg-type reflection from moving diffraction gratings created from the refractive-index variations caused by acoustic waves propagating axially in the fiber material. These acoustic waves can be generated spontaneously by thermal excitation and, when this is the case, the resulting scattering effect on optical waves is known as 'spontaneous' Brillouin scattering. (The alternative, 'stimulated' Brillouin scattering, will be described in the next section).

Thus, the Stokes scattered wave will be from an axially propagating acoustic wave moving away from an optical "pump" pulse and the anti-Stokes wave from one moving towards it. The Brillouin effect leads to a Stokes and anti-Stokes frequency shift in the optical fiber which is given by[13]:

$$\nu_B = \pm \frac{2nV_a}{\lambda} \tag{1}$$

where n is the refractive index of the fiber material, V_a is the acoustic velocity and λ is the free-space pumping wavelength. For silica fiber at a pumping wavelength of 1.55 μm we have $\nu_B \approx 10.8$ GHz.

As Brillouin frequency shift depends on both the optical refractive index and the acoustic wave velocity, it changes whenever these quantities change in response to local environmental variations and can be used to deduce the temperature and strain along the fiber. Several experiments have demonstrated an excellent linearity of the Brillouin frequency shift with respect to both fiber strain and temperature, for a wide range of these quantities. Figure 3 shows the dependence of the Brillouin frequency shift on temperature (a) and strain (b) measured for a pump wavelength of 1.32 μm, from which a linear temperature coefficient of 1.36 MHz/°C and strain coefficient of 594.1 MHz/% can be derived.[14] In spontaneous Brillouin scattering based systems, the scattered power level is also dependent on the temperature and longitudinal strain, since the scattering cross-section depends upon these measurand parameters. Hence it is possible to exploit the simultaneous dependence of Brillouin shift and power from temperature and strain, in order to measure both measurands. A considerable amount of work has been done in this area.[15-18] The technique has been designated Brillouin optical time-domain reflectometry (BOTDR).

The primary advantages for measurement of temperature using BOTDR are that the backscattering level is larger than that for the Raman effect and that the detection requires measurement of a frequency shift. The former allows distributed measurements over very long lengths (up to 100 km sensing length was demonstrated[19]), whilst the latter allows a variety of sophisticated opto-electronic techniques to be used in the detection process.

Disadvantages, however, are that the persistence of the acoustic wave (the phonon lifetime in the quantum description) limits, in principle, the spatial resolution to, at best, around 1 m, which is too large for many applications; and that the frequency shift is simultaneously dependent both upon the temperature and upon the longitudinal strain, making it difficult to discriminate between them.

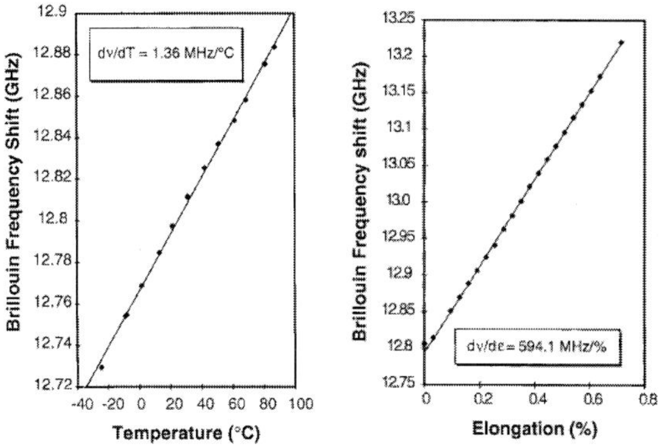

(a) (b)

Figure 3. Brillouin frequency shifts as a function of (a) temperature and (b) strain. (After M. Niklés, L. Thévenaz, and Philippe A. Robert, J. Lightwave Technol., 15, 1842, 1997).

These limits also apply to sensors working on stimulated Brillouin scattering, and a discussion on the main techniques proposed to overcome them will be given in the next section.

4. Non-Linear Forward-Scattering Systems

A forward-scattering DOFS system is based on the interaction, via a non-linear optical effect, between two beams counter-propagating along a fiber.[20] When this non-linear interaction is influenced, in some deterministic way, by an external measurand field, then that measurand value can be mapped along the fiber to comprise the DOFS measurement (Fig. 1(c)). In most cases the counter-propagating radiations comprise an optical pulse and an optical CW. The positional information is provided via knowledge of the pulse's position at any time (in common with backscattering systems) and the non-linear interaction is mapped along the fiber by observing the magnitude of the non-linear effect on the emerging CW, from the pulse's launch end of the fiber, as a function of time.

Another spatially-resolving technique makes use of a sinusoidally-intensity modulated beam and an optical CW. In this case, the complex amplitude of the modulation induced on the CW beam, as a result of the nonlinear interaction, is measured for a range of modulation frequencies.

The so measured base-band transfer function represents the harmonic response of the fiber, and is equivalent, from a theoretical point of view, to the pulse response measured in the time domain. This approach is referred to as optical frequency-domain reflectometry (OFDR).[21-22] Generally speaking, OFDR systems offer higher accuracy with respect to OTDR systems, thanks to synchronous detection, but also require longer acquisition times.

Forward-scattering schemes have been demonstrated both for Raman and Brillouin effect. In the next sub-sections, we will discuss the principle and the main features of both configurations.

4.1. The Forward-Scattering Raman Gain DOFS

Figure 4 shows a possible arrangement for distributed sensing based on stimulated Raman scattering.[23] A Nd:YAG-pumped dye laser was used as the pump laser, whereas light from a He–Ne laser was used as the probe. The latter was amplified during its propagation along the fiber, due to interaction with the pulsed pump. The effectiveness of the stimulated Raman process depends upon the relative polarization states of the two counter-propagating beams, being maximum when they coincide and minimum when they are orthogonal. Thus any measurand which can affect the polarization properties of the fiber will so affect the stimulated Raman interaction and is, in principle, capable of being measured in a distributed manner using the method.

In Ref. 23, the sensing fiber was subjected to a stress field which altered the polarization properties and thus the detected time dependence of the gain received by the He–Ne laser light. However, it is clear that a stress at any one point on the fiber (see Fig. 4) will affect the polarization states and thus the Raman gain at all other points.

Consequently, the signal processing will be extremely complex, in order to compensate for such nonlocal effects.

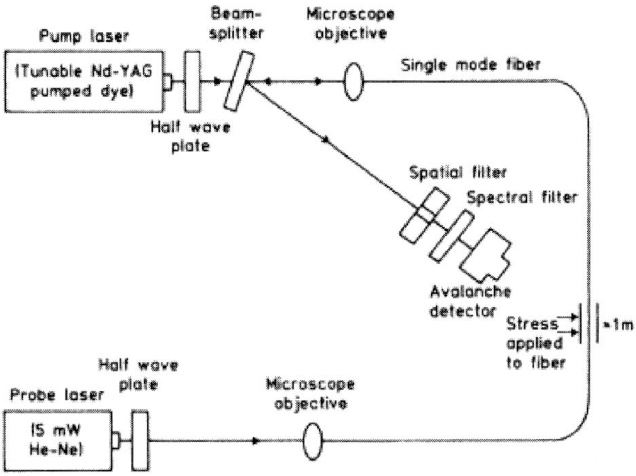

Figure 4. The forwards-scattering Raman gain DOFS arrangement (After A. Rogers, Meas. Sci. Technol., 10, R75 (1999)).

4.2. Brillouin Optical Time-Domain Analysis Sensors

In Brillouin optical time-domain analysis (BOTDA) configurations, the effect of stimulated Brillouin scattering is employed in order to perform distributed temperature and strain measurements along a standard single-mode optical fiber. If an acoustic wave propagates in a medium, the variations in pressure give rise to variations in the refractive index of the medium, via the strain-optical effect. Some acoustic waves will always be present in a medium, above the absolute zero of temperature, since the molecules are in motion and will couple some of their energy into the dynamic modes of the structure. Optical scattering from these thermally excited acoustic waves comprises, as was noted in Section 3, the phenomenon of spontaneous Brillouin scattering. However, as the optical pump power is increased the wave scattered backwards from an acoustic wave will increase in amplitude and will interfere significantly with the forward-traveling pump wave. An optical beat signal arises within the fiber, which generates a pressure wave having the same frequency as the optical beat signal, via the phenomenon of electrostriction (Fig. 5); this pump-induced index grating scatters the pump light through Bragg diffraction. Scattered light is down-shifted or up-shifted in frequency because of the Doppler shift associated with a grating moving at the acoustic velocity V_A. This positive feedback, backscattering process is known as the stimulated Brillouin scattering (SBS).

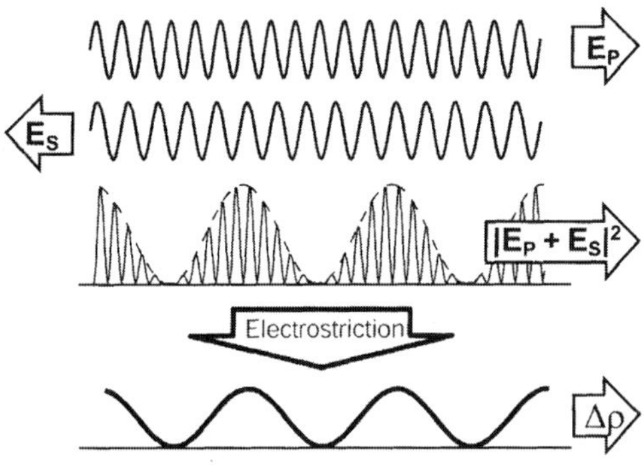

Figure 5. Principles of stimulated Brillouin scattering in optical fibers.

It leads to much larger backscattering at the Stokes and anti-Stokes frequencies than in the spontaneous case and, indeed, causes depletion problems in narrow-band optical-fiber telecommunications systems[24], although several schemes have been proposed to overcome this limit. These schemes include the use of fibers with a Brillouin frequency shift distribution,[25] Bragg gratings which are used for reflection of the scattered light,[26] and the simultaneous amplification of more than one lasing frequency. [27-28]

DOFS systems make use of this phenomenon by employing a pump–probe arrangement similar to that of the Raman forward-scattering system of section 4.1. The basic arrangement is shown in Fig. 6. A coherent pulse of light acts as the pump, and a counter-propagating CW is scanned in frequency around the Stokes line. When it coincides with the Stokes line it will receive gain from the pump via the SBS process. Essentially what is happening in this case is that the CW is giving rise to a large-amplitude interference with the pump, thus generating the acoustic wave from which the pump is strongly reflected. By observing the probe level as a function of time and frequency as the pump propagates, the Stokes frequency can be mapped as a function of position along the fiber.

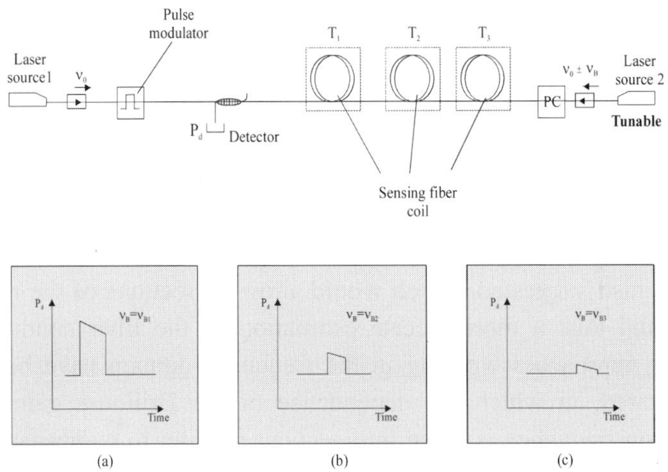

Figure 6. Basic configuration for BOTDA. (a), (b) and (c) show the Stokes signals acquired when the frequency offset between the two lasers is tuned to the Brillouin frequency shift at regions 1, 2 and 3, respectively.

The first experiments along these lines achieved a strain resolution of $2\cdot10^{-5}$, equivalent to a temperature resolution of 3 °C, with a spatial resolution of 100 m over 1.2 km.[29,30] Later systems have improved the spatial resolution to 1 m over a 22 km fiber.[31] Further developments of the system have attempted to overcome the problem of pump depletion, as a result of the strong backscattering, by using the anti-Stokes, rather than the Stokes, line. In this case the CW probe is attenuated and the pump is amplified. This technique, designated “loss BOTDA” has achieved a 1°C temperature resolution, a spatial resolution of 5 m and a total length of 51 km.[32]

Spatial resolution in SBS-based sensors is limited by the phonon lifetime: Brillouin profile is broadened to a Gaussian-like profile as the pulse width decreases below the phonon lifetime (i.e. when sub-meter spatial resolution is needed), thus reducing the accuracy of temperature/strain measurements.[33] An intense research activity has been devoted to overcome this limit. As regards time-domain schemes, many proposed approaches are based on the use of opportunely shaped pump pulses. Specifically, it has been demonstrated that using a pump beam with a small dc component, results in a spectral narrowing of the Brillouin gain spectrum. Such a narrowing arises from the background acoustic intensity generated by interaction of the cw beam with the baseline of the pulsed beam.[34-35] Other approaches based on pre-pump[36] or dark pulses[37] have been also demonstrated. However, the use of complex pulse shapes may also result in a distortion of the Brillouin gain curves, leading to errors in the determination of the Brillouin frequency.[34,38] The opportunity to use accurate reconstruction algorithms has been also suggested, which would allow corrections of the recorded spectra and thus a more-precise estimation of the fiber condition.[34,39] Different approaches working in the frequency-domain have been also demonstrated, in which the dependence of the Brillouin gain on the modulation frequency is taken into account in order to compensate from Brillouin spectrum distortion.[40]

The major advantage of BOTDA, with respect to spontaneous scattering based systems, is that of a strong signal, thus an easing of detection problems, with the associated beneficial spatial resolution trade-off. This provides valuable performance over very large distances. The major disadvantage, apart from a more demanding requirement on the source coherence, is that there is now no dependence of the signal power on temperature or strain, since the scattering process is now controlled by the wave interference rather than by the intrinsic fiber properties. Consequently, it is no longer possible to measure strain and temperature simultaneously, as in the spontaneous case. Each can only be measured if variations in the other are known to be absent, or are independently determined. The most practical way to achieve this consists in deploying the fiber in such a way that half of it is subjected to only temperature changes, whereas the other half is mechanically attached to the structure to be monitored, so that both temperature and strain changes are detected. Brillouin frequency shift measurement in the first region allows subtracting temperature effects from the measurement taken in the second half.[41] Another solution exploits the weak dependence of the Brillouin gain peak on temperature, so that the combined measurement of Brillouin frequency shift and Brillouin power can provide the additional information required for temperature/strain discrimination.[42] Finally, another interesting approach involving the use of dispersion-shifted fibers (DSF) is worthy to be reported.[43] In DSF fibers, the non-uniform refractive index transverse profile gives rise to multiple peaks in the Brillouin gain spectrum, each resonance peak being associated to a definite acoustic mode of the fiber. In Ref. 43, the Authors report that simultaneous temperature and strain sensing can be achieved by monitoring the frequency position of more than one peak in the Brillouin spectrum.

Temperature sensing capabilities of Brillouin-based distributed sensor can be exploited for electrical power cables monitoring, in particular for hot-spot localization.[44] For this purpose the fibers can be bundled in the screen layer of the power cable: The perfect dielectric properties of the fiber, makes the light propagation insensitive to the extreme electromagnetic environment of such a cable. Brillouin-distributed temperature sensing is also of use for temperature monitoring in lakes, oceans, for environmental issues, or in tunnels for realizing an alarm detection system. As an example, a demonstration of the use of a Brillouin-distributed temperature sensor for monitoring the thermal gradients in boreholes in volcanic areas has been recently reported.[45]

As regards strain sensing capabilities, fiber strain is an important parameter to be measured for assessing the reliability of optical-fiber cables, because strain can cause degradation in fiber strength (stress corrosion), leading eventually to fiber failure. Therefore, Brillouin-based strain measurement has found many applications in the research and development of optical fiber and cables and their related technologies[46]. Reports have been published on the strain evaluation of optical-fiber communications cables.[47]

Strain sensing also permits pipeline monitoring. A structural health monitoring (SHM) system for pipeline networks would permit to monitor continuously their structural integrity, reducing the overall risks and costs associated with current methods. Pipeline monitoring could be of use also in geotechnical applications, in which the deformation of a buried pipeline can be correlated to landslide movements. In Ref. 48, the authors show that high-spatial-resolution Brillouin sensing can be adopted in order to detect the formation of pipeline buckling, resulting from excessive concentric and bending loads. A technique for monitoring pipeline dislocation has been also demonstrated,[49] in which a Brillouin sensor is used in order to measure the strain distribution along three longitudinal directions running along the pipeline.

5. Conclusions

Distributed optical-fiber sensor systems will have a large part to play in the monitoring and diagnostics of critical extended structures. This is especially true for the new generation of self-adjusting, self-monitoring "intelligent" or "smart" structures.

Despite the unique opportunities offered by fiber-optics sensor technology, commercialization of the ideas emerging has been slow. One of the main obstacles to more effective deployment is the interface problem, i.e. that of ensuring an optimized interaction, between the fiber system and the measurand field. For example, in case of strain sensing, the problem of optimal strain transfer from the host structure to the fiber is still an open issue.[50-52] On the more positive side, the rapid advance of optical fiber telecommunications has given rise to a large range of high-performance and low-cost components and fiber types which have assisted considerably in the advance of the fiber sensors technology. As the requirement for ever greater understanding, monitoring and control of large structures increases its demands on sensor technology, more and more technical and commercial attention will be paid to the powerful advantages offered by distributed optical-fiber sensing methods. Also distributed chemical sensing may represent an opportunity to extend the number of application fields which may take advantage of fiber-optics distributed sensing systems.[53]

References

1. J. K. Barnoski and S. M. Jensen, *Appl. Opt.* **15**, 2112 (1976).
2. A. Rogers, *Meas. Sci. Technol.* **10**, R75 (1999).
3. A. H. Hartog, *IEEE J. Lightwave Technol.* **1**, 1983, 498 (1983).

4. M. C. Farries, M. E. Fermann, R. I. Laming, S. B. Poole and D. N. Payne, *Electron. Lett.* **22,** 418 (1986).
5. D. K. Gifford, B. J. Soller, M. S. Wolfe and M. E. Froggatt, ECOC Technical Digest, Glasgow, Scotland, 2005, paper We4.P.5.
6. A. J. Rogers, *Electron. Lett.* **16,** 489 (1980).
7. A. J. Rogers, *Appl. Opt.* **20,** 1060 (1981).
8. J. N. Ross *Appl. Opt.* **21,** 3489 (1981).
9. J. P. Dakin, D.J. Pratt, G. W. Bibby and J. N. Ross, *Electron. Lett.* **21**, 569 (1985).
10. J. P. Dakin, D. J. Pratt, J. N. Ross and G. W. Bibby, *Proc. Conf. on Optical-Fibre Sensors 3 (San Diego)* postdeadline paper (1985).
11. R. Feced, M. Farhadiroushan, V. A. Handerek and A. J. Rogers, *Proc. IEE Opto-electronics* 144, 183 (1997).
12. M. A. Farahani and T. Gogolla, *IEEE J. Lightwave Technol.* **17**, 1379 (1999).
13. G.P. Agrawal, Nonlinear Fiber Optics (Academic Press, San Diego, 2001).
14. E. P. Ippen and R. H. Stolen, *Appl. Phys. Lett.* **21** 539, (1972).
15. T. R. Parker, M. Farhadiroushan, V.A. Handerek and A.J. Roger, *IEEE Photon. Technol. Lett.* **9**, 979 (1997).
16. T. R. Parker, M. Farhadiroushan, R. Feced, V. A. Handerek and A. J. Rogers, *IEEE Journal of Quantum Electronics* **34,** 645 (1998).
17. P. C. Wait and A. H. Hartog, *IEEE Photon. Technol. Lett.* **13**, 508 (2001).
18. M. Alahbabi, Y. T. Cho and T. P. Newson, *Opt. Lett.* **29**, 26 (2004).
19. M. N. Alahbabi, Y. T. Cho and T. P. Newson, *Meas. Sci. Technol.* **15,** 1544 (2004).
20. A. J. Rogers, *Opt. Lasers Eng.* **16** 179 (1992).
21. D. Garus, K. Krebber, F. Schliep and T. Gogolla, *Opt. Lett.* **21,** 1402 (1996).
22. R. Bernini, A. Minardo and L. Zeni, *Opt. Lett.* **29**, 1977 (2004).
23. M. C. Farries and A. J. Rogers, Proc. Conf. on Optical-fiber Sensors 2 (Stuttgart) 121, (1984).
24. D. Cotter, *J. Opt. Commun.* **4** 10 (1983).
25. K. Shiraki, M. Ohashi and M. Tateda, *IEEE J. Lightwave Technol.* **14**, 50 (1996).
26. H. Lee and G.P. Agrawal, *Opt. Express* **11**, 3467 (2003).
27. M. Tsubokawa, S. Seikai, T. Nakashima and N. Shibata, *Electron. Lett.* **22**, 472 (1986).
28. P. Weßels, P. Adel, M. Auerbach, D. Wandt, and C. Fallnich, *Opt. Express* **12**, 4443 (2004).
29. T. Horiguchi and M. Tateda, *Opt. Lett.* **14** 408 (1989).
30. M. Tateda, T. Horiguchi, T. Kurashima and K. Ishihara, *IEEE J. Lightwave Technol.* **8** 1269 (1990).
31. T. Horiguchi, T. Kurashima and Y. Koyamada, *IEEE Technical Digest of Symp. on Optical Fibre Measurements*, Boulder *CO* 73 (1994).
32. X. Bao, J. Dhliwayo, N. Heron, D. J. Webb and D. A. Jackson, *IEEE J. Lightwave Technol.,* **13**, 1340 (1995).

33. T. Horiguchi, K. Shimizu, T. Kurashima and Y. Koyamada, Proc. SPIE 2507, 126-135 (1995).
34. X. Bao, A. Brown, M. DeMerchant and J. Smith, *Opt. Lett.* **24**, 510 (1999).
35. V. Lecoeuche, D. J. Webb, C. N. Pannell and D. A. Jackson, *Opt. Lett.* **25**, 156 (2000).
36. K. Kishida, C.-H Lee and K. Nishiguchi, Proc. SPIE **5855**, 559 (2005).
37. A. W. Brown, B. G. Colpitts and K. Brown, *IEEE Photon. Technol. Lett.* **17**, 1501, (2005).
38. X. Bao, Q. Yu, V. P. Kalosha and L. Chen, *Opt. Lett.* **31**, 888 (2006).
39. A. Minardo, R. Bernini and L. Zeni, *Opt. Express* **15**, 10397 (2007).
40. R. Bernini, A. Minardo and L. Zeni, *IEEE Photon. Technol. Lett.* **18**, 280 (2006).
41. J. Smith, A. Brown, M. DeMarchant and X. Bao, *Appl. Opt.* **38**, 5382 (1999).
42. X. Bao, J. Smith, A. W. Brown, Proc. SPIE, Vol. 4920, Advanced Sensor Systems and Applications, Shanghai, China, p. 311 (2002).
43. C. C. Lee, P. W. Chiang and S. Chi, *IEEE Photon. Technol. Lett.* **13**, 1094 (2001).
44. D. Villacci, R. Vaccaro, R. Bernini, A. Minardo and L. Zeni, *IET Generation Transmission and Distribution* **1**, 912 (2007).
45. L. Zeni, A. Minardo, Z. Petrillo, M. Piochi, M. Scarpa and R. Bernini, European Geosciences Union (EGU 2007), Vienna, Austria, (2007).
46. T. Horiguchi, K. Shimizu, T. Kurashima and M. Tateda, *IEEE J. Lightwave Technol.,* **13**, 1296 (1995).
47. M. Tateda, T. Horiguchi, T. Kurashima and K. Ishihara, *IEEE J. Lightwave Technol.* **8**, 1269 (1990).
48. L. Zou, X. Bao, F. Ravet and L. Chen, *Appl. Opt.* **45** 3372 (2006).
49. R. Bernini, A. Minardo and L. Zeni, *Smart Mater. Struct.,* **17**, 015006 (2008).
50. G. Duck and M. Leblanc, *Smart Mater. Struct.* **9** 492 (2000).
51. W. R. Habel and A. Bismark, *Journal of Structural Control*, **7**, 51, (2006).
52. H. N. Lee, G. D. Zhou, L. Ren and D .S. Li, *Opt. Eng.*, **46**, 054402, (2007).
53. W. C. Michie, B. Culshaw, I. McKenzie, M. Konstantakis, N. B. Graham, C. Moran, F. Santos, E. Bergqvist and B. Carlstrom, *Opt. Lett.* **20** 103 (1995).

LIGHTWAVE TECHNOLOGIES FOR INTERROGATION SYSTEMS OF FIBER BRAGG GRATINGS SENSORS

D. Donisi,[a,b,*] R. Beccherelli[b] and A. d'Alessandro[a]

[a]*Dipartimento di Ingegneria Elettronica, "La Sapienza" Università di Roma, Via Eudossiana, 18, 00184 Rome, Italy*

[b]*Istituto per la Microelettronica ed i Microsistemi, CNR Via del Fosso del Cavaliere 100, 00133 Rome, Italy*

[*]*E-mail: domenico.donisi@uniroma1.it*

A review of Fiber Bragg Grating (FBG) features and the different schemes of fiber sensors interrogation are reported. The interrogation system represents the key element of monitoring systems in terms of both performance and cost as it has to measure relatively small shifts in Bragg wavelength of FBG elements.

An innovative interrogation system prototype for structural sensing based on a high-performance electro-optic edge filter on glass is also presented here. It provides a wavelength-dependent transmittance with a linear relationship between the Bragg wavelength shift and the output intensity change of the filter. The resulting device is the demonstration of a simple and inexpensive technology to implement low cost FBG sensors monitoring system based on innovative integrated optic functional component on glass.

1. Introduction

In 1978, Hill et al. reported[1] the possibility to form refractive index variation patterns in germano-silicate optical fibers. Hill's gratings were written in the fiber core by standing wave of 488 or 514.5 nm argon laser light. Afterwards, Meltz et al. demonstrated how to produce Bragg gratings by exposing the fiber core, through the side of the cladding, to a coherent UV two-beam interference pattern.[2] With this technique, gratings with a wide range of bandwidths and reflectivities can be formed in times between 20 ns (the duration of a 248 nm excimer laser pulse) to a few minutes. Indeed, the refractive index modulation of Fiber Bragg Gratings (FBGs) can be achieved with different schemes:[3,4] the conventional free-space two-beam holographic method, the diffractive phase mask technique and the point-by-point method. Argon fluoride (193 nm) or krypton fluoride (248 nm) excimer lasers are often efficiently used as a light source.

In recent years, FBG have been used extensively in the telecommunication industry for dense wavelength division demultiplexing, dispersion compensation, laser stabilization, and erbium amplifier gain flattening. In addition, FBGs have been used for a wide variety of sensing applications including structural health monitoring (SHM) of civil structures (highways, bridges, buildings, dams, etc.),[5] smart manufacturing and non-destructive testing (composites, laminates, etc.),[6-9] remote sensing (oil wells, power cables, pipelines, space stations, etc.),[10] smart structures (airplane wings, ship hulls, buildings, sports equipment, etc.),[11-13] nuclear power plants,[14-16] medical industry,[17,18] as well as traditional strain, pressure and temperature sensing.[19–23]

FBG offer attractive characteristics that make them very suitable and, in some cases, the only viable sensing solution. Some of the key attributes of FBG are their relatively high immunity to electromagnetic interference, excellent resolution and range, remote access, absolute measurement, long term measurements stability, small size, light weight, low cost, easy cabling and operation in harsh environments. Indeed, in nuclear power plants and refineries, they are often the most appropriate solution as use of electronic sensors is impractical or dangerous.

When the FBGs are used in the structural integrity monitoring field as structural sensors (strain gauges), they are mechanically linked to the structure whose dynamical-structural behavior needs to be monitored. Depending on the kind of monitored structure the FBG sensor can be either glued on the structure surface (usually with epoxy resin), or incorporated in the structure during its production (e.g. in the case of concrete or glass/carbon fiber composite materials). Once the sensors are fixed onto the object to be monitored, every deformation or vibration of the structure is transferred to the FBG sensors. With adequate dynamic-structural analysis techniques, the FBG sensors signal permits to detect the onset of an abnormal behavior of the structure, thus anticipating overload damages and structural breakdown. Moreover, the FBG sensors can measure, at the same time, both the structure temperature and the strain in the point where the sensor is located. In this case, a reference

FBG sensor is often used to distinguish mechanical from thermal stress and to increase the overall accuracy of the system.

FBG have become attractive optical components for sensing because of their robust wavelength-encoding capability. Due to the wavelength-encoded nature of the signals in optical FBG, there are no problems associated with transmission or bending losses in the fiber.

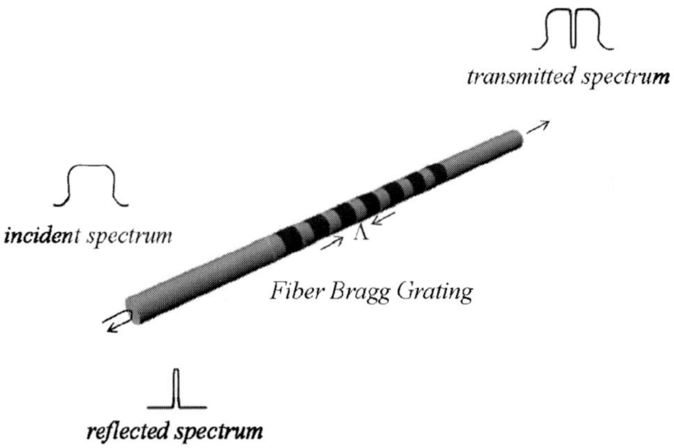

Figure 1. Schematic drawing of structure and spectral response of FBG.

The monitored measurands are determined only by detecting the Bragg wavelength shift of the light back-reflected from an FBG sensors array. This operation is achieved by transmitting the FBG sensors optical signal to an optoelectronic analysis system, which represents the core of the entire monitoring system. While much research is focused on developing and applying new optical sensing technologies, there is also a great need to develop performing compact and cost efficient fiber sensor interrogators.

In this chapter, we review the principles of FBG interrogation systems. After a brief description of the FBG optical properties in Sec. 2, consolidated FBG interrogation techniques will be reviewed in the Sec. 3. We also describe in Sec. 4 an innovative compact and cost-effective light-wave technology for FBG sensors wavelength demodulation and analyze its performance in Sec. 5.

2. Operating Principle of the Fiber Bragg Grating Sensor

An FBG is a highly wavelength-selective reflection filter formed by a modulated periodically refractive index structure within the core of an optical fiber. The amount of change induced in the refractive index[24,25] of the core ranges from 10^{-5} to 10^{-2}.

Whenever a broad-spectrum light beam impinges on the grating, it will have a portion of its energy transmitted through, and another reflected off as depicted in Fig. 1.

When the Bragg condition is satisfied, the contributions of reflected light from each grating plane add constructively in the backward direction to form a back-reflected peak with a center wavelength defined by the grating parameters. The Bragg wavelength is given by

$$\lambda_B = 2n_{eff}\Lambda, \tag{1}$$

where n_{eff} is the effective refractive index of the fiber core and Λ is the grating period.

A typical FBG has a physical length of a few mm and can provide virtually 100% peak reflectivity, with a reflection bandwidth which ranges from 0.05 to 0.3 nm. A general expression for the approximate full width at half-maximum bandwidth of a grating is given by[26]

$$\Delta\lambda_B = \lambda_B \alpha \sqrt{\left(\frac{\Delta n}{2n_{co}}\right)^2 + \left(\frac{1}{N}\right)^2}, \tag{2}$$

where N is the number of the grating planes. The parameter α is ~ 1 for strong gratings (for grating with near 100% reflection) whereas α is ~ 0.5 for weak gratings.

The sensing function of an FBG derives from the sensitivity of both the effective refractive index of the guided mode in the fiber and grating period to externally applied mechanical or thermal perturbations. Perturbation of the grating results in a shift in the Bragg wavelength of the device[27,28] which can be detected in either the reflected or transmitted spectrum. The shift in the Bragg grating center wavelength due both to strain (ε) and temperature changes (ΔT) can be calculated by differentiating Eq. 1:

$$\Delta\lambda_B = 2\left(\Lambda\frac{\partial n}{\partial l} + n\frac{\partial\Lambda}{\partial l}\right)\Delta l + 2\left(\Lambda\frac{\partial n}{\partial T} + n\frac{\partial\Lambda}{\partial T}\right)\Delta T, \quad (3)$$

where the first term represents the strain effect on an optical fiber. This corresponds to the combination of a change in the grating spacing and the strain-optic induced change in the refractive index. For axial loads, the wavelength change is typically 1.2 pm/με at 1550 nm (12 nm for 1% strain). The second term represents the temperature effect on an optical fiber, which corresponds to the dependence of the index of refraction of the glass on temperature and the thermal expansion of the glass. Typically, the fractional wavelength change in the Bragg wavelength is of the order of 10 pm/°C.

Figures 2 and 3 show an experimental measurement of the Bragg wavelength shift in the reflection and transmission spectral response for a FBG, respectively, with a designed nominal working wavelength of ~ 1550 nm, strained with tensile stress. Strain applied to an FBG elongates it (compresses it, for negative strain); hence, the grating period is increased (decreased), which results in a shift of the Bragg wavelength to longer (shorter) wavelengths.

For FBG sensors applications, the side lobes in the spectral response are very undesirable. For this reason FBGs are apodized. The spatial apodization smooths the refractive index of the core over the transition from the homogeneous region to the periodic region along the propagation axis by using a proper smoothing function. In Figs. 2 and 3, it is possible to notice the effect of FBG apodization on the spectral responses. Moreover the FWHM and the spectral shape are stable under deformation.

3. FBG Interrogation Techniques

In this section we will briefly discuss the ways in which optical fiber Bragg grating sensors can be individually interrogated and collectively multiplexed in order to be able to perform multi-point sensing.

The interrogation system, which processes the back-reflected FBG sensors optical signal, represents the key element of monitoring systems in terms of both performance and cost.[29] The requirements for FBG interrogation systems are low power, high resolution, high speed, small size and capability to deliver real-time measurements. FBG sensors require expensive optical sensing interrogator to achieve all these performances.

Ideally, one would desire a simple and low cost fiber sensor monitoring system for embedded instrumentation applications and for long-term

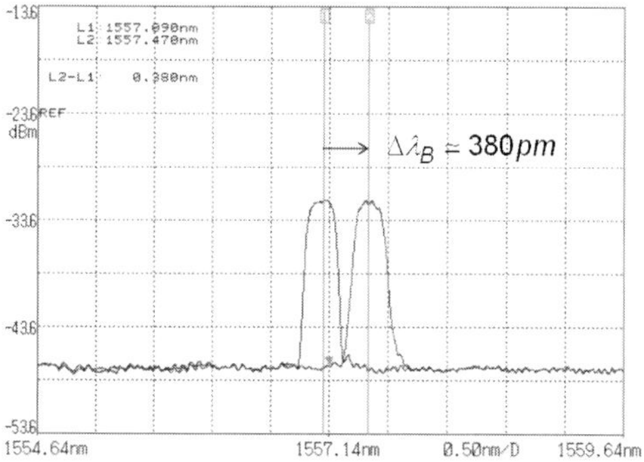

Figure 2. Experimental measure of a Bragg wavelength shift in the reflection response.

operation in harsh environments.[30] Moreover, data collection and analysis has to be fast and easy by exploiting existing communication protocols. FBG have been extensively accepted by engineers and have become the most prominent sensors for structural health monitoring (SHM), because of their high accuracy. The typical resolutions and the measure ranges of the FBG sensors are the ones required by civil engineering: (i) resolution as low as 1 με and 0.1 °C, which translates into a wavelength resolution of about 1 pm, (ii) strain measurement ranges in the order of 10 mε and (iii) more than 200 °C as temperature operating range. Whereas this wavelength resolution is easily achieved with expensive laboratory instrumentation, the ability to resolve changes on this order using small, packaged electro-optics units able to operate on the field is more of a challenge. The choice of fiber Bragg interrogation method depends on the available optical component technology suitable for a specific application. The most straightforward method for interrogating a FBG sensors array is based on passive broadband illumination in the telecom

C band (1530-1565 nm). A linear sensor array can be created in a single long optical fiber by writing a set of Bragg gratings with different and unique Bragg wavelengths or by bonding stubs of common optical fiber to different FBGs. FBG sensors wavelength spacing can be 1-2 nm, allowing up to a few tens to be multiplexed in a single fiber operating in the C band. Each FBG can be localized at any position along the optical fiber. However, minimum spacing and maximum number of gratings are ultimately ruled by cross-talk coming from multiple reflections and spectral shadowing. Thus the same optical fiber behaves as an array of stress sensors, as a multiplexing system and as the transmission medium. This makes it possible to have multi-point, as well as quasi-distributed sensing.

Light with a broadband spectrum which covers that one of an FBG sensor feeds the system, and the narrowband component reflected by the FBG is routed to a wavelength detection system. The two most important interrogation schemes are: WDM and TDM. For a fixed level of acceptable cross-talk, one can increase number of sensors by combining time division multiplexing (TDM) with WDM. In this configuration, a short light pulse from the broadband source is launched into this system and the response is measured with controlled delays proportional to the distance of each subset of adjacent FBGs.

Many general purpose optical sensor monitoring systems are large and bulky, costly and have high power consumption. For example a typical optical spectrum analyzer (OSA) has lots of functionalities that are unnecessary for standard sensor monitoring. Various FBG interrogation schemes have been so far implemented by means of set of discrete optical beam splitters and precisely engineered interferential filters. Such filters are hard to design and expensive to make with the desired spectral specifications. Commonly, they are realized by means of multiple coatings in vacuum and lack of integrability and compactness.

The state of the art of commercially available fiber Bragg grating interrogator is represented by Micron Optics products[31] which are designed specifically for fiber sensor applications. Anyway, the FBG interrogator dimensions are still incompatible with aerospace vehicles.[32] Thus while the sensor itself can be extremely compact and adaptable to a variety of situations, the interrogation system is not. Therefore, the key to a practical and low cost monitoring system based on FBG sensors lies in the development of innovative integrated devices capable of determining the relatively small shifts in Bragg wavelength of FBG elements. This area has received significant attention over the past three or four years, with various approaches demonstrated so far. Several options[33–39] exist for measuring the wavelength of the optical signal reflected from a FBG element. These include the use of a wavelength interrogation scheme with a scanning Fabry–Perot filter,[40] a tunable acousto-optic filter,[41] an interferometric detection,[42] a diffraction grating,[43] an AWG,[44,45] a frequency-locking circuit[46,47] or frequency-modulated multimode laser.[48,49] Usually the wavelength measurement is not very simple; thus, the general principle is to convert the wavelength shift into some easily measurable parameter, such as amplitude or phase. Amplitude measurement is the most common and direct technique used in optical fiber sensors. Converting wavelength shift to amplitude change makes the interrogation operation simple and cost-effective. Several approaches can be related to amplitude measurement. The most simple and low-cost wavelength-amplitude conversion technique for the measurement of the wavelength shift caused by FBG sensors is based on an edge filter.[50]

4. An Integrated Tunable Filter using Composite Holographic Grating

We present here a simple and high-performance interrogation system prototype for optical fiber structural sensors, based on an innovative integrated electro-optic edge filter on glass as an alternative to other optical filter technologies (MEMS, MOEMS, acousto-optics,...).[51-53] Indeed, currently even the mature waveguide technology based on $LiNbO_3$, characterized by highly efficient electro-optic effect and acousto-optic effect, does not allow to overcome problems of too high insertion losses and high fabrication costs.

This innovative filter provides a wavelength–dependent transmittance, offering a linear relationship between the FBG wavelength shift and the output intensity change of the filter. This hybrid filter is based on a double ion-exchanged channel waveguide and a holographic composite Bragg grating (see Fig. 4). The grating consists of polymer slices alternated with films of regularly aligned Nematic Liquid Crystal (NLC), known as POLICRYPS (POlymer LIquid CRYstal Polymer Slices).[54] Such a composite grating is used as overlayer of a single mode optical channel waveguide. We use a reliable and reproducible K^+-Na^+/Ag^+-Na^+ double ion-exchanged process in BK7 glass to obtain low loss (< 1 dB/cm) and high index-contrast ($\Delta n \sim 0.04$) optical waveguides.[55]

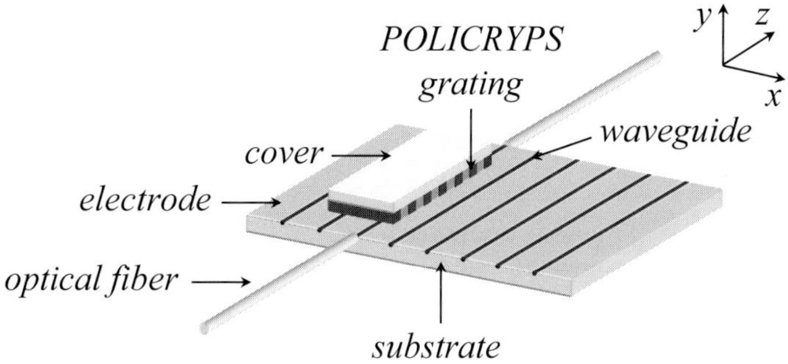

Figure 4. Schematic illustration of the integrated optical filter with POLICRYPS morphology.

The image in Fig. 5 is collected by means of an optical microscope and shows a 6 μm wide optical waveguide perpendicularly aligned to the overwritten grating.

The filter structure also includes coplanar electrodes which allow in-

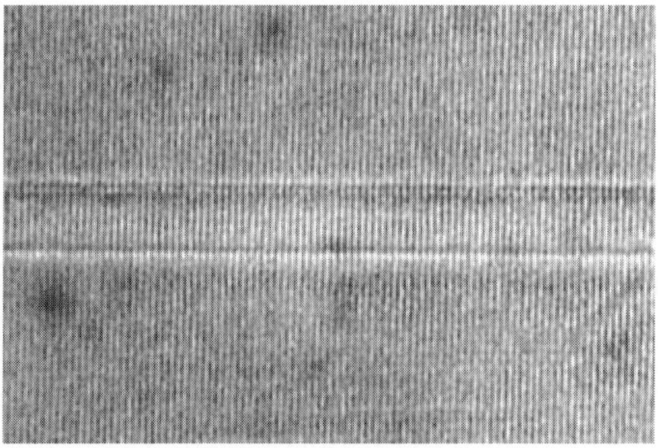

Figure 5. Grating - waveguide alignment.

plane reorientation of the NLC molecules between the polymer slices by exploiting the electro-optic effect of the POLICRYPS grating.[56] When the external electric field is absent, the director of the LC molecules is aligned normally to the polymer/LC interface and along the direction of propagation because of the boundary conditions imposed by the confining walls of the polymers. This is schematically illustrated in Fig. 6a which shows the top view of the grating structure. The desired tilt of the LC molecules is obtained by applying a suitable control voltage.

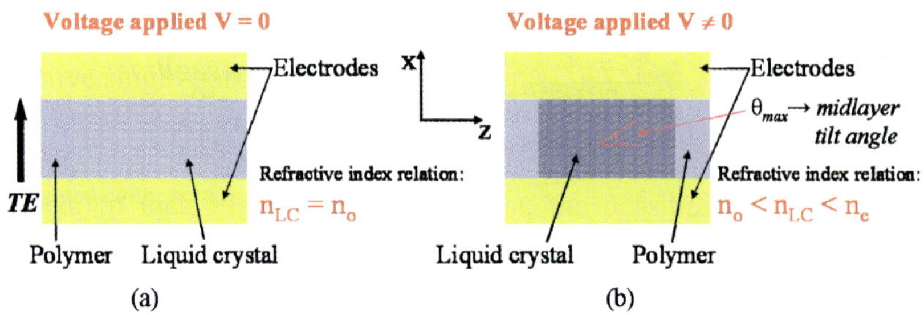

Figure 6. Top view of the device and sketch of the working principle. (a) Without applied voltage and (b) with applied voltage.

Such LC reorientation lets only guided light with TE polarization to "see" a refractive index modulation of the overlaying hybrid cladding. In particular, a TE-like optical field will see the ordinary refractive index n_o of the LC when no external signal is applied. Applying an external electrical field by means of coplanar electrodes the molecules rotate in the plane (xz) and the TE optical field will see a higher LC refractive index (see Fig. 6b). In this way, the Bragg wavelength of the guided wave optical filter can be tuned by controlling the effective refractive index of the guided mode.

5. POLICRYPS Filter–based FBG Sensors Interrogation

In our system, the POLICRYPS filter acts as a wavelength demodulation device, working with the same principle of an edge filter, for measuring the wavelength shift of the FBG sensors.

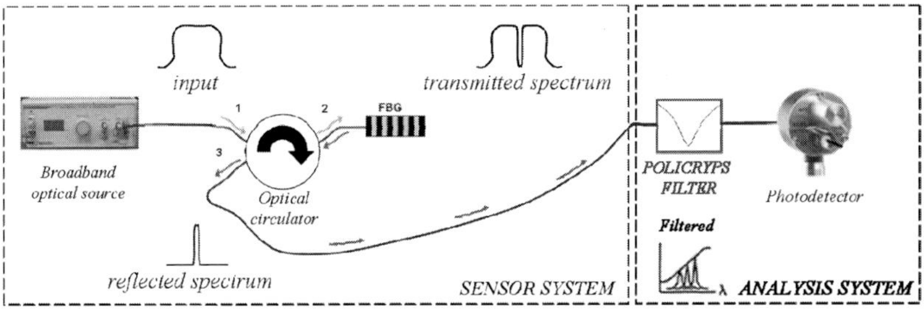

Figure 7. Schematic of POLICRYPS filter-based FBG sensor interrogation.

In the experiment illustrated in Fig. 7, the back-reflected light of a FBG sensor is launched into the input of a POLICRYPS filter and its output intensity measured by means of a photo-detector. A simple optical setup is used in order to evaluate the response of the filter for different values of the mechanical stress acting on the FBG sensor.

The broadband light source used to illuminate the Bragg grating sensor is an Erbium-Doped-Fiber-Amplifier (model EBS-4015/EFA, from MPB technologies Inc.). The model of this Broadband Source and Power Amplifier utilizes amplified spontaneous emission (ASE) from diode-pumped Erbium-doped fiber and a cleverly-conceived spectral-shaping scheme to produce more than 15 dBm of unpolarized output centered at

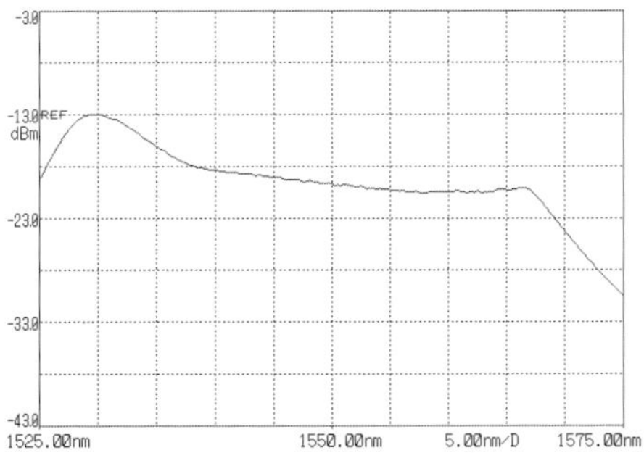

Figure 8. Typical input source spectrum to interrogate FBG sensors.

1548 nm and with a near flat-top spectrum (see Fig. 8); maximum ripple does not exceed 2.0 dB over most of the spectrum, namely 39 nm, while 3-dB bandwidth is larger than 40 nm. The commercially available FBG sensor used in this work has a nominal working wavelength close to 1548 nm.

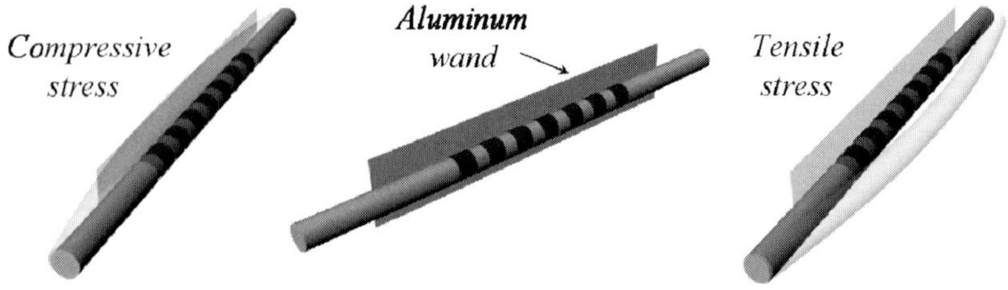

Figure 9. FBG under compressive or tensile stress.

The FBG sensor was aligned parallel the main axis of an aluminum wand and glued in such a way to permit the grating to be strained with either tensile or compressive stress (see Fig. 9). The POLICRYPS filter transmittance is shown in Fig. 10.

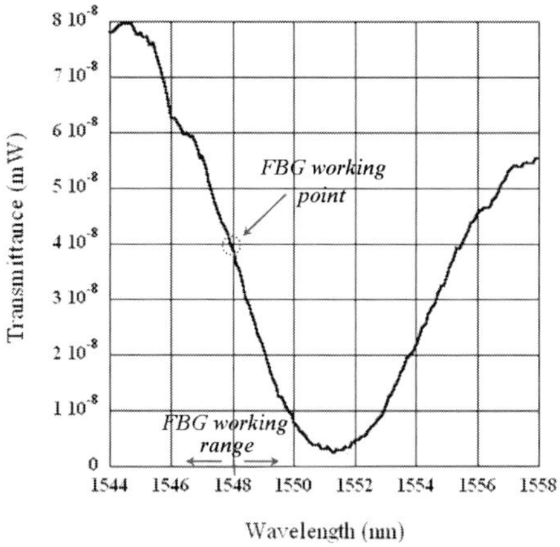

Figure 10. Wavelength-amplitude conversion with a POLICRYPS filter.

The POLICRYPS filtering function is electrically pre-biased so that the linear regime of its transmittance curve coincides with the FBG working range (± 1.5 nm). A linear, wavelength-dependent, filtering function is observed over the wavelength range from 1547.5 to 1550 nm.

The calibration is performed by applying a controlled strain to the FBG sensor with a nano-positioning stage. The integrated output response of the FBG versus wavelength and deformation as filtered by the POLICRYPS filter is plotted in Fig. 11. As expected from the spectral shape of the POLICRYPS filter transmittance, this integrated response linearly decreases with increasing wavelength, i.e., when stress varies from compressive to tensile.

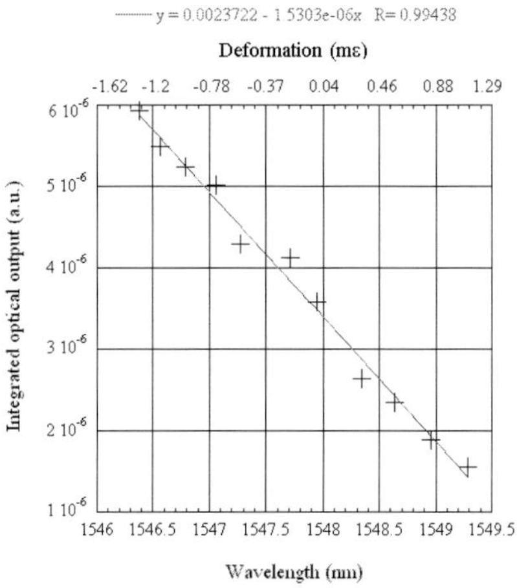

Figure 11. Integrated response of FBG as filtered by POLICRYPS.

By substituting the photodetector with a spectrum analyzer in the optical setup, we have performed a spectral scan of the POLICRYPS filter output as validation. Experimental result of this spectral scan is shown in Fig. 12. We point out the quasi-linear decrease of the output peak power from lower to higher wavelengths with respect to the center working wavelength.

5. Conclusions

In this chapter the most important optical properties of the FBG sensors have been reviewed. Examples of FBG interrogation techniques are referenced. The implementation of a monitoring system based on FBG sensors by using an innovative analysis system has been presented. This is based on a POLICRYPS filter in a wavelength demodulation system. The filter electrical tunability is exploited to pre-bias the POLICRYPS

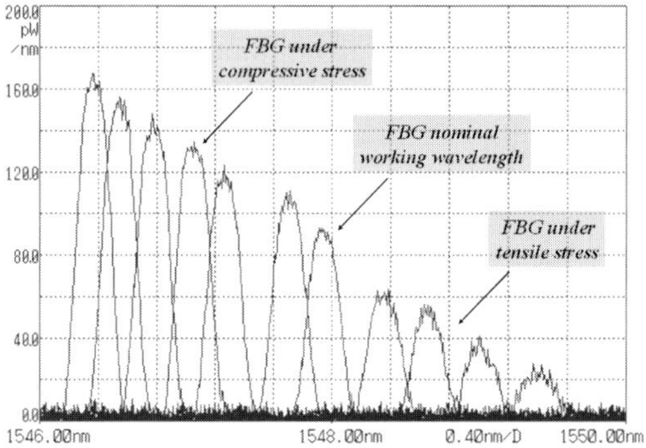

Figure 12. Response of FBG as filtered by POLICRYPS.

filtering function to operate in a linear regime. Hence the wavelength shift is translated into a linear intensity change which can be detected with a simple photodiode.

Acknowledgments

The authors are grateful to Dr. M. Caponero and Prof. C. Umeton for enlightening discussions.

References

1. K. O. Hill et al., *Applied Physics Lett.*, 32 (1978).
2. G. Meltz, W. W. Morey and W. H. Glenn, *Opt. Lett.*, 14 (1989).
3. K. O. Hill, B. Malo, F. Bilodeau, D. C. Johnson and J. Albert, *Appl. Phys. Lett.*, 62 (1993).
4. A. Othonos and K. Kalli, Eds., in *Fiber Bragg Gratings—Fundamentals and Applications in Telecommunications and Sensing* (Artech House, Boston, 1999).
5. Z. Zhou, T. W. Graver, L. Hsu and J. Ou, *Pacific Science Review*, 5 (2003).
6. X. Li, C. Zhao, J. Lin and S. Yuan, *Optics and Lasers in Engineering*, 45 (2007).
7. S. Takeda, S. Minakuchi, Y. Okabe and N. Takeda, *Composites Part A (Applied Science and Manufacturing),* 36 (2005).

8. Y. Okabe, S. Yashiro, R. Tsuji, T. Mizutani and N. Takeda, *Proceedings of the SPIE*, 4704 (2002).
9. W. S. Kim, S. H. Kim and J. J. Lee, *Key Engineering Materials*, 297 (2005).
10. L. Liu, P. Long and T. Liu, *Proceedings of the SPIE*, 5579 (2004).
11. Weimin Chen, Yi Jiang and Shanglian Huang, Proceedings of SPIE, 3241 (1997).
12. Jinsong Leng and A. Asundi, *Sensors and Actuators **A** (Physical),* 103 (2003).
13. Xiyuan Chen and Lin Fang, *Key Engineering Materials*, 336 (2007).
14. A. Gusarov, F. Berghmans, O. Deparis, A. Fernandez, Y. Defosse, P. Mégret, M. Decréton and M. Blondel, *IEEE Photonics Technology Letters* (1999).
15. A. Fernandez, B. Brichard, F. Berghmans and M. Decréton, *IEEE Trans. on Nuclear Science,* 49 (2002).
16. A. Gusarov, A. Fernandez, S. Vasiliev, O. Medvedekov, M. Blondel and F. Berghmans, *Nucl. Instr. Methods in Phys. Res. **B**,* (2001).
17. Y.-J. Rao, D. J. Webb, D.A. Jackson, L. Zhang *J.of Lightwave Technology*, 15 (1997).
18. Y.J. Rao, D.J. Webb, D.A. Jackson, L. Zhang and I. Bennion, *J. of Biomedical Optics*, 3 (1998).
19. W.W. Morey, G. Meltz and W. H. Glenn, in *Fiber Optic & Laser Sensors, VII,* Proceedings of SPIE, 1169 (1989).
20. W. W. Morey et al., Proceedings of OFS '89 (Paris, 1989).
21. A. D. Kersey, M. A. Davis, H. J. Patrick, M. LeBlanc, K. P. Koo, C. G. Askins, M. A. Putnam and E. J. Friebele, *J. Lightwave Technol.*, 15 (1997).
22. Y. J. Rao, *Opt. Lasers Eng.*, 31 (1999).
23. I. C. Song, S. K. Lee, S. H. Jeong and B. H. Lee, *Appl. Opt.* , 43 (2004).
24. K. O. Hill, B. Malo, F. Bilodeau and D. C. Johnson, *Annu. Rev. Mater. Sci.*, 23 (1993).
25. W. W. Morey, G. A. Ball and G. Meltz, *Opt. Photonics News* , 8 (1994).
26. P. St J. Russell, J. L. Archambault and L. Reekie, *Phys. World*, 41 (1993).
27. A. D. Kersey, M. A. Davis, H. J. Patrick, M. LeBlanc, K. P. Koo, C. G. Askins, M. A. Putnam and E. J. Friebele, *IEEE J. Lightwave Technol.*, 15 (1997).
28. Y. J. Rao, *Opt. Lasers Eng.*, 31 (1999).
29. I. C. Song, S. K. Lee, S. H. Jeong and B. H. Lee, *Appl. Opt.*, 43 (2004).
30. S. W. Lloyd, J. A. Newman, D. R. Wilding, R. H. Selfridge and S. M. Schultz, *Review of Scientific Instruments*, 78 (2007).
31. G. B. Tait, *Appl. Opt.,* 46 (2007).
32. http://www.micronoptics.com/sensing.htm.
33. G. B. Tait and R. S. Rogowski, *Proceedings of 2005 Quantum Electronics and Laser Science Conference (OELS)*, (IEEE, 2005).
34. Y.-J. Chiang, L. Wang, H.–S. Chen, C.-C.Yang and W.-F. Liu, *Appl. Opt.*, 41 (2002).
35. Yi Jiang, *Appl. Opt.,* 47 (2008).

36. A.G. Simpson, K. Zhou, L. Zhang and I. Bennion, in *Bragg Gratings Photosensitivity and Poling in Glass Waveguides*, (2003).
37. T. Farrell, P. O'Connor, J. Levins, D. McDonald, Proceedings of SPIE, 5826 (2005).
38. Y. Xiufeng, Z. Chun-Liu, P. Qizhen, Z. Xiaoqun and L. Chao, *Optics Communications,* 250 (2005).
39. C.Z. Shi, C.C. Chan, M. Zhang, J. Ju, W. Jin, Y.B. Liao, Y. Zhang and Y. Zhou, Proceedings of SPIE , 4920 (2002).
40. M. D. Todd, G.A. Johnson and B. L. Althouse, *Measurement Science & Technology*, 12 (2001).
41. M. G. Xu, H. Geiger and J. P. Dakin, *J. of Lightwave Technology*, 14 (1996).
42. A. D. Kersey, T. A. Berkoff and W. W. Morey, *Electron. Lett.*, 28 (1992).
43. A. Ezbiri, A. Munoz, S. E. Kanellopoulos and V. A. Handerek, in *IEE Colloquium on Optical Techniques for Smart Structures and Structural Monitoring,* Digest 1997/033 (Institute of Electrical Engineers, London, 1997).
44. P. Niewczas, A. J. Willshire, L. Dziuda and J.R. McDonald, Proceedings of the 20th IEEE Instrumentation Technology Conference, 2 (2003).
45. H. Su and X. Guang Huang, *Optics Communications*, 275 (2007).
46. A. Arie, B. Lissak and M. Tur, *J. Lightwave Technol.*, 17 (1999).
47. S. Yamashita, A. Inaba, *Measurement Science & Technology*, 15 (2004).
48. G. Gagliardi, M. Salza, P. Ferraro and P. De Natale, *J. of Optics A: Pure and Applied Optics,* 8 (2006).
49. L. A. Ferreira, E. V. Diatzikis, J. L. Santos and F. Farahi, *J. Lightwave Technol.*, 16 (1998).
50. S. M. Melle, K. Liu, R. M. Measures, *Appl. Opt* , 32 (1993).
51. K. Hirabayashi, H. Tsuda, T. Kurokawa, *J. of Lightwave Technology,* 11 (1993).
52. D. C. Abeysinghe, S. Dasgupta, H. E. Jackson and J. T. Boyd, *J. of Micromechanics and Microengineering*, 12 (2002).
53. D. A. Smith, R.S. Chakrawarthy, Z. Bao, J. E. Baran, J. L. Jackel, A. d'Alessandro, D. J. Fritz, S. H. Huang, X. Y. Zou, S. M. Hwang, A. E. Willner and K. Li, *J.of Lightwave Technology*, 14 (1996).
54. R. Caputo, L. De Sio, A. Veltri and C. Umeton, *Opt. Lett.*, 29 (2004).
55. J. Zou, F. Zhao and R. T. Chen, *Appl. Opt.*, 41 (2002).
56. A. d'Alessandro, R. Asquini, C. Gizzi, R. Caputo, C. Umeton, A. Veltri and A.V. Sukhov, *Opt. Lett.*, 29 (2004).

SURFACE PLASMON RESONANCE: APPLICATIONS IN SENSORS AND BIOSENSORS

Roberto Rella* and Maria Grazia Manera

Istituto per la Microelettronica e i Microsistemi, CNR Via per Monteroni "Campus Universitario" 73100 Lecce, Italy.
** E-mail: roberto.rella@imm.cnr.it*

Surface Plasmon Resonance (SPR) is an optical technique that uses evanescent waves as a valuable tool to investigate chemical and biological interactions taking place at the surface of a thin sensing layer. SPR offers a real time analysis of dynamic adsorption and desorption events for a wide range of surface interactions. After a brief theoretical introduction, examples of a wide range of applications of SPR are presented. Main application areas involve the detection of biological analytes and study of biomolecular interactions in liquid phase. Applications in chemical sensors will be also illustrated by using different classes of organic and inorganic material as sensing layers.

1. Introduction

The potentiality of Surface Plasmon Resonance as optical transduction technique for the optical characterization of thin films and monitoring processes taking place at metal interfaces has been recognized in the late twenties. Since the pioneering work of Otto[1] end Kretschmann[2] in the late 1960s, considerable progress has been devoted to the development of surface plasmon resonance as powerful tool for the optical characterization of thin films,[3,4] gas sensing,[5,6] biosensing techniques,[7,8] immunosensing,[9,10] SPR microscopy,[11,12] etc.. In 1983, Nylander and Liedberg exploited surface plasma waves excited in the Kretschmann geometry for gas detection and biosensing.[13,14]

Since then, the numerous possibilities opened in this field have attracted the interest of a wide spectrum of scientists, ranging from physicists, chemists and materials scientists to biologists.

2. SPR Theory

Surface Plasmon Spectroscopy (SPS) is an optical method based on the excitation of an evanescent electric field that is enhanced by electron plasmon resonance at a metal/dielectric interface. The use of evanescent fields for the detection of surface reactions is advantageous because of the limited field distribution near the interface where excited. The decay length of a surface plasmon enhanced wave outside the metal interface is no more than a few hundred nanometers, and depends on the wavelength of the light.[15] An evanescent field is created when light is totally reflected at the interface between media with different optical constants. The light energy is then reflected at the interface although the electromagnetic field penetrates into the second medium. The electric field exponentially decays normal to the boundary surface and is actually propagating parallel to it. The propagating wave along the interface is called an evanescent wave.[16] Surface Plasmon Spectroscopy (SPS) uses evanescent waves to excite so-called surface plasmons in a thin metal film (Fig. 1).
A surface plasmon is an electromagnetic wave associated with longitudinal oscillation of the free electron gas on the interface of a dielectric medium (such as water or air) and a metal with corresponding constant ε_d and ε_m. The electrical field of a plasmon propagating along the interface $x=0$ (fig.1) in the z-direction is given by: $\mathbf{E}(x, z, t) = \mathbf{E}_0(x) \exp i(\omega t - k_z z)$, where ω is the angular frequency and $k_z = k_z' + ik_z''$ is the propagation constant.

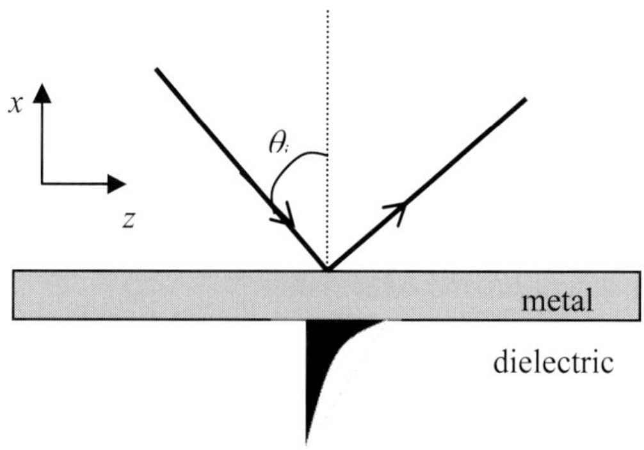

Figure 1. Schematic illustration of how an evanescent field is induced upon total reflection.

In order to excite the SP oscillation we need an electric field component, E_x, perpendicular to the interface. Consequently, p-polarised (TM) light is used, while s-polarised (TE) light yield no SP-excitation. The propagation constant $k_{z,sp}$ of a surface plasma wave propagating at the interface between a metal and a dielectric is given by the following expression:

$$k_{z,sp}= \frac{\omega}{c}\sqrt{\frac{\varepsilon_m \varepsilon_d}{\varepsilon_m + \varepsilon_d}} \tag{1}$$

where c is the velocity of light, ε_m and ε_d are the complex dielectric constants of the metal and the dielectric layer, respectively.

The propagation constant is generally a complex number because the dielectric function of metal ε_m is a complex function of angular frequency. The real part of the propagation constant is related to the effective refractive index whereas the imaginary part describes the modal attenuation due to the electron finite mass that damp their electronic oscillations. The damping is dependent on the exciting wavelength.

In order to excite a surface plasmon, the wave vector of the incident light must coincide with the unique wave vector of surface plasmon at the particular metal/dielectric interface:

$$k_z= \frac{\omega}{c} sen\theta_i \sqrt{\varepsilon_d} = \frac{\omega}{\mathrm{c}}\sqrt{\frac{\varepsilon_m \varepsilon_d}{\varepsilon_m + \varepsilon_d}} \tag{2}$$

For the solution of Maxwell's equation to exist, the dielectric constants should satisfy several conditions:

- ε_d should be real and positive
- the real part of the metal dielectric constant ε_m must be negative and its absolute value should be greater than the imaginary part.

At optical wavelengths, this condition is fulfilled by several metals[17] of which gold and silver are the most commonly used.

As the propagation constant of a surface plasma wave is always larger than that of a light wave in the dielectric, a surface plasmon wave (SPW) at a planar metal-dielectric interface cannot be excited directly by an optical wave from the dielectric. One possible solution to solve this problem is to let the exciting light beam pass through a high refractive index dielectric before hitting the metal surface. There are some different technical solutions for how to arrange the SP excitation conditions.[18] The most used method to enhance the momentum of an optical wave to allow coupling between a light wave and an SPW at the metal is the Attenuated Total Reflection method (ATR). A light beam is first propagated through the glass prism and then totally reflected at a glass/metal interface generating an evanescent field in the metal film in the Kretschmann configuration (Fig. 2). The wave vector along the interface in this case is given by:

$$k_{xp} = k_x \sqrt{\varepsilon_p} = \frac{\omega}{c} \sin\theta_i \sqrt{\varepsilon_p} \qquad (3)$$

If the metal layer is not too thick, the evanescent field can extend through it and couple to the plasmon resonance frequency at the metal/dielectric interface if the condition $k_{xp} = k_{x\,SP}$ occurs.

For gold and silver layers the evanescent field has in this configuration a maximum range of about 200 nm normal to the metal surface when the layer has an optimal thickness of about 50 nm.[19] If the metal layer is too thin the plasmon will reach back into the prism and thereby suffer intensity losses.

If we scan the angle of incidence of light onto a metal film of the right thickness, then at a certain angle the reflected light intensity will go sharply to almost zero, indicating resonant coupling to surface plasmons (Fig. 3). This angle is always greater than the angle of total internal reflection of the prism/outer dielectric interface (the so-called critical angle), and is called the Attenuated Total Reflection (ATR) angle. The position and the width of this 'resonance point' angle are very sensitive to the properties of the surface and the media next to it. It makes it possible to use surface plasmon resonance techniques for chemical and biological sensing.

Exploiting the sensitivity of the propagation constant of a surface plasmon wave to refractive index, SPR sensors allows us to measure changes in the refractive index produced by a change in the propagation constant of the plasmon wave.

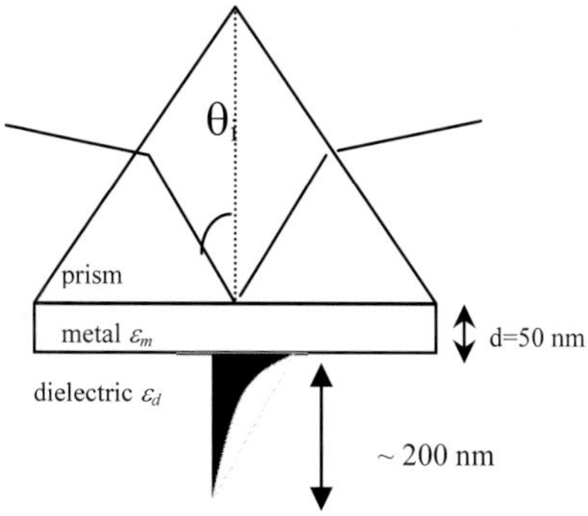

Figure 2. Schematic of excitation of a surface plasmon wave using a prism coupler in the Kretschmann configuration.

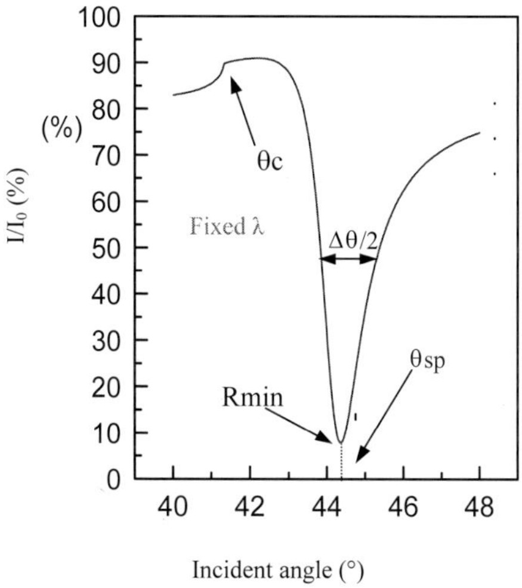

Figure 3. Reflectance curve of a p-polarised light beam incident on a thin Au film (50 nm) depositied on a BK7 prism (n=1.514) in air. The absoption peak in correspondence of θ_{sp} angle is due to the plasmon resonance. The position of the critical angle is evidenced too.

3. Optical Sensors based on Surface Plasmon Resonance

Recently, the applications of the SPR approach in sensors devices are becoming frequent. They involve the study of optical properties in metal layers, film thickness characterization, measurements of the optical parameters of organic layers deposited onto metal surfaces,[20-22] adsorption and desorption mechanism involving biomaterials, applications in gas sensing and biosensors.[23,24] There are a variety of subject areas ranging from physical applications to biological applications. Generally, optical sensors based onto surface plasma waves are referred to the measure of the changes in the refractive index or changes in non-optical quantities that can produce changes in the refractive index. In other words, a change in the refractive index can be measured by monitoring the change in the propagation constant of the SPW (surface plasmon wave) as a result of the change in the characteristic of the light wave interaction with the SPW. It is possible to classify SPR sensors on the basis of which characteristic of the light wave interacting with SPR is measured.

3.1. SPR Sensor with Angular Modulation

The coupling strength between the incident light wave vector and the surface plasmon wave is monitored at multiple angles of incidence of the light wave onto the metal surface, the light wavelength being fixed. When the surface plasmon wave is excited by the optical wave, resulting in a resonant transfer of energy into the surface plasmon wave, SPR manifests itself by resonant absorption of the energy of the optical wave at a particular incident angle. Variations in the optical parameters of the transducing medium can be detected by monitoring the shape and the angular position of the reflectance minimum.[25,26]

3.2. SPR Sensors with Wavelength Modulation

The wavelength modulation technique utilizes a fixed angle of incidence and modulates the wavelength. By measuring the reflected light intensity in the wavelength domain there will be a resonant minimum that satisfies the coupling condition.[27,28] However, unlike the angle modulation technique, as the incident wavevector varies, also the surface plasmon wavevector is modulated. Indeed there is a dependence of the plasmon wavevector on the metal and dielectric complex permittivities which also vary as a function of wavelength. But the variation of refractive index of the dielectric medium is much smaller than that of the dielectric medium, so the latter contribution is neglected.

3.3. SPR Sensors with Intensity Modulation

Both the angle of incidence of the light onto the metal film and its wavelength are kept constant. This technique measures the change in intensity of the light wave interacting with the surface plasmon wave.[29,30]

3.4. SPR Sensors with Phase Modulation

Also in this case, both the angle of incidence of the light onto the metal film and its wavelength are kept constant. The shift in phase of the light wave interacting with the surface plasmon wave is measured near the resonance angle.[31] A rapid jump in the phase is produced and it's more pronounced when the minimum approaches zero.

3.5. SPR Sensors with Polarization Modulation

Under SPR condition, both the amplitude and the phase of a p-polarized light wave change dramatically with the angle of incidence, while similar parameters for an s-polarized component remain almost constant. Therefore, if the exciting beam has an elliptical polarization (which comprises the components of both s- and p-polarization), the polarization state of the incident light will be sensitive to the variation in the propagation constant of the surface plasmon wave.[32]
Today, measurements of the resonant momentum of the optical wave are the most prevalent approaches because of the possibility of inherent simultaneous multiple data measurement which offers better signal to noise figures.

4. Application of SPR in Chemical Sensors and Biosensors

Because of the strong concentration of the electromagnetic field in the dielectric, the propagation constant of the surface plasma wave is very sensitive to variations in the optical properties of the transducing medium. Therefore, variations in its optical parameters can be detected by monitoring the interaction between the surface plasma wave and the optical wave. In this sense, SPR technique has been largely used in monitoring a variety of chemical and biochemical processes occurring on the metal surface.

Particularly, the Kretschmann geometry of ATR method has been found to be very suitable for sensing and has become the most widely used geometry in SPR sensors and biosensors.

In chemical SPR sensors the transducing layers are often polymers. Thin optically homogeneous polymers layers can be produced using spin coating or dip coating technique.[33,34] Thin films of organic materials such as phthalocyanines have been prepared using spin coating[6] and Langmuir-Blodgett technique[35] which allows the preparation of very ordered thin (monomolecular) layers and provides good thickness control.

Major areas of applications of surface plasmon resonance techniques are: the measurements of physical quantities, the monitoring of chemical sensing and biosensing. In the first area, SPR sensors for the measurement of displacement,[36] angular position[37] and temperature[38] have been exploited. On the contrary, chemical SPR sensors are based on the measurements of SPR variations due to the adsorption or a chemical binding of an analyte with a biomolecular layer which results in changes in its optical properties. Applications include monitoring of the concentration of hydrocarbons[39] vapors, aldehydes,[40] alcohols[41] and also molecular hydrogen[42] or NO_2[43] molecules or NH_3.[44]

As concerns SPR biosensors, they are devoted to monitor biomolecular interactions for rapid and parallel detection. The biomolecular recognition elements are immobilized onto the SPR sensor surface. When a solution containing analyte molecules is brought into contact with the SPR sensor, analyte molecules in solution bind to the biomolecules on the sensor surface, producing an increase in the refractive index at the sensor surface. This change produces a change in the propagation constant of the Surface Plasmon wave and it is eventually measured by the change in one of the characteristics of the light wave interacting with the surface plasmon wave.

Respect to the majority of current optical transducers based on fluorescence, SPR technique has the advantage to be a label-free technology for monitoring biomolecular interactions. Since the first application of SPR to biosensing, demonstrated in 1983,[14] the detection of bio-specific interaction was developed by different groups.[45-47]

To this purpose several immobilization chemistries that provide desired chemical properties for stable and defined binding of ligands have been developed. Surface immobilization techniques have been considered for optical sensors by Koller and Wolfbeis and, in particular, mechanical (adsorption and entrapment), electrostatic (or ionic), and chemical (covalent) methods have been used. Surface modification for immobilization includes mechanical polishing and sometimes electrochemical modification or chemical etching of the surface as a preliminary step, but these techniques are not confined exclusively to optical sensors and are not described here. For application in optical sensors devices the most widely used immobilization methods are physical-chemical techniques, including adsorption, sol-gel, lipophilic membranes, and chemical (both electrostatic and covalent); here we focused our attention describing covalent binding onto suitable gold surfaces for SPR imaging application. Typical approach is based on covalent binding of the biological recognition element to a metal thin film suitable for SPR application via linker layer. In this case self assembled monolayers (SAMs) are generated when organic molecules spontaneously chemisorb on the surfaces of metal layer, typically organic thiols and disulfides on gold or silver surfaces. The most robust and best characterized SAMs are those comprising alkanethiolates on gold.[48] By variation of the length of the alkane chain and the identity of the functional group at its terminus, the thickness of the organic layer and the chemical properties of the exposed interface can be controlled with great precision. For example, thio-oligonucleotides can be immobilized onto gold surfaces by SAMs in ethanolic solution and successively passivated with thioalkane to contrast some undesiderable a-specific adsorptions. A different approach uses surface preactivation of gold surface by SAMs of 11-Mercaptoundecanoic acid (MUA); this polyanionic MUA surface should be capable of binding polycationic

molecules such as Polylisine, PDTC (paraphenylenediisothiocyanate) and successive covalent attachment of thio/amino DNA sequences.
Alternatively, metal surfaces may be functionalized by thin polymer films to which ligands may by coupled via amino-groups.

5. SPR Instrumentation: From Traditional SPR Instrument to SPR Imaging

In addition to the traditional scanning SPR techniques,[50,51,52] fixed-angle SPR imaging can also be employed to monitor the adsorption of organic monolayers onto a suitable surface[53] and to simultaneously monitor the molecular probes affinity against a target molecule, for example in the analysis of DNA and RNA oligonucleotide hybridization[54,55] and DNA-protein interactions.[56] By setting the incident angle near the resonance angle, the refractive index can be measured by the variation in the reflection intensity. Under a parallel light illumination at a fixed incident angle, the reflected light intensity represents the refractive index distribution of the active layer surface.
The principle of SPR imaging was demonstrated first by Rothenhäussler and Knoll.[57] Two configurations were proposed to achieve lateral resolution. In the first configuration the incident beam was focused in order to minimize its spot size. Then, the focused beam was scanned across the sample surface and the reflected energy measured by a single channel detector.[58] In the second configuration (Fig. 4), a plane wave was employed to illuminate the complete analyte layer and the reflected image was observed by an array detector.[59] Recently, SPR imaging using CCD camera have provided a method for overcoming the limitation on the number of pixels acquired.[60]
The parallel beam of monochromatic light with a certain angle of incidence is coupled into surface plasmons. The collimated light source can be realized with a 1 mW HeNe laser equipped with a spatial filter and a beam expander.
Samples can be introduced into the imaging apparatus by attaching the gold-coated glass slides to a coupling prism with index matching fluid.

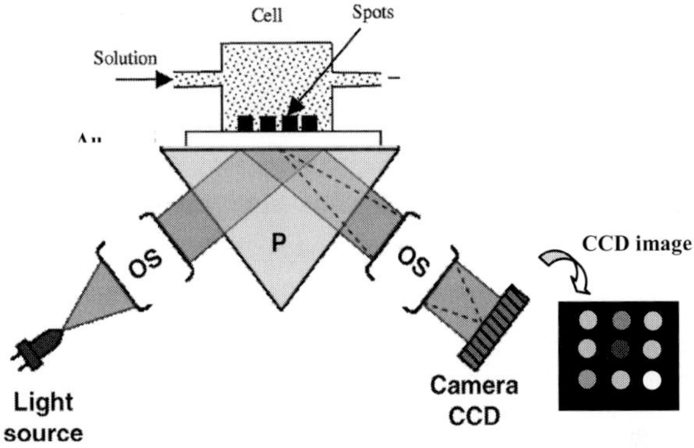

Figure 4. SPR imaging set-up in Kretschmann geometry. Light source can be a laser or a LED (λ=630 nm); OS is the optical system devoted to expand the incident beam and to direct the beam onto the CCD camera. P is the prism. SPR image arises from variations in reflected light intensity due to deposition of different spots of biomolecules onto the Au surface.

For in situ measurements, a Teflon flow cell can be attached to the prism/sample assembly so that a selected area of the chemically modified gold surface is in contact with solution. The reflected light is focused by a simple glass lens, captured with the CCD camera and then transferred to a computer for analysis. For a given sample arrangement and a given wavelength of the light, the resonance coupling appears as a sharp minimum in the angular distribution of the reflected light. The minimum shifts towards higher angles upon slightest changes of the refractive index or layer thickness of the sample. Adsorption of molecules such as nucleic acids onto the surface affects the index of refraction, thereby causing a change in reflectivity of incident light which can be monitored with the CCD camera in order to have a map of refractive index distribution. If the dielectric medium in contact to the metal layer is patterned, the resonance angles will be different for different areas of the metal film. This feature forms the contrast mechanism in SPR imaging, as shown in Fig. 5.

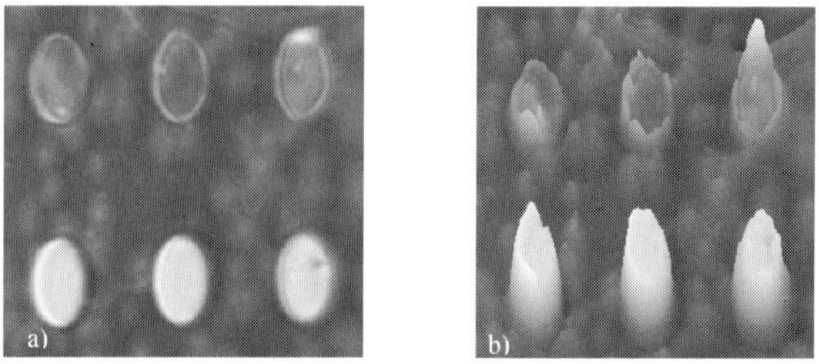

Figure 5. SPR contrast 2D (a) and 3D (b) images (λ=650 nm) representing the immobilization of different spots (with different thickness) of a material (n=1.6) onto a gold (d=52 nm) layer (in dark in the image). The images have been acquired in liquid phase.

This fact can be particularly useful in the analysis of DNA or DNA-RNA hybridization. Using UV-photopatterning techniques,[61] it is possible to create DNA arrays on gold surfaces for use with SPR detection by imaging. Areas on the gold surface destined for spotting with DNA probes are surrounded by regions modified with protecting groups that confines each DNA probe to its respective array position on the surface. Changes in the index of refraction where hybridization adsorption occurs, affects the reflectivity of incident light showing an SPR image of the DNA array ready for the subsequent analysis.
Improvements in the design of the SPR imaging system have yielded improved image contrast and sufficient sensitivity to clearly detect interactions between biological molecules without amplification.[62-64]

6. Future Capabilities

Food and environmental analysis can benefit greatly from the real-time aspects of biosensor analysis. There are many other important areas including medicine, biotechnology, drug and hazardous agents monitoring where SPR biosensors can play an important role. In these fields SPR biosensors devices have to compete with other types of biosensors.[9,65-67] In order to support this competition, improvements in the detection capabilities of SPR biosensors are desired to enable direct detection of biomolecular interactions improving sensitivity and resolution. For this reason, now research and development of SPR sensing devices are devoted to improvements of detection limits and specificity of SPR biosensors, to multichannel approaches in order to enhance throughput of SPR sensors and provide them with multi-analyte detection capability, and also to the miniaturization of SPR sensing devices.

As regards SPR imaging technique, recently, several methods have been introduced for image contrast enhancement; the simplest seems to be the dark-field technique.[68] The method based on SPR interferometry too[69] has demonstrated to enhance sensitivity of SPR imaging.

Given their extremely wide capabilities and ever-evolving applications, we envision the use of SPR biosensor technology will continue to expand as a modern bioanalytical tool.

References

1. A. Otto, *Z. Phys.*, **216**, 398 (1968).
2. E. Kretschmann, *Z. Phys.*, **241**, 313 (1970).
3. S. Szunerits and R. Boukherroub, *Langmuir*, **22**, 1660 (2006).
4. D.Roy, *Optics Communications*, **200**, 119 (2001).
5. K. Yoshinori, S. Masahiro, N. Takayuki, I. Hiroshi and U. Norihiro, *Sensors, IEEE* 628 (2007).
6. R. Rella, A. Rizzo, A. Licciulli, P. Siciliano, L. Troisi and L. Valli, *Mater. Sci. Eng. C* **22**, 439 (2002).
7. F. A. Tanious, B. Nguyen, W. D. Wilson, *Methods Cell. Bio.,* **84**, 53 (2008).
8. R. Wang, M. Minunni, S. Tombelli and M. Mascini, *Biosens. Bioelectron.* **20**, 598 (2004).
9. F. Ricci, G. Volpe, L. Micheli and G. Palleschi, *Anal. Chim. Acta*, **605**, 111 (2007).
10 F. Yu and W. Knoll, *Anal. Chem.*, **76**, 1971 (2004).
11. X. Li, K.Damada, A. Baba, W. Knoll and M. Hara, *J. Phys. Chem.* **B110**, 15755 (2006).
12. G. Stabler, M.G. Somekh and C. W. See, *J. Microsc.,* **214**, 328 (2004).
13. C. Nylander, B. Liedberg and T. Lind, *Sens. Act.* **B 3**, 79 (1982).
14. B. Liedberg, C. Nylander and I. Lundstrom, *Sens. Act.* **B 4**, 299 (1983).
15. T. Liebermann and W. Knoll, *Colloids Surf.* **A 171**, 115 (2000).
16. P. Yeh, in *Optical waves in layered media*, John Wiley and Sons, USA (1988).
17. M.A. Ordal, L. L. Long, R.J. Bell, S. E. Bell, R. R., Bell, R. W. Alexander, J. Ward and C. A. Ward, *Appl. Opt.* **11**, 1099 (1983).
18. H. Raether, *Springer Tracts in Modern Physics*, Vol. 11, Germany (1988).

19. T. Liebermann, W. Knoll, P. Sluka, and R. Herrmann, *Colloids Surf.* **A**, 169,337 (2000).
20. M. G. Manera, G. Leo, M. L. Curri, R. Comparelli, R. Rella, A. Agostiano and L.Vasanelli, *Sens. Act.* **B 115**, 365 (2006).
21. C.M. Pettit, D. Roy, *Analyst.*, **132**, 524 (2007).
22. S. Patskovsky, S. Bah, M. Meunier and A. V. Kabashin, *Appl. Opt.* **45**, 6640 (2006).
23. M. Schneider, A. Andersen, P. Koelsh and H. Motschmann, *Sens. Act.* **B104**, 276 (2005).
24. M. A. Plunkett, Z. Whang, M.W., Rutland and D. Johannsmann, *Langmuir,* **19**, 6937 (2003).
25. S. W. Kim, M. G. Kim, J. Kim, H. S. Lee andH. S. Ro, *J. Virol. Methods*, **148**, 120 (2008).
26. S. Conoci, M. Palumbo, B. Pignataro, R. Rella, L. Valli and G. Vasapollo, *Colloids Surf.* **A**, 198-200, 869 (2002).
27. J. Dostálek, J. Pribyl, J. Homola and P. Skládal, *Anal. Bioanal.Chem.* **389**, 1841 (2007).
28. J. Mavri, P. Raspor, M. Franko *Biosens. Bioelectron.* **22**, 1163 (2007).
29. M. G. Manera, P. D. Cozzoli, M. L. Curri, G. Leo, R. Rella, A. Agostiano and L.Vasanelli, *Synth. Met.* **148**, 25 (2005).
30. Y. Sakao, F. Nakamura, N. Ueno and M. Hara, *Colloids Surf. B*: **40**, 149 (2005).
31. A. K. Sheridan, R. D. Harris, P. N. Bartlett and J. S. Wilkinson, *Sens. Act. B* **97**, 114 (2004).
32. M. Piliarik, H. Vaisocherovà and J. Homola, *Biosens. Bioel.* **20**, 2104 (2005).
33. R. Capan, A. K. Ray, T. Tanrisever and A. K. Hassan, *Smart Mater. Struct.* **14** N11-N15 (2005).
34. J. F. Masson, T. M. Battaglia, Y. C. Kim, A. Prakash, S. Beaudoin and K. S. Booksh, *Talanta* **64**, 716 (2004).
35. S. Mukhopadhyay and C. Hogart, *Adv. Mater.* **6**, 162 (2004).
36. M. H. Chiu, B.Y. Shih, C.H. Shih, L. C. Kao and L. H. Shyu, *Proc. SPIE 6038*, 315 (2006).
37. H. P. Chiang, J. L. Lin, R. Chang, S. Y. Su, *Optics Letters*, **30**, 2727 (2005).
38. S. K. Ozdemir and G. Turhan-Sayan, *J. Lightwave Tech.* **21**, 805 (2003).
39. N. M. Aguirre, L. M. Pérez, J. A. Colin and E. Buenrosto- Gonzales, *Sensors* **7**, 1954 (2007).
40. S. Miwa, T. Arakawa, *Thin solid films*, **281**, 466 (1996).
41. M. G. Manera, G. Leo, M. L. Curri, P. D. Cozzoli, R. Rella, P. Siciliano, A. Agostiano and L. Vasanelli, *Sens. Act.* B, **100**, 75 (2004).
42. P. Tobiška, O. Hugon, A. Trouillet and H. Gagnaire, *Sens. Act. B*, **74**, 168 (2001).
43. K. Kato, C.M. Dooling, K. Shinbo, T.H. Richardson, F. Kaneko, R. Tregonning, M. O. Vysotsky andC. A. Hunter, *Colloids Surf.* **A**, **198**, 811 (2002).
44. Y. C. Kim, S. Banerji, J. F. Masson, W. Peng, K. S. Booksh, *The Analyst*, **130**, 838 (2005).

45. H. Vaisocherová, K. Mrkvová, M. Piliarik, P. Jinoch, M. Steinbachová and J. Homola, *Biosens. Bioelectron.*, **22**, 1020 (2007).
46. Y. Li, H. J. Lee and R. M. Corn, *Anal. Chem.*, **79**, 1082 (2007).
47. M. Ito, F. Nakamura, A. Baba, K. Tamada, H. Ushijima, K. H. A. Lau, A. Manna and W. Knoll, *J. Phys. Chem. C*, **111**, 11653 (2007).
48. Y. Arima and H.Iwata, *J. Mater. Chem.*, **17**, 4079 (2007).
49. C. E. Jordan, B. L. Frey, S. Kornguth and R. Corn, *Langmuir*, **10**, 3642 (1994).
50. Z. H. Zhang and C. L. Feng, *Biotechnol. J.* **2**, 743 (2007).
51. B. Snopok, M. Yurchenko, L. Szekely, G. Klein and E. Kashuba, *Anal. Bioanal.Chem.*, **386**, 2063 (2006).
52. R. Slavìk, J. Homola and E. Brynda, *Biosens. Bioel.*, **17**, 591 (2002).
53. J.M. Brockman, B. P. Nelson, R. M. Corn, *Annu.Rev.Phys.Chem.*, **51**, 41 (2000).
54. I. Mannelli, L. Lecerf, M. Guerrouache, M.Goossens, M.C. Millot and M. Canva, *Biosens. Bioelectron.*, **22**, 803 (2006).
55. H. J. Lee, A.W. Wark, Y. Li and R. M. Corn, *Anal. Chem.*, **77**, 7832 (2005).
56. B. H. Garcia and R. M Goodman, J. *Virol. Methods*, **147**, 18 (2008).
57. B. Rothenhäussler and W. Knoll, *Letters to Nature*, **332** 615 (1988).
58. E. Yeatman and E. A. Ash, *Electron. Lett.*, **23**, 091 (1987).
59. W. Hickel, W. Knoll, *Nature*, **332**, 615 (1988).
60. D. Boecher, A.Zybin, K.Niemax, C, Grunwald, V. M. Mirsky, *Rev. Sci. Instrum.*, **79**, 023110 (2008).
61. E. A. Smith, M. G. Erickson, A. T. Ulijasz, B. Weisblum and R. M. Corn, *Langmuir*, **19**, 1486 (2003).
62. L. Malic, B. Cui, T. Veres and M. Tabrizian, *Opt. Lett.*, **32** 3092 (2008).
63. K. S. Philips, Q. Cheng, *Anal. Bioanal. Chem.*, **387**, 1831(2007).
64. G. Steiner, *Anal. Bioanal. Chem.*, **379**, 328 (2004).
65. K. Länge, B. E. Rapp and M. Rapp, *Anal. Bioanal. Chem.*, (2008) in press
66. G. Gauglitz, G. Proll, *Adv. Biochem. Eng. Biotechnol.*, **109**, 395 (2008).
67. C.A. Marquette and L.J. Blum, *Anal. Bioanal. Chem.*, **390**, 155 (2008).
68. P. Jain, X. Huang, I. El- Sayed, M. El- Sayed, *Plasmonics*, **2**, 107 (2007).
69. X. Yu, X. Ding, F. Liu, X. Wei and D. Wang, *Meas. Sci. Technol.*, **19**, 015301 (2008).

MICRORESONATORS FOR SENSING APPLICATIONS

Simone Berneschi,[a] Gualtiero Nunzi Conti,[a,b] Stefano Pelli[a] and Silvia Soria[a,b,]*

[a]*Istituto di Fisica Applicata "Nello Carrara , CNR*
50019 Sesto Fiorentino, (Fi) Italy
[b]*Centro Studi e Richerche "Enrico Fermi"*
00184 Rome, Italy
**E-mail: s.soria@ifac.cnr.it*

Nowadays sensing represents a very active area of research due to many possible applications. A particular need exists for miniature sensors for the detection of several biochemical species and tracking mechanical changes. Several optical techniques have proven to be quite effective. Here we provide a quick overview of the recent progresses in the development of optical biosensors based on resonant cavities, where light propagation occurs through whispering-gallery modes (WGMs). The effect of any perturbation to the optical resonance structure of a WGM resonator is such that a very high sensitivity can be achieved.

1. Introduction

In this chapter, we address sensing applications of passive devices based on Whispering Gallery Modes (WGMs) resonators. We focus mainly on biochemical sensors and briefly summarize some relevant results of chemical and mechanical sensing.

WGM resonators have different geometries with different confining principles and unique spectral properties, including narrow line-width, high stability, and tunability. High quality factor Q and long recirculation of light in compact WGM devices are the most important features for sensing applications, where the change in Q or resonant wavelength can be used for measuring the change in ambient parameters in the surrounding environment or binding phenomena on the WGM resonator surface.

In chapter 2 we describe the theory of WGM spherical resonators, but it can be extended to all types of WGM. In chapter 3 we summarize the results obtained in sensing.

2. Whispering Gallery Modes in a Microsphere

WGMs were first observed in the gallery of the cupola of St Paul's Cathedral in London: a whisper spoken close to the wall can be heard all the way along the gallery, some 42 m to the other side. From that, the term "whispering gallery" was introduced.[1] Some authors have also referred to these modes as "morphology dependent resonances" (MDR), however this terminology has not been widely adopted.

These optical modes are confined in the microcavity by total internal reflections (TIR) at the dielectric air interface. If scattering losses by TIR at the boundary of the microsphere are minimal and absorption of light in the transparent material is very low, thus, the photons are able to circulate on their orbit several thousand times before exiting the microcavity by loss mechanism. This long lifetime of the confined photon is associated to a long optical path length because of the resonant nature of the phenomenon. When a micro or nanoscopic object like a bacterium or a molecule is brought in contact with the confined circulating light, the interaction will be resonantly reinforced.

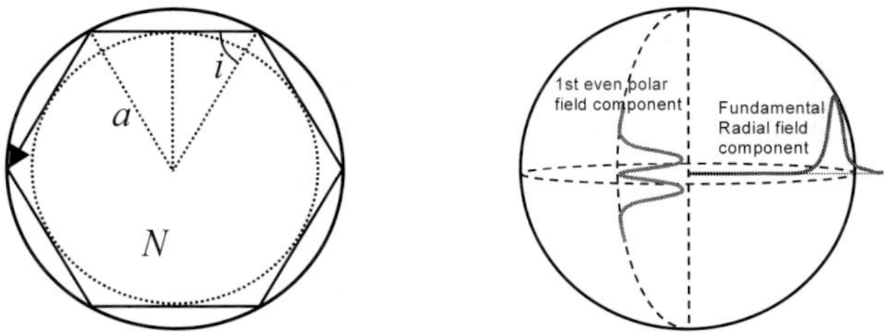

Figure 1. Schematic of the total internal reflection of rays. Right: Radial and polar field distribution for different mode numbers.

Simply using geometrical optics we can carry out a quick analysis of the propagation. With reference to Fig. 1, where a indicates the radius of the sphere and N its refractive index, a ray of light will undergo total internal reflection if the angle of incidence i is higher than the critical angle $i_C = arcsin(1/N)$. A dimensionless size parameter is generally introduced, defined as $x = 2\pi a/\lambda = k\,a$, where k is the wave number. Let us suppose that the radius of the sphere is much larger than the wavelength of radiation ($a \gg \lambda$, or $x \gg 1$) and that rays are at glancing incidence ($i \approx \pi/2$): at the sphere surface, the condition to have resonance is that the optical path length, which is approximately equal to the circumference of the sphere , should correspond to an integer number of wavelengths in order to keep in phase:

$$2\pi a \approx l(\lambda/N) \tag{1}$$

with l integer number. In this way it is easy to understand that light may be confined in a band around a great circle of the sphere and that a caustic region can be defined, comprised between the outer sphere and the inner sphere to which the propagating and bouncing rays are tangent.

Quite obviously, the geometrical optics description has severe limitations: as an example, it cannot explain how the light can couple into a WGM (or escape from a WGM) in a perfect sphere, nor it can take into account the polarization of light.

A complete description can be provided by the electromagnetic theory, and the resonances can be analyzed using the generalized Lorenz-Mie theory.

The optical modes of a dielectric microsphere can be calculated by solving Helmholtz equation in spherical coordinates. Polarization is assumed to be constant along the optical trajectories since the microspheres are made of homogeneous dielectric and the optical modes reflect with grazing incidence at the dielectric-air boundary. The fields can be express in terms of either TM or TE mode polarizations and solutions are found by solving the scalar equation by the separation of variables approach. The radial field can be described by spherical Bessel functions inside the sphere and an exponential tail outside, while the polar component is described by Legendre polynomials, and the equatorial behavior is sinusoidal. A given WGM is identified, thus, by mode numbers n, m, and l, and by the polarization mode (TE or TM).

The value of n gives the number of maxima in the radial component, l depends on the equatorial length, expressed in number of wavelengths, and $l-|m|+1$ gives the maxima in the polar component. Polar modes are often referred to as even or odd, based on number of lobes. For each angular number l, the allowed azimuthal mode numbers are in the range of $-l<m<+l$, leading to degeneracy in the azimuthal modes of $2l+1$.

The position of the resonances, in term of the size parameter, can be approximated by the following equation:

$$Nx_{n,l}^{P} \approx l + 1/2 + \left(\frac{l+1/2}{2}\right)^{1/3}\left[\frac{3\pi}{2}(n-1/4)\right]^{2/3} - \frac{P}{\sqrt{N^2-1}} \qquad (2)$$

where $P=N$ for TE modes and $P = 1/N$ for TM modes. A number of characteristics of the WGM spectrum can be obtained from this equation including the quasi-periodicity for WGMs with same n value and $\Delta l=1$ corresponding to a pseudo-free spectral range Δ_0 given by:

$$\Delta_0 = c/(2\,\pi\,N\,a) \qquad (3)$$

and the spacing between modes having the same modal numbers but different polarizations Δ_P given by:

$$\Delta_P = \Delta_0\sqrt{\frac{N^2-1}{N^2}} \qquad (4)$$

The modes with smallest mode volume, which are those of interest in many applications, are those with low n values and with $m \approx l$ and they are most closely confined to the surface of the sphere.

In this case the propagation constant β for the WGMs can be simply written as $\beta = k\,l/\,x_{n,l}$.

The most important parameter of microspherical resonators in sensing applications is their quality factor Q. It can be considered as an indication of the fraction of the light lost during each cycle around the sphere. The intrinsic Q of a microsphere is determined by contributions from several types of losses:

$$Q^{-1} = Q_{rad}^{-1} + Q_{scat}^{-1} + Q_{mat}^{-1} + Q_{sens}^{-1} \qquad (5)$$

where Q_{rad} denotes intrinsic curvature losses, Q_{scat} scattering losses on residual surface inhomogeneities, Q_{mat} intrinsic material losses, and finally Q_{sens} indicates the losses introduced by analytes to be detected. Q_{rad}^{-1} vanishes exponentially with increasing size, so with $2a/\lambda \geq 15$, $Q_{rad} > 10^{11}$. Calculations based on the model of Rayleigh scattering by molecular-sized surface clusters under grazing incidence and total internal reflection yield the following estimate for Q_{scat}:

$$Q_{scat} = \frac{\lambda^2 d}{2\pi^2 \sigma^2 B} \tag{6}$$

where $d = 2a$ is the sphere diameter, and σ and B are the rms size and the correlation length of surface inhomogeneities, respectively. Experimental values reported for fire polished glass surfaces are $\sigma = 0.3$ nm and $B = 3$ nm; using these figures, $Q_{scat} > 10^{10}$ may be expected for $2a > 50$ µm (if $\lambda > 1$ µm). Thus, in the absence of contaminants, the Q factor of large spheres may reach the limit defined by material losses:

$$Q_{mat} = 2\pi\, n_{eff} / \alpha\, \lambda \tag{7}$$

where n_{eff} is the effective index of refraction for the WGM being tested, α is the absorption coefficient of the sphere's material, and λ is the wavelength of propagating light. As the optical attenuation α in standard fused silica fiber for telecom systems is around 0.2 dB/km at 1.55 µm, Q_{mat} may reach a value of 1.3×10^{11}.

One of the methods to measure the Q-factor is based on the transient response: one observes how quickly the output power decays after an input of short pulse, and Q is determined from the decay time τ according to the equation:

$$Q = 2\pi\nu\tau, \tag{8}$$

where ν is the resonance frequency.

The Q-factor may also be derived from the measurement of the spectral line-width $\Delta\nu$ of the mode:

$$Q = \nu/\Delta\nu \tag{9}$$

where $\Delta\nu$ is the full width of the resonance at the half-maximum points. It is obvious that in this case it is the line-width of the laser used in the measurement to set the upper limit for the measurable Q-factor.

In low-loss fused-silica microspheres, with diameters in the range 50 to 500 μm, Q's in excess of 10^{10} have been demonstrated.[2] Smaller microresonators have a wider free spectral range (FSR) and a smaller number of modes; they also exhibit a very small mode volume (low-order WGMs have smaller mode volume, e.g. V_{eff} ~1000 μm^3 when $2a$ ~ 40 μm) and high finesse.

The efficient coupling of light in or out of a microsphere is a key issue and requires the use of near field coupling: the evanescent field of a phase-matched optical waveguide should overlap with the evanescent field of the whispering gallery mode. Selective excitation of high-Q WGMs (lowest n values) is possible through the use of phase-matched wave coupling from an adjacent waveguide, a prism under total internal reflection, or a tapered fiber.

Coupling can be characterized by the fractional depth K of the resonance dip in intensity transmittance through the coupler. K is observed upon varying the frequency of the exciting wave around the resonance and can be expressed in the following way as a function of the intrinsic quality-factor of the WGM Q_0:

$$K = 4\, Q_0 Q_c \Gamma^2 / (Q_0 + Q_c)^2 \tag{10}$$

where Q_c describes loading, i.e., it is proportional to the inverse of the transmittance of the coupler, and the coefficient Γ describes mode-matching (a single mode coupler is always mode-matched). The quality factor of the system Q_S is related to Q_0 and Q_c by the following equation:

$$1/Q_S = 1/Q_0 + 1/Q_c \tag{11}$$

Unlike the case of Fabry–Perot cavities with their fixed coupling to external beams, the sphere-coupler system provides a unique opportunity to easily control the bandwidth of the cavity. In fact Q_S can be adjusted by increasing the gap between the coupler and the sphere from the over-

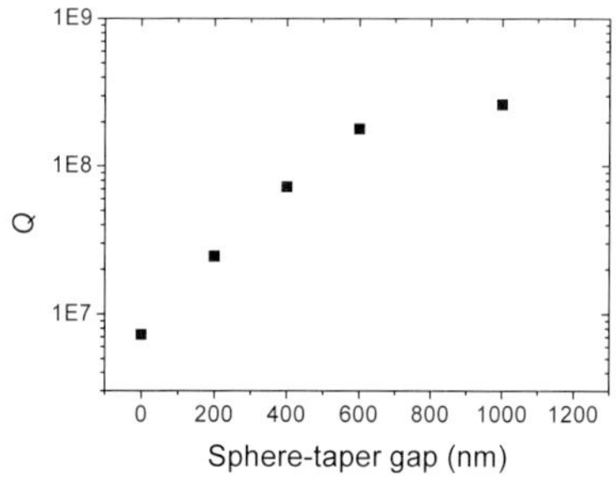

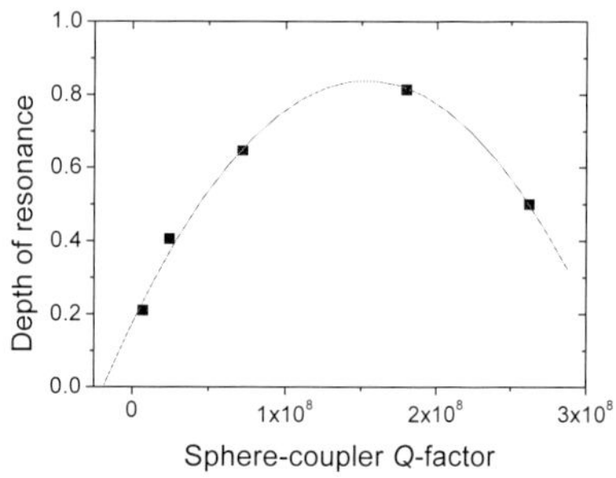

Figure 2. Left: Measured Q values vs fiber-microsphere gap. Q factor saturates to its intrinsic value $Q \sim 3{\cdot}10^8$. Right: Depth of resonance K vs sphere-coupler system Q factor.

coupled regime ($Q_c << Q_0$) to the under-coupled one which permits a clear observation of saturation of the measured Q_S up to its intrinsic (unloaded) value Q_0 (fig.2). Maximum contrast is achieved when coupling losses equal intrinsic cavity losses, i.e. $Q_c = Q_0$ or $Q_S = 1/2\ Q_0$, and the entire coupled power is lost inside the resonator (fig.2). This regime is usually called critical coupling.

A frequency shift of the resonances when the radius and/or the refractive index of the sphere change makes possible the utilization of the WGMs of microspheres in detecting trace amounts of chemical and biological molecules. In other words, when the analyte aggregates at the surface, it interacts with the evanescent part of the WGM field inducing a change in the Q factor or a shift in the wavelength (fig.3). The latter can be then quantitatively predicted using a perturbation theory.

$$\delta\lambda/\lambda = \alpha_{ex}\sigma/\ [\varepsilon_0(n_s^2 - n^2_m)a] \tag{12}$$

where α_{ex} is the excess of polarizability and σ the surface density of molecules forming a layer. This equation is exact if the layer is considerably thinner than the evanescent field depth. It is worth pointing that also rings and toroids can be considered WGM resonators and that rings and capillaries are also included in this type of resonators, though their modes are not strictly WGM.

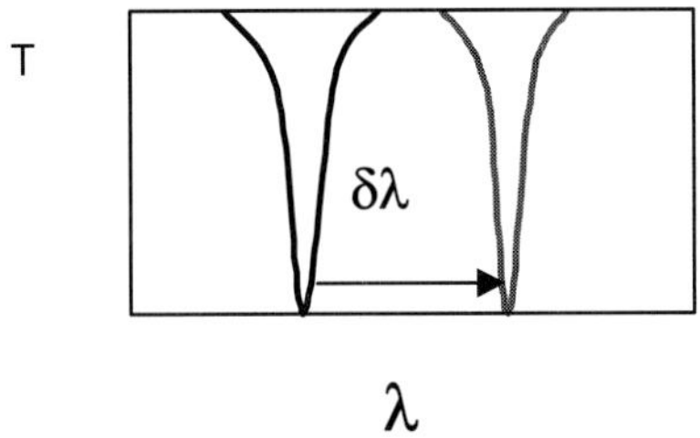

Figure 3. Resonance shift after analyte binding to the surface of a WGM sensor. Transmittance vs wavelength.

3. WGM Resonators: Applications in Sensing

3.1. Sensing of Biological and Chemical Agents

The International Union of Pure and Applied Chemistry (IUPAC) defined a biosensor as "a self-contained integrated device, which is capable of providing specific quantitative or semi-quantitative analytical information using a biological recognition element (biochemical receptor) which is retained in direct spatial contact with a transduction element". The interaction produces an effect measured by the transducer, which converts the information into a measurable effect, in our case an optical signal. An important aim of much of the work done in (bio)sensing has been to develop disease detection systems with the longer term goal of prevention or cures.

The quality of biosensors, thus, does depend on the total sensor system defined by the transducer, the sensitive layer, and the electronics and on evaluation of the acquired data[3]. The requirements for a biosensor are selectivity, sensitivity, stability and reversibility. These are mostly provided by the biochemical receptor, the sensitivity is also provided by the quality of the transducer. Of the same importance are high signal-to-noise-ratio (SNR), short response time, low limit of detection (LOD), high sensitivity at low cost and in real samples. Thus, the main goals of a (bio)sensor are to obtain the maximum amount of information from the smallest amount of sample, and to detect many binding events simultaneously.

The optical techniques used are based on phase changes (change in the index of refraction), amplitude changes (absorption) or frequency changes (fluorescence). The phase and amplitude changes are a direct monitoring technique whereas the frequency changes often include and indicator or marker (labeled system). The disadvantage of the labeled system (cost, expenditure, possibly reduced reactivity) are normally compensated by lower LOD, while in the case of direct monitoring, the disadvantage lies on the ineffectiveness of detecting small molecular weights analytes and the sensitivity to non specific binding. However, recently the ability of single molecule detection has been proved.[4]

Of the different optical sensor principles, none is generally superior but rather depends on the application. Most of the optical sensors are based on detection by evanescent field at the transducer surfaces. Some examples of low LOD in planar geometries and label-free detection are given in Clerc et al.[5] where the authors measured changes in the index of refraction of $\Delta n_{eff} = 3 \times 10^{-6}$. This translates into the possibility to detect coverage of 10 pg/mm^2, if the molecule to be detected is an IgG (M_w=150 kDa) the numbers could be translated in a concentration as low as 0.3 nM. Recently, Schmitt et al. were able to detect a lower LOD of $\Delta n_{eff} = 9 \times 10^{-9}$, coverage of 13 fg/ mm^2, IgG concentration of 0.39 pM.[6] For labeled detection in planar geometry, Plowman et al. were able to detect concentrations of about 3 fM.[7]

A crucial step for producing effective biosensors is the surface functionalization or chemical modification of the transducer's surface. Protein adheres to any glass surface rising the unwanted non-specific effects. There are several ways of functionalizing the surface of a biosensor, among them the most common are based on the silanization of the glass surface through covalent binding of the silane groups with the glass surface and the use of biotin and/or streptavidin layers. The first one shows a reduced non-specificity and enables a further functionalization with ligands or receptors, whereas the second one is based on a high affinity binding with streptavidin and biotinylated molecules. This crucial layer, however, has to be very thin between 10-100 nm (below the evanescent field tail) and homogeneous in order to preserve the high quality of the transducer.

WGMs based sensors are devices that aim at ultra-low detection of binding events (either biological or chemical) at the interfaces of the cavities.[8] Figure 4 shows the surface functionalization steps.

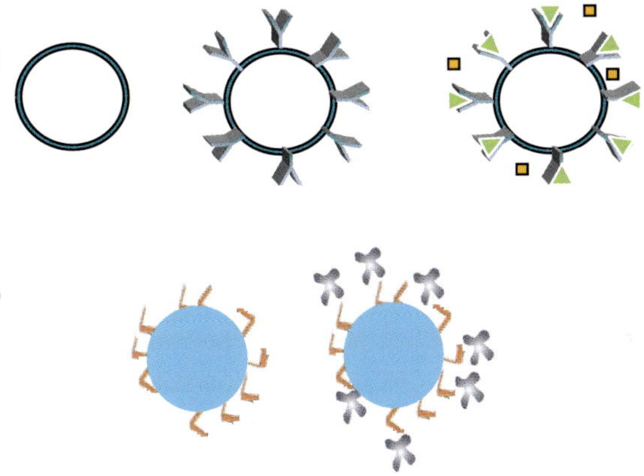

Figure 4. Functionalization of a WGM sensor: a) i) deposition of the precursor layer (green color) on the bare WGM resonator, ii) covalent binding of the antibody (bio-receptor) to the precursor, iii) selective binding of the antigen (analyte) to the antibody (bio-receptor); b) deposition of a biotin layer (red), ii) selective binding of streptavidin (blue) to biotin.

WGM sensors can be classified according to the geometries (spheres, disks, rings or capillaries) or to their degree of planar integration: optically integrated WGM sensor (in/out-coupling system and cavities are both integrated), hybrid WGM sensors (either the cavity or the in/out-coupling are integrated), and non-integrated systems (neither the cavity nor the in/out-coupling system are integrated). Common properties of WGM sensors are high Q factors, tunability, high stability and sensitivity and small sample volume needed to detect a given analyte. They can be considered a miniaturized and updated version of the conventional optical cavity ring down spectroscopy (CRDS), where long and bulky conventional Fabry-Perot resonators are used to obtain an effective optical path length sufficient to enable high resolution. They can be used for label-free[9-11] and/or labeled detection.[8,11,12-15] In the latter case, long photon lifetime enhances dramatically the absorption quantum efficiency of the fluorophores, resulting in an enhanced fluorescent emission. In WGM resonators, the Q can be partially spoiled during the various steps of surface functionalization (see figure 5). As the analyte binds to the bio-receptor, the resonance shifts. In fact, the frequency (wavelength) locations of the resonances depend on the size and refractive index of the WGM resonators. This is generally the working principle of label-free WGMs sensors and by measuring these changes, the amount of analyte bound can be quantified.

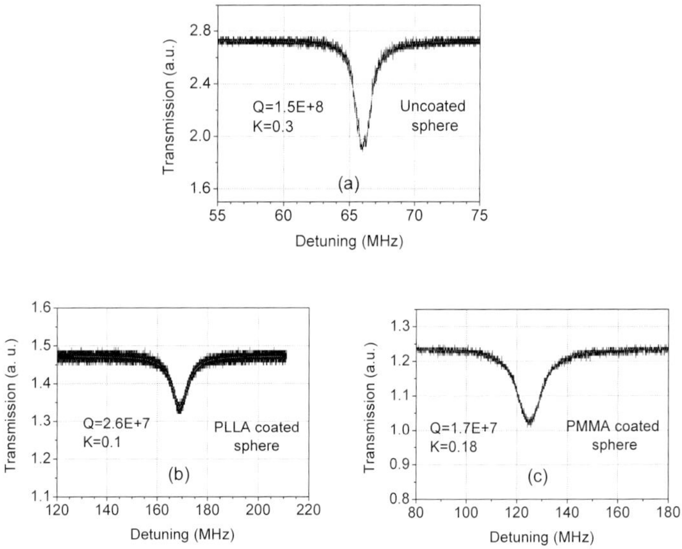

Figure 5. Measurement of *Q*-factors of a: a) Bare microsphere of 250 μm diameter (Q= 1.5x10^8), b) microsphere coated with a thin layer of polylactic acid (PLLA) (Q= 2.6x10^7) and c) microsphere coated with a thin layer of PMMA (Q= 1.7x10^7).

Optically integrated WGM sensors are usually microrings or microdisks with two adjacent single mode port waveguides for in and out coupling the light to and from the resonator; the port waveguides could be arranged horizontally or vertically. In the first case, both waveguides and WGM sensor are exposed to the analyte, whereas in the second one the waveguides are shielded from the analytes by a separation layer (see Fig. 6). An approach with a MEMS actuated coupling waveguide allowing to change the coupling efficiency has been developed by Yao et al.[17] Microdisks and microrings can be easily produced using lithographic or imprint technologies (see Fig. 7), in SiO_xN_y materials[16] or in polymers[10,18,19] but they show low Q factors of about 10^5 in aqueous environment. For glucose sensing, Krioukov et al.[16] were able to measure changes in index of refraction below 10^{-4} whereas Chao et al.[10] measured changes of about $3x10^{-5}$. For protein detection, it has been shown that for Δn_{eff}=10^{-5} concentrations of 0.3 nM (or 10 ng/ml) of avidin solutions can be detected.[19-21] Recently Ramachandran et al.[22] have shown that microrings can be also used for bacterial and nucleic acid detection.

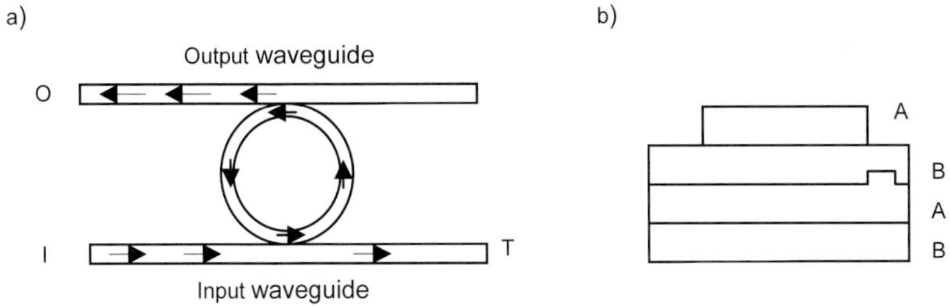

Figure 6. Sketch of the two basic geometries for optically integrated WGM sensors: a) top view of a microring with two port waveguides: O output port, I input port, T through port; b) cross section of a microdisk with vertical coupling, A: Si_3N_4layer, B: SiO_2 layer.

Hybrid WGM sensors can be of two types, Fig. 8a shows the type that consists of an integrated WGM sensor and a discrete in-output coupler[4,23] while figure 8b and 8c show the one that consists of an integrated in-output coupler and a non integrated 3-D WGM sensor.[24,25]

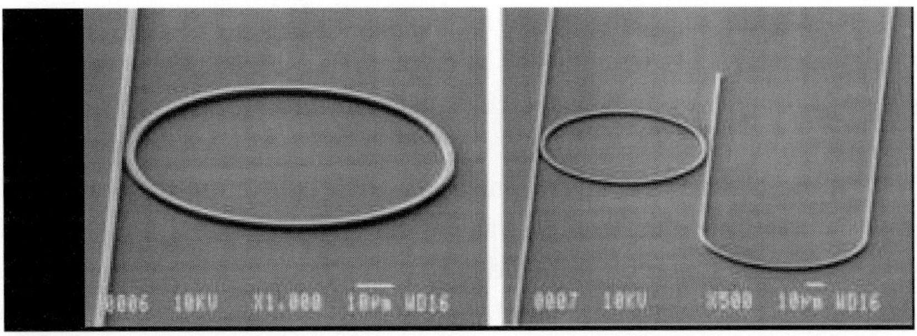

Figure 7. Scanning electron microscope picture of nanoimprinted polystyrene microrings for a single coupled case (right) and doubly coupled case (left) (© 2003 IEEE, reprinted with permission from[10].)

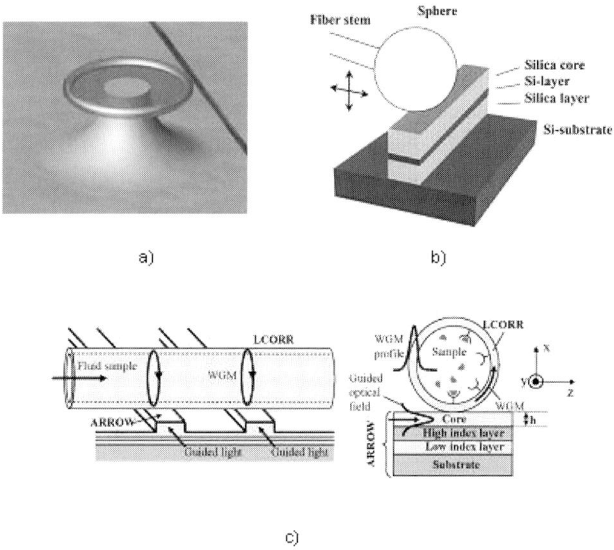

Figure 8. Sketch of hybrid WGM sensors: a) microtoroid coupled to a tapered optical fiber (© 2005 AIP, reprinted with permission from[23].); b) microsphere on top of a SPARROW (strip-line pedestal ARROW) (©2000, IEEE, reprinted with permission from[30].); c) LCORR-ARROW system and cross section viewed from the LCORR (liquid core optical ring resonator) on top of an ARROW (anti-resonant reflecting optical waveguides). (©2006, AIP, reprinted with permission from[28].)

Armani et al. used planar arrays of microtoroids coupled to a tapered fiber for detecting the difference between two similar chemical species,[26] like H_2O and D_2O and for single molecule detection of interleukin-2 (concentrations of 100 aM)[4]. Microtoroid-based-WGM sensors are specially suited for single molecule detection due to their ultra high Q factor,[27] which is above 10^8.

Fan et al. have developed a new sensing architecture that combines anti-resonant reflecting optical waveguides (ARROW) and a 2-D WGM sensor, which is based in a liquid core optical ring resonator (LCORR). An LCORR is a thin-walled quartz capillary that acts simultaneously as a microfluidics channel and a ring resonator. The ARROW is brought into contact perpendicularly to the LCORR, in- and out-coupling the light to and from the ring cross-section of the capillary. The evanescent field interacts with the analyte that is flown inside the capillary. Thus, as the analyte binds to the inner surface of the LCORR, the WGMs shift spectrally.[24] This approach is quite promising for lab-on-a-chip and sensor arrays technology.[28,29] A similar architecture was used to couple the light in a strip-line pedestal ARROW to a microsphere.[30]

Discrete micro-optical or non-integrated WGM sensors are 3-D or 2-D structures like microspheres or microcapillaries (Fig.9) in which the light is coupled by means of an optical component like a taper fiber. Microspheres have ultra high *Q* factors of about 10^9, but they lack of robustness and integration capability.

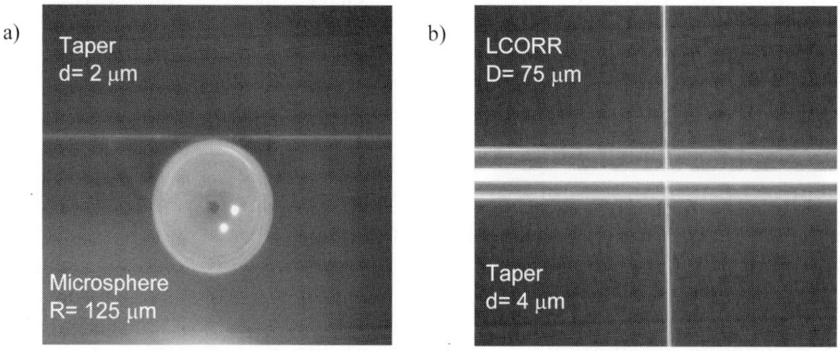

Figure 9. Discrete (non-integrated) WGM sensors: a) microsphere coupled to a tapered optical fiber of diameter below 2 μm; b) LCORR coupled to a taper of diameter of 4 μm (©2006, OSA, reprinted with permission from[29]).

Zhu et al. have been able to detect bovine serum albumina (M_w= 66 kDa) below 10 pM concentration with 0.5 pg/mm^2 mass detection limit.[31] The authors used an LCORR of 100μm diameter with a *Q* factor above 10^6, thermoelectrically cooled because for such a *Q* factor the noise caused by temperature fluctuations is a dominant factor. Ling and Guo[33] studied an LCORR coupled by prism and showed a record sensitivity of 600 nm/RIU for 32 μm thick capillary. The authors attributed this high sensitivity to a new type resonance mode. Such mode has the highest optical field in the low index region fluid. Suter et al. have demonstrated real time detection of DNA sequences quantitatively.[33] The LCORR used in that work has achieved the detection of DNA in bulk, with a strand length of 25 bases and a concentration of 10 pM. Sumetsky et al. have imbedded an LCORR into a polymer matrix (n = 1.384) in an attempt to integrate and increase the robustness of the WGM sensor. The authors suggested the use of Teflon AF (n = 1.291) for improving the sensing capabilities of the liquid ring resonator optical sensor (LRROS) .[34]

Microspheres were first proposed as gas sensors.[35-37] The authors described the construction of a prototype system for atmospheric and acetylene trace-gas detection. Microspheres were and are still one of the most studied WGM sensors due to their intrinsic ultra high Q factor[38] of about 10^8-10^9. However, the Q factor of microspheres immersed in aqueous solutions[9,39] is just in excess of 10^6. In the last years, it has been shown that the LOD for proteins like BSA[9], DNA[39] and protease[40] deposited on the surface of the microspheres can be as low as 1-10 pg/mm^2. In Vollmer et al.[39] it was also demonstrated that multiplexed DNA detection by using two microspheres was possible, and that a single nucleotide mismatch in an 11-mer oligonucleotide could be discriminated. Recently, microspheres based sensors able to detect small molecules at subfemtomole level[41] were also demonstrated. The same authors tested their WGM sensor system for mercuric ions in water.[42] Keng et al. detected nanoparticles like viruses undergoing Brownian from resonance fluctuations.[43] The size of the nanoparticles could be estimated from the analysis of the noise. The authors incorporate a microfluidic system (see figure 10). Ren et al. used microspheres to detect quite large particles, like rod-like bacteria. They were able to measure down to 34 pg/mm^2 mass loading of E. coli (44 bacteria bound randomly to the surface).[44]

Another example of microfluidics is given in Levy et al.[45] In there the authors have designed and fabricated a microfluidics chip for an optical microring resonator in which they could mix up to two liquids with different refractive index. Figure 11 shows a detail of the integrated microfluidics chip with water and green dye injected into the inlets.

Active WGM resonators have been also proposed for sensing applications. Fang et al used hybrid zinc oxide/silica microdisk lasers for sensing volatile organic compounds such as toluene and nitrobenzene.[46] They monitored the red shift of the laser line caused by the increase of the refractive index of the microdisk surface after adsorption of the organic molecules.

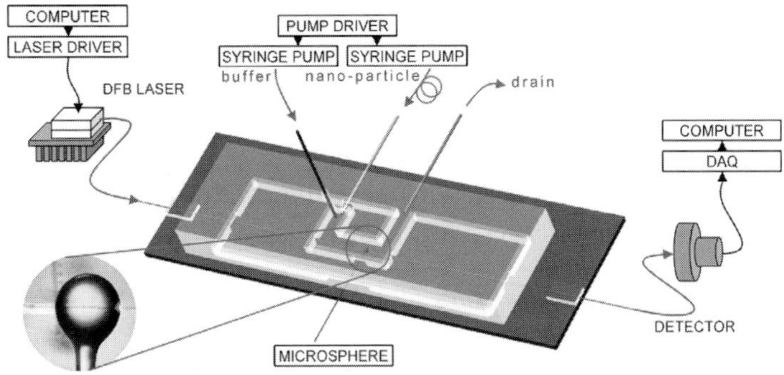

Figure 10. Microfluidic system incorporating a WGM sensor and a coupling fiber. (©2007, OSA, Reprinted with permission from[43].)

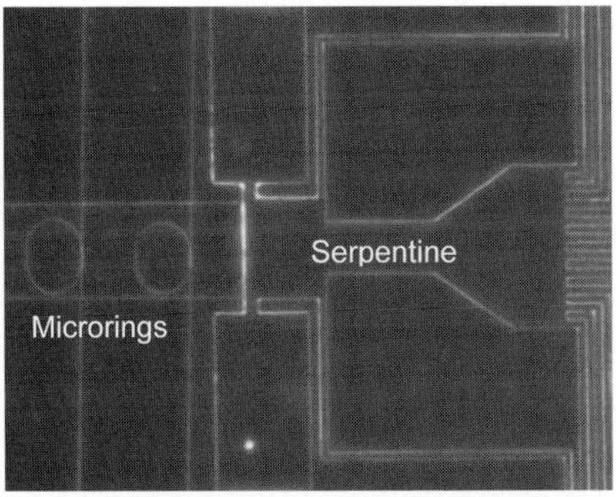

Figure 11. Microfluidic chip integrated with a WGM sensor based on microrings coupled to waveguides. (©2006, AIP reprinted with permission from[45].)

Yang et al.[47] proposed active polymeric microspheres for sensing applications. The Q factor of such WGM sensor is ultra high ($Q = 10^{10}$) and could provide a low LOD of $\Delta n_{eff} = 10^{-10}$.
Polymeric microspheres were recently proposed as sensors. Lutti et al.[48] measured Q factors as high as 10^6 on polystyrene microspheres of 30 μm held by optical tweezers in aqueous solutions. They estimated a protein detection sensitivity of 0.25 pg/mm^2.

Another trend in this field is the use of arrays of WGM resonators. Schweiger et al.[49] used an array of microspheres as a miniaturized spectroscopic device. The array consisted of 16 polymethyl methacrylate microspheres placed on top of a microscope slide serving as optical waveguide. Francois et al. have recently proposed the use of clusters of dielectric particles as a novel detection scheme for optical biosensing and they were able to detect concentrations of analyte below the femtomole.[50]

3.2. Mechanical and Other Sensors

Laine et al.[51] proposed a hybrid WGM sensor as an accelerometer. The fiber stem that holds the microsphere is displaced and a variation of the coupling gap is created. With this system, sensitivities better than 1 mg at 250 Hz bandwidth and a noise floor of 100 μg were demonstrated. Armenise et al.[52] used an optically integrated active WGM sensor for gyroscope systems. The authors obtained a very good performance of the device in terms of gyro quantum limit, thermal range of operation, power consumption and detectable velocity. On the other hand, Matsko et al. proposed an array of crystalline WGM resonators coupled to a waveguide. The authors demonstrated that the composite structure would allow for several orders of magnitude enhancement.[53]

WGM sensors could be also used for measuring changes of temperature. Brenci et al.[54] were able to detect changes in the resonance wavelength of 14.2 ± 0.4 pm/°C around 27 °C and 16.4 ± 0.8 pm/°C around 54 °C. Figure 12 shows the experimental set-up and the linear wavelength shift with temperature.

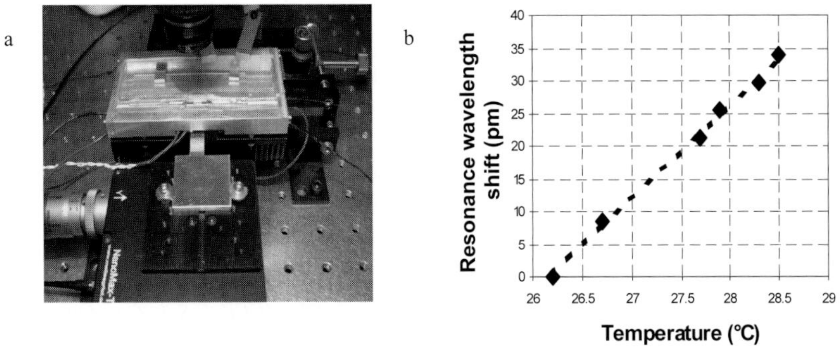

Figure 12. a) Photograph of the thermostatic cell where the microsphere sensor is placed, b) Resonance wavelength shift vs. temperature, for a microsphere of 350 μm diameter.

Acknowledgments

We would like to thank a number of colleagues who have been collaborating with us since many years: F. Baldini, M. Brenci, R. Calzolai, F. Cosi, A. Giannetti, G.C. Righini (IFAC CNR, Florence), A. Chiasera and M. Ferrari (IFN CNR, Trento) and P. Feron (ENSSAT-Laboratoire d'Optronique, Lannion).

References

1. Lord Rayleigh, *Scientific Papers*, **5**, 617 (1912).
2. D. W. Vernoy, V.S. Ilchenko, H. Mabuchi, E. W. Streed and H. J. Kimble, *Opt. Lett.* **23**, 247 (1998).
3. G. Gauglitz, *Anal. Bioanal. Chem.*, **381**, 141 (2005).
4. A. M. Armani, R.P. Kulkarni, S. E. Fraser, R. C. Flagan and K. J. Vahala, *Science*, **317**, 783 (2007)
5. D. Clerc and W. Lukosz, *Sens. Actuators* **B**: *Chem.*, **40**, 53 (1997).
6. K. Schmitt, B. Schirmer, C. Hoffman, A. Brandenburg and P. Meyrueis, *Biosens. Bioelectron.*, **22**, 2591 (2007).
7. T. E. Plowman, W. M. Reichert, C. R. Peters, H. K. Wang, D. A. Christensen and J. N. Herron, *Biosens. Bioeletron.*, **11**, 149 (1996).
8. J. L. Nadeau, V. S. Ilchenko, D. Kossakovski, G. H. Bearman and L. Maleki, Proceedings of SPIE **4629**, 172 (2002).
9. F. Vollmer, D. Braun, A. Libchaber, M. Khoshsima, I. Teraoka and S. Arnold, *Appl. Phys. Lett.*, **80**, 4057 (2002).

10. C.Y. Chao and L. J. Guo, *Appl. Phys. Lett.*, **83**, 1527 (2003).
11. E. Krioukov, J. Greve and C. Otto, *Sens. Actuators* **B 90**, 58 (2003).
12. S. Blair and Y. Chen, *App. Opt.*, **40**, 570 (2001).
13. R.W. Boyd and J. E. Heebner, *App. Opt.*, **40**, 5742, (2001).
14. E. Krioukov, D. J. W. Klunder, A. Driessen, J. Greve and C. Otto, *Opt. Lett.*, **27**, 1504 (2002).
15. E. Krioukov, D.J.W. Klunder, A. Driessen, J. Greve and C. Otto, Talanta, 65, 1086 (2005).
16. E. Krioukov, D.J.W. Klunder, A. Driessen, J. Greve, and C. Otto, *Opt. Lett.*, **27**, 512 (2002).
17. J.Yao, D. Leuenberger, M-C. M. Lee and M. C. Wu, *IEEE J. Sel. Top. Quantum Electron.*, **13**, 202 (2007)
18. C.Y. Chao, W. Fung, L.J. Guo, *IEEE J. Sel. Top. Quantum Electron.*, **12**, 134, 142 (2006).
19. A. Yalçin, K.C. Popat, J. C. Aldridge, T. A. Desai, J. Hryniewicz, N. Chbouki, B. E. Little, O. King, V. Van, S. Chu, D. Gill, M. Anthes-Washburn, M. S. Unlu and B. B. Goldberg, *IEEE J. Sel.Top. Quantum Electron.*, **12**, 148 (2006).
20. A. Ksendzov and Y. Lin, *Opt. Lett.*, **30**, 3344 (2005).
21. K. De Vos, I. Bartolozzi, E. Schacht, P. Bienstman, and R. Baets, *Opt. Express* **15**, 7610-7615 (2007).
22. A. Ramachandran, S. Wang, J. Clarke, S. J. Ja, D. Goad, L. Wald, E. M. Flood, E. Knobbe, J. Hryniewicz, S. Chu, D. Gill, W. Chen, O. King and B. E. Little, *Biosens. Bioelectron.*, **23**, 939 (2008).
23. A. M. Armani, D. K. Armani, B. Min, S. M. Spillane, and K. J. Vahala, *Appl. Phys. Lett.*, **87**, 151118 (2005).
24. X. Fan, I.M. White, H. Zhu, J. D. Suter and H. Oveys, Proceedings of SPIE, **6452**, 1 (2007) and references therein.
25. B. E. Little, J. P. Laine, D. R. Lim, H. A. Haus, L. C. Kimerling and S. T. Chu, *Opt. Lett.*, **25**, 73 (2000)
26. A. M. Armani and K. J. Vahala, *Opt. Lett.*, **31**, 1896 (2006).
27. D. K. Armani, T. Kippenberg, S. M. Spillane and K. J. Vahala, *Nature*, **421**, 925 (2003).
28. I. M. White, H. Oveys, X. Fan, T. L. Smith and J. Zhang, *Appl. Phys. Lett.*, **89**, 191106 (2006).
29. I. M. White, H. Oveys and X. Fan, *Opt. Lett.*, **31**, 1319 (2006).
30. J. P. Laine, B. Little , D. R. Lim, H. C. Tapalian L. C. Kimerling and H. Haus, *IEEE Phot. Technol. Lett.*, **12**, 1004 (2000)
31. H. Zhu, I. M. White, J. D. Suter, P. S. Dale and X. Fan, *Opt. Express*, **15**, 9139 (2007).
32. T. Ling and L. J. Guo, *Opt. Express*, *15*, 17424 (2007)
33. J. D. Suter, I. M. White, H. Zhu, H. Shi, Ch. W. Caldwell and X. Fan, *Biosens. Bioelectron.*, **23**, 1003 (2008).

34. M. Sumetsky, R.S. Windeler, Y. Dulashko and X. Fan, *Opt. Express*, **15**, 14376 (2007).
35. A. T. Rosenberger and J.P. Rezac, Proceedings of SPIE, **3930**, 186 (2000).
36. A. T. Rosenberger and J.P. Rezac, Proc. SPIE 4265, 102 (2001).
37. G. Farca, S. I. Shopova and A.T. Rosenberger, *Opt. Express*, **15**, 17443 (2007).
38. M. L. Gorodetsky, A. A. Savchenkov and V. S. Ilchenko, *Opt. Lett.*, **21**, 453 (1996).
39. F. Vollmer, S. Arnold , D. Braun, I. Teraoka and A. Libchaber, *Biophys. J.*, **85**, 1974 (2003).
40. N. M. Hanumegowda, I. M. White, H. Oveys and X. Fan, *Sens. Lett.*, **3**, 315 (2005).
41. I. M. White, N. M. Hanumegowda and X. Fan, *Opt. Lett.*, **30**, 3189 (2005).
42. N. M. Hanumegowda, I. M. White and X. Fan, *Sens, Actuators* **B 120**, 207 (2006).
43. D. Keng, S. R. MacAnanama, I. Teraoka and S. Arnold, *Appl. Phys. Lett.*, **91**, 103902 (2007).
44. H.C. Ren, F. Vollmer, S. Arnold and A. Libchaber, *Opt. Express*, **15**, 17410 (2007).
45. U. Levy, K. Campbell, A. Groisman, S. Mookherjea, and Y. Fainman, *Appl. Phys. Lett.* **88**, 111107 (2006).
46. W. Fang, D. B. Buchholz, R. C. Bailey, J. T. Hupp, R. P. H. Chang and H. Cao, *Appl. Phys. Lett.*, **85**, 3666 (2004).
47. J. Yang and L. J. Guo, *IEEE J. Sel. Top. Quantum Electron.*, **12**, 143 (2006).
48. J. Lutti, W. Langbein and P. Borri, *Appl. Phys. Lett.*, **91**, 141116 (2007).
49. G. Schweiger, R. Nett and T. Weigel, *Opt. Lett.*, **32**, 2644 (2007).
50. A. Francois, S. Krishnamoorthy and M. Himmelhaus (6862-27), Photonics West (2008).
51. J. P. Laine, C. Tapalian, B. Little and H. Haus, *Sens. Actuators* **A 93**, 1 (2001).
52. M. N. Armenise, V. M. N. Passaro, F. De Leonardis and M. Armenise, *J. Lightwave Technol.*, **19**, (10), 1476 (2001).
53. A. B. Matsko, A. A. Savchenkov, V. S. Ilchenko and L. Maleki, *Opt. Comm.*, **233**, 107 (2004).
54. M. Brenci, R. Calzolai, F. Cosi, G. Nunzi Conti, S. Pelli and G. C. Righini, Proceedings of SPIE, **6158**, 61580S (2006).

PHOTONIC CRYSTALS: TOWARDS A NOVEL GENERATION OF INTEGRATED OPTICAL DEVICES FOR CHEMICAL AND BIOLOGICAL DETECTION

Armando Ricciardi,[a] Caterina Ciminelli,[b] Marco Pisco,[c,*]
Stefania Campopiano,[a] Carlo Edoardo Campanella,[b] Emanuele Scivittaro,[b]
Mario Nicola Armenise,[b] Antonello Cutolo[c] and Andrea Cusano[c]

[a] Università degli Studi di Napoli "Parthenope", Facoltà di Ingegneria, Centro Direzionale Napoli, Isola C4, 80143 Napoli, Italy
[b] Laboratorio di Optoelettronica, Dipartimento di Ingegneria Elettrica ed Elettronica, Politecnico di Bari, Via Re David 200, 70125 Bari, Italy
[c] Dipartimento di Ingegneria, Divisione di Optoelettronica, Università del Sannio, Corso Garibaldi 107, 82100 Benevento, Italy
**E-mail: pisco@unisannio.it*

A new class of materials, called Photonic Crystals (PhCs), affects a photon's properties in much the same way that a semiconductor affects an electron's properties. PhCs, in fact, possess a photonic bandgap, which means that light of certain wavelengths cannot propagate through them. These structures have very interesting properties of light confinement and localization together with the strong reduction of the device size, orders of magnitude less than the conventional photonic devices, allowing a potential very high scale of integration. Due to their unique features, these structures can possess unique characteristics that enable the structures to behave like optical waveguides, high Q resonators, selective filters, lens or superprism, just to name a few. The ability to mold and guide light leads naturally to novel applications in several fields including optoelectronics, telecommunications. The authors present in this chapter an introductory survey of the basic concepts of this new technology with particular emphasis on their applications for chemical and biological sensing.

1. Introduction

During the last decades, the effort of many research groups in the world has been focused on the study and the development of a new generation of photonic devices based on photonic band gap (PBG) structures, also called photonic crystals (PhCs). These structures have very interesting properties of light confinement and localization together with the strong reduction of the device size, orders of magnitude less than the conventional photonic devices, allowing a potential very high scale of integration.

The authors present in this chapter an introductory survey of the basic concepts of this new and emerging technology starting form the fundamental principles of operation. A number of optical functional devices based on photonic band gap structures have been reported including optical microresonators and lasers, waveguides, filters, super-collimators and super-prisms.

Particular attention has been focused on the exploitation of the fascinating properties of PhCs for the development of promising integrated and multifunctional technological platforms to be employed in chemical and biological applications. PhC sensors have been reviewed including configurations based on the band-gap effect, defect engineering and crystal fibers. New intriguing solutions exploiting their unique dispersion properties have been also analyzed outlining the envisaged advantages, potentialities and limitations that lie ahead.

2. Photonic Crystals — Fundamental Principles

In 1987 E. Yablonovitch[1] proposed a three dimensional structure that could have the capability to completely inhibit spontaneous emission within its electromagnetic band gap. S. John[2] in the same year demonstrated the possibility of strongly localizing the light. The focus point of the above mentioned works is the idea that a periodic arrangement of either dielectric or metallic elements can exhibit polarization and/or direction dependent band gap regions for certain frequency ranges, where the propagation of the electromagnetic waves is forbidden.[3] This property depends on the material and on the crystal lattice properties. The band gap regions are analogous to the electronic band gap in semiconductor crystals.

Starting from the pioneering works in late 80s the research activity in this field has strongly advanced, seeing a very large amount of theoretical and experimental results on one-dimensional (1D), two-dimensional (2D) and three-dimensional (3D) PhC structures both in dielectric and metallic materials. The simplest PhC is the configuration formed by a multi-layer of two alternating materials, well known as Bragg reflector.[4-5] 2D PhCs are more difficult to be fabricated that 1D PhCs but their increased technological complexity, even if still quite less than that 3D PhCs, is largely compensated by the potential applications in integrated photonic circuits. Two different types of structure can be considered. The first one consists of dielectric rods in air while the second one is composed by holes in a dielectric medium.

Depending on the physical and geometrical characteristics of the structure, such as the refractive index, the radius of holes/cylinders, the lattice periodicity, waves propagating inside the crystal may interfere each other, in such a way that a photonic band gap can be created. The band gap can be defined "complete" when it exists independently of the polarisation and the angle of incidence of the light. Since 2D PhCs do not have a periodic structure in the third dimension, the immediate consequence for that is still the issue to prevent the light from escaping out-of-plane also in presence of a dielectric slab.

The translational symmetry of the periodic lattice is disturbed when defects are introduced and the consequence of this is that Bloch mode can not still be considered a solution of the Maxwell's equations. If the defect is created by adding extra dielectric material to one or more cells, one ore more localized evanescent mode can be created within the photonic band-gap. The parts of the crystal on both sides of the defect behave like mirrors where modes exponentially decay. Any light propagating in the space between the mirrors bounds back and forth and, thus, is trapped. Since the distance between the mirrors is of the order of the light wavelength, the modes are quantized. Therefore, due to the presence of defects corresponding to frequencies inside the photonic band gaps, localized mode can exist. In particular, states with frequencies close to the middle of the band gap can be localized more tightly than states near to the band edge. The type and the size of the defect define the shape and the properties of the localized states, such as frequency, polarization, symmetry, field distribution. Mainly, the defects can be classified in point defects and extended defects. Point defects determine the presence of e.m. modes at discrete frequencies that can be considered analogous to isolated electronic states while extended defects result in the presence of transmission bands inside the photonic band gap of the unperturbed PhC.

3. Functional Photonic Band Gap Components and Devices

The importance of the PhC structures in the development of the integrated photonics strongly depends on the large variety of devices that can be realized on a single chip, with a better performance with respect to the conventional photonic devices. In this section the main applications of the PhCs will be briefly described.

3.1. Microcavities and Lasers

PhCs allow realizing resonant cavities with a high quality factor and a small mode volume and controlling the characteristics of the resonant cavity by means of the geometrical parameters of the lattice.[6] In particular, the localization of light and, then, the realization of sub-micrometric resonant cavities can be done by introducing a point defect inside a PhC lattice.

The structures reported in literature are particularly suitable in telecommunication and sensing devices. As examples, multiplexer/ demultiplexer devices[7] and add/drop filters,[8] based on the use of waveguides and resonant microcavities in PhC have been proposed. The high value of Q is also the best condition for realizing lasers with extremely low threshold and high quantum efficiency.[9]

3.2. Waveguides in Photonic Integrated Circuits

Since their introduction, PhCs have been proposed as ideal candidates for realizing photonic integrated circuits. PhCs allow to realize optical waveguides based on physical effects different from the conventional total internal reflection.[10-11]

The simplest waveguide can be formed by a 2D PhC lattice, without one or more rows of holes/columns, so creating a linear defect. An optical beam with a wavelength within the photonic band gap can not penetrate in the PhC and will be forced to propagate along the axis of the linear defect. The structure behaves as an optical waveguide based on PBG effects. The PhC slab is characterized by a vertical confinement due to total internal reflection in combination with the PBG in-plane confinement.[12] The advantages of the PhC waveguides with respect to the conventional waveguides consist of the capability of guiding optical signals along paths with very large curvature angles[13,14] and of the possibility to easily couple PhC resonant cavities to optical waveguides. This latter advantage opens interesting possibilities of realizing components that are building blocks of the photonic circuits.

3.3. Superprism and Supercollimator

Dispersion effects occurring in PhCs can find several applications in the field of passive devices. Supercollimators and superprisms are well known examples. To understand their operating principle the concept of isofrequency curve in the **k** plane has to be introduced. The isofrequency curve is the locus of points having the same frequency $\omega(\mathbf{k})$, that can be determined by intersecting the dispersion diagram with a plane normal to the frequency axis at ω. The gradient of the function $\omega(\mathbf{k})$ in each point of an isofrequency curve determines the direction, the sense and the velocity at which the e.m. energy travels inside the crystal.

In isotropic bulk materials the isofrequency curves are circles having radius changing with the frequency. The situation is totally different in the PhCs, where regions with a curve gradient constant with respect to ω and k, and regions with a propagation direction for the e.m. energy changing very rapidly can both exist.

Regions with gradient constant with respect to ω and k, can be exploited for realizing the supercollimators.[15] If an un-collimated optical beam, that can be represented as a combination of plane waves with different propagation directions, impinges on the surface of a PhC, the refracted waves, associated to different components, can propagate in parallel directions, giving an optical beam that appear to be perfectly collimated.

Different is the operating principle at the basis of the superprisms.[15-16] It is well know that the prism is a structure able to separate the spectral components of an optical beam, exploiting the chromatic dispersion of the materials, i.e. the variation of the refractive index as a function of the wavelength. The crystal can be designed so that in the spectral region of interest the isofrequency curves show a strong sensitivity to small variations in frequency, changing in shape from concave to convex forms. Under these conditions when a polychromatic beam impinges of the crystal surface, the refracted waves generated from the spectral components, propagate along directions that can be also very different each other even for small changes of the wavelength.

3.4. Photonic Crystal Fibers

PhC fibers are a new class of optical fibers based on the use of periodic dielectric structures.[17]

Two types of PhC fibers can be considered: bandgap guiding fiber (PBF) and index guiding fibers (PCF). In both types, a periodic pattern is realized in the cross section of the fiber.

In case of PBF, the PhC lattice has the property of confining the light by means of the mirroring function realized by the crystal, while in PCF the lattice is made by air holes within a guiding dielectric material, with a resulting effect of reducing the cladding refractive index.

In this last case, the guiding effect is similar to that one of the conventional fibers, but the periodic structure allows a fine control of the cladding refractive index.

4. Photonic Crystals for Chemical and Biological Sensing

PhCs have inspired a lot of interest and many research efforts have been devoted to their possible applications in communications and information fields due to the opportunity they offer to efficiently manipulate the light on wavelength and sub-wavelength scale. The outstanding potential of photonic bandgap structures encourages their employment also in sensing applications. As a matter of fact, the microstructure of the crystal opens up for a large degree of freedom in optical waveguides design, enabling the implementation of novel and intriguing transduction principles for sensing applications basically exploiting the dependence of the PhCs' optical properties on the physical and geometrical features of the crystal itself. Furthermore, the possibility to realize a PhC through holes-patterned in a dielectric would allow the integration with sensitive materials in order to improve the functionality of the final device for chemical and biological sensing, either tailoring the sensing system performance or conferring selectivity capability. The proper filling of the PhC air holes with a sensitive material, in fact, leads to an enhancement of the interaction between the light and the sensing material that can be strongly improved and optimized by properly designing the guiding structures. Alternatively, dielectric structures patterned with holes can be exploited to directly detect target molecules by infiltrating analytes solutions or gas mixtures into the holes.[18]

On these bases, it can be noted that PhCs offer a new possibility of realizing effective and compact sensors and open the way for the development of 'lab-on-chip' portable devices which allow several chemical and biological analysis to be performed in parallel onto the same platform, by taking advantage of the large scale integration and wavelength multiplexing capabilities of the PhCs. To this aim, a significant advance would be represented by the integration of the PhCs, with micro-fluidic or nano-fluidic circuits in order to provide compact and multifunctional systems to be employed as valuable technological platform for modern chemical and biological applications.[19] In spite of the outlined potential of the PhC for chemical and biological sensing applications, the PhCs fabrication processes, the defects introduction, as well as the integration with additional materials enabling sensing capabilities imply several challenges of physical realization and process availability, that still prevent PhCs to be fully exploited in the sensing fields. In the following the first successfully attempts of exploitation of the fascinating PhC properties for chemical and biological sensing are reviewed.

Historically, the main PhC research trend has concerned the design and the development of structures able to exhibit a photonic band-gap and thus to enhance or suppress emission of light. Successively, many researchers have focused their attention on the study of other unique and unusual dispersive properties characterizing PhCs such as slow light,[20] self-collimation and superprism.[15] On this line of argument, here, we have chosen to start with the PhC sensors based on the band-gap effect and defect engineering and then to deal with sensors exploiting crystals dispersion properties. Finally, the PhC fiber based structures are presented.

4.1. Photonic Crystal Defect based Sensors

The introduction of proper defects in a PhC lattice in terms of geometrical or physical parameters modifications, acting as defect states inside the PBG, gives to the resulting device interesting functionalities such as resonant states and unique guidance capabilities,[21] as already mentioned. The appealing feature of such defected structures for chemical and biological sensing applications relies on the dependence of the photonic device's spectral response on the defect properties.

PhC cavities in fact allow a strong localization of the optical field in a very small volume at a characteristic resonant wavelength, related to the geometrical and physical features of the cavity itself. Furthermore, as the cavity is designed to have a high Q-factor and a small mode volume, a sharper defect band into the PBG is provided.[22] Such cavities can be exploited as chemical and biological sensors because the line-width associated to the defect location and thus the spectral response of the whole device is sensitive to the cavity refractive index. In particular, if the cavities are designed to trap light in the air pores, an interaction between light and analyte solution that fills the holes occurs, and the refractive index change, induced by the liquid or gas insertion, causes a spectral shift or an intensity variation at a fixed wavelength of the cavity resonances. In principle, the presence of a sensitive material in the cavities or constituting the cavities could enhance selectiveness of the PhC defect based sensor.

The first attempt of using planar PhCs for chemical sensing was carried out by Loncar et al in 2003.[23] They proposed a planar PhC laser source based on a single defect in triangular lattice. The cavity was fabricated in InGaAsP strained quantum-well material grown on an InP substrate using metal-organic chemical vapour deposition. The laser fabrication procedure consists of EBL, followed by dry-and wet-etching steps. Optical gain was provided by four compressively strained quantum wells. The scanning electron microscope (SEM) image of the realized structure is reported in Fig. 1(a). As observable, the laser cavity consists of a defect hole smaller than the surrounding ones and a row of holes that is elongated along one direction. The elongation is able to lift the degeneracy between the two dipole modes and to increase the Q-factor of one of these modes. In Fig. 1(b) the field spatial distribution of the resonant mode is reported demonstrating that the energy of the mode is mostly confined into the central defect hole. In order to increase the interaction between light and the material within the central hole, a larger defect diameter would be preferred. However, increasing the central hole size reduces the gain provided by the light-emitting quantum wells within the laser cavity. Therefore, a trade-off between the field optical overlap with the analyte and the optical gain is needed. In the configuration proposed by Loncar et al. the interaction with the analyte to detect is not limited to the central hole but the whole structure is immersed in a liquid solution with refractive index n_{ext}. Consequently, a linear red-shift of the band-gap edges and a reduction of the band-gap width occur as n_{ext} increases. Thus, by evaluating shifts in the emission wavelength of the laser, it is possible to optically detect refractive index change of the material surrounding the cavity. However, the introduction of liquid solution degrades the cavity Q-factor which decreases from about 6000 down to 1000 when the refractive index passes from 1 up to 1.4. The Q-factor decreases because the vertical confinement of the slab becomes weaker due to the reduced index contrast between the slab and the environment.

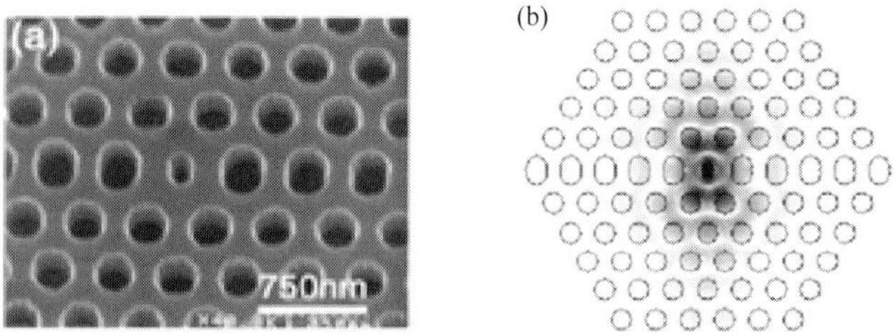

Figure 1. Photonic nanocavity laser sensor: (a) scanning electron micrograph, (b) field distribution.[23]

In order to overcome this problem an optical gain is introduced into the cavity to narrow the width of the cavity resonance and consequently improve the sensor sensitivity. In passive devices, in fact, this sensitivity is directly dependent only on the width of the cavity resonance peak which in turn is determined by the cavity Q-factor; the introduction of gain into the cavity can lead to significantly narrower linewidths. Refractive index resolution of $\Delta n \approx 10^{-3}$ within samples of femtoliter volumes has been experimentally demonstrated. Two years later the same authors proposed a microfluidic integration of the same structure,[24] obtaining a more precise control on the liquid delivery but not achieving a selective hole filling. By embedding the structure into polydimethylsiloxane (PDMS) microfluidic flow channels, they showed how it is possible to deliver picoliter quantities of reagents. The PDMS chip was fabricated using multilayer soft lithography.[25] This new capability opens the way for a new class of highly integrated optofluidics devices.

Chow et al.[26] proposed a planar PhC refractive index sensor employing a passive structure constituted by single defect cavity, obtained by introducing one hole with a smaller radius in the centre of a slab patterned with a triangular lattice of holes (see Fig. 2). The sample is fabricated on SOI substrate with a silicon slab 260 nm thick separated from the Si substrate with 1 μm SiO_2. A 0.75 mm long conventional ridge waveguide on each side of the structure is used for coupling light in and out of the PhC cavity. The realized structure is able to sense a surrounding refractive index variation through the peak shift of transmission spectra when immersed in a liquid solution. These shifts are measured by focusing and then detecting a TE polarized collimated laser into the crystal when it is embedded in different liquid solutions at different densities. The main difference with respect to the work proposed by Loncar et al. relies in the absence of optical gain into the cavity. This implies a less sensitive refractive sensor able to detect index changes of 2×10^{-3} RIU.

close to 1.49 μm for TE-like modes with a quality factor of about 2000. The defect size has been chosen in such a way that single target particle of proper dimension can fall within the defect hole. The biosensor is capable of detecting single latex sphere that is trapped inside the micro-cavity causing the typical resonance red-shift. However, it is clear that this configuration, while suffers from the challenge of the selective delivery of the bio-particles only into the defect hole, still remains based on a refractive index changes without employing sensitive materials able to enhance the selective properties of the sensing operation; in this case in fact, the selectivity is only based on size exclusion principle leading to a weak specificity mainly for classes of different molecules. In the aforementioned configuration, the cavity is basically formed by a point defect, created by omitting or changing the radius of one o more holes in the centre of the slab.

A new sensing device based on a 2D PBG filter in polymeric slab has been proposed by Ciminelli et al.[30] The filter is a Fabry-Perot cavity having a self-sustained membrane configuration. The parametric analysis, carried out on that device taking also into account the fabrication tolerances, proved that best performance can be achieved in case of square lattice of holes in polystyrene material.

Nevertheless, a defect in the PBG can be also created as well by using 'hetero-structure' configurations, consisting of joining two or more PhC structures with slightly different lattice constants as shown in Refs. 31-32. Tomljenovic-Hanic et al.[33] exploited this concept to numerically design a new PhC hetero-structure substituting the air in the holes with materials at different refractive index, such as liquid crystals, nano-porous silica or polymers, rather then changing the lattice constant.

The main advantage of this approach, capable to provide a high Q-factor, even if with a typically lower modal volume, relies on the regularity of the basic structure not more requiring nanometre precision. Even though the hetero-structure has not been demonstrated experimentally, it can be considered an interesting starting point for the development of new PhC platform for chemical and biological sensing.

4.2. Photonic Crystal Guided Resonances Based Sensors

As described in previous section, in a planar PhC, light is confined within the plane of the slab by a two-dimensional periodic dielectric structure, and in the third dimension by total internal reflection resulting from the refractive index contrast between the slab and the cladding. Because of the internal reflection condition, this confinement in the third dimension may be not complete. Some modes are guided, while others can couple to external radiation modes. These latter modes, with finite lifetimes inside the slab, have been described as guided resonances.[34] In a band structure, confined modes and guided resonances are separated by the light line: modes that cannot couple to external radiation are below the light line, while guided resonances that have finite lifetime due to coupling to external propagating modes are above the light line. This coupling provides a way to excite these guided resonances using external waves. Since these discrete resonances couple to a continuum of free-space modes, their transmission and reflection spectra exhibit a Fano resonance line shape[35] characterized by a very sharp variation of the transmission spectrum from 0% to 100% over a narrow frequency range, as depicted in Fig. 4. Resonant line shape width strongly depends on the radius of holes of the PhC structure: the Q-factor can vary from tens of thousands to a few tens if, for instance, the radius is 0.2 a or 0.4 a (where a is the lattice constant) respectively. It is also worth noting that a relative low refractive index contrast of the PhC slab is sufficient to excite the guided resonances phenomena. From all these concepts, it follows that guided resonances, being standing electromagnetic waves strongly confined in the slab but that can couple to the external radiation, provide an effective way to sense external environment and thus are well suited for sensing application. In fact, the transmission through the slab is deeply influenced by the refractive index change of the environment surrounding the slab which results in a frequency shift of transmission peak.

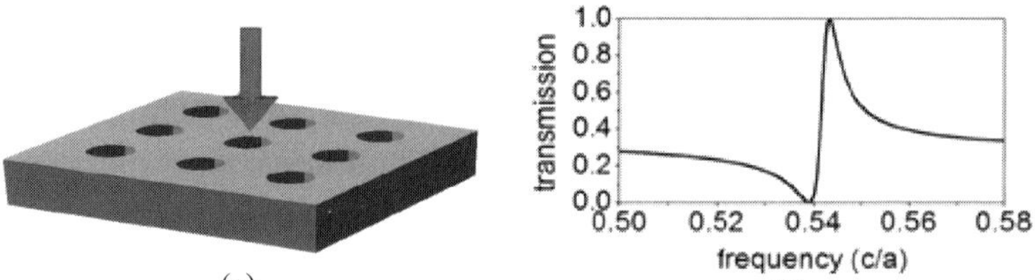

(a) (b)

Figure 4. (a) PhC structure: the arrow represents the direction of the externally incident light, (b) typical transmission spectrum with a Fano line shape indicating the existence of a guided resonance.[36]

On these bases, in 2007 Levi et al. proposed[36] a sensitivity analysis of a PhC sensor based on guided resonances capable to detect refractive index changes in aqueous solution. The guided-resonance sensor was fabricated in SiO_2/SiN_x dielectric materials by using EBL to pattern a square array of holes on a resist and subsequently Fluorine-based dry etching to transfer the structure from the resist to the SiN_x layer. The sensor consisted of a silica (n = 1.46) slab of thickness d = 250 nm perforated by a square array of air holes of radius r =100 nm with lattice constant a = 500 nm. The entire slab was immersed in different liquid solution (Iso-propanol/Water mixtures) of refractive indices varying in the 1.330-1.338 range. The excitation of the PhC slab guided resonances occurs perpendicular to the sample. The local refractive index change results in a variation in the guided resonance condition and consequently generates a shift in the resonance peak location. They were able to measure a transmission peak shift of 2 nm corresponding to a detectable refractive index change of 1.5×10^{-3}.

Another very interesting research about PhC sensors has been conducted by the group of Cunningham at the University of Illinois. They developed a category of biosensors based on PhC structure that can be used for both fluorescence-based (labeled) and label-free detection.[37] The sensor they proposed consisted of a low refractive index plastic material with a periodic surface structure coated with a thin layer of high refractive index dielectric (titanium dioxide). If used for label-free detection, the device was designed to reflect only a relatively narrow band (2 nm) when illuminated with a white light source at normal incidence; red shifts of the reflected peak wavelength value are a sign of the detected material adsorption on the sensor surface. Concerning the labeled fluorescence detection, the sensor structure was in turn designed to enhance the intensity of emission of fluorescent samples that are placed close to the regions where the PhC guided resonances concentrate their energy. The excitation enhancement was obtained since the samples were absorptive at the same resonant wavelength of guided resonances. Fluorescence amplification factor of about 500 have been experimentally demonstrated.[38]

4.3. Photonic Crystal Opal Based Sensors

Before presenting some devices based on dispersive properties of PhCs, it is important to mention other sensor configurations based on PhC fabricated by self-assembly of uniform particles, resulting in the formation of artificial opals. Self-assembly of colloidal particles represents a simply and effective way to create three dimensional PhC with respect to the layer-by-layer lithographic method. When monodisperse and highly-charged colloidal particles are placed in low ionic strength medium, in fact, they arrange to form crystalline colloidal array (CCA) that exhibits interesting optical properties[39]. The application of CCA for sensing purpose, though, is significantly limited due to its fluidic nature and high sensitivity against ionic impurities.

To overcome these limitations, in 1997 the Asher Research Group at the University of Pittsburgh developed[40] a method to solidify CCA by polymerizing acrylamide hydrogel network around CCA to obtain PCCA (polymerized CCA). With this technique, they combined the distinct optical properties of colloidal crystals coupled with the sensitivity of polymer hydrogels to their environment, resulting in an effective PhC nanomaterial which can be utilized for sensing application. Due to its crystalline structure, PCCAs exhibit a photonic band-gap for visible light. In addition, hydrogel responds to changes of chemical environments with swelling or shrinking. As the PCCA volume changes, the diffraction wavelength changes accordingly. For example, when the PCCA swells, the spacing between the planes of the CCA increases, leading to a diffraction at higher wavelength red-shift. Likewise, when the gel shrinks, the spacing decreases and a blue-shift is verified. This volume change could be related to the analyte concentration using the diffraction wavelength shifts of PCCA. Based on this principle, Asher and coworkers developed several novel sensors which can detect variations in temperature, metal concentration, pH value, ionic strength, glucose concentration among others.[40-41] In this context, Cheng-Yu Kuo et al. proposed[42] a work that reports on the chemical sensing ability of defect-free 3D PhC structures such as polystyrene (PS) opals and gold and titania inverse opals. In order to sense ethanol and water mixture concentrations at ratios of 20, 40, and 80% they were simply passed through the fluidic cell containing the opal/inverse opal, fixed on a stage, at a constant feeding rate. Since the gold-coated slide glass is opaque, they measured the reflectance instead of transmittance to detect the stop band shifts. Peak position shifts of the resulting stop bands were identified to differentiate the species contained in the samples flowing through. Ethanol and water mixture utilized in their experiment have a refractive index difference as small as 7.5×10^{-3} that produces a 1.5-2 nm shift in stop band position with the polystyrene opal structure. By using gold and titania inverse opal structures which have larger void space, stop band shifts of 4.5-6 nm were obtained. Detection of the presence of bound analytes was also demonstrated because the analyte adsorption on the sphere surface increases the total solid volume fraction of the structure. Shifts of 3.5 and 4.5 nm in stop band position induced by binding of $C_8H_{18}S$ (n = 1.452) and $C_{16}H_{34}S$ (n = 1.464) respectively onto the surface of the gold inverse opal, were also obtained.

4.4. Photonic Crystal Sensors based on Dispersive Properties

While the existence of a photonic band-gap in PhC structures provides an excellent way to manipulate flow of light, and thus, allows for a variety of sensing applications as discussed in previous sections, PhCs, due to their unique dispersion characteristics, can also carry out a lens (self-collimation effect) or superprism function (see Section 3.3). According to the self-collimation effect for which collimated light propagation is insensitive to the divergence of the incident beam, electromagnetic waves can be efficiently guided within a planar PhC without the use of channel defects or structural waveguides. In addition, PhCs exhibit the superprism phenomenon: the light path shows a drastic wide swing with a slight change of the incident light angle owing to the strong modification of group velocity.

In 2006 Martin et al.[43] exploited the self-collimation effect to numerically simulate a compact PhC refractive index sensor based on a Mach-Zehnder interferometer (MZI). A conventional MZI based sensor requires perfect alignment of two beam splitters and two mirrors that compose it. The alignment process is a challenging task, especially if the components dimensions are rather small. Moreover, MZI requires long light-matter interaction length (typically of the order of centimeter) to achieve high sensitivity, and so micro-sized devices can not be realized. These problems can be overcome by realizing the interferometer embedded in a self-collimated PhC structure. The new MZI configuration includes a beam splitter and mirrors embedded in a self-collimating two-dimensional PhC structure consisting of a square lattice of air columns in silicon. The beam splitter is designed to equally split the optical signal. Two mirrors (one in the upper right corner and another in the lower left) direct the beams to the output ports. If half of the interferometer is exposed to a gas or liquid and the other half is protected from the material under test, the MZI becomes a relative refractive index sensor. The authors also suggested that by adding further beam splitters, but using the same optical source, multiple Mach-Zehnder sensors could be fabricated in parallel to sense multiple biological or chemical agents.

In the same year, Prasad et al presented[44] a numerical sensitivity study of a three-dimensional PhC sensor based on the superprism effect. In their work, they proposed an optical sensor architecture based on the angular deviation of light, rather than on its spectral properties. It is shown that the propagation direction of a light ray inside a PhC can be extremely sensitive to the material parameters of the crystal such as the dielectric constant. They computed the beam displacement as a function of the refractive index of the polymer medium constituting the crystal for several different wavelengths at a fixed incidence angle. A slight change in the refractive index contrast leads to a large change in the internal propagation angle, which produces a displacement on a position-sensitive detector. From calculated data the authors were able to extrapolate a minimum detectable refractive index shift of the order of

10^{-5} that represents an improvement over previous sensor but its practical realization seems to be a very hard task.

Another very attractive property of PhC platforms and that can be exploited for sensing applications is the so called slow light.[45] For some particular values of frequencies and wavenumbers of the propagating electromagnetic radiation, PhC structures are able to slow down the wave group velocity. As a consequence, slow-light propagation can enhance the energy density of the electromagnetic field within the structure and thus improve the light-matter interaction. This phenomenon is very exciting because can compensate for the reduced optical path in typical lab-on-a-chip systems for bio-chemical sensing applications.

At the same time, the enhanced local fields have the potential to increase the overall sensitivity of chemical and biological sensors.

5. Photonic Crystal Fibers Sensors

PhC fibers have attracted a high interest in the scientific community in the last years for their exceptional properties enabling new optical phenomena and hence new applications. The microstructuration of the cladding, in fact, offers a high degree of freedom in the fabrication of the PhC fibers and allows also to develop new sensing configurations by filling the PhC fibers with proper sensitive materials[46] or directly by infiltrating gases or liquids into their holes.[47-48]

5.1. Index-Guiding Photonic Crystal Fibers based Sensors

The first sensor configuration based on PhC fibers was theoretically and numerically envisaged by Monro et al.[48] by exploiting the in-holey-cladding evanescent wave interaction with filling analytes.

In principle, in order to obtain a strong light-analyte interaction, a significant percentage of modal power should be located in the fiber holes at the wavelength range of the analyte absorption.[48] The amount of power in the holes increases with wavelength and decreases for larger core sizes because the mode-field diameter increases with the wavelength and thus the mode sees the cladding holes for smaller air-filling fractions. In general, index-guiding PCFs exhibit less than 1% of the modal power in the holes but with an optimized fiber design the overlap at a particular wavelength can be increased up to 30%.[49]

The first experimental demonstration of evanescent wave gas detection with PCFs was reported successively by Hoo et al. in 2002.[50] The fiber taken into account consisted of a silica core surrounded by a single ring of large air holes. They measured the absorption spectrum of acetylene by inserting one end of a 75-cm long PCF into a pressure chamber filled with 100% acetylene gas. The fiber was subsequently removed and the absorption spectrum was measured quickly in order to minimize out-diffusion of the gas. The fiber used in this experiment had a relatively weak penetration of the optical field into the air holes, but the long interaction length provided by the PCF was used to compensate the weak evanescent wave interaction.

Jensen et al. detected[51] the presence of fluorescently labeled antibodies, α-streptavidin-Cy3 and α-CRP-Cy3, in less than 1 pL sample volume. They considered in their experiment a microstructured polymer optical fiber reported in Fig. 5 (a). The fiber core was composed by the 6 air holes arranged in a circle and possessed an outer diameter of 300 μm and an air hole diameter of 60 μm. A fiber section of 10 cm was filled with the sample using capillary forces and the transmission spectrum was measured. When the sample contained the labeled DNA, the transmission spectrum showed dips at wavelengths corresponding to the absorption bands of the Cy3 molecule. An alternative and very interesting use of PCFs for sensor applications relies on two-photon fluorescence detection,[52] where these fibers can improve detection efficiency compared to conventional single mode fibers. In 2003 Myaing et al. experimentally demonstrated[53] that a double-cladding PCF (see Fig. 5(b)) serves to improve the efficiency of two-photon fluorescence detection of biomolecules. In particular, they showed that two-photon excitation and collection of fluorescence could be accomplished using the same optical

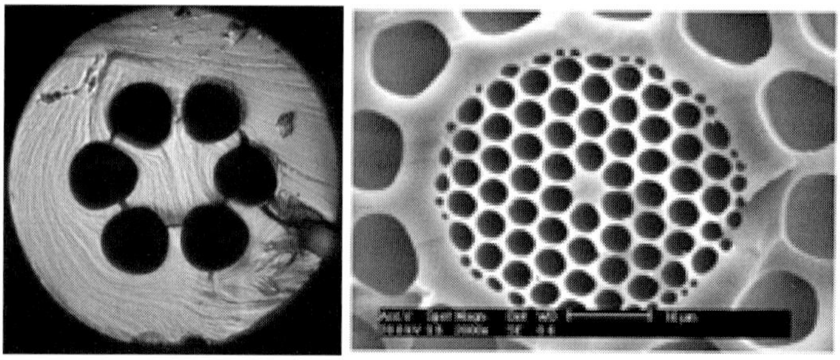

Figure 5. (a) Micrograph showing the end-facet of the microstructured fiber used in the biosensor experiments[50]. (b) Example of dual cladding PCF.[53]

fiber. In fact a dual-cladding PCF allows one to independently adjust the fiber parameters to have single-mode propagation of excitation light in the centre core while having a large numerical aperture of the inner cladding for efficient fluorescence collection.

For the interrogation system, a dual-cladding PCF was used to sense luminescent species in a distant liquid sample; laser radiation was delivered to a sample through the central core of the fiber with a diameter ranging in their experiments from 1 up to 10 μm. The fiber cladding with a substantially larger diameter, and consequently a higher numerical aperture, was used to collect the fluorescent response from the sample and to guide it in the backward direction to a detector.

Koronov et al. demonstrated[54] that a dual-cladding PCF can be used in two different optical sensing interrogation schemes, simultaneously serving, whenever necessary, for the collection and on-line monitoring of liquid-phase samples. PCFs used in their experiments were fabricated in fused silica and soft glasses by using the standard fabrication technique. In the first scheme, a laser radiation was delivered to a sample through the central core of a dual-cladding PCF with a diameter of a few micrometers, while the large diameter fiber cladding was responsible to collect the fluorescent response from the sample and to guide it to a detector in the backward direction.[54] In the second scheme, the liquid sample was collected by a micro-capillary array in the PCF cladding and interrogated by laser radiation modes guided in PCF. Several dye molecules dissolved in water, alcohol, and dimethyl sulfoxyl (DMSO) were used to illustrate the above-described protocols of optical sensing

with PCFs. Their measurements demonstrated the equivalence of spectroscopic data achievable with the two PCF-sensing methods.
Finally, it is worth reporting that in the solid core of PCFs a modulation of the refractive index can be impressed to create a grating in the fiber. In 2006 Rindorf et al. presented[55] a label-free technique for detection of biomolecules using a 18.2 mm long-period grating (26 pitches of 700 μm) written with a CO_2 laser in a PCF (PCFLPG). They demonstrated experimentally that the PCFLPG was able to detect the average thickness of a layer of biomolecules, deposited on the sides of the PCFs holes, within a few nm. The PCFLPG was shown to exhibit a refractive index sensitivity of approximately 10^{-4} RIU. By measuring the adsorbed molecules thickness the technique may thus be used for label-free detection of selective binding of biomolecules such as DNA and proteins.

5.2. Hollow Core PCFs: Improving the Sensor Performances

In the previous section, it has been shown that the index-guiding PCFs provide a quite strong interaction between the guided light and molecules present in the air holes through evanescent wave interaction. On the other hand, PBFs show a significant advantage for sensor applications because, in comparison to evanescent field devices, the overlap between the molecules and the mode field of light is considerably improved. As a matter of fact, such fibers can guide more than 98% of the power in the air-regions of the fiber, thus, reducing the influence of material parameters on the optical properties of the fiber. A PBF can evidently provide a stronger interaction over several tens of centimeters, while using only a few micro liters of sample volume.
A comparison between the performances of a PCF and a PBF as liquid sensing devices was carried out by Smolka et al.[56] By selectively filling the hollow core of the PBF with a dye solution an interaction of the guided light with the sample material of nearly 100% was achieved, while only 1% was found for the completely filled PCF. Their studies clearly showed that selectively infiltrated PBFs outperform existing evanescent sensing devices. Especially with regard to fluorescence sensing the detection limit is improved by four orders of magnitude

compared to the investigated PCF. Although PCFs are usually much more cost effective and easier to fabricate, the clear advantages of PBFs make them ideal tools for the most demanding sensing applications where high sensitivities are needed.

Ritari et al. studied[57] gas sensing in air-guiding PBFs by filling them in turn with hydrogen cyanide and acetylene at low pressure and measuring absorption spectra using a LED and a tunable laser. The obtained results indicated, also in this case, that the absorption strength per equal fiber length can be improved by using air-guiding PBFs instead of index-guiding PhC fibers. They also showed that gas sensing is feasible using low-power light sources, allowing for the realization of compact cost-effective single-point measurement systems.

Furthermore, PBFs based sensors are well suited for monitoring poisonous gases in places where space is confined and high sensitivity is needed. In fact, PBFs offer, in addition, the possibility of fabricating hybrid devices with sensitive materials like polymers or high index fluids filling the air holes of the fibers.[58]

For the first time, Cusano et al. proposed[46] the integration of hollow-core PhC fibers with single walled carbon nanotubes (SWCNTs) as sensitive material for chemical sensing, by exploiting the PBG modifications induced by SWCNTs - target molecules interaction.

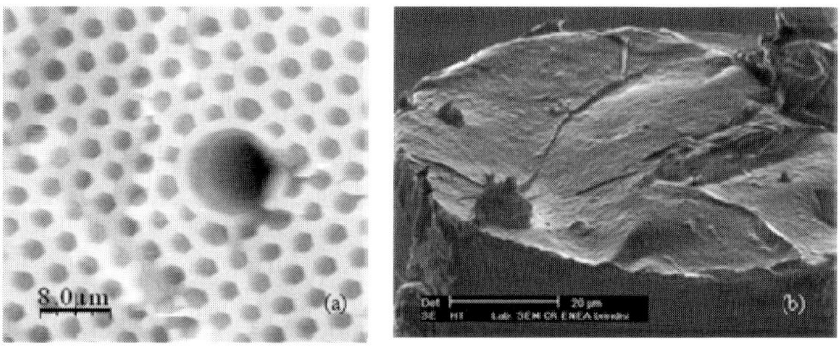

Figure 6. (a) AFM image of a hollow fiber; (b) SEM image of the hollow fiber interface coated by 10 monolayers of SWCNTs.

The SWCNTs covered the PhC fiber (see Fig. 6) and were slightly infiltrated within the fiber holes to create an interferometric configuration able to convert the SWCNTs physical changes in reflectance modifications. The sensing capability of the HOF sensors was firstly investigated by exposure in a test chamber to traces of Tetrahydrofuran.[46] Successively, the same authors demonstrated the sensing capability of the proposed sensors towards Volatile Organic Compounds (VOCs).[59] Experimental results demonstrated both the success of the SWCNTs partial filling within the HOF holes and the sensor capability to perform VOCs detection with a good sensitivity and fast response times achieved through a suitable control of the fabrication process conferring to the functionalized HOF the capability to perform chemical detection.[59]

6. Perspectives and Challenges

While several functionalities have been demonstrated in principle, the PhC devices (for sensing but for telecommunication too) designed and fabricated so far still lack the requirements to impose itself on the industrial scale application.

In general the requirements for next generation of chemical and biochemical sensors can broadly be classified into three different categories: sensitivity and specificity (detect rarer targets with greater precision), multiplexing (many reactions have to be tested at the same time on the same platform) and reduction in measurements complexity and costs (reduce the number of required sample processing steps as well as the amount of on-chip or off chip infrastructure). The reasons for this slow entry into the optics industrial market are manifold. First of all, PhC based components for the optical industry meet the strong competition of already established products which can be fabricated with high quality for a reasonable price with currently well developed processes and they are capable to sufficiently fulfill almost all demands of today's optics industry. In turn, the fabrication of PhC components for the technologically interesting wavelength range of 1.3 μm and 1.5 μm is still working at the limits of currently available production systems. As a consequence, nowadays PhCs found industrial attention only in niche markets. One example to be mentioned here are the PhC fibers by Russel et al.[47] that are currently the only commercially successful application of PhC.

Up to now, great effort has been carried out by the scientific community to develop photonic devices, however, the weak integration of competencies required to address this challenge, intrinsically multidisciplinary, limits the capability to achieve high performances devices. A highly integrated approach involving continuous interactions of different backgrounds aimed to optimize each single aspect with a continuous feed-back, would enable the definition of an overall and global design concept.

A novel generation of photonic devices for chemical and biological sensing is expected being based on the concurrent addressing of the issues related to the different aspects of their global design, such as: dielectric properties definition, materials identification, functionalization and activation, novel optical transduction principle development. This new class of photonic sensors should be driven by an integrated design and development that accounts for shaping the optical responsive materials to molecular analytes and simultaneously for selecting the better transduction schemes provided by PhCs technology.

References

1. E. Yablonovitch, *Phys. Rev. Lett.*, **58**, 2059, (1987).
2. S. John, *Phys. Rev. Lett.*, **58**, 2486, (1987).
3. J. D. Joannopoulos, R. D. Meade and J. N. Winn, Princeton University Press, (1995).
4. A. Yariv, M. Nakamura, *IEEE J. of Q. Electr.* Vol. QE-13, No.4, 233 (1977).
5. K. J. Kasunic, *IEEE J. of Lightwave Techn.*, Vol. 18, No. 3, 425 (2000).
6. R. Coccioli, M. Boroditsky, K.W. Kim, Y. Rahamat-Samii, E. Yablonovitch, *IEE Proceedings of Optoelectronics*, Vol. 145, No. 6, 391 (1998).
7. K. H. Hwang and G. H. Song, *Optics Express*, Vol. 13, No. 6, 1948 (2005).
8. B.-K. Min, J.-E. Hwang and H.Y. Park, *Optics Comm.*, Vol. 237, No.1-3, 59 (2004).
9. H.-G. Park, J.-K. Hwang, J. Huh, H.-Y. Ryu, S.-H. Kim, J.-S. Kim and Y.-H. Lee, *IEEE J. of Quantum Electronics*, Vol. 38, No. 10, 1353 (2002).
10. A. Adibi, Y. Xu and R. K. Lee, A. Yariv, *IEEE J. of Lightwave Techn.*, Vol. 18, No.11, 1554 (2000).
11. M. Lončar, J. Vuckovic, A Scherer, *J. of Optical Society of America **B***, Vol.18, No. 9, 1362 (2001).
12. S. G. Johnson, S. Fan, P. R. Villeneuve, J. D. Joannopoulos and L. A. Kolodzejski, *Phys. Rev. **B***, Vol. 60, No. 8, 5751 (1998).

13. A. Chutinan and S. Noda, *Phys. Rev.* ***B***, Vol. 62, No.7, 4488 (2000).
14. M. Augustin, H.-J. Fuchs, D. Schelle, E.-B. Kley, S. Nolte, A. Tunnermann, R. Iliew, C. Etrich, U. Peshel and F. Lederer, *Optics Express*, Vol. 11, No. 24, 3284 (2003).
15. L. Wu, M. Mazilu and T. F. Krauss, *IEEE J. of Lightwave Techn.*, Vol. 21, No. 2, 561 (2003).
16. A. Lupu, E. Cassan, S. Laval, L. El Melhaoui, P. Lyan and J. M. Fedeli, *Optics Express*, Vol. 12, No. 23, 5690 (2004).
17. J. C. Knight, *Nature*, **424**, 847 (2003)
18. D. Erickson, S. Mandal, A. H. J. Yang and B. Cordovez, *Microfl. Nanofl.*, Vol. 4, 33 (2008).
19. C. Monat, P. Domachuk and B. J. Eggleton, *Nature Photonics*, Vol. 1, 106 (2007).
20. T.F. Krauss, *J. Phys. D. Appl. Phys.*, Vol. 40, 2666 (2007).
21. P.R. Villeneuve, S. Fan and J. D. Joannopoulos, *Phys. Rev.* ***B***, Vol. 54, No. 11, 7837 (1996).
22. Y. Akahane, T. Asano, B. S. Song and S. Noda, *Nature*, Vol. 425, 944 (2003).
23. M. Lončar, A. Scherer and Y. Qiu, *Appl. Phys. Lett.*, Vol. 82, 26, (2003).
24. M.L. Adams, M. Lončar, A. Scherer and Y. Qiu, *IEEE J. on Select. Areas in Comm.*, Vol. 23, 7, (2005).
25. Y. Xia and G.M. Whitesides, *Annu. Rev. Mater. Sci.*, Vol. 28, 153 (1998).
26. E. Chow, A. Grot, L.W. Mirkarimi, M. Sigalas and G. Girolami, *Opt. Lett.*, Vol. 29, 1093 (2004).
27. P.S. Cremer, *Nature Biotech.*, Vol. 22, 2 (2004).
28. M.R. Lee and P.M. Fauchet, *Opt. Expr.*, Vol. 15, No. 8, 4530 (2007).
29. M.R. Lee and P.M. Fauchet, Group IV Photonics, 2007 4th IEEE International Conference, 234 (2007).
30. C. Ciminelli and M. N. Armenise, EWOFS 2007, Napoli, (2007).
31. B.S. Song, S. Noda, Y. Akahane and T. Asano, *Science*, **300**, 1537 (2003).
32. B.S. Song, S. Noda, T. Asano and Y. Akahane, *Nature Materials*, Vol. 4, 207 (2005).
33. S. Tomljenovic-Hanic, C. Martijng de Sterke and M.J. Steel, *Opt. Expr.*, Vol. 14, 25, (2006).
34. S. Fan and J.D. Joannopoulos, *Phys. Review* ***B***, Vol. 65, 235112 (2002).
35. U. Fano, *Phys. Rev.*, Vol. 124, 1866 (1961).
36. O. Levi, M. M. Lee, J. Zhang, V. Lousse, S. R. J. Brueck, S. Fan and J. S. Harris, *Proc of SPIE*, 6447, 0P1-9, (2007).
37. B. Cunningham, *SPIE Newsroom*, (2008).
38. P.C. Mathias, N. Ganesh, L. L. Chan and B. T. Cunningham, *Appl. Optics*, Vol. 26, 2351 (2007).
39. I. M. Krieger and F. M. O'Neill, *J. Am. Chem. Soc.* Vol. 90, 3114 (1968).
40. J. H. Holtz and S. A. Asher, *Nature*, Vol. 389, 829 (1997).
41. K. Lee and S. A. Asher, *J. Am. Chem. Soc.*, Vol. 122, 9534 (2000).

42. C.-Y. Kuo, S.Y. Lu, S. Chen, M. Bernards and S. Jiang, *Sensors and Act.* ***B***, Vol. 124, 452 (2007).
43. R. Martin, A. Sharkawy and E. Kelmelism, *SPIE Newsroom*, (2006).
44. T. Prasad, D.M. Mittleman and V.L. Colvin, *Opt. Mat.*, Vol. 29, 56 (2006).
45. M.A. Fiddy, *SPIE Newsroom*, (2006).
46. A. Cusano, M. Pisco, M. Consales, A. Cutolo, M. Giordano, M. Penza, P. Aversa, L. Capodieci and S. Campopiano, *IEEE Phot. Tech. Lett.*, Vol. 18, No. 22, 2431 (2006).
47. F. Benabid, J.C. Knight, G. Antonopoulos and P. St. J. Russell, *Science*, Vol. 298, 399 (2002).
48. T.M. Monro, W. Belardi, K. Furusawa, J.C. Baggett, N.G.R Broderick and D.J. Richardson, *Meas. Sci. Technol.*, Vol. 12, 854 (2001).
49. T. Ritari, G. Genty and H. Ludvigsen, *Opt. Lett.*, Vol. 30, 3380 (2005).
50. Y. L. Hoo, W. Jin, H. L. Ho, D. N. Wang and R. S. Windeler, *Opt. Eng.*, Vol. 41, 8 (2002).
51. J. Jensen, P. E. Hoiby, G. Emiliyanov, O. Bang, L. H. Pedersen and A. Bjarklev, *Opt. Express*, Vol. 13, 5883 (2005).
52. W. Denk, J. Strickler and W. Webb, *Science*, Vol. 248 (4951), 73 (1990).
53. M. T. Myaing, J. Y. Ye, T. B. Norris, T. Thomas, J. R. Baker, W. J. Wadsworth, G. Bouwmans, J. C. Knight and P. SJ. Russell, *Opt. Lett.*, Vol. 28, No. 14, 1224 (2003).
54. S.O. Koronov A. Zheltikov and M. Scalora, *Opt. Expr.* Vol. 13, 9, (2005).
55. L. Rindorf, J.B. Jensen, M. Dufva, L.H. Pedersen, P.E. Høiby and O. Bang, *Opt. Expr.*, Vol. 14, 18, (2006).
56. S. Smolka, M. Barth and O. Benson, *Opt. Expr.*, Vol. 15, No. 20, 12783 (2007).
57. T. Ritari, J. Tuominen, H. Ludvigsen, J. Petersen, T. Sørensen, T. Hansen and H. Simonsen, *Opt. Expr.*, Vol. 12, No. 17, 4080 (2004).
58. C. Kerbage et al., *Opt & Phot. News*, Vol. 13, No. 9, 38 (2002).
59. M. Pisco, M. Consales, A. Cutolo, M. Penza, P. Aversa and A. Cusano, *Sensors and Actuator **B**: Chemical*, **129**, 163 (2008).

MICROMACHINING TECHNOLOGIES FOR SENSOR APPLICATIONS

Pasqualina M. Sarro,[a] Andrea Irace[b,*] and Paddy J. French[a]

[a]*DIMES - TU Delft Feldmannweg, 172600 CT Delft, The Netherlands*
[b]*Dipartimento di Elettronica e dell'Ingegneria delle Telecomunicazioni, Università "Federico II", Via Claudio, 21, 80125 Napoli, Italy*
[*] *E-mail:a.irace@unina.it*

In this contribution we review some of the more conventional micromachining technologies, underlining their advantages and disadvantages. Then, new developments in some of the existing technologies as well as new technologies will be illustrated.
Finally, some examples of integrated micro-machined sensors and devices are presented to underline the expectations of these technologies.

1. Introduction

The fabrication of micromechanical structures with the aid of etching techniques to remove part of the substrate or deposition techniques to add a thin film is usually called *micromachining*. Silicon has excellent mechanical properties[1] making it an ideal material for machining. In the 60s the first micro-machined silicon sensors using isotropic etching[2] or mechanical milling[3] were reported. Crystal orientation dependent etchants lead to more precise definition of structures and increased interest.[3,4] In 1976 anisotropic etching of silicon was introduced. This process, making use of crystal orientation dependent etchants, leads to a more precise definition of structures and increased interest in micromachining[3]. An early silicon pressure sensor, based on anisotropic etching was made by Greenwood in 1984.[5] Surface micromachining also dates back to the 1960s. Basically, surface micromachining involves the formation of mechanical structures from thin films on the surface of the wafer. Early examples included metal mechanical layers.[5] The early 1980's saw the growth of silicon based surface micromachining using a polysilicon mechanical layer.[6] In recent years, a number of new technologies have been developed using both silicon and also alternative materials. These include the epi-processes where the epi-layer is used as a mechanical layer and a number of deep plasma etching processes. This chapter concentrates on silicon based micromachining processes.

In the past decade a large number of micro-machined sensors have been developed and more recently, the importance of microsystems, i.e. systems combining electronic functions with mechanical, optical and others and that employ miniaturization in order to obtain high complexity in a small space, has been acknowledged. The advantages deriving by the use of conventional IC processes to fabricate the various components of a microsystem has further stressed the importance of the integration of micro-machined sensors. This integration is challenged by the constraints of the IC processes and the very tight control on material properties required to produce functioning electronic devices. This requirement, generally referred to as *IC compatibility*, forms one of the most important issues in the development of process flows for integrated micro-machined sensors. In short, we can say that the major requirements for successful integration of micro-machined devices are:

- process *compatibility* with conventional *IC* processes, which limits the kind of materials and the temperature budget allowed;
- a certain *flexibility* in the additional processing to allow more types of sensors/structures in the same process flow (no custom-made process);
- process *complexity:* the amount of additional steps should be kept minimal and the level of complexity low, to favor commercial development.

There are essentially two approaches to satisfy the above mentioned requirements, when using a standard baseline *IC* process: pre-processing or post-processing. In the first case, all micromachining process steps are done before the circuitry and in the latter are done after the circuitry. Conventional silicon micromachining techniques are essentially divided into two categories: bulk and surface micromachining. Bulk micromachining covers all techniques which etch the substrate (bulk) material and the bulk is part of the micro-machined structure. Surface micromachining, on the other hand is referred to techniques which use a stacked thin-layer structure, created on top of the substrate.

2. Bulk Micromachining

Bulk micromachining generally encompasses techniques that remove significant amounts of the substrate (bulk) material and the bulk is part of the micro-machined structure. This microstructuring of the substrate is done to form structures that can physically move, such as floating membranes or cantilever beams. Other types of structures that can be realized by bulk micromachining are wafer-through holes. Typical bulk micro-machined structures are shown in Fig. 1

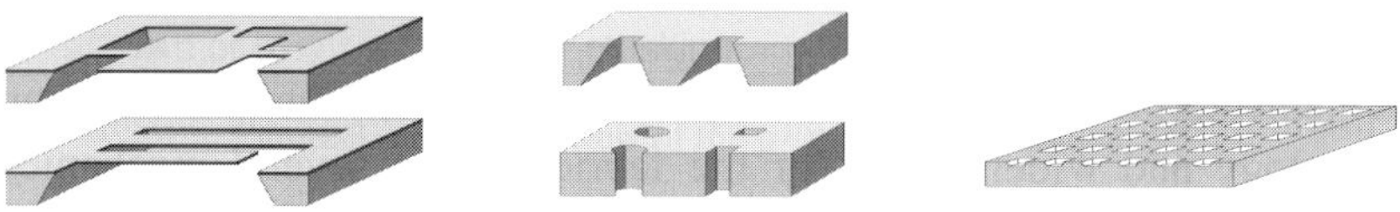

(a) (b) (c)

Figure 1. Typical bulk micro-machined structures: a) membranes and beams, b) wafer-through holes, c) microwells.

The substrate removal can be done using a variety of methods and techniques. In what follows, a number of currently available processes will be introduced and their potentials and limitation indicated. Various aspects, such as etch characteristics, compatibility to conventional *IC* processes, complexity and costs, will be illustrated in order to evaluate the suitability of each technique for a specific application.

2.1. Wet Etching

Wet etching of silicon[7] is often used if large amounts of the silicon bulk have to be removed and it is more diffused than dry etching, the preferred method used in *IC* technology. The reasonably fast etch rates that can be achieved, the low cost of wet etching, due to the low complexity equipment, the availability of masking materials to perform the process selectively are among the major reasons for the large use of wet etching of silicon. Chemical solutions that remove the silicon anisotropically (orientation dependent etch rates) or isotropically (etch rate is equal in all directions), are available. These are shown in fig. 2. These two types of etching will be discussed separately below.

where the etch process stops consists of a *p*-type silicon wafer with an *n*-type epilayer on top or if devices are fabricated in a CMOS process, the *n*-type well in a *p*-type epi or substrate. A positive voltage is applied to the *n*-side of the junction with respect to a Pt counter electrode. The p-side of the junction is etched and when the etch-front reaches the junction the etch stops as the *n*-side is passivated by the applied voltage. The special wafer holder required and the proper contact pattern to passivate all n-regions on the wafer, are often indicated as factors limiting the use of this technique. Recently, wafer holders and computer-controlled systems have been made available, contributing to reduce some of the concerns. A more recent technique is the *Photovoltaic etch-stop*.[16-18] This technique does not require external electrodes as the external power source is a high intensity light source. Contrary to the ECE technique, the *p*-side of a junction is passivated and the *n*-side is etched. A platinum/titanium film is sputtered on the wafer backside and acts as a masking layer as well as a contact to the *n*-type silicon.
The wafer is illuminated by a strong light source to ensure an etch-stop at the *p*-type epilayer. The *n*-type substrate and the platinum film interact galvanically. This method still presents some difficulties as a large platinum electrode is needed and the required high-power light source complicates the set-up. A new etch-stop technique, *galvanic etch-stop* that does not use any external power source has been introduced a few years ago.[19] A very similar structure as the one used for ECE *p*/*n* etch stop is used. The electrical power used to stop the etching process is generated within the structure itself. The gold/silicon combination forms a galvanic cell[20]. The reduction of oxygen at the gold electrode generates the cell current. When the structure is immersed in the etching solution, the gold electrode and the *p*-type Si bulk are insulated by the reverse-biased *p*/*n* junction. If the etch front reaches the junction, the insulation is destroyed and the galvanic cell is formed. When a sufficiently large gold electrode is used, etching stops and an n-type membrane is obtained. A disadvantage of this technique is that often a relatively large gold electrode is required. Further investigation to accurately establish the potentials of this technique is needed.

2.1.2. Isotropic Etching

Silicon can also be etched isotropically in HF based solutions. The main dissolution mechanism is anodic for which valence band holes are required. The reaction can be controlled by changing the surface hole concentration. This can be achieved with an electrochemical cell (anodic dissolution); by using an oxidizing agent (electroless or chemical dissolution) or by high intensity illumination (open-circuit light-assisted dissolution).[20] This last one is generally not used for micromachining of silicon and will not be discussed here.

In the case of *anodic dissolution*, the silicon wafer is the anode in an electrochemical cell. A Pt electrode is used as counter electrode. The set-up used is very similar to the one used for anisotropic ECE and it is shown in Fig. 5. More often a three-electrode configuration with a Ag/AgCl reference electrode is used. The important process variables are the current density and the solution concentration. There is a critical current density for each value of HF concentration. Above this value silicon is uniformly etched (electropolishing) while below this value a porous silicon layer is formed. Doping and illumination determine the general type of porous film, but the morphology depends on current density and etchant concentration.[20,21] Often an additive is added to the HF solution to enhance the formation process.[22] These additives act as surfactants, reducing the surface tension of the solution and allowing the hydrogen formed as by-product of the process to escape freely. This prevents it sticking to the silicon and erroneously masking it, causing a non-homogeneous layer.

HF concentrations range from 1% to 40% although recently some higher concentrations (73%) have been employed as the etch rate of aluminum is strongly reduced at this high concentration. LPCVD silicon nitride or noble metals can be used as etch masks. Less aggressive with respect to photoresist, but with lower etch rates, are mixtures of ammonium fluoride, HF and water, generally addressed as buffered HF solutions.

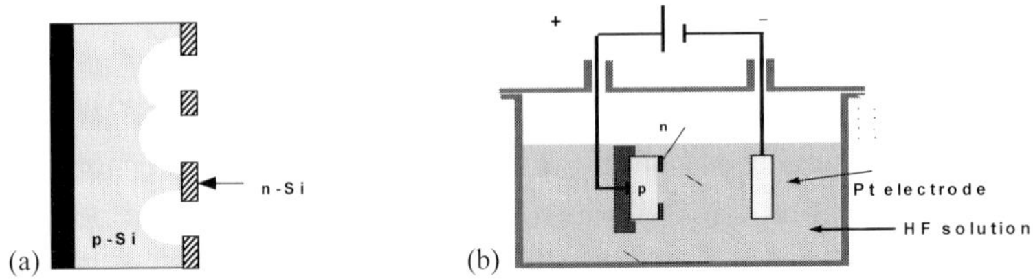

Figure 6. Anodic etching of silicon in HF solutions: a) schematic cross section of etched wafer; b) etch set-up).

Next to being used as dielectric in humidity sensors, porous silicon is also an interesting material as sacrificial layer in epi-micromachining (see section 3.3.5.) and to realize deep trenches and 3D-free standing silicon microstructures by using the effect of light on porous silicon formation (see section 2.2).

Electroless dissolution requires a strong oxidizing agent added to the HF solution. The oxidising agent is reduced and thereby injects holes into the valence band.[20] The holes are then consumed in the silicon dissolution reaction. The etch rate can be increased by increasing the oxidizing agent concentration responsible for an increase in hole injection current. If the reduction reaction is mass-transport controlled,[23] agitation of the solution generally results in an enhanced reaction rate as it has a considerable effect on the hole injection current. Two types of surface morphology are possible: a uniformly etched surface or a porous layer is formed (stain etching). If the dissolution reaction is limited by hole injection a porous layer is formed, while if it is limited by diffusion of HF species to the surface electro-polishing occurs. Commonly used etchants are HNA, i.e. a mixture of HF, nitric acid (HNO_3) as oxidizing agent and acetic acid (CH_3COOH) to stabilize the oxidizing agent concentration[24,26] and mixtures of HF, HNO_3 and H_2O. For both systems the etch rates of silicon and the quality of the etched surfaces strongly depend on the proportion of the acids in the mixture. A mixture of HNO_3:H_2O:NH_4F (126:60:5) has also been used.[27] The use of NHF instead of HF results in a buffer action, keeping the HF and HF_2^- concentrations from changing rapidly with use. Moreover, photoresist can be used as masking layer.

Etch-stop techniques: The most commonly known etch-stop techniques in isotropic etchants are of the extrinsic type. The only intrinsic one is lightly doped etch-stop technique. The HNA system etches heavily doped silicon preferentially with respect to lightly doped silicon.[28] For some solution composition etch rates between 0.7 and 3 μm/min have been reported for 10^{-2} Ωcm silicon with no appreciable etch rate for 6.8×10^{-2} Ωcm silicon. Several extrinsic etch-stop techniques are available. The major characteristics of these techniques are summarized in table I.

The *p/n etch stop* is the most commonly used in porous silicon micromachining applications.[20,29] The *n*-type mechanical structure is realized in a *p*-type bulk wafer. The *p*-type is anodized in dark (otherwise enough holes are generated to make the *n*-type porous as well) and made porous, while the *n*-region is not. Then the porous silicon is removed in a weak alkaline solution.

The *lightly doped etch stop*[35] uses *p*-type layers for the mechanical structures. These regions are protected form the solution by a masking layer and the *n*-type bulk is positively biased. Only the *n*-type becomes porous because the *p/n* junction provides a barrier for the charge carriers. Low-doped *p*-type structures can be fabricated with this method, that is therefore complementary to the *p/n* etch stop. The last two techniques, the *photovoltaic etch stop*[36] and the galvanic etch-stop[37] are contactless. For the photovoltaic etch stop, *n*-type regions in a *p*-type wafer are defined. The *n*-type regions will form the mechanical structures. When the wafer is immersed in the HF solution in the presence of illumination, the *p/n* junction works as a photovoltaic cell, short-circuited by the HF solution. The photocurrent flows from the positive *n*-side to the negative *p*-side resulting in its anodic dissolution.

The etch rate is dependent on the light intensity. The *galvanic etch stop* uses a similar structure to that of the *p/n* etch stop. Anodization is accomplished without an external power source. A platinum/chromium or gold/chromium layer is used to make (backside) contact to the *p*-type silicon wafer. The oxygen (or another strong oxidizing agent) is reduced at the metal surface.

The holes generated at the metal/HF solution interface flow to the *p*-type Si/HF solution interface where the *p*-type is anodically dissolved.

Etch-stop technique	**Thickness control**	**External power source**	**Stop on *p*- or *n*-type**	Ref.
Intrinsic	poor	no	*p* and *n*	30
p/*n*	Reasonable/good	yes	*n*	31
Resistivity gradient (n/n^+)	poor	yes	*n*	30, 32-33
Lightly doped	Good	yes	*p*	37
Photovoltaic	good	yes (contactless)	*n*	36
Galvanic	good	no	*n*	37

Table I. Comparison of etch-stop techniques for isotropic wet etching of silicon[20].

The *n*-type regions are not attacked in HF in dark conditions, resulting in intact freestanding structures.

3. Surface Micromachining

Surface micromachining is a quite different technology from the bulk micromachining processes described above. Basically surface micromachining involves the deposition of thin films on the wafer surface and selectively removing one or more of these layers to leave free standing structures. In recent years a number of new processes have been developed which use the epitaxial layer, or the upper few microns of the substrate as a mechanical layer. These technologies will also be discussed in this section and have been given the collective title of epi-micromachining.

3.1. Basic Process Sequence

Surface micromachining techniques can be traced back to the 1960's. The resonant gate structure of Nathanson et al. showed how free standing structures could be fabricated with thin film (in this case metal).[5,38] In the 1970's surface micro-machined devices were fabricated where the sacrificial layer was the epitaxial layer. An example of this was the deflectable aluminum coated oxide mirrors.[39]

In the 1980's came the first of the micromachining processes using entirely chemical vapor deposited layers (CVD).[6,40] In this case polysilicon and oxide were used as the mechanical and the sacrificial layers, respectively. This early work showed the potential of this new process and it was the first examples of moving mechanical parts. The basic principle of surface micromachining is given in Fig. 6. Two types of layers can be seen, the sacrificial layer and the mechanical layer. The sacrificial layer is so called because it is removed during subsequent processing. In this example, first the sacrificial layer is deposited and defined, followed by the same process for the mechanical layer. At the end of the process the sacrificial layer is removed to leave the free-standing mechanical structure. As shown in Fig. 7ii, the sacrificial layer is accessed from the side of the structure or through access holes.

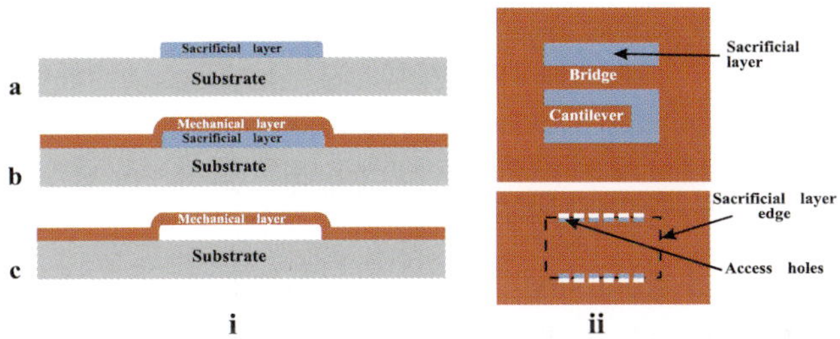

Figure 6. Basic surface micromachining process.

There are many options for both the sacrificial and mechanical layers. The combination is chosen to ensure good mechanical properties of the mechanical layer and ease of removal of the sacrificial layer. Frequently used mechanical layers include LPCVD polysilicon silicon nitride and a number of metals. More recently PECVD silicon carbide has been shown to be a highly suitable material as a mechanical layer. Silicon dioxide is probably the most commonly used sacrificial layer due to its ease of etching, although other materials include polyimide and a range of metals.

3.2. Epi-Micromachining

Epi-micromachining is a group of micromachining technologies which use the epitaxial layer or a similar thickness of the upper part of the substrate as the mechanical layer and use a buried layer, or simply the substrate, as the sacrificial layer. Recently a modification to the bulk micromachining technique, where the micromachining is performed in the top few microns of the substrate, has been receiving some attention. These techniques also called "*epi micromachining*" since the mechanical devices are often formed in the epilayer, have many of the advantages of both bulk and surface micromachining without most of the disadvantages.[41] A basic epi-micromachining process sequence is schematically depicted in Fig. 7.

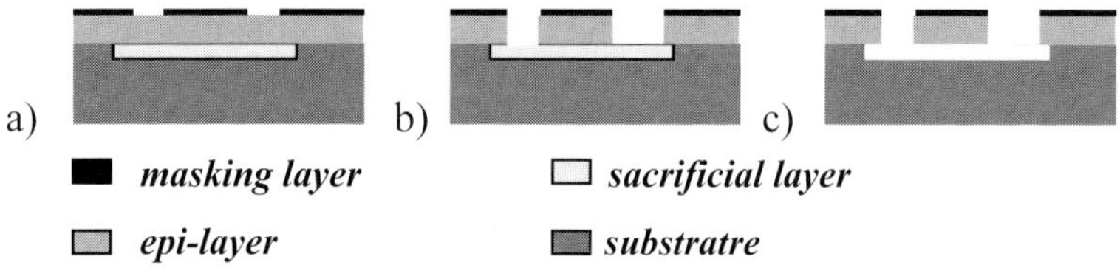

Figure 7. Schematic epi-micromachining process sequence: a) the wafer is ready for the post-processing to start; b) an opening is made through the epilayer to reach the sacrificial layer; c) the sacrificial layer is removed.

It is essentially a front-side bulk micromachining, using wet or dry etching (or a combination of both) to from the mechanical structures. The structures realized in this way maintain the good mechanical properties of single crystal material with lateral dimensions similar to surface micromachining.

There are various techniques which are used to implement epi-micromachining. After the basic *IC* process is performed, a last mask step is applied. Then a two-step etch consisting of a dry etch step to expose the sacrificial layer and, just as for surface micromachining, an etch step to remove the sacrificial layer. For this last etch step either dry or wet etching can be used. Among the techniques using dry etching to release the microstructures are *SCREAM* (Single Crystal Reactive Ion Etching And Metallization), *SIMPLE* (SIlicon Micromachining by PLasma Etching), *BSM* (Black Silicon Method). In this case, the sacrificial layer consists of single crystal silicon itself. For other techniques, such as the use of *SIMOX* wafers or the epitaxial lateral overgrowth of silicon, the buried oxide will be the sacrificial layer and will be removed by wet etching. Another type of epi-micromachining is based on the use of a porous silicon layer as sacrificial layer.

3.2.1. SIMPLE

The *SIMPLE* process (Silicon Micromachining by Plasma Etching), forms micro-machined structures using a single etch step. This process makes use of a Cl_2/BCl_3 chemistry which etches low doped material anisotropically and *n*-type material above a threshold of about $8x10^{19}$ cm^{-2}, isotropically. The basic process sequence is shown in Fig. 9. The first additional step, to the bipolar process, is a heavily doped buried layer since the bipolar buried layer has a doping level which is too low to be under-etched. This buried layer is formed by ion implantation of arsenic. The implantation required is thus $1x10^{16}cm^{-2}$ at 180 keV followed by a 1200 °C, 4 hour anneal and drive-in. This is followed by the formation of the standard bipolar buried layer using implantation of antimony. The standard bipolar epitaxial layer is then grown (Fig. 8a) followed by an additional deep diffusion where the mechanical structure will be formed (Fig. 8b). After this a full standard bipolar process is performed (Fig. 8c). A thick PECVD oxide is deposited which serves as a mask for the plasma etching of the epi and buried layer. The ratio of lateral to vertical etch-rate and selectivity over mask etching is dependent upon the gas ratio, power and pressure. A limiting factor on under-etching is the drop in etch-rate as a function of time. This is due to the fact that the etching results from spontaneous chemical reactions, which are related to the concentration of Cl atoms at the reaction surface and the rate of desorption of reaction products. Typically, for an epi thickness of 4 μm and mechanical structures of width 3 μm, and masking oxide layer of 1 μm would be required. The final structure is given in figure 15d. This shows clearly how the vertical etching continues in the trench during the lateral etching of the buried layer. This technique has the advantage of simplicity. It is performed after the aluminum deposition, therefore it is fully compatible with the electronics. The high doped buried layer, required for under-etching has some influence on the electronics due to auto-doping. The increased epi-doping concentration results in a small increase in vertical *npn* transistor gain and a fall for lateral *pnp* devices.

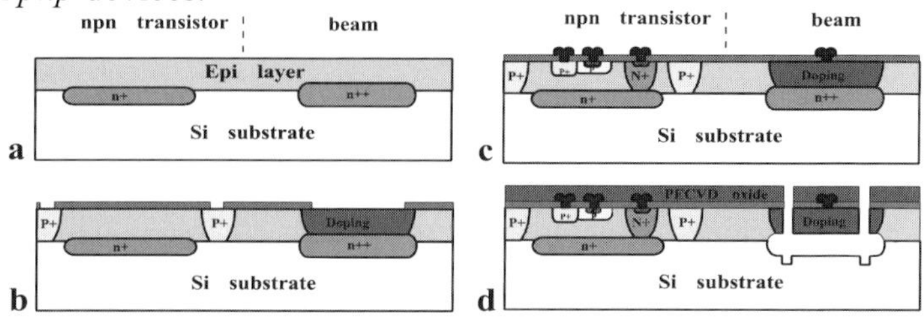

Figure 8. Main steps for the *SIMPLE* process.

3.2.2. SCREAM

The process known as *SCREAM I* (Single Crystal Reactive Etching And Metallization),[42] uses a combination of anisotropic and isotropic plasma etching. The basic process sequence is shown in Fig. 9. After patterning the oxide, trenches are etched which define the sidewalls of the structure (a). An oxide layer is deposited (Fig. 9b) and etched-back using the plasma etching, resulting in the structure given in Fig. 9c. A further plasma etch deepens the trench beyond the sidewall protection (Fig. 9d). The trenches now need to be under-etched; therefore an isotropic etch is required. This structure is shown in Fig. 9e. Finally, aluminum is sputtered to form a contact to the mechanical structures, as shown in Fig. 9f. High aspect ratios can be achieved although this is limited by the thickness of the masking oxide which must survive all the etch steps. Maximum reported beam height was 20 μm with a total width (including metal) of 6 μm. If the devices are not to be integrated with electronics, processing finishes at this point and the devices can be contacted with bond wires. The cross-section shows how isolation is achieved through the poor step coverage of the aluminum. Integration with electronic can be achieved using thick resist layers to pattern the wafer after micromachining. In this integrated version the micromachining is performed after the processing of the electronics, including metallization. First, an oxide layer is deposited to protect the electronic circuitry during subsequent processing. This oxide also forms the masking layer for the micromachining. After the formation of the free-standing structures (Fig.

9e), the oxide is etched back to reveal the metal pads of the electronics. A metal layer is then deposited as in the non-integrated version. In this case the metal serves for three purposes; 1) capacitor electrodes, 2) interconnect and 3) contact the *IC* metallization. Finally a thick resist layer is used to pattern the metal. An OCG 895I 90cs resist was used for this purpose. Since the detailed metallization of the electronics has already been defined before the micromachining, only large structures need to be defined using the thick resist. Similar to SCREAM is the Black Silicon method[43] with the important difference that the sidewall passivation is achieved in the plasma etcher and thus a multi-step one-run process can be achieved.

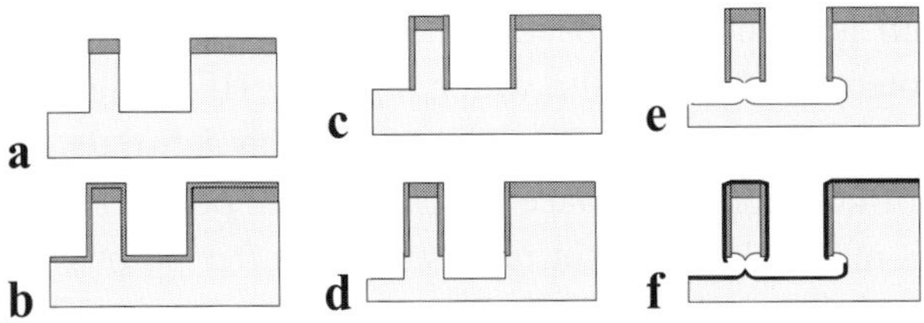

Figure 9. *SCREAM* process sequence

3.2.3. MELO

The *MELO* process (Merged Epitaxial Lateral Overgrowth) is an extension of selective epitaxial growth SEG. Selective epitaxial growth uses HCl added to the dichlorosilane. The HCl etches the silicon. However, if there is a pattern of bare silicon and silicon dioxide, the silicon deposited on the oxide will have a rough grain like structure, with a large surface area, and will therefore be removed more quickly. Thus, although the growth rate will be lower than a normal deposition, a selective growth can be achieved. Once the silicon layer reaches the level of the oxide, both vertical and lateral growth will occur yielding lateral overgrowth.

If two of these windows are close enough together they will merge giving the MELO process. As a result we have buried silicon dioxide islands. This lends itself well to micromachining [45-49]. This process has the advantage of producing single crystal structures, but the disadvantages that beams must be orientated in the <100> direction (due to growth mechanisms of silicon[46]) and since growth continues both laterally and vertically, the lateral dimensions are limited.

3.2.4. Porous Silicon

A process using silicon as both mechanical and sacrificial layer is the sacrificial porous silicon technique.[30,33,46] Basically regions of silicon are selectively made porous and this region is used as sacrificial layer. The porous formation set-up is very similar to the electrochemical KOH etching, although in this case the etchant is HF. A positive voltage is applied to a platinum electrode and negative to the backside of the wafer. If aluminum is used as the back contact then a special holder must be used to protect the back of the wafer from the HF etchant. The current flow between the platinum electrode and silicon substrate enhances the formation of holes on the surface which result in pores. The high surface area of this materials results in rapid etching, after porous formation, in KOH. This makes the material highly suitable as a sacrificial material. The porous silicon formation rate is highly dependent upon the current density, HF concentration, lighting and the doping in the substrate. The process sequence is given in figure 10. In this case the selective etching of *p*-type material over the *n*-type epi is used. First a plasma etch is used to etch through the epi-layer to reveal the substrate (Fig. 10b). The porous layer is then formed using the process described above as shown in Fig. 10c, and finally the porous layer is removed in KOH at room temperature (Fig. 10d).

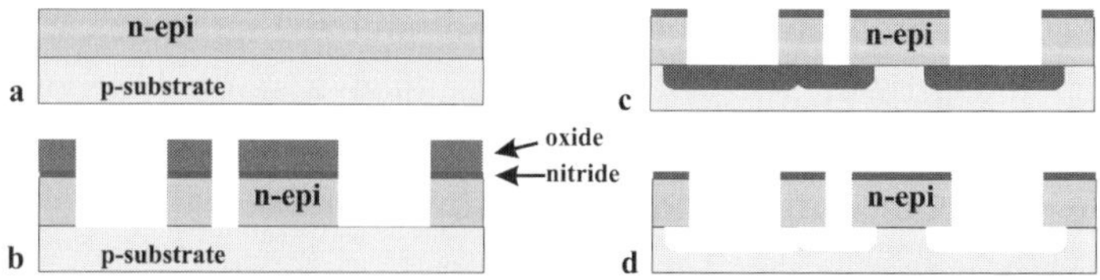

Figure 10. Basic process steps for sacrificial porous silicon based micromachining.

The porous silicon technique is extremely simple and can be applied as a post processing step and it is therefore fully compatible with the electronic circuitry. The only remaining problem is to protect the areas of electronics and metallization from the HF etchant, although this can be achieved by using an alternative etchant. One disadvantage of this technique is the added process complexity introduced by the requirement of a backside electrical contact during etching.

Porous silicon has been known for quite some time, and used as sacrificial layer in some sensor structures. This material can be formed by selective electrochemical etching of silicon in HF. Doping and illumination determine the general type of porous film, but the morphology depends on current density and etchant concentration. Due to its porosity, it is a fast etching material, therefore quite suitable to be used as sacrificial layer in an epi-micromachining process. By properly selecting the doping of the area indicated as sacrificial layer in Fig. 1b, this region can be made porous after openings in the epilayer have been plasma etched. Finally, the porous layer is removed in KOH at room temperature as the last processing step. Although this technique is quite simple and a post processing one, it has the problem that metallization must be protected from the HF etchant. Recently, porous silicon has been formed in an ammonium fluoride etch mixture. A micro-machined membrane for a vertical accelerometer, is shown in figure 11. After the porous silicon formation in the ammonium fluoride mixture, no further etching was performed on this structure.

Figure 11. Vertical accelerometer membrane freed from the substrate using a porous layer formed in ammonium fluoride mixture at 10 mA/cm^2, for 600 sec. (Ref. 28).

3.2.5. *SIMOX*

An alternative technique is to start with a wafer with a buried sacrificial layer, such as a SIMOX (separation by implantation of oxide) wafer.

SIMOX wafers have a buried oxide layer above which is a single crystal silicon layer. The substrates are prepared by implantation of oxygen with a typical dose of $1.8x10^{18}$ cm^{-2}. The implantation is performed at temperatures above 500 °C to avoid amorphization of the silicon. A high temperature anneal (> 1300 °C) is used to eliminate defects generated by the implantation. A typical resulting structure is a 2000 Å upper silicon layer with a 4000 Å buried oxide. The wafers may then be further processed with an additional epitaxial growth. A plasma etch through the epi is used to reveal the sacrificial oxide and this layer is then removed in the same manner as with surface micromachining. This process has the advantages of substrate industrially available (SIMOX), CMOS compatibility and single crystalline silicon as surface layer. Moreover, it offers freedom on the surface structure thickness by standard epitaxial processing and SiO_2 buried layer as a sacrificial and insulating layer. However, the higher costs of the start material may present problems for some applications. A further problem which may arise is unwanted under-etching at the epi-oxide interface, although in many applications this does not present a significant problem.

3.2.6. Epi-Poly

An interesting alternative to the above described processes is the epi-poly technique. In this case polysilicon layers are grown in the epitaxial reactor. Although this technique departs from using single crystal silicon as a mechanical material it has greater flexibility in terms of lateral dimensions. It has been used as a thick surface micromachining process. Alternatively, the mechanical layers can be formed at the same time as the single crystal epi required for the electronics]. The basic process is shown in figure 12. After the formation of the sacrificial oxide, a polysilicon seed is deposited (Fig. 12a). This polysilicon seed ensures a uniform growth. A standard epi growth will then form epi-poly on the seed and single crystal where the substrate is bare (Fig. 12b). The epi growth rate on the polysilicon seed is about 70% of that on the single crystal silicon. Therefore the total thickness of the sacrificial layer and seed can be adjusted to ensure a planar surface after epi growth. The mechanical layer is then patterned (Fig. 12c) and released through sacrificial etching as shown in Fig. 12d. The process can be seen to be extremely simple in that the mechanical layer is formed at the same time as the epi-layer for the electronics. One potential problem with epi-poly is the compressive stress which can be generated during oxidation. This problem can be eliminated by protecting the epi-poly with a thin silicon nitride throughout the bipolar processing. This technique has been successfully applied, resulting in zero or low tensile stress.[46]

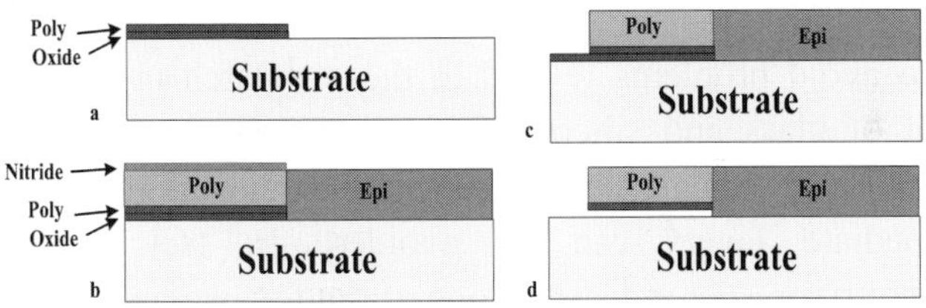

Figure 12. Basic epi-poly process.

The epi-poly process requires minimal additional processing before epitaxial deposition and no detrimental effect on the electronics' characteristics has been found. The process is therefore fully compatible with the electronics processing.

A few examples to illustrate the potentials of this technique are a thermally actuated indicator (extremely suitable as process control device) and a lateral accelerometer structure which benefit from the large lateral capacitance attainable by using the thick polysilicon layer. An SEM photograph of an accelerometer structure is shown in figure 13. In the close-up the suspended beams and the overforce stops and are clearly visible. The thickness of the epi-poly layer is 4 μm and the distance from the substrate is 2 μm.

3.2.7. Low Temperature Si-Si Bonding

The most mature bonding technology is still glass-to-silicon anodic bonding. Although glass wafers are currently available in many sizes and thicknesses, and a certain degree of processing can be performed on glass, this is generally limited to the etching of recesses to create cavities and to the patterning of a metal layer.

Figure 13 a), b). SEM micrograph of an accelerometer structure fabricated with the epi-poly process: a) the whole device (about 1 mmx 1 mm in size); b) a close-up of the area indicated in figure 14a, showing the suspended beams and the overforce stops.

Further, to avoid problems related to the mismatch in the temperature coefficient of glass and silicon, silicon-to-silicon bonding could be a more suitable technique if the high temperature annealing steps or the use of intermediate layers can be avoided. In fact, the processed device/circuitry present at least on one of the two wafers, cannot tolerate high temperature steps. Recently, a truly IC-compatible silicon-to-silicon bonding process has been developed. By proper handling of the wafer surface, room temperature pre-bonding can be achieved and an anneal step at temperatures well below 400 °C, results in quite strong bonding. An ammonium fluoride etch mixture, which does not attach the aluminum metallization or the scratch protection layer, is used instead of HF to remove the native oxide. In this way, samples containing electronic circuitry have been successfully bonded, without affecting the electronic or device performance.

4. Characterization of Thin Film Membranes

Micro-machined sensor industry makes an extensive usage of thin dielectric or conducting films such as silicon nitride or polysilicon both as supporting structures, passivation/isolation layers or active components.[50] Since the first appearance of these sensors, mainly for calibration purposes, there has been an urgent need to characterize the physical properties of the materials. In fact, it is widely demonstrated that they can be significantly different from the bulk material ones and, moreover, that they are often process dependent.[51, 52] In what follows we discuss in some more detail an experimental method to measure the thermal conductivity of an amorphous SiC thin film deposited by PECVD technique on SiN freestanding membranes.

4.1. The Measurement Principle

With reference to figure 14, where a schematic overview of the test structure is depicted, we use a polysilicon (PS) resistor to create the required heat flux and several PS-Al thermocouples, which sensitivity are in the range of few hundreds microvolts per Kelvin degree, as a temperature difference sensor.

The usage of a thermocouple instead of a thermistor (as it had been in many previous works) makes possible to work with a higher signal to noise ratio and eliminates some concern about the calibration and linearity of the temperature sensor itself.

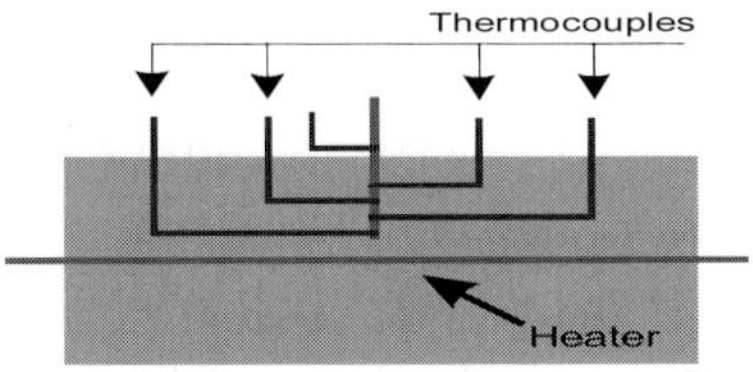

Figure 14. Schematic overview of the heater-thermocouple system.

with $\rho_0=5\cdot10^{-6}$ [$\Omega\cdot$m] and m $\approx$ 2.5 as constants. Both the heater and the thermocouple are realized over a thin film membrane or, if needed, over a multi-layered stack deposited over a SiN membrane which acts as a mechanical support (this is particularly useful when there are no techniques useful to pattern the material under test as a membrane). For a detailed discussion of the measurement technique see Ref. 52.

The experimental procedure has been successfully used to characterize the thermal conductivity of 500 nm-thick PECVD SiC thin films deposited over a 4mmx2mm, 500 nm-thick SiN membrane. Since it is reasonable to assume that all the heat transfer is taking place inside the SiC (because the thermal conductivity of the SiN is 1.5 W/mK), we simply have to double the equivalent thermal conductivity. In fact, the thermal resistance is given by a layer that is half the thickness of the entire system. The experimental results are resumed in Table II.

	SiC #1	SiC #2
Thickness [nm]	500	500
Deposition temperature [°C]	400	350
Stress[a] (compressive) [MPa]	890	780
k_T [W m^{-1} K^{-1}]	160	130
SiN thickness [nm]	500	500

Table II. Summary of experimental results

[a] *stress measured on as-deposited film*

The retrieved values of the thermal conductivity are lower than the reported value for the same material in its crystalline form. This is not surprising, as it is widely known that properties of materials in thin-film form can be significantly different from the bulk material ones.

5. Conclusions and Outlook

The microminiaturization of sensors and instruments is bringing about a revolution in instrumentation technology. The technological development of the last decades in micromachining technologies, thin film deposition and bonding technology, along with continued advances in silicon semiconductor circuitry, have resulted in miniature sensing devices whose performance is equal to and sometimes better than their full-sized counterpart. In order to be able to move microsystems from research prototyping to industrial development, some requirements need to be fulfilled. Among them the compatibility of silicon micromachining with IC processing has often been stressed. A few upcoming technologies, all aiming at respecting this compatibility are reviewed. They focus on a small amount, low complexity additional processing to be placed before or preferably after the electronic process. Some of them have already demonstrated their applicability, while others are at an early development stage. Particularly interesting are some epi-micromachining techniques due to the limited, low temperature, additional processing. Further, the very encouraging preliminary results in low temperature silicon to silicon bonding can have a significant impact primarily for the integration of closed microstructures. Although most of the research/development effort concentrates on the fabrication processes for integrated micro-machined sensors, some progress in the other areas involved, such as design, modeling and packaging must be acknowledged too.

References

1. K. E. Petersen, *Proc. IEEE*, 70, (1982).
2. O. N. Tufte and G.D. Long, *Journal of Applied Physics*, **33**, (1962).
3. K. E. Bean, *IEEE Trans Electron Devices*, ED-25, (1978).
4. H. C. Nathanson and R.A. Wickstrom, *Appl. Phys. Lett.*, 7, (1965).
5. J. C. Greenwood, *J. Phys. E. Sci. Instrum.*, 17, (1984).
6. R.T. Howe and R.S. Muller, *Sensors and Actuators*, 4, (1983).
7. H. Seidel et al. *J. Electrochem. Soc.* 137 (1990).
8. D. Zielke and J. Frühauf, *Sensors & Actuators **A**,* 48 (1995).
9. K. Sato et al., *Sensors & Actuators **A**,* 64 (1998).
10. L. D. Clark Jr., J. L. Lund and D. J. Edell, *Techn. Digest IEEE Solid State Sensor and Actuator Workshop*, USA, June 6-9, 198.
11. Y. Gianchandani and K.Najafi, *IEDM Techn. Digest* (1991).
12. A. Perez-Rodriguez, A. Romano-Rodriguez, J. R. Morante, M. C. Acero, J. Esteve and J. Montserrat, *J. Electrochem. Soc.* 143 (1996).
13. H. A. Waggener, *Bell Sys. Tech. J.,* 49 (1970).
14. B. Kloek, S. D. Colllins, N. F. de Rooij and R. L. Smith, *IEEE Tran. Electron Dev.*, 36 (1989).
15. P. M. Sarro, S. Brida, C. M. A. Ashruf, W .v .d. Vlist and H. v. Zeijl, *Sensors & Materials*, 10 (1998).
16. A. W. van Herwaarden, D. C. van Duyn, B. W. van Oudheusden, P. M. Sarro, *Sensors and Actuators **A*** 21-23 (1989).
17. E. Peeters, D. Lapadatu, R. Puers, W. Sansen, PHET*, J.Microelectromech. Sys.*, 3 (1994).
18. D. Lapadatu, M. de Cooman and R. Puers, *Sensors & Actuators A*, 53 (1996).
19. P. J. French, M. Nagao and M. Esashi, *Sensors & Actuators A*, 56 (1996).
20. C.M.A. Ashruf, P. J. French, P. M. M. C. Bressers, P. M. Sarro and J. J. Kelly, *Sensors & Actuators A*, 66 (1998).
21. H. L. Offereins, H. Sandmaier, K. Marusczyk, K. Kuhl and A. Plettner, *Sensors & Materials*, 3 (1992).
22. G. M. O'Hallaran, M. Kuhl, P. J. Trimp and P. J. French, *Sensors & Actuators **A***, (1997).
23. G.T. Kovacs et al., *Proc of the IEEE*, vol.86, (1998).
24. P.T.J. Gennissen and P.J. French, *Proceedings Transducers 97*, (1997).

25. H. Robbins, B. Schwartz, , *J. Electrochem. Soc.* 123 (1976).
26. S. D. Collins, *J. Electrochem. Soc.* 144 (1997).
27. K. R. Williams and R. S. Muller, *J.MEMS*, 5 (1996).
28. K. C. Lee, *J. Electrochem. Soc.* 137 (1990).
29. T. Bischoff, G. Muller, W. Welser and F. Koch, *Sensors & Actuators **A***, 60 (1997).
30. C. Ducso et al., *Sensors & Actuators A*, 60 (1997).
31. C. J. M. Eijkel, J. Branebjerg, M. Elwenspoek and F. C. M. v. d. Pol, *IEEE Electr. Dev. Lett.* 11 (1990).
32. G. Kaltsas and A. G. Nassiopoulou, *Sensors & Actuators A*, 65 (1998).
33. W.Lang, P.Steiner, H.Sandmaier, *Sensors and Actuators A*, Vol.51 (1995).
34. P.T.J. Gennissen and P.J. French, *Proceedings SPIE Micro-machined Devices and Components* vol.3876, (1999).
35. T. E. Bell and K. D. Wise, *Proceedings of IEEE MEMS*, (1997).
36. T. Yoshida, T. Kudo and K. Ikeda, *Sensors Mater.*, 4-5 (1993).
37. C. M.A. Ashruf, P. J. French, P M. M. C. Bressers and J. J. Kelly, *Sensors & Actuators A*, 74 (1999).
38. H. Wensink, J. W. Berenschot, H. V. Jansen and M. C. Elwenspoek, *Proc. IEEE MEMS* (2000).
39. H.C. Nathanson, W.E. Newell, R.A. Wickstrom and J.R. Davis Jr., *IEEE Trans. Electron Dev.*, 14, (1967).
40. R.N. Thomas, J. Guldberg, H.C. Nathanson and P.R. Malmberg, *IEEE Electron Dev.*, ED-22, (1975).
41. A. Merlos, M. Acero, M. H. Bao, J. Bauselles, J. Esteve, *Sensors & Actuators **A***, 37-38 (1993).
42. K.A. Shaw and N.C. MacDonald, *Proceedings IEEE MEMS*, (1996).
43. M. de Boer, H. Jansen and M. Elwenspoek, *Proceedings Transducers 95*, (1995).
44. M. J. Dececlerq, L. Gerzberg, J. D. Meindl, *J. Electroch. Soc.*, 122 (4) (1975).
45. A.E. Kabir, G.W. Neudeck and J. A. Hancock, *Proceedings Techcon 93*, (1993).
46. P.T.J. Gennissen, Delft University Press. ISBN 90-407-1843-1.
47. P.T.J. Gennissen, P.J. French, D.P.A. De Munter, T.E. Bell, H. Kaneko and P.M. Sarro, *Proceedings* ESSDERC'95, (1995).
48. B. Diem, P. Rey, S. Renard, S. Viollet Bosson, H. Bono, F. Michel, M. T. Delaye and G. Delapierre, *Sensors and Actuators,* (1995).
49. P.T.J. Gennissen, M. Bartek, P.M. Sarro and P.J. French, *Sensors and Actuators* (1997).
50. S. Lee and D.G. Cahill *J. Appl. Phys.,* (1997).
51. X. Ziang and C.P. Grigoropoulos, *Rev. Sci. Instrum.* (1995).
52. A. Irace and P. M. Sarro, *Sensors and Actuators **A***, (1998).

SPECTROSCOPIC TECHNIQUES FOR SENSORS

Stefano Pelli,[a,*] Alessandro Chiasera,[b]
Maurizio Ferrari[b] and Giancarlo C. Righini[a,c]
[a]*MDF-Lab, Istituto di Fisica Applicata "N. Carrara", CNR*
Via Madonna del Piano 10, 50019, Sesto Fiorentino (Fi), Italy
[b]*Istituto di Fotonica e Nanotecnologie, CNR*
Via Sommarive 14, 38050, Povo(Tn), Italy
[c]*Dipartimento Materiali e Dispositivi, CNR*
Via dei Taurini 19, 00185 Roma, Italy
[*]*E-mail: s.pelli@ifac.cnr.it*

The aim of this chapter is to give some basic elements concerning the utilization of absorption, luminescence, Raman and Brillouin spectroscopies in the field of optical sensors. Particular attention is paid to the diagnostics of thin films and to the applications of activated optical waveguides.

1. Introduction

The spectroscopic and optical techniques are currently used in the development of sensor technology. The application of optical fibers for chemical, biochemical, biomedical, temperature and pressure sensing is well established, and it is discussed in several papers and books.[1-7] Recently, several kinds of optical waveguides in planar format have been successfully applied to develop a great number of sensors based on the changes of the optical features of indicator layers. Application of planar waveguides as sensors based on refractive index,[8] surface plasmon resonance,[9] absorbance,[10] fluorescence,[11] Raman,[12] pressure[13] and interferometric[14] measurements have been reported. The main advantages of planar format are the high integration and the relatively large optical path length for attenuated total reflection spectrometry, characteristic of the integrated optical waveguide geometry.

As an example, the typical experimental configuration used to study the spectroscopic properties of the planar waveguide sensors is shown in Fig. 1. This set-up is practically the same one used in *m-line* spectroscopy, where the light is injected in the film by prism coupling.[15,16]

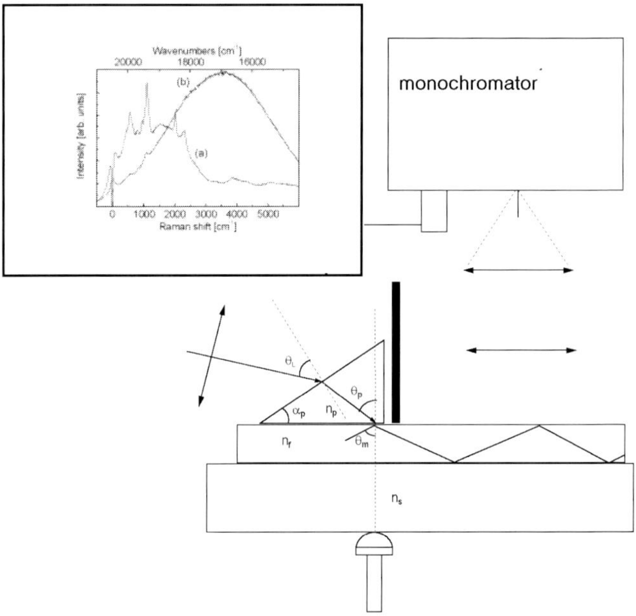

Figure 1. Experimental setup used for Raman, Brillouin and luminescence measurements in waveguide configuration. The laser light is injected into the guide by prism coupling as in the typical arrangement used for *m-line* spectroscopy.

This configuration allows an appreciable increase of the contrast as well as selectivity of both mode and polarization. Detailed discussions about the wave-guiding geometry are reported in several books and review articles.[15-21] Raman and Brillouin spectroscopies, combined with the values of the optical parameter obtained by modal measurements, are a powerful non destructive tool for the structural characterization of the sample, in particular in the case of the multicomponent systems. As a more specific example, the dependence of Brillouin intensity with strain has been applied to distributed fiber sensing.[22,23] In this context, we have to note that although the temperature sensitivity of the Brillouin backscattered signal is roughly 0.3% K^{-1}, compared to the Raman sensitivity of 0.8% K^{-1}, the intensity of Brillouin backscattered light is an order of magnitude greater than Raman scattering, which is already used in commercial instruments.[22, 23] The interest of absorption and luminescence spectroscopy is obvious in photonics. In particular, absorption and luminescence techniques allow to determine the spectroscopic properties of the optical species embedded in the matrix.[24,25] Papers concerning the application of the lùminescence in the field of optical sensors are practically countless. Among these applications, we recall the technique of time resolved fluorescence spectroscopy[11] and the fundamental role played by rare-earth doped optical sensor materials.[6,26,27] The fundamentals of optical spectroscopy, both theoretical and experimental, can be found in several textbooks and scientific papers, for instance in Refs. 24, 25, 28-31 and references therein.

2. Absorption, Reflectance and Transmission Measurements

Absorbance spectra are fundamental to determine the factors governing the sensitivity of optical sensors. The simplest schema is to measure the decrease of the optical power of the beam travelling in the material as a function of wavelength, λ. In the linear behavior, the absorption coefficient α for a sample of thickness L is given by the Lambert-Beers law.[31] In the discussion of the absorption spectra other quantities such as transmittance, optical density, extinction coefficient, and absorption cross section are currently used. Details concerning the definition of these quantities are reported in Ref. 31.

As an example, we discuss a chalcogenide-based fiber optical sensor.[7] The compound selected for testing the sensitivity was ethanol, which has absorption bands between 9 and 12 μm, corresponding to the transparency window of the used fiber. The experimental apparatus consisted in a FTIR spectrometer coupled to an IR glass fiber which was tapered in the middle, therefore allowing a very small bending radius and increased sensitivity.

This probe is introduced into the liquid or the gas to be analyzed after being immobilized on a metallic or glass support. The IR signal which has been absorbed by the chemical species in contact with the fiber was analyzed by a suitable detector.[7] The ideal optical situation for this class of sensors is to have a quasi-single mode regime in the sensing zone. This single mode would correspond to a fiber diameter two or three times larger than the analytical wavelength. In this case, most infra red light will propagate on the surface of a fiber, giving a more evanescent absorption. This effect was experimentally demonstrated by Hocdé et al.[7] performing infrared analysis using tapered fibers of different diameter. The absorbance results indicated that the sensitivity of the sensor was inversely proportional to the fiber diameter, as expected from the theory of the evanescent wave propagation.[7] Other examples of sensors based on selective absorption can be found in Refs.1, 2, 7, 8, 10, 32-36.

Reflectance and transmission measurements are employed in several sensing applications, in particular for photonic crystals-based systems where tunability in the optical response is used to detect physical or chemical modifications. John Ballato and Andrew James reported on the temperature dependence of refractive index (dn/dT) used to tune the optical response of a sol-gel-derived photonic structure.[37] Authors observed photonic bandgap effects in a disordered array of sol-gel-derived SiO_2 particles prepared via centrifugation of a colloidal suspension. The tunability in the optical response was achieved by infiltrating into the interparticle void space an organic liquid (1-methylnaphthalene) that possesses a moderately high temperature coefficient of the refractive index ($\mathrm{d}n/\mathrm{d}T$). The varying $\mathrm{d}n/\mathrm{d}T$ value between SiO_2 and 1-methylnaphthalene was used to tune the temperature dependent contrast ratio, resulting in an optical temperature sensor based on adaptive ceramic photonic crystal.

Methylnaphthalene was chosen because it possesses a relatively large temperature coefficient of the refractive index (approximately -5×10^{-4}/K). As the temperature was increased, the refractive index of methylnaphthalene, nominally $n = 1.607$ at a wavelength of 656 nm and 25 °C temperature, decreased ~ 50 times more than that of SiO_2 for a given temperature change. The result is an increase in transmission with increased temperature, because the relative difference in refractive index (Δn) between methylnaphthalene and SiO_2 is decreasing.

Figure 2 shows the measured relative change in transmission as a function of temperature at wavelengths of 550 and 600 nm. About 35%

relative percent change in transmission is observed over a temperature range of 12 K. As expected, the transmission of the methylnaphthalene-infiltrated SiO_2 composite increases as the temperature increases. More importantly, the temperature dependence is quite high, which enables greater precision for optical temperature sensors that are made according to this configuration.

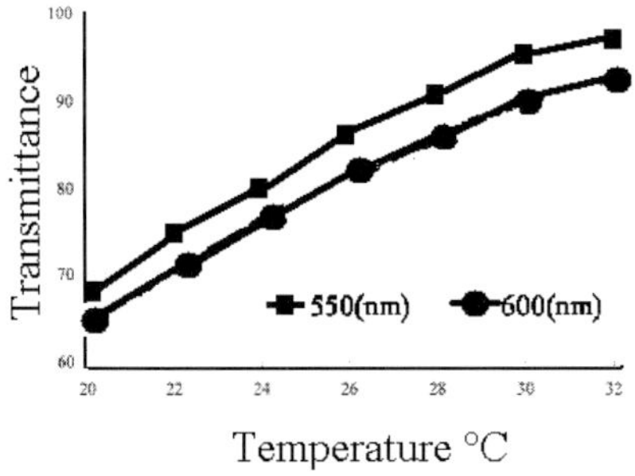

Figure 2. Relative change in percent transmission with temperature of sol gel-derived, 1-methylnaphthalene-infiltrated SiO_2 photonic crystal (Reprinted from J. Ballato, and A. James, J. Am. Ceram. Soc., 82 (1999) pp. 2273-2275).[37]

The unique properties of colloidal opals were investigated for potential applications as optical sensing of strain. Otto Pursiainen et al. fabricated an opal structure consisting of hard polystyrene (PS) cores, covered by a polymethylmethacrylate (PMMA) interlayer, and a polyethylacrylate (PEA) shell.[38] Self-assembly processes occur during the shearing by uniaxial compression of the precursor melt resulting in fcc crystallization of the PS PMMA cores, with the soft PEA shell material filling the spaces between the PS-PMMA lattice sites thus forming an elastic film with the (111) plane of the fcc lattice parallel to the sample surface.

The final shape of the system presented by Pursiainen et al. was a disk with a diameter of 10 cm and a thickness of around 250 μm.[38] In order to evaluate the properties of the flexible absorptive photonic crystals for sensor applications, authors measured reflection and transmission properties as a function of strain. The shifting of the (111) plane band gap is clearly visible in both transmission and reflection spectra reported in Fig. 3, as the strain is increased from 0% to 2%, 4%,

8%, and 13%. The reflection spectra show an identical shift of the (111)-plane Bragg-peak towards shorter wavelengths, reaching a − 5% wavelength shift at the maximum + 13% strain.

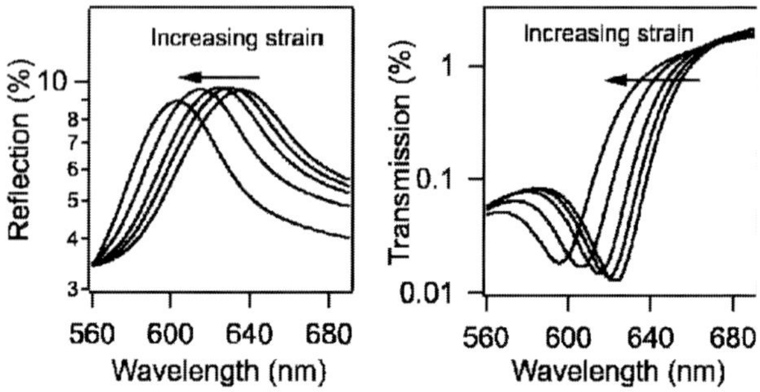

Figure 3. Strain-induced shifts in reflection (left) and transmission (right) of the absorbing photonic crystal, as the strain is increased from 0% to 2%, 4%, 8%, and 13% (Reprinted with permission from Otto L. J. Pursiainen, Applied Physics Letters, 87, 101902 (2005). Copyright 2005, American Institute of Physics.).[38]

Fudouzi and Sawada improved the system presenting an innovative silicone rubber sheet that exhibits tunable and reversible structural color due to mechanical strain.[39] Periodically arranged latex particles were embedded in a silicone rubber so that they do not come in contact with each other, thus assuring that the interparticle distance can sensitively and reversibly change with the elastic deformation of the rubber matrix.

Figure 4 illustrates well the concept of tuning the structural color of the composite film. The polystyrene spheres are closely packed and they were spaced out by filling an elastic polymer. When the elastic polymer is stretched by mechanical stress, the lattice constants increase slightly in the horizontal direction but decrease in the vertical direction. Light is selectively diffracted by the array of planes of the spheres and the stop band position λ shifts according to the Bragg equation $\lambda = 2d\sqrt{(n_e^2 - \sin^2\theta)}$, where d is the interplanar spacing of the planes, n_e is the effective refractive index , and θ is the angle of incident light. The structural color of the sheet due to Bragg diffraction and its peak position correspond to the elongation of the photonic rubber sheet.

Figure 5 shows the relationship between diffraction peak and mechanical strain. The peak position shifts from 590 to 560 nm as a function of mechanical strain $\Delta L/L$ up to 20%.

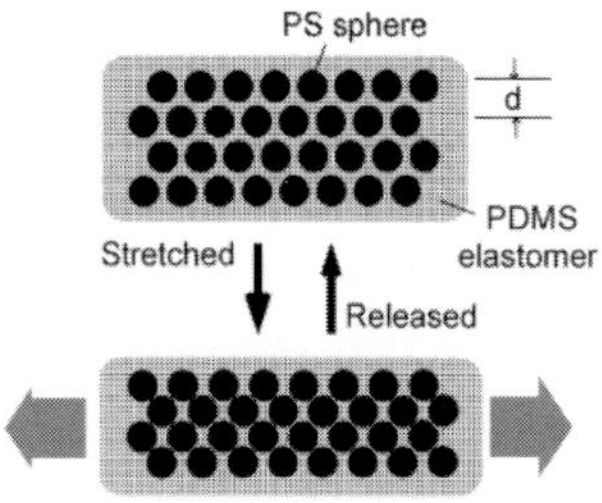

Figure 4. Elastic deformation of the colloidal crystal composite film. Concept of reversible tuning lattice distance of polystyrene, PS, array embedded in a poly (dimethylsiloxane), PDMS, elastomer matrix due to stretching and shrinking. (Reprinted from Fudouzi et al. Langmuir 22 (2006) pp. 1365-1368).[39]

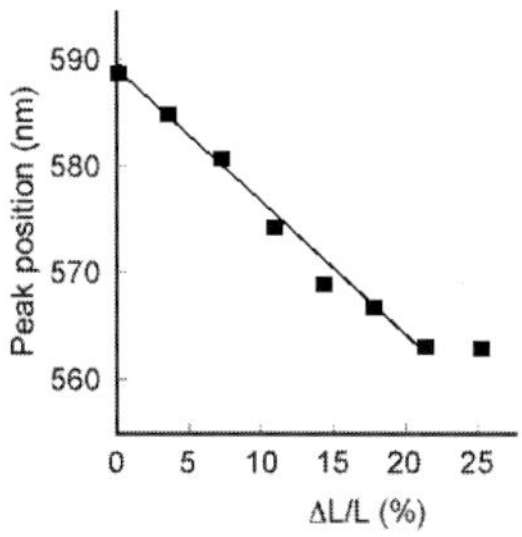

Figure 5. Relationship between the peak position and elongation of the silicone rubber as a function of the mechanical strain. (Reprinted from Fudouzi et al. Langmuir 22 (2006) pp. 1365-1368).[39]

3. Luminescence Measurements

The spectroscopic properties of the systems activated by cromophores ions, such as emission quantum efficiency, lifetime of the excited electronic states, dynamical processes i.e. non-radiative relaxation mechanism, up conversion and cooperative processes, are investigated by luminescence spectroscopy. In particular, rare-earth doped materials are used in a large number of optical sensors because of the large number of absorption and emission bands available using various rare-earth elements.[27] For instance, the variation in the green intensity ratio between $^2H_{11/2}$ and $^4S_{3/2}$ energy levels to the ground state of Er^{3+} ions has been used for temperature sensors.[6, 40-42] Many optical sensors described so far use luminescence spectroscopy for sensing chemical species, in which the luminescence intensity of an indicator is dependent on the concentration of the respective analyte.[27,35,43,44] However, the disadvantage of intensity-based techniques is that they suffer from variations in the intensity of the light sources and the sensitivity of the detector. A possibility for overcoming these problems is to use luminescence decay time of indicator cromophores for sensor purposes and time-resolved fluorescence spectroscopy has been successful applied in chemical sensors.[11] A detailed discussion about luminescence measurements is given in reference 31.

In practice, fluorescence experiments are performed under continuous or pulsed excitation. In the first case the system is considered in equilibrium i.e. the population density of the excited state is constant.[31] By recording the fluorescence signal at a certain frequency, while varying the frequency of the exciting radiation, the so called excitation spectrum is obtained. Such a spectrum can be correlated to the absorption spectrum of the system. The excitation spectrum is identical to the absorption spectrum if the molecules decay rapidly from any higher excited state to the emitting state, which is known as Kasha's rule.[45] If the excitation spectrum shows deviation from the absorption spectrum the measure is an indication of inhomogeneity of the spectral behavior. Measurements of the excitation spectrum are important in multi-site systems, and can uncover the bands responsible for the energy storage and subsequent emission of the radiative energy.

Under pulsed excitation interesting information about the relaxation mechanisms can be obtained. In particular, the fluorescence decay time or lifetime of the emitting state can be determined.[31]

The total luminescence energy emitted in a sufficiently long time interval is proportional to the total energy of the exciting light in the same interval, and the proportional constant corresponds to the luminescence yield of the observed state. The correct definition for the quantum yield Φ of a luminescent system is the ratio of the spontaneous emitted photons per unit time N_{em} divided by the total number of absorbed photons N_{abs} per unit time.[31] The absolute measurement of the quantum yield is very difficult because of the geometry of the emission and the presence of re-absorption.

It is usually measured in direct comparison to samples with known quantum yields or by using an integrating sphere.

As an example, Table 1 gives the quantum yields of some materials.

The trivalent lanthanide ions, in particular Eu^{3+}, Tb^{3+} and Yb^{3+}, have been used as luminescence sensors for pH, pO_2, OH, NH and CH using the radiative rate constants for depopulation of their excited states.[43,44]

Material	λ_{exc} (nm)	λ_{emi} (nm)	τ_{meas} (ns)	Φ
p-terphenil in cyclohexane[29]	275	340	0.95	0.93
Anthracencee in cyclohexane[29]	340	400	4.9	0.27
Perylene in cyclohexane[29]	410	470	6.4	0.94
Acridine yellow in ethanol[29]	480	500	5.1	0.86
Acridine red in ethanol[29]	550	590	3.8	0.33
$[EuL^1]$[44]	300-375	594	710 x10^3	0.011
$[EuL^1H]^+$[44]	300-375	594	720 x10^3	0.03
$[TbL^1]$[44]	300-375	547	980 x 10^3	0.025
$[TbL^1H]^+$[44]	300-375	547	100 x 10^3	9.1 x 10^{-4}

Table 1. Quantum yields, excitation and emission wavelengths λ_{exc} and λ_{emi}, and measured fluorescence lifetime τ_{meas} of some materials (Reprinted from Menzel R. *Photonics Linear and Nonlinear Interactions of Laser Light and Matter*, Springer-Berlin, 2001, p. 513[29] and Parker D., Senanayake P. K. and Williams J. A. G. *Luminescence sensors for pH, pO_2, halide and hydroxide ions using phenanthridine as a photosensitizer in macrocyclic europium and terbium complexes* J. Chem. Soc., Perkin Trans. 2 (1998) pp. 2129-2139[44]).

The technique is based on the coupling of the electronic states of the rare-earth ion to the vibrational overtones of proximate oscillators. For instance, in the case of the Eu^{3+} ion the energy gap between the $^5D_0$ state and the ground state manifold is approximately 12 000 cm^{-1} which corresponds to three vibrational quanta of the OH stretching (3500 cm^{-1}).

Following the suggestion of Horroks and Sudnik,[46] the rate of non-radiative relaxation of the $^5D_0$ state can give a measure of the number of molecules of water coordinated to Eu^{3+} in a liquid environment. The average number $N(H_2O)$ of water molecules can be obtained using the phenomenological relation

$$N(H_2O) = \frac{A_{Ln}}{\tau_{O-H}} \tag{1}$$

where $\tau_{O-H}^{-1} = \tau_{meas}^{-1} - \tau_r^{-1}$ is the non-radiative transition probability due to the OH vibrational modes and A_{Ln} is a constant which depends on the lanthanide ion (A_{Eu}=1.05 and A_{Tb}=4.2).[46] τ_{O-H} can be obtained from the measured lifetime τ_{meas} if the radiative rate τ_r^{-1} is known. The latter can be estimated by the luminescence spectra by measuring the ratio of magnetic dipole emission intensity ($^5D_0 \rightarrow {}^7F_1$) I_{DM}, to the total emission intensity I_{Tot}, and by taking for the magnetic dipole radiative lifetime τ_{MD}^{R} the value of 20 ms, assumed to be independent on the environment:

$$\frac{I_{DM}}{I_{Tot}} = \frac{\tau_r}{\tau_{MD}^{r}} \tag{2}$$

Another example of time-resolved spectroscopy-based sensors is given in Ref. 11. In general, the fluorescence decay function $\phi(t)$ is not exponential because of the local environment inhomogeneities. In the more general case, when the relaxation refers to a number of locally active channels varying throughout the system, the stretched exponential or Kohlrausch model[47,48] is used:

$$\phi(t) = \phi(0)\left[\exp\left(\frac{t}{\tau}\right)^{\beta}\right] \tag{3}$$

where $0 < \beta < 1$. In the simplest case, where the distance between the fluorophore and the interaction sites is homogeneously distributed, the total decay function is calculated by the integration of the differential equation that describes the time evolution of the excited state and the summing up over all sites:[49]

$$\phi(t) = \phi(0)\exp\left[-\frac{(1+c)t}{\tau} - a\left(\frac{t}{\tau}\right)^{\frac{1}{2}}\right] \tag{4}$$

where a is the interaction parameter between cromophore and matrix and c is the quenching parameter and depends on the concentration of the chemical species to detect. In the case reported in Ref. 11, c was dependent on the concentration of oxygen, the system was pyrene in PVC with $\tau = 293$ ns and $a = 0.73$. The proposed oxygen sensor had a resolution of 5 hPa. If two different components of $\phi(t)$ can be observed and the relation

$$\phi(t) = A\phi_A(t) + B\phi_B(t) \tag{5}$$

is valid, then A and B describe the relative amounts of the two chemical species present in the system. Note that the difference between the decay times corresponding to the forms A and B must be significant, a factor 100 being a good value of reference.

We conclude this section reminding that one of the most widespread applications of rare-earth doped materials as sensors is in the field of temperature sensing.[27] The classical method consists in detecting the green intensity ratio between $^2H_{11/2} \rightarrow {}^4I_{15/2}$ and $^4S_{3/2} \rightarrow {}^4I_{15/2}$ transitions of Er^{3+} ions as a function of the temperature. In this context Maurice et al. presented an erbium-doped silica fiber for intrinsic fiber-optic temperature sensor.[40] The sensor principle is based on the thermalization among the $^2H_{11/2}$ and $^4S_{3/2}$ energy levels of the Er^{3+} ions, which has been observed in a variety of hosts. With the thermalization of the population of the $^2H_{11/2}$ and $^4S_{3/2}$ energy levels and avoiding as well as possible self-absorption processes,[50] the ratio of the integrated fluorescence intensities $^2H_{11/2} \rightarrow {}^4I_{15/2}$ and $^4S_{3/2} \rightarrow {}^4I_{15/2}$ is given by

$$R = \frac{I_H}{I_S} = \frac{c_H(\upsilon)}{c_S(\upsilon)} \frac{p_H^r}{p_S^r} \frac{g_H}{g_S} \frac{h\upsilon_H}{h\upsilon_S} \exp\left(-\frac{\Delta E}{kT}\right) \tag{6}$$

where I_H and I_S are the measured intensities, g_H and g_S are the degeneracies (2J+1) for a given $^{2S+1}L_J$ level, and $p_{H,S}^r$ are the total spontaneous emission rates of the $^2H_{11/2}$ and $^4S_{3/2}$ energy levels, respectively. The response of the detection system in the employed frequency region is given by $c_{H,S}(\upsilon)$. $h\upsilon$ is the photon energy, ΔE is the energy difference between the involved levels, with k the Boltzmann constant and T the temperature in K.

Considering only temperature dependence and with the hypothesis that the spontaneous emission rates are temperature independent, Eq. (6) can be simplified as:

$$R = a \exp\left(-\frac{\Delta E}{kT}\right) \quad \text{or} \quad \ln(R) = c - \frac{b}{T} \tag{7}$$

where a, b, and c are constants for a particular host. Consequently, it is possible to determine the temperature from a measurement of the green intensity ratio based on two calibration points using:

$$T = \frac{b}{c - \ln(R)} \tag{8}$$

Using this approach and the experimental set-up shown in Fig. 6, Maurice et al. obtained a temperature sensor with a dynamic of 11 dB and a sensitivity of 0.016 dB/°C from room temperature to 600 °C.

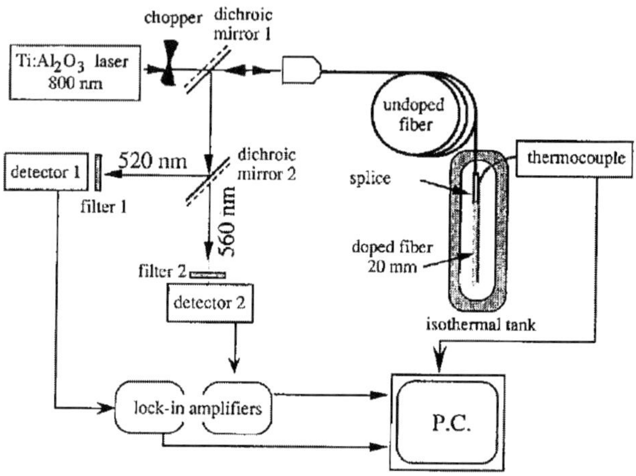

Figure 6. Schematic experimental arrangement for Er^{3+}-based fiber-optic temperature sensor. (Reprinted from Maurice et al., Applied Optics 34 (1995) pp. 8019-8025).[40]

4. Raman and Brillouin Measurements

Raman spectroscopy is a powerful tool for providing structural information. The basic schema and the energy relations for Raman scattering are shown in Fig. 7. The application of Raman spectroscopy as both a qualitative and quantitative analytical detector is limited by its inherent lack of sensitivity: approximately 1 in 10^7 photons is scattered at an optical frequency different from that of source excitation.[51] However, the combination of dielectric waveguides and Raman spectroscopy yields, for example, a useful and sensitive method with which to analyze thin indicator layers present on top of these waveguides.
Moreover, a graded-index waveguide allows to perform Raman (and luminescence) depth-selective measurements.[52]
Surface-Enhanced Raman Scattering (SERS) is largely employed in Raman-based sensors. SERS is a phenomenon in which Raman scattering cross sections are dramatically enhanced for molecules

adsorbed on nanostructured metal surfaces.[53] Silver is the most SERS-active, followed by gold, copper, and transition metals. Regarding the enhancement mechanism, both a long-range electromagnetic and a short-range chemical effect are thought to be simultaneously operative; factors of 8-10 orders of magnitude can arise from electromagnetic surface plasmon excitation, while the enhancement factor due to chemical effects is of the order of 10^1-10^2.

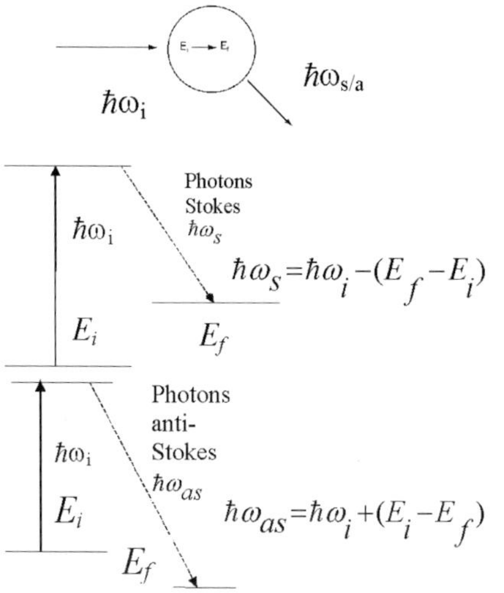

Figure 7. Basic schematic and energy relations for Raman scattering.

There is a huge amount of literature about SERS and here, as an example of Raman-based sensor we mention a liquid core Raman waveguide detector which was proposed by Marquardt et al. for liquid chromatography.[54] Thanks to a waveguiding approach they achieved detection limit enhancements of over 1000-fold for the measurement of alcohols in aqueous phase.

Arjyal and Galiotis reported on a laser Raman sensor for stress monitoring in composites.[55] This technique is based on the fact that most Raman backbone vibrational modes of crystalline fibers shift to lower values in tension and to higher values in compression. Bond extension or contraction changes the bond stiffness and hence the atomic vibrational frequencies. The magnitude of this Raman frequency shift can be related to external stress or strain, hence making possible stress strain measurements in composites. Raman sensors provide independently values of fiber stress and strain from composite sample volumes as small as 1 mm^3. In addition, this is the only technique that can directly measure stress in composites as most of the currently available non-destructive methods can only provide strain measurements. The experimental procedure consists by plotting the shift of Raman frequency as a function of axial stress or strain. Using this approach Arjyal and Galiotis observed a linear dependence of the Raman frequency shift $\Delta\omega$ on the applied tensile stress ε in the bulk of a composite laminate with a resolution of 3.65 cm^{-1}/GPa in the range 0-3 Gpa.[55]

Xia et al. used Raman backscattering to determine the temperature distribution along a fiber.[56] Raman scattered light is caused by thermally influenced molecular vibrations. It is sensitive to temperature but not to strain, consequently the backscattered light carries the information on the local temperature where the scattering occurred, thus the temperature along the fiber can be determined. To accurately predict the temperature changes, the Raman signal has to be referenced to a temperature independent signal measured with the same spatial resolution. Xu et al. retrieved the temperature along the fiber by measuring the power ratio of the Stokes to anti-Stokes Raman backscatters in the time domain. The equation relating the power ratio Γ_R to the temperature is given by

$$\Gamma_R = \frac{P_{AS}}{P_S} = \left(\frac{\lambda_S}{\lambda_{AS}}\right)^4 \exp\left(-\frac{h\Delta\upsilon}{kT}\right) \tag{9}$$

where P_{AS} and P_S are anti-Stokes and Stokes power measured at the corresponding wavelengths λ_{AS} and λ_S. Xia et al. measured at room temperature a Raman shift $\Delta\upsilon = 11.9\,THz$ THz and a $\Gamma_R = 0.22$ along a single mode fiber.

An exhaustive discussion of the Brillouin scattering in solid is reported in Ref. 57. The Brillouin scattering results from the interaction between the incident light beam and thermally generated acoustic waves. The sensing fiber temperature may derived from the ratio of the elastic (Rayleigh) scattering to Brillouin scattering which is known as the Landau Placzek Ratio and is given by:[22,23]

$$LPR = \frac{T_f}{T}\left(\beta_T\, \rho\, v_a^2 - 1\right) \qquad (10)$$

were T_f is the fictive temperature, β_T is the isothermal compressibility, ρ the material density, v_a the acoustic velocity. Reasonable values for commercially available fiber sensors, i.e. T_f = 1943 K, ρ=2200 Kg m^{-3}, β_T = 7x10^{-11} m^2 N^{-1}, v_a = 5960 m s^{-1}, given LPR ≈ 30. The temperature sensitivity of the Brillouin scattering intensity is 0.3% K^{-1} (Refs. 22, 23).

Xia et al. employed Brillouin scattering to analyze the strain information within measurement range of 2000 με strain and 400 °C temperature.[56] The Brillouin shift, spectrum width and intensity of Brillouin backscatter are sensitive to temperature and strain. To avoid this problem Xia et al. considered the relationship between the strain experienced by the sensor fiber and the transmittance of a Fabry-Perot interferometer for the Brillouin backscatter with knowledge of the temperature profile of the fiber. With the look-up table of response functions of the detected temperature, the strain profile along the fiber can be retrieved using the strain response function at known temperature defined by

$$R(\varepsilon)_{TD} = \frac{\zeta(\varepsilon)_{TD}}{1-\xi(\varepsilon)_{TD}} \qquad (11)$$

where $\xi(\varepsilon)_{T_D}$ is the Brillouin transmission that at a given temperature T_D varies only with the strain ε experienced by the fiber.

It is reasonable to think that the Brillouin technique can be employed in a planar device. However, for planar waveguide excitation the discussion of Brillouin scattering measurements requires an *ad hoc* model which is reported in Ref. 58. The wave propagation in the planar waveguide is shown in Fig. 8.

Brillouin scattering from acoustic vibration depends on the exchanged q vector of the scattered photon and is a coherent process which involves the whole illuminated region. In a homogeneous waveguide the light propagates along a zig-zag path so that at a fixed angle of detection two values of q are sampled. Since the two q values change with the mode index, Brillouin spectra present doublets at different frequencies for different excitation modes.[58]

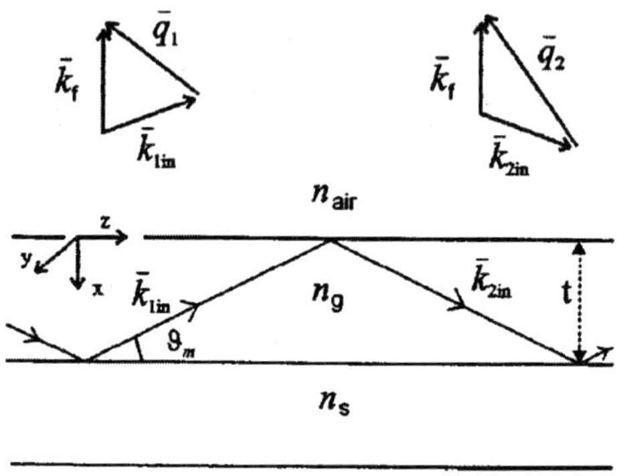

Figure 8. Wave propagation in the planar waveguide. $\vec{q}_1$ and $\vec{q}_2$ are the exchanged wave vectors of scattered light in the zig and zag paths. (Reprinted figure with permission from Montagna et al., Phys. Rev. B 58 pp. R547-R550, 1998. Copyright (2008) by the American Physical Society).[58]

Under these conditions the Brillouin scattering intensity I_{BS} is given by:[58]

$$I(\vec{q}_1, \vec{q}_2, \omega) \propto \frac{T}{\omega^2} \left| \int q_1 \, e^{i(\vec{k}_p - \vec{q}_1)\vec{r}} + e^{i\gamma} \, q_2 \, e^{i(\vec{k}_p - \vec{q}_2)\vec{r}} \, d\vec{r} \right|^2 \qquad (12)$$

where T is the temperature, $\omega = \mathrm{v}_{\mathrm{L,T}} \, k_p$, and $\mathrm{v}_{\mathrm{L,T}}$ is the longitudinal, transverse sound velocity, $\vec{k}_p$ the wave vector of the phonons which scatter the light and γ is the relative phase of the zig and zag fields in the mth mode of the guide. The linear dependence of I_{BS} on T is always present in Eq. 12, however the Brillouin spectrum of a glass waveguide will show four peaks in the Stokes and four in the anti-Stokes spectrum. Two peaks are due to longitudinal phonons and the other two to transverse phonons.[58]

5. Conclusions

The present chapter gives a very short survey of the possibilities of application in sensing, offered by well consolidated optical spectroscopy techniques such as absorption, photoluminescence, Raman and Brillouin scattering. Some innovative aspects related to the exploitation of photonic crystals and colloidal self-assembled systems have been mentioned.

References

1. K.T.V. Grattan, and B.T. Meggit, *Optical Fiber Sensors Technology* London, Chapman & Hall (1995).
2. J. Lin, , *Trends in analytical chemistry,* **19**, 541 (2000).
3. K.T.V. Grattan, and T. Sun, , *Sensors and Actuators*, **82**, 40 (2000).
4. A.G. Mignani, and F. Baldini, *Rep. Prog. Phys.*, **59**, 1 (1996).
5. H. H. Gao, Z. Chen, J. Kumar, S. K. Tripathy and D. L. Kaplan, *Opt. Eng.*, **34**, 3465 (1995).
6. E. Maurice, G. Monnom, B. Dussardier, A. Saïssy, D.B. Ostrowsky, and G.W. Baxter, *Appl. Opt.*, **34**, 8019 (1995).
7. S. Hocdé, C. Boussard-Plédel, G. Fonteneau, D. Lecoq, H. L. Ma and J. Lucas, *J. Non-Cryst. Solids*, **274**, 17 (2000).
8. C. R. Lavers, K. Itoh, S. C. Wu, M. Murabayashi, I. Mauchline, G. Stewart and T. Stout, *Sensors and Actuators* **B 69**, 85 (2000).
9. J. Ĉtyroký, J. Homola, P. V.Lambeck, S. Musa, H. J. W. M. Hoekstra, R. D. Harris, J. S. Wilkinson, B. Usievich and N. M. Lyndin, *Sensors and Actuators* **B 54**, 66 (1999).
10. L. Yang, S.S. Saavedra and N. R. Armstrong, *Anal. Chem.*, **68**, 1834 (1996).
11. S. Draxler and M. E. Lippitsch, *Appl. Opt.* **35**, 4117 (1996).
12. J. B. Marquardt, P. G.Vahey, R. E. Synovec and L.W. Burgess, *Anal. Chem.*, **71**, 4808 (1998).
13. U. J. Gibson and M. Chernuschenko, *Optics Express*, **4**, 443 (1999).
14. K. Tsunoda, S. Kikuchi, K. Nomura, K. Aizawa, K. Okamoto and H. Akaiwa, *Analytical Science*, **15**, 241 (1999).
15. S. Pelli and G. C. Righini, in *Advances in Integrated Optics*, Edited by S. Martellucci, A. N. Chester and M. Bertolotti, New York: Plenum Press, pp. 1-20 (1994).
16. P. K. Tien, *Reviews of Modern Physics*, **49**, 361 (1997).
17. S. B. Mendes and S. S. Saavedra, *Optics Express*, **4**, 449 (1999).
18. R. E. Kuntz in *Integrated Optical Circuits and Components Design and Applications*, Edited by E. J. Murphy, New York, Marcel Dekker Inc., 335 (1999).
19. M. Ferrari, F. Gonella, M. Montagna and C. Tosello, *J. Appl. Phys.*, **79**, 2055 (1996).

20. M. Ferrari, F. Gonella, M. Montagna and C. Tosello, *J. Raman Spectrosc.*, **27**, 793 (1996).
21. C. Duverger, S. Turrell, M. Bouazaoui, F. Tonelli, M. Montagna and M. Ferrari, *Phil. Mag.* **B 77**, 363 (1998).
22. P. C. Wait and T. P. Newson, *Optics Communication*, **122**, 141 (1996).
23. P. C.Wait, K. De Souza and T. P. Newson, *Optics Communication*, **144**, 17 (1997).
24. *Laser Spectroscopy of Solids* Edited by M. W. Yen and P. M. Selzer, Berlin: Springer (1981).
25. *Optical Spectroscopy of Glasses* Edited by J. Zschokke, Dordrecht, Reidel, (1986).
26. F. J. McAleavery, J. O'Gorman, J. F. Donegan, B. D. MacCraith, J. Hegarty and G. Mazé, *IEEE J. of Selected Topics in Quantum Electronics*, **3**, 1103 (1997).
27. B. G. Jr. Potter and M. B. Sinclair, *J. of Electroceramics*, **2**, 295 (1998).
28. W. Demtröder, in *Laser Spectroscopy, Basic Concepts and Instrumentation* Berlin, Springer (1996).
29. R. Menzel, in *Photonics Linear and Nonlinear Interactions of Laser Light and Matter* Berlin, Springer, 2001.
30. B. Di Bartolo, *Optical Interaction in Solids* New York: John Wiley & Sons Inc., (1967).
31. G. C. Righini and M. Ferrari, *Rivista del Nuovo Cimento*, **28**, 1 (2005).
32. P. J. Skrdla, S. S.Saavedra, N. R. Armstrong, S. B. Mendes and N. Peyghambarian, *Analytical Chemistry*, **71**, 1332 (1999).
33. K. Tóth, G. Nagy, B. T. T.Lan, J. Jeney and S. J. Choquette, *Analytica Chimica Acta*, **353**, 1 (1997).
34. J. Lin and C. W.Brown, *Trends in analytical chemistry*, **16**, 200 (1997).
35. P. L.Edmiston, C. L.Wambolt, M. K. Smith and S. S. Saavedra, *J. of Colloid and Interface Science,* **163**, 395 (1994).
36. X. M. Chen, K. Ithoh, M. Murabayashi and C. Igarashi, *Chemistry Letters*, **2**, 103 (1996).
37. J. Ballato and A. James, *J. Am. Ceram. Soc.*, **82**, 2273 (1999).
38. O. L. J. Pursiainen, J. J. Baumberg, K. Ryan, J. Bauer, H. Winkler, B. Viel and T. Ruhl, *Appl. Phys. Lett.,* **87**, 101902-1/3 (2005).
39. H. Fudouzi and T. Sawada, *Langmuir* **22**, 1365 (2006).
40. E. Maurice, G. Monnom, B. Dussardier, A. Saïssy and D. B. Ostrowsky, *Opt. Lett.*, **19**, 990 (1994).
41. G. S. Maciel, L. De S. Menezes, A. S. L. Gomes, C. B. de Araújo, Y. Messaddeq, A. Florez and M. A Aegerter, *IEEE Photonics Technology Letters*, **7**, 1474 (1995).
42. P. V. Dos Santos, M. T. de Araujo, A. S. Giuveia-Neto, J. A. Medeiros Neto and A. S. B. Sombra, *Appl. Phys. Lett.*, **73**, 578 (1998).

43. A. Beeby, I. M. Clarkson, R. S. Dickins, S. Faulkner, D. Parker, L. Royle, A. S. de Sousa, J. A. G. Williams and M. Woods, *J. Chem. Soc., Perkin Trans.*, **2**, 493 (1999).
44. D. Parker, P. K. Senanayake and J. A. G. Williams, *J. Chem. Soc., Perkin Trans.*, **2**, 2129 (1998).
45. M. Kasha, *Disc. Faraday Soc.*, **9**, 14 (1950).
46. W. De W. Jr. Horrocks and D. R. Sudnik, *Acct. Chem. Res*, **14**, 384 (1981).
47. D. L. Huber, *Mol. Cryst. Liq. Cryst.* **291**, 17 (1996).
48. D. L. Huber, *Phys. Rev.* **B 31**, 6070 (1985).
49. S. Draxler and M. E. Lippitsch, *Sensors Actuators* **B 29**, 199 (1995).
50. M. Mattarelli, M. Montagna, L. Zampedri, A. Chiasera, M. Ferrari, G. C. Righini, L. M. Fortes, M. C. Gonçalves, L. F. Santos and R. M. Almeida, *Europhys. Lett.*,**71**, 394 (2005).
51. L. A. Woodward, *Raman Spectroscopy*, **4**, New York, Plenum Press (1967).
52. M. Ferrari, M. Montagna, S. Ronchin, F. Rossi and G. C.Righini, *Appl. Phys. Lett.*, **75** 1529 (1999).
53. R. K. Chang and T. E. Furtak, in *Surface Enhanced Raman Scattering*, Plenum Press, New York (1982).
54. B. J. Marquardt, P. G. Vahey, R. E. Synovec and L. W. Burgess, *Anal. Chem.*, **71**, 4808 (1999).
55. B. Arjya and C. Galiotis, in *Application of a laser Raman sensor for stress monitoring in composites,* SPIE Proceedings, **2779**, 142 (1996).
56. H. Xia, H. Mu, Y. Yang and D. Sun, SPIE Proceedings , 683021-1/6 (2007).
57. J. D. Dil, *Rep. Prog. Phys.*, **45**, 285 (1982).
58. M. Montagna, M. Ferrari, F. Rossi, F. Tonelli and C. Tosello, *Phys. Rev.* **B 58** R547-R550 (1998).

LASER DOPPLER VIBROMETRY

Paolo Castellini, Gian Marco Revel and Enrico Primo Tomasini*

Dipartimento di Meccanica, Università degli Studi di Ancona,
Via Brecce Bianche, Località Montedago, 60131, Ancona, Italy
**E-mail: ep.tomasini@univpm.it*

In this chapter the potentials of Laser Doppler Vibrometry (LDV) technique for the measure of vibration velocity on solid surfaces are presented. LDV allows to perform vibration measurements without contact, eliminating the intrusivity of traditional devices. This makes LDV suitable for many applications, from noise&vibration analysis in automotive domains up to works of art diagnostics. Different LDVs configurations are discussed (i.e. single-point, differential in fibre, scanning system, in-plane and rotational). A description of the optical schemes and signal processing strategies is given.

1. Introduction

During the last decade, laser-based techniques for measurement and analysis of mechanical vibrations have become widely and intensively investigated because of vibrations relevance in a very large number of industrial applications, from process monitoring to on line diagnostics and quality control.
Moreover the problems connected with noise emission, vibrations induced fatigue, vibration isolation are of great importance in the automotive and aerospace areas, and this has given a new boost to research. In addition, the possibility of measuring vibrations is becoming of significant importance also in many fields outside industry, such as the diagnostic of artworks and the biomedical applications.
Laser technology has attracted a special interest because of the increasing request for non-intrusive measurement techniques, which guarantee the absence of any kind of alterations on the measurand under investigation.
In fact, the non-intrusivity is becoming a fundamental feature in the study of mechanical vibrations of very small and light objects (e.g.

microchips, hard disks, human tympanic membranes, etc.), of highly damped non linear materials (as rubber), where the conventional transducers, based on the use of accelerometers, strain gages, triangulation or reflection sensors and proximity, become difficult to be employed and always involve complex and costly installations, and often cannot provide the desired information.

Nowadays, the most investigated and used laser technique is the Laser Doppler Vibrometry (LDV); this instrument is basically an interferometric device which measures the instantaneous velocity of a target through the measurement of the Doppler shift of scattered laser light coming from the vibrating object.

The first commercial LDV system was launched on the market during the 1970s by DISA[1] as a result of primitive studies presented in the late 1960s;[2-4] the device was based on an optical heterodyne detection of the Doppler shift and it was the first prototype to those available at present; its major limit was a very low optical sensitivity which allowed measurements only on very diffusive surfaces.

LDV techniques have proved to be a unique measurement instrument suited to many applications; these allowed to overcome problems related to vibration measurements, such as intrusivity, frequency response, and to implement measurements under harsh conditions (high temperature surfaces and noisy environments) or when hard-to-reach, small, or weak structures are analyzed.

Furthermore, they are considered a primary reference for the calibration of other velocity or displacement sensors (e.g. Ref. 5).

In the state of art, these devices are effectively used in structural dynamic testing, biological and clinical diagnostics, fluid-structure interaction study, on-line monitoring of industrial plants, acoustics fields, etc.

Moreover the enormous potential of LDV is being applied for new and non-conventional uses to open up new possibilities, e.g. in the field of measurements in tracking mode on rotating or arbitrarily-moving structures (e.g. Ref. 6).

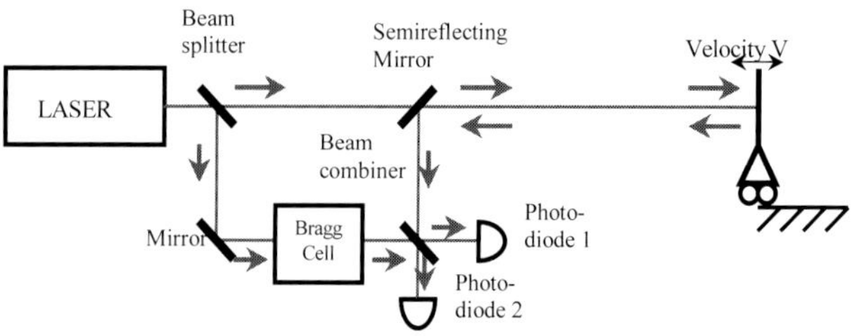

Figure 1. Optical schematic of the single-point vibrometer (Mach-Zehnder).

2. Laser Doppler Vibrometry Technique

2.1. The Laser Doppler Vibrometer: Basic Configurations for the Single Point out-of-Plane Measurement

The Laser Doppler Vibrometer is a non-contact velocity transducer, based on the analysis of the Doppler effect on a laser beam scattered from a solid moving surface. As shown in Figure 1, the vibrometer is basically composed of a laser source and an interferometer. The laser beam is focused on the vibrating surface, which diffuses the light with a frequency shift Δf_D proportional to the velocity along the laser axis, according to the equation:

$$\Delta f_D = \frac{2v}{\lambda} \tag{1}$$

where λ is the laser wavelength and v is the object velocity.

If a He-Ne (λ=632.8 nm) laser is used for the measurement of a velocity of 1 m/s, the frequency f = 4.7 10^{14} Hz is shifted by 3.16 10^{6} Hz. A dynamic range of about 6.7 10^{-9} (164 dB) is required for the detection.

The interferometric approach can be used in order to measure such relatively small frequency shift and to have a reference in the phase assessment.

The so-called "Single-point Vibrometer" is the first system developed and the most diffused vibrometer; this device is basically able to measure the velocity component along the direction of the incident laser beam.

For the extraction of the Doppler frequency information, different interferometric configurations can be adopted, in particular the most used are the Michelson and the Mach-Zehnder schemes. Usually in all schemes, the laser beam is split in two, with one beam acting as a stationary reference (reference beam), while the other beam is directed at the vibrating surface (measuring beam) and carries the velocity information through the frequency shift. The two beams are finally mixed at the photodiodes, where beating phenomena will be observed.

2.2. Signal Processing Scheme

The signal obtained from photodiodes can be a heterodyne or a homodyne signal, depending on the optical arrangement of the interferometer.

Such signal can be processed in different ways in order to extract velocity or displacement. The demodulation of the signal from photodiodes (in the heterodyne at frequency Δf_D usually in the range of 40 MHz for surface velocity up to 10 m/s) is realized basically as a down mixing of interference signal with the reference signal driving the Bragg cell. The Bragg cell is an acousto-optic device employed to impose a fixed frequency shift Δf_{BC} in the reference beam, in such a way as to eliminate the direction ambiguity in the velocity measurement. In fact, the final frequency shift emerging from the beating of the two beams will be $\Delta f_{TOT} = \Delta f_D + \Delta f_{BC}$ giving the possibility to distinguish the velocity direction.

While the frequency demodulation supplies the instantaneous velocity, the direct displacement assessment can be implemented by counting (through digital counters) the number N of "fringes", i.e. of constructive and destructive interference. In fact, when a displacement d of the target surface occurs, a path length variation $2d$ in the arm of the interferometer corresponds and the interference signal presents a phase shift (see Eq.2) directly proportional to N, given by

$$\Delta\phi = \frac{4 \cdot \pi \cdot d}{\lambda} = 2\pi \cdot N \tag{2}$$

Each fringe corresponds to a displacement of half of the wavelength λ.

If a frequency shifter, as a Bragg cell, is employed, the displacement signal is obtained through the comparison between two counters, one working on a reference signal from the shifting device and one working on the interference signal.
The interference signal is usually interpolated; in this way an improvement of resolution performances up to $\lambda/80$ is obtained.

2.3. Differential Fiber Vibrometer

The differential vibrometer is used to measure the relative vibration between two surfaces, instead of two independent measurements of single velocity. Such device is realized using both arms as reference and measurement beams at the same time. If an optical fiber is incorporated in both arms, the interferometer measures the differential movement between the two targets at which the two probe beams are focused.
The fiber optic allows to extend the applicability of this system to the analysis of surfaces which are difficult to reach and if one target is a "steady" reference object, this differential vibrometer can be used as a fiber optic single-point system.
On the other hand, the application of fibers, the reduction of the optical aperture and the fact that both arms have diffusive target yield a strong decrease in the optical signal energy and in the signal to noise ratio; therefore, it is, often, necessary to increase the optical quality of the target surface.

2.4. Scanning Laser Doppler Vibrometer (SLDV)

Scanning laser Doppler vibrometer, invented by Stoffregen and Felske in 1981,[8] is basically a coupling of a Laser Doppler single point vibrometer and a Scanning system (two orthogonal scanning mirrors and a co-ordinate controllers), which performs the directing of the laser beam (Fig. 2). In this way a single point vibrometer sensor can be used to scan across a surface, gathering multi-point data from large objects vibrating in stationary state.
To aim the measuring beam at specified locations, two voltages have to be applied to the co-ordinate controllers (motors), so that the two

scanning mirrors rotate of the desired angles: galvanometer scanners are usually employed.

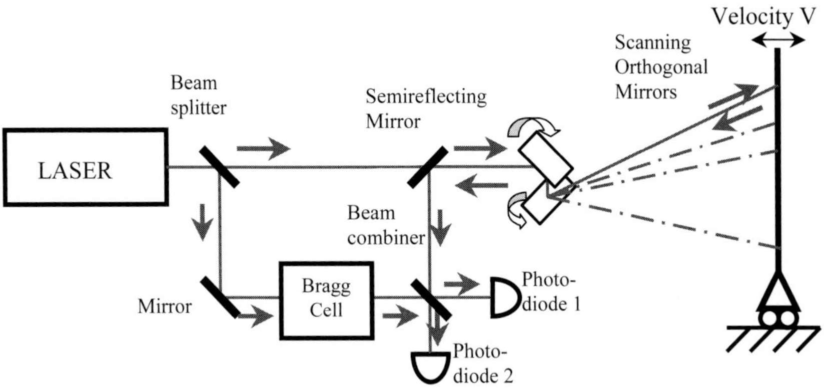

Figure 2. Optical schematic of the Scanning vibrometer.

The mirrors orientation is accurately obtained only if the spatial relationship between the test structure and the SLDV system is precisely determined.[9] In order to uniquely define a position on the test object, a calibration has to be performed by directing the laser spot to specified calibration points.

One of the features, which turned out to be extremely helpful for the calibration and the application of the SLDV, was the integration in the scanning head of a video camera. This solution allows the operator to monitor the location of the laser beam on the actual test item and to overlay the measured results on an image of the test item.

The most common scanners, which are incorporated into commercial SLDVs, have a closed loop design. A built-in capacitive rotation sensor provides position information to a servo amplifier.

An important problem in the development of SLDV is the accuracy in the positioning of the measurement spot.

Some authors[10] developed a procedure for scanners evaluation and calibration in a fast and accurate way. Martarelli et al.[11] analyzed different control algorithms and their sensitivity to alignment procedure and operative conditions.

2.5. In-Plane Vibrometer

The in-plane vibrometer (Figure 3) is suitably utilized to experimentally determine in-plane or tangential vibrations.

Its functioning is based on the focusing of two laser beams in a measurement point and on the analysis of interaction between the surface roughness and the interference fringes area (measurement volume) formed in correspondence with the intersection, where an optical heterodyne phenomenon is produced. This arrangement is similar to the one commonly used in laser Doppler anemometry which measures fluid velocity. The determination of the velocity component, perpendicular to the optical axis, occurs by Doppler effects in the scattered light, which is collected by the optics to the photodetector.

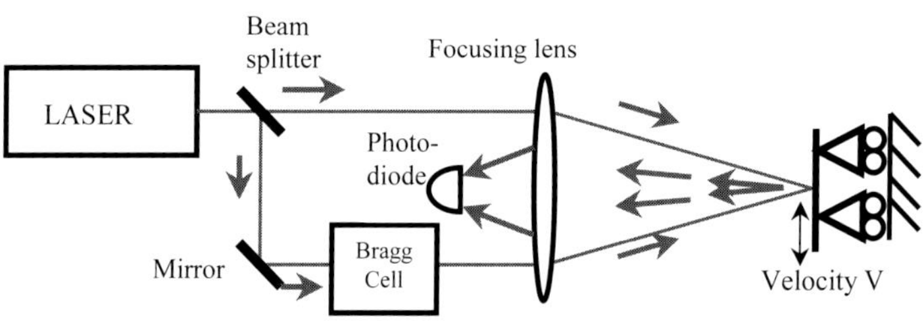

Figure 3. Optical schematic of the in-plane vibrometer.

If λ is the laser wavelength and θ is the angle between the two beams, the relation between the in-plane velocity v and the measured Doppler shift Δf_D is:

$$v = \frac{\lambda/2}{\sin(\theta/2)} \Delta f_D \tag{3}$$

A signal proportional to velocity is obtained through a frequency-voltage converter.

In order to eliminate the direction ambiguity, a Bragg cell can be introduced to induce an optical frequency shift in one beam.

Typical performances are a bandwidth from 0 to 10 kHz and a velocity range up to 100 m/s DC. The typical calibration accuracy is of about ± 0.5%.

The problems involved in the measurement principle are mainly connected with the reflecting power and the morphological characteristics of the illuminated surface of the vibrating structure.[12]
The reflective power and the roughness influence respectively the signal-to-noise ratio on the signal to be demodulated and the spatial scattering pattern; the in-plane vibrometer suffers also from speckle phenomenon.[13]
The knowledge of in-plane velocity fluctuations is important for a large number of mechanical and industrial applications (e.g. paper production plant or printing processes) and therefore such instruments are becoming widely studied (e.g. Ref. 14).

2.6. Rotational Vibrometer

The rotational vibrometers are based on the design presented by Halliwell[15] to optically measure the angular velocity and to analyze the torsional vibrations (Figure 4). Two parallel interferometric measurement beams, with separation *d*, acquire the velocity components, v_A and v_B, in the direction of the beams in two different points of a rotating object; from the knowledge of these velocity components, the instantaneous value of the rotational speed is easily computed.
In fact, each point on the perimeter of a rotating part of any shape has a tangential velocity v_t depending on the rotational radius R and on the angular velocity, Ω.

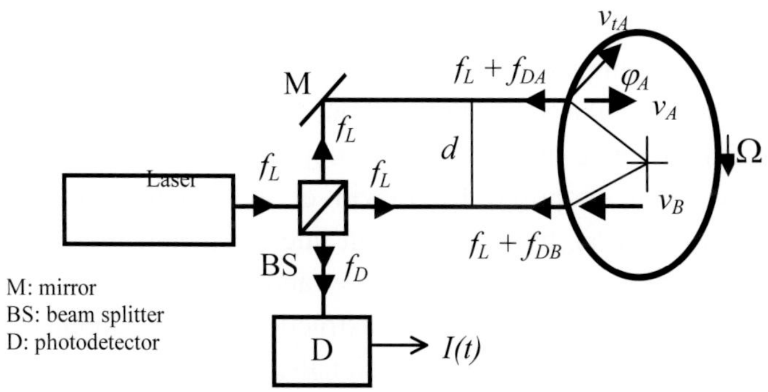

Figure 4. Schematic of the differential mode interferometer for rotational vibration measurements by Halliwell (1983).

The Doppler frequencies Δf_{DA} and Δf_{DB}, produced in the back scattered beams, are:

$$\Delta f_{DA} = \frac{2v_A}{\lambda} \qquad \Delta f_{DB} = \frac{2v_B}{\lambda} \tag{4}$$

Where

$$v_A = \Omega \cdot r_A \cdot \cos\varphi_A \quad v_B = \Omega \cdot r_B \cdot \cos\varphi_B \tag{5}$$

The geometrical relationship for the distance d between the measurement beams and the angles φ_A and φ_B at given radii r_A and r_B is given by:

$$\Omega \cdot r_A \cdot \cos\varphi_A + \Omega \cdot r_B \cdot \cos\varphi_B = d \tag{6}$$

Thus, combining Equation (4), (5) and (6), the following formula is obtained for the Doppler frequency shift:

$$\begin{aligned}\Delta f_D &= \Delta f_{DA} + \Delta f_{DB} = \frac{2v_A}{\lambda} + \frac{2v_B}{\lambda} \\ &= \frac{2}{\lambda} \cdot \Omega \cdot (r_A \cdot \cos\varphi_A + r_B \cdot \cos\varphi_B) = \frac{2}{\lambda} \cdot \Omega \cdot d\end{aligned} \tag{7}$$

where Δf_D depends only on the vibrometer constructive parameters (d and λ) and on the angular velocity Ω.

In order to improve the SNR (also without a retro-reflective paint) on the surface and to measure stationary objects, a dual interferometer configuration was proposed by Lewin et al.[16] Here, instead of the standard configuration, two independent interferometers are used for the acquisition of the velocity components v_A and v_B, but only one laser and one Bragg cell are employed. The extraction of the Doppler shift is performed by mixing of the electrical signals.

This technique presents two relevant advantages: it seems to be basically insensitive to bending motions of the rotating object not in the plane of beams[17] and no particular surface preparation is usually required. Typical

velocity range is from -7000 to 11000 rpm. The frequency range in vibrational angle measurement is from 1 Hz to 10 kHz and the calibration accuracy is of about ± 0.5%.
The use of rotational vibrometers is helpful to extend the applicability of torsional modal analysis (e.g. Ref. 18) and rotational velocity measurements, also for on-line continuous processes, thanks to their easy and non-contacting set-up.

3. Applications

In recent years, Laser Doppler Vibrometry (LDV) has gained a technical level suitable for the solution of many practical and industrial problems. Different system configurations allow to vary the measurement performances, depending on the specific application to be approached.
Due to the sensitivity, accuracy and versatility of laser Doppler vibrometers, different systems based on this principle are now largely diffused for a very broad variety of applications.[19] In fact, these systems were applied not only in "mechanics", traditional field of application of vibration measurement techniques, but also where other techniques demonstrate important limits.
In addition, the portability of LDV systems offers a unique possibility of in-field tests, without complex and costly installations, and makes it possible to operate directly where the structure is usually working or installed.
A large number of applications in the field of system identification of light or very small structures have been developed thanks to the non-contacting nature of LDV measurements. It is well known that effects of mass loading due to accelerometers are often significant also in large or rigid structures (e.g. Ref. 20). When mass or dimension of the measurand is comparable to those of the transducer, this effect becomes important.

The versatility and easiness in installation and control allow the employment of Doppler systems in several industrial applications and Quality Control systems.

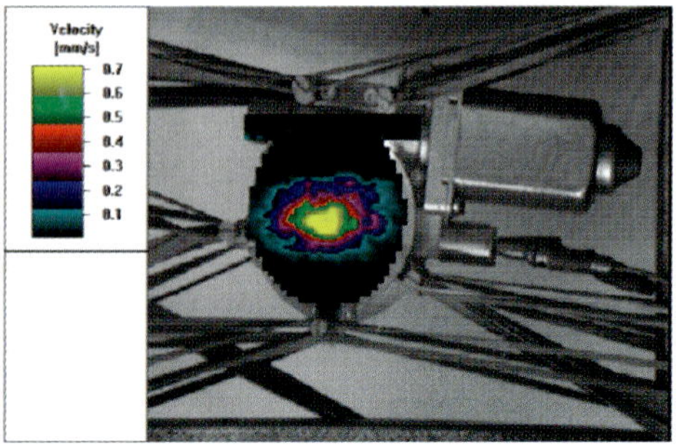

Figure 5. Vibration map of the cover of an automotive window lift system.

In automotive industry, LDV has been proposed to analyze the behavior of mechanical and structural components, as window lift system.[21] In Figure 5 an example of vibration map measured by a SLDV on an electric motor cover surface is reported. The resonance of the gear cover is well highlighted with high spatial resolution. In Quality Control, laser Doppler sensors have been employed in bench for automatic testing and selection of washing machines[22] at the end of the production line. Data obtained during operating conditions are analyzed using neural network algorithms.

Another interesting application is the vibration measurement across a flame (e.g. in the inner surface of a burner). Among the different "non-contact remote" optical measurement techniques, laser Doppler vibrometers may be usefully employed for this task, but interactions between laser beam and flame have to be carefully considered.[23]

A field which demonstrates an increasing interest in laser Doppler techniques is the biomedical engineering, in particular for the vibration measurement and analysis of different human body parts (tooth, membranes etc) or conditions.[24]

Laser vibrometer has been successfully employed for structural diagnostics of artworks[25] such as frescoes, mosaics, ceramics and easel painting. The main purpose has been to implement a novel diagnostic procedure based on non-intrusive measurement instruments: the investigated surfaces are very slightly vibrated by mechanical or acoustic actuators and a SLDV scans the objects measuring surface velocity.

The fields, which present higher velocity than neighboring areas, identified a structural defect. Moreover, the application of laser vibrometer allowed to recognize structural resonance frequencies and to reach a complete characterization of defects. An example is shown in Figure 6.

Scanning Laser Doppler Vibrometry systems have been applied also to test and verify the resistance to shock and vibrations of electronic devices, such as personal computer, power supply units, lamp, etc.; the main idea is to improve the efficiency of the traditional on-line "go-no-go" tests generating feedback information for the design process.

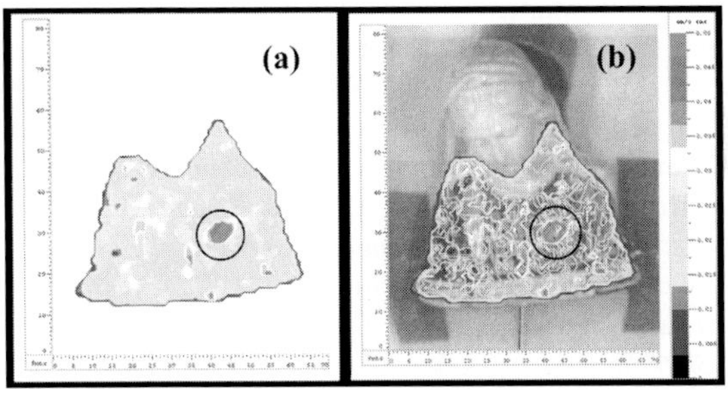

Figure 6. Vibration map supplied by a SLDV system of a ceramic sample with a defected area.

The use of this kind of electro-optic technique has allowed to measure vibration behavior of single components of above mentioned devices under known excitation: an example of obtained results is presented in Figure 7, where the acceleration vibration map and relative frequency spectrum of the capacitor of a power supply unit are visualized (e.g. Ref. 26).

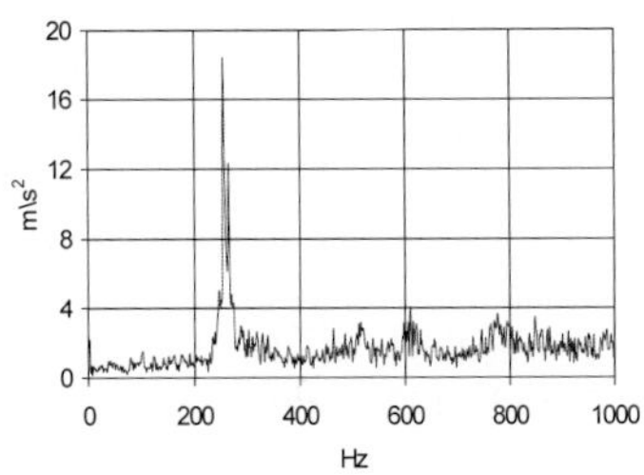

Figure 7. Resonance of a capacitor at 255 Hz: acceleration vibration map and relative frequency spectrum.

References

1. P. Buchhave, Optics and Laser Technology, 11 (1975).
2. G: A. Massey in *Study of vibration measurement by laser methods*, Accession Number: 66N27953; Document ID: 19660018663; Report NASA-CR-75643 (1965).
3. G. A. Massey, in *Laser vibration analyzer*, Accession Number: 68N17069; Document ID: 19680007600; Report Number: NASA-CR-73167 (1967).
4. R. R: Carter G. A. and Massey, in *A portable laser instrument for vibration analysis and transducer calibration*, Accession Number: 68N17052; Document ID: 19680007583; Report Number: NASA-CR-73137 (1967).
5. A. Link, H.J von Martens and W. Wabinski, Proceedings of 2nd International Conference on Vibration Measurements by Laser Techniques: Advances and Applications, SPIE No.3411, 224-235 (1998).
6. P. Castellini and N. Paone, in *Review of Scientific Instruments*, vol.71, n°12, p. 4639, Ed. American Institute of Physics, Woodbury, NY (2000).
7. H. Marguerre, in *Optical Sensors*, Ed. W Göpel et al., vol. 6321(1992).
8. B. Stoffregen and A. Felske, in *Scanning laser Doppler vibration analysis system*, SAE Technical Paper Series 850327 (1981).
9. W. X. Li and L.D. Mitchell, in *Error Analysis and Improvement for Using Parallel-Shift Method to Test a Galvanometer-Based Laser Scanning System*, Proceedings of the 1st International Conference of Vibration Measurement by Laser Techniques, SPIE Vol. 2358, 13 (1994).
10. X. Zeng, A. L. Wicks and L.D. Mitchell, in The Determination of the Position and Orientation of a Scanning Laser Vibrometer for the Laser-Based Mobility Measurement System, Proceedings of the 1st International Conference of Vibration Measurement by Laser Techniques, SPIE Vol. 2358, 81, Ancona, Italy (1994).
11. M. Martarelli, G. M. Revel and C. Santolini, in *Mechanical Systems and Signal Processing*, Academic Press, **15**(3), 581 (2001).
12. L. E. Drain, in *The laser Doppler technique,* Ed., J. Wiley & Sons (1980).
13. M. Gasparetti, G. M. Revel and E. P. Tomasini, in *Theoretical modeling and experimental evaluation of an in-plane laser Doppler vibrometer in different working conditions*, Proceedings of 3rd International Conference on Vibration Measurements by Laser Techniques: Advances and Applications, SPIE No. 3411, 317, (1988).
14. D. Calhoun and J. Tangren, in *A Three Axis Modal Analysis of a Magnetic Head Gimbal Assembly using a Laser Doppler Vibrometer and an In-plane Vibrometer*, Proceedings of the 15th International Modal Analysis Conference, Orlando, Florida, 1727 (1997).
15. N. A. Halliwell, C. J. D. Pickering and P.G. Eastwood, *J. Sound Vib.*, **93**, 588 (1984).
16. A. C. Lewin, V. Roth and G. Siegmund, in *New Concept for Interferometric Measurement of Rotational Vibrations*, Proceedings of 1st International

Conference on Vibration Measurements by Laser Techniques: Advances and Applications, SPIE No.2358, 24-36 (1994).

17. G. M. Revel and E.P. Tomasini, in *Torsional Vibrations: a Laser Vibrometry Approach*, Proceedings of Vibration, Noise & Structural Dynamics, 448, Venice, Italy (1999).
18. S. Seidlitz, in *Using modal analysis to verify a torsional mass elastic model*, Proceedings of the 16th International Modal Analysis Conference, Santa Barbara, 1505 (1998).
19. P. Castellini, G. M. Revel and E. P. Tomasini, in *Laser Doppler Vibrometry: A review of advances and applications*, The Shock and Vibration Digest, Vol. 30, No. 6, 443 (1998).
20. G. D' Emilia, C. Santolini and E. P. Tomasini, in *Comparison among modal analysis of axial compressor Blade using experimental data of different measuring systems*, Proceeding of Stress and Vibration Analysis conference, 191 (1989).
21. G. M. Revel, C. Santolini and E. P. Tomasini, in Laser Doppler Vibration and Acoustic Intensity Measurements for Dynamic Characterization and Noise Reduction of a Car Window Lift System, Proceedings of the 15th International Modal Analysis Conference, 1636 (1997).
22. N. Paone and L. Scalise., in *Non Invasive Laser Measurement Techniques in On-line Diagnostics of House-hold Appliances*, Proceedings of the 5th European Congress on Intelligent Techniques & Soft Computing (EUFIT), 1738(1997).
23. N. Paone and G. M. Revel, Optics *& Lasers in Engineering*, Vol.30, 163 (1998).
24. R. Deboli, G. Miccoli and G. L. Rossi, in *Pedestrian Controlled Tractor Operator Hand Vibration: Laser Measurement Technique Application*, Proceedings of the 7th International ISTVS Conference, 470 (1997).
25. P. Castellini, E. Esposito, B. Marchetti, N. Paone and E. P.Tomasini, in New applications of Scanning Laser Doppler Vibrometry (SLDV) to non-destructive diagnostics of artworks: mosaics, ceramic, inlaid wood and easel painting, Proceedings of Laser in the Conservation of Artworks-LACONA IV, 203, Paris 11-14 September 2001.
26. G. M. Revel, in A new vibration measurement procedure for on-line Quality Control of Electronic Devices, Shock and Vibration, ed. IOS Press, 9, 3 (2002).

LASER DOPPLER VELOCIMETRY

Nicola Paone*, Lorenzo Scalise and Enrico Primo Tomasini

Dipartimento di Meccanica, Università Politecnica delle Marche
Via Brecce Bianche, Ancona, Italy
**E-mail: e.p.tomasini@univpm.it*

A short historical survey of laser Doppler velocimetry is presented, followed by a description of the measurement principle, basic configurations and main components used in existing instrumentation. The measurement volume is described. Aspects and problems related to optical access and fluid seeding are discussed. Some hints are given on signal processes.

1. Introduction

Soon after the invention of the laser, the first paper about laser Doppler velocimetry appeared in 1964 by Yeh and Cummins[1] from Columbia University. In the four following decades Laser Doppler Velocimetry (LDV) has evolved from a complex laboratory technique into a well established measurement technique, implemented into different instrumentation lay-outs having many application areas. In 1965, Foreman et al.[2] reported the first experiment implementing the heterodyne detection and reporting the first of the known LDV optical configurations named reference-beam. Although this configuration is still used, the dual beam configuration proposed in 1969 by Penney et al.[3] (also known as differential Doppler) has almost completely supplanted it. Jackson and Paul[4] in 1970 have shown that LDV can be used at very high flow velocities without heterodyne detection, but with direct spectral analysis. Reference and dual beam systems have been estensively analysed by Edwards et al.[5] and Adrian and Goldstein.[6] In 1969, Rudd[7] introduced the fringe model, successively discussed by Durst and Whitelaw[8] in 1971. Charactheristics of the Doppler signal, its relation with the particle location and the stochastic proprieties of the measured frequency have been analysed by Lading,[9] Drain[10] and Greated and Durrani.[11] An important paper from Grant and Orloff,[12] regarding a two-color optical system was presented for solving the problem of the simulateneous measurement of two velocity components.

LDV has found many application areas, the main one is fluid dynamics. Research in hydrodynamics has been conducted with LDV system measuring velocity fields over a large velocity range. Subsonic, transonic and supersonic aerodynamics in the range up to 1000 m/s is possible with Doppler anemometers. Special version of the instrumentation has been developed for convective flows and medicine in the μm/s and mm/s velocity range. Combustion problems, turbomachinery, two-phase flows (sprays) are other explored application areas.

The main advantage of LDV is that it is a non intrusive technique, because the probe volume is realised by laser ligth. Therefore, expecially in transonic or supersonic flows as well as in combustion process, there is a minimum modification of the natural conditions of the measurand. Another important advantage of LDV consists in the linearity of the instrument response to velocity. These are important features of LDV compared to hot wire anemometers, usually characterised by high sensitivity to the combination of flow velocity and temperature and therefore undergoing problems of interfering inputs. At present, LDV is considered a reference measurement technique, allowing accurate measurements, for highly turbulent flows or in reverse flows, where recirculations zones and vortices cause rapid modifications of the instantaneous velocity vector proprieties. LDV systems typically allow the measurement of one or two components of the velocity vector istantaneouslly and locally; three component systems existed as well.

The main problem with LDV systems is that the instrument does not measure the fluid velocity, but it measures the velocity of the particles immersed in the flow, which are supposed to follow the flow. In many cases flow seeding is necessary and this could affect the original flow conditions. Particle arrival rate cannot be controlled and this causes a random sampling of the velocity in time and it can randomly affect signal-to-noise ratio of the instrument. Some limitations, due to stray light, have been dimonstrated for boundary layer investigations close to solid walls. Finally, the instrumentation complexity (fluid-particle interaction, optical and mechanical lay-out and signal processing) makes it an high-cost measurement system which provides high-quality data only if properly used.

2. Basic Optical Components in LDV Systems

Laser is the acronym of *L*ight *A*mplification by *S*timulated *E*mission of *R*adiation. Historically the laser has been realised in the summer of 1960 by T.H. Maiman of the Hughes Aircraft Company Laboratories, following the studies on the maser, a similar device using microwaves instead of visible light waves. A laser is an electro-optical device able to generate an intense, low-divergence, monocromatic beam of coherent light. The principle for the emission of coherent light in a laser is extensively presented in Smith et al.[13] For LDV pourposes the most important laser property is the degree of coherence, mathematically expressed by a three-dimensional coherence function. This function describes the phase coherence in the direction of propagation (*temporal coherence*) and the phase correlation across the wavefront (*spatial coherence*). Coherence length, which is equal to velocity of light times the coherence time, is an important parameter for LDV. In fact, in some LDV configurations, the laser beam is splitted in two beams and then recombined in order to form the interference fringe pattern; the maximum difference between the beam path length must be significantly less then the coherence length of the laser source. As the path length difference along the beams increases, the fringe visibility decrease to zero. For LDV system the main laser sources are continuous wave (cw) lasers even if some interesting LDV configurations have been proposed utilising pulsed lasers. Multimode He-Ne (λ = 632 nm) and Argon-ion (λ = 514, 488 and 442 nm) lasers are the most common sources in LDV.

Two kinds of photo-detectors are commonly utilised in LDV system: photomultipliers and photodiodes. Photomultipliers are used in case of low light level or when large signal bandwidth is needed; limits in the use of such detectors, are due to high-voltage power supply and additional shot noise. Photodiode detectors can be used in normal measurement conditions (normal bandwith and sufficient signal-to-noise ratio), being carachterised by high amplification efficiency, low noise, small size and relatively low cost. Sensitivity of photodetectors is function of wavelength, most of the detectors have the maximum in the blue-green range (argon-ion lasers) rather than in the red (He-Ne lasers).

Beam splitters are used in LDV set-ups for beam separation. The simplest beam splitter is composed by a thin glass plate aligned at an angle to the incident laser beam. Two beams can be more easly obtained utilising a prism beam splitter.[14,15] Parallel beams can be obtained by means of special designed lateral displacement beam splitter.

3. Doppler Effect

In the following we will present the basic equation for the Doppler shift from a moving target (the particle). Let us consider figure 1, where *S* is the emitting source and *P* the moving particle, with a velocity $\vec{w}$. The laser beam propagates in direction $\overrightarrow{SP}$, the observer detects scattered light along direction $\overrightarrow{PO}$; the following angles will be used in the analysis of Doppler effect: θ_1 is the angle between $\overrightarrow{PS}$ and $\vec{w}$; θ_2 is the angle between $\vec{w}$ and $\overrightarrow{PO}$; α is the angle between $\overrightarrow{PS}$ and $\overrightarrow{PO}$ and it is $\alpha = \pi - \theta_1 - \theta_2$.

Light originating from the laser S and having frequency f_0 and wavelength λ impinges the particle P, whose motion in the direction of propagation of the light beam causes the particle to detect a frequency f_{D1} smaller than f_0 (Doppler effect on moving target).

$$f_{D1} = f_0 - \frac{|\vec{w}|\cos(\alpha + \theta_2)}{\lambda} = f_0 - \frac{|\vec{w}|\cos(\pi - \theta_1)}{\lambda} = f_0 + \frac{|\vec{w}|\cos\theta_1}{\lambda} \qquad (1)$$

The moving particle scatters light towards the fixed detector in direction $\overrightarrow{PO}$; a second Doppler shift occurrs, due to the particle velocity component along direction $\overrightarrow{PO}$, which increases the frequency detected by the observer at:

$$f_{D2} = f_{D1} + \frac{|\vec{w}|\cos\theta_2}{\lambda} = f_0 + \frac{|\vec{w}|}{\lambda}(\cos\theta_1 + \cos\theta_2) = \\ = f_0 + \frac{|\vec{w}|}{\lambda} 2\cos\frac{\theta_1 + \theta_2}{2}\cos\frac{\theta_1 - \theta_2}{2} \tag{2}$$

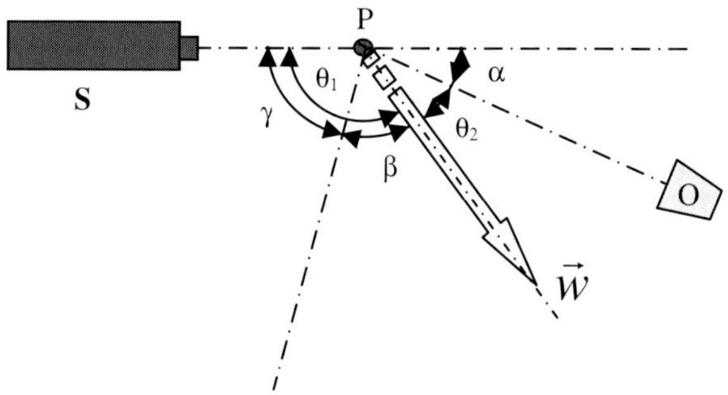

Figure 1. Doppler shift on scattering from a moving target.

Therefore, the overall Doppler shift is:

$$\Delta f_D = f_{D2} - f_0 = 2\frac{|\vec{w}|}{\lambda}\cos\frac{\theta_1 + \theta_2}{2}\cos\frac{\theta_1 - \theta_2}{2} = \\ \cong 2\frac{f_0}{c}|\vec{w}|\cos\frac{\theta_1 + \theta_2}{2}\cos\frac{\theta_1 - \theta_2}{2} \tag{3}$$

In this equation it is assumed that $\Delta f_D << f_0$, i.e. that the particle velocity is much smaller than the velocity of light c; this is the classical expression reported in literature for Doppler shift on moving scatterer.[14]

This equation, after some geometrical consideration, can be written in a useful form which allows to understand which velocity component determines the Doppler shift. In fact:

a) being $\alpha = \pi - \theta_1 - \theta_2$, then $\alpha/2 = \pi/2 - (\theta_1 - \theta_2)/2$ it can be observed that the term $\cos\frac{\theta_1 + \theta_2}{2} = \sin\left[\frac{\pi}{2} - \left(\frac{\theta_1 - \theta_2}{2}\right)\right] = \sin\frac{\alpha}{2}$; this term therefore depends only on the configuration of the instrument (the angle α);

b) it can be demonstrated that the angle $\beta = \frac{\theta_1 - \theta_2}{2}$ is the angle which forms between the velocity vector $\vec{w}$ and the bisector of the angle $(\theta_1+\theta_2)$; in fact, as it can be seen in the drawing, the bisector forms an angle γ with $\overrightarrow{PS}$, therefore:

$$\gamma = \frac{\theta_1 + \theta_2}{2}$$

$$\gamma + \beta + \theta_2 + \alpha = \pi$$

from which $\beta = \dfrac{\theta_1 - \theta_2}{2}$.

The angle β indicates in which direction the velocity vector is projected, it defines the velocity component $(|\vec{w}|\cos\beta)$ to which the Doppler Δf_D shift is proportional. The resulting equation is:

$$\Delta f_D = \frac{2|\vec{\mathrm{w}}|}{\lambda}\cos\beta\sin\frac{\alpha}{2} \tag{4}$$

Eq. 4 shows that Doppler shift is proportional to $|\vec{w}|\cos\beta$, the velocity component along the bisector of $(\theta_1+\theta_2)$, and depends on scattering angle α. In mediums different from air, the wavelength λ differs from its value in air λ_0 and depends on the refractve index of the medium *n*, according to the well known expression $\lambda = \lambda_0/n$.

3.1 Optical Heterodyne

Apart from some early applications in very high velocity flows, the Doppler shift cannot be detected directly with a spectrometer. For He-Ne laser (λ = 632 nm) back-scattered radiation (i.e. $\theta_1 = -\theta_2$) provides a

Doppler shift equal to 3.16 MHz/(m/s); such a shift is too small with respect to the laser frequency (f_0 =4.7 10^{14} Hz) to be resolved by any available spectrometer. Therefore most LDV systems make use of optical interference to produce a beat signal at the Doppler frequency Δf_D . For the purpose, the light beam of the laser source is made to interfere with the scattered radiation on a photodetector.

Defining E as the amplitude of the electric field generated by the laser source and E' the amplitude of the electric field associated to the Doppler shifted radiation, it is possible to express these fields at time t as:

$$E = E_1 \cos(2\pi f_1 t + \phi_1) \tag{5}$$

$$E' = E_2 \cos(2\pi f_2 t + \phi_2) \tag{6}$$

where ϕ_1 and ϕ_2 are the arbitrary phases and f_1 and $f_2 = f_1 + \Delta f_D$ are the frequencies associated to the waves. Optical heterodyning of these two electric fields on a photodetector generates a current i proportional to the square of the total electric field:

$$i(t) = B(E_1 \cos(2\pi f_1 + \phi_1) + E_2 \cos(2\pi f_2 + \phi_2))^2 \tag{7}$$

where B is the sensitivity of the photodetector. Equation 7 contains terms characterised by frequencies f_1, f_2 and $(f_1 + f_2)$ higher then the photodiode bandwidth; the corresponding photodiode output will be a low frequency term proportial to the mean value of such terms. Such term is called *Pedestal* and is proportional to the light beam intensity. The only term which generates a time dependent signal at the photodiode contains the difference frequency $(f_2 - f_1) = \Delta f_D$, i.e. the Doppler frequency shift. The photodiode current is therefore proportional to:

$$i(t) \propto ((Pedestal + E_1 E_2 \cos(2\pi (f_1 - f_2) t + (\phi_1 - \phi_2)) \tag{8}$$

Apart from the *Pedestal*, the signal $i(t)$ at the output of the photodectector results to be frequency modulated by the difference $(f_2 - f_1) = \Delta f_D$ which is the Doppler shift proportional to the target velocity component $|\vec{w}|\cos\beta$ according to Equation 4. The signal forms only when a particle crosses the laser beam, whose intensity profile is gaussian, and scatters light. Therefore each particle produces a so called burst signal, whose frequency Δf_D is related to velocity $|\vec{w}|\cos\beta$; the burst is made of the *Pedestal* due to the gaussian profile of laser beam intensity and an oscillating part having amplitude proportional to E_1E_2. The depth of modulation of the signal is optimal when E_1 and E_2 are comparable.

4. Laser Doppler Velocimeters: Main Optical Configurations

An LDV system relies on collection of scattered laser light from moving particles, that after interference with a reference beam on the photodetector, provides an electric signal whose frequency is related to particle velocity. This is achieved by different optical configurations, described hereafter.

4.1 Forward-Scatter (Reference Beam)

Historically, this has been the first optical configuration proposed. In Figure 2 it is shown that a laser beam is focussed at a point in the flow, and a photodetector is arranged in such a way to collect scattered light from particles that cross the focal region and mix it with a reference laser beam coming directly from the laser.

Figure 2 shows an example where the laser source and the observer are situated in such a way that the velocity vector is in the direction of the bisector of the angle between the laser beam and the direction of observation; in this case $\theta_1 = \theta_2$ and $\beta = 0$. In order to maximize signal to noise of the beat signal, it is necessary that the two waves which interfere have similar amplitudes. Therefore, the reference beam is usually attenuated, in order to match the low intensity of scattered light from a small particle.

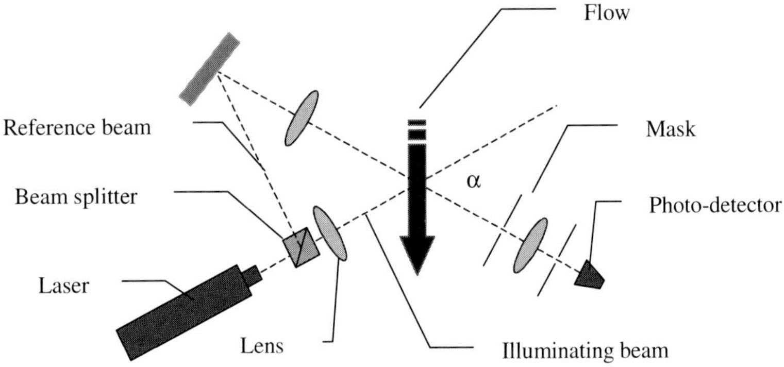

Figure 2. Forward-scatter heterodyne system.

4.2 Forward-Scatter (Differential)

In a differential configuration, two scattered waves are used for optical interference, instead of a scattered wave and a reference one. This leads to the differential configuration reported in Fig. 3.

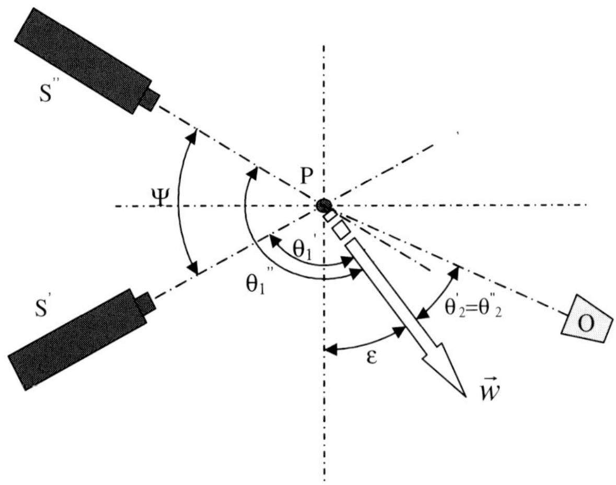

Figure 3. Interference between scattered waves in differential configuration.

Two beams are crossed in the measurement volume at an angle ψ. The observer collects two scattered waves, one from each beam, having different Doppler shifts; their frequencies are:

$$f_D^I = f_0 + \frac{|\vec{w}|}{\lambda}\left(\cos\theta_1^I + \cos\theta_2^I\right)$$
$$f_D^{II} = f_0 + \frac{|\vec{w}|}{\lambda}\left(\cos\theta_1^{II} + \cos\theta_2^{II}\right) \tag{9}$$

The two scattered waves interfere and the photodetector outputs an electric signal at the difference frequency Δf_D:

$$\Delta f_D = f_D^{II} - f_D^I = \frac{|\vec{w}|}{\lambda}(\cos\vartheta_1^{II} + \cos\theta_2^{II} - \cos\theta_1^I - \cos\theta_2^I) \tag{10}$$

This expression can be simplified, taking into account that:

- $\psi = \theta_1^{II} - \theta_1^I$ is the angle between the two crossed beams;
- $\theta_2^I = \theta_2^{II}$
- $\dfrac{\theta_1^I + \theta_1^{II}}{2} = \theta_1^I + \dfrac{\psi}{2} = \varepsilon + \dfrac{\pi}{2}$

where ε is the angle between the velocity vector $\vec{w}$ and the direction orthogonal to the bisector of the illuminating beams, which is called the optical axis of the differential Doppler configuration.

It becomes:

$$\Delta f_D = \frac{|\vec{w}|}{\lambda}\left[\cos(\vartheta_1^I + \psi) - \cos\theta_1^I)\right] = \frac{|\vec{w}|}{\lambda}(-2)\sin\frac{\theta_1^I + \psi - \theta_1^I}{2}\sin\frac{\theta_1^I + \theta_1^{II}}{2} =$$
$$= \frac{|\vec{w}|}{\lambda}(-2)\sin\frac{\psi}{2}\sin(\varepsilon + \frac{\pi}{2}) = \frac{|\vec{w}|}{\lambda}(-2)\sin\frac{\psi}{2}\cos\varepsilon \tag{11}$$

where a direct optical access is not available. Usually in fiber systems laser beam separation and frequency shift units are put before the beams are launched into the fibers. The system operates in differential mode, in back scatter configuration. These systems utilise polarization preserving monomode fiber, so that at the probe volume beams are still fully coherent and can produce effective interference. Back-scattered light is collected by a multimode fiber, which brings the optical signal back to the photodetector.

A schematic of a typical fiber optic LDV system is reported in figure 6. Typical problems with fiber optics are due to vibro-acoustic perturbation of the fiber. These phenomena can induce periodical strain excitation on the fibers and consequent phase modulation, resulting in noise added to the Doppler bursts. Commercial fiber links for LDV systems are therefore usually composed by optic fiber cables immersed in a special damping cover and protected by an external metallic shield. Temperature effect on fibers, due to environmental conditions, have no significant influence on beams relative phase, being the time constant of these phenomena much larger than the time scale of observation.

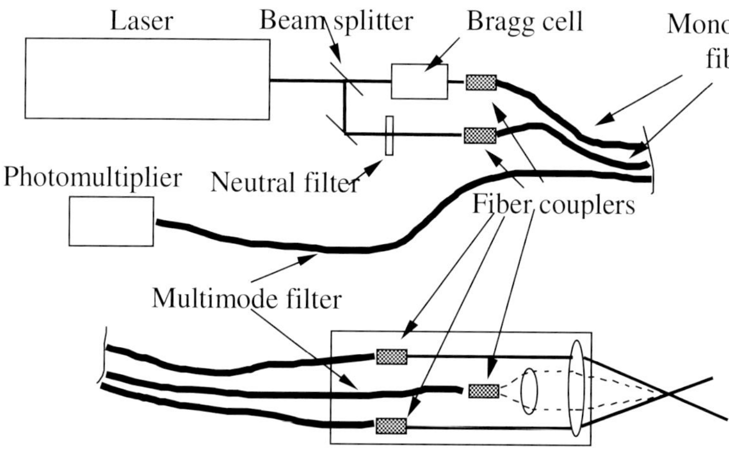

Figure 6. Fiber optic laser Doppler velocimeter.

4.5 Two Velocity Component Systems

The need to measure simultaneously two velocity components is typical of many fluid dynamic flow analysis. LDV systems allows 2D velocity measurement on a plane orthogonal to the optical axis of the LDV system. Different techniques have been proposed for the realisation of 2D LDV systems. The most common is the wavelenght division multiplexing. This technique utilises two spectral lines of the laser source (for example: 514 and 488 nm for Argon-ion lasers source), for the generation of two different interference patterns. These two patterns are set perpendicular to each other. Back reflected signal is then collected by a lens. The two Doppler shifts corresponding to the two orthogonal components of velocity are then separated and filtered by ineterference filters before photodetecors, so that two signals are produced.

4.6 Three Velocity Component Systems

For some specific application a 3D LDV is needed. The 3D components can be measured focousing a set of three sets of interference fringes of different colour on the same measurement volume.[15] The fringes having known relative angles allow the reconstruction of the velocity vector in space. Argon-ion lasers are generally used; they provide power at: 514, 488 and 442 nm. These systems are usually complex in alignment and require larger access windows to the flow with respect to 2D systems.

5. Measurement Volume

The characteristics of the measurement volume are of extreme inportance for the determination of LDV system performance. The two coherent laser beams (having a Gaussian beam profile) form an interference fringe pattern schematically reported in figure 7.

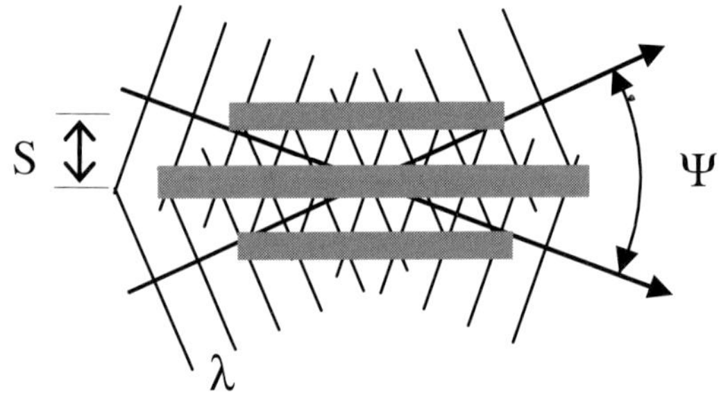

Figure 7. Interference fringe pattern as generated by two coherent laser beams.

Fringe spacing S is proportional to the laser wavelength λ and inversely proportional to half the angle ψ between the two beams and can be calculated by the formula:[7]

$$S = \frac{\lambda}{2\sin\frac{\psi}{2}} \tag{12}$$

Particles inside the flow will travel trough the fringe region and will scatter light to the photo-detector. Signals at photo-detectors will have a frequency equal to Δf_D.

The fringe pattern is three-dimensional region having an ellipsoidal shape and a gaussian intensity profile, as reported in Fig. 8. Dimensions of the measurement volume are determined by geometrical optics and diffraction:

$$d_x = \frac{d_f}{\cos\frac{\psi}{2}}; \quad d_y = d_f; \qquad d_z = \frac{d_f}{\sin\frac{\psi}{2}} \tag{13}$$

where d_f is the diameter of the focused laser beam defined as the region where intensity is larger than $1/e^2$ of the maximum at centre:

$$d_f = \frac{4}{\pi}\frac{F\lambda}{Ed_l} \tag{14}$$

being, Ed_l the expanded laser beam diameter entering the front lens which forms the probe volume, E the expansion factor and d_l is the laser beam diameter before expansion and F the focal length. Beams are often expanded so to reduce the diffraction limited dimension of the probe volume. Number of fringes N_f in the probe volume can be calculated by the following formula:

$$N_f = \frac{8F\tan\frac{\psi}{2}}{\pi Ed_l} \tag{15}$$

Measurement volume tipically has the following dimensions: $d_x \cong 0.1$ mm, $d_y \cong 0.1$ mm, while $d_z \cong 1 \div 3$ mm. These dimensions determine spatial resolution of the LDV. In literature measurement volume of 35 μm x 35 μm x 66 μm are reported for small-scale turbulent motions, obtained with short focal length optics.[16]

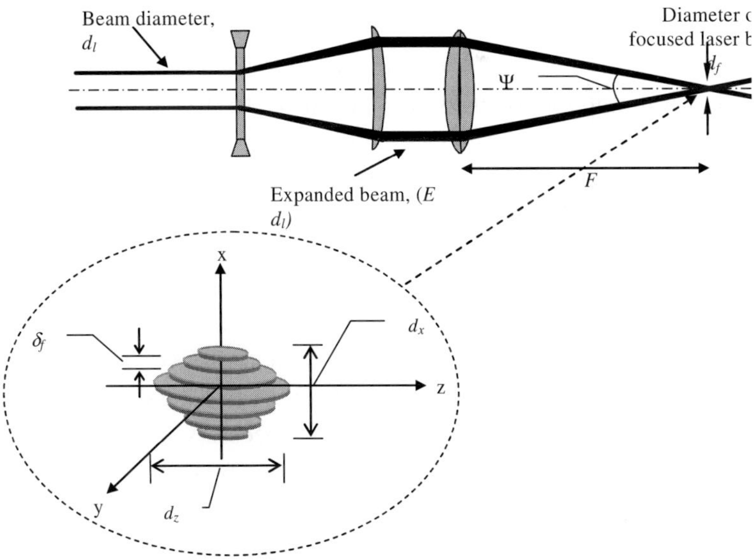

Figure 8. Measurement volume for a LDV system.

The use of very short focal length lenses allowed Tieu et al.[17] to reach measurement volume of 5 μm x 5 μm x 10 μm.

6. Optical Access

As in any optical technique, the question of optical access is fundamental and it may pose very difficult problems in practical applications. In order to outline the problems that one may have to face in laser Doppler velocimetry, it should be beared in mind that: 1) the fluid must be transparent to the laser light of the LDV system; 2) the fluid must be optically accessible to the LDV laser beams. The first condition is generally met in air flows, in many gases, in clear water, but limitations may arise when dealing with oils, smoke, slurries, and biological fluids and in general with fluids heavily loaded with particles. If the fluid is not transparent, LDV measurements cannot be done and the experimenter needs to address the problem by substituting the original fluid with a transparent one and then using the concepts of fluid dynamic similitude.

As an example, this approach is typical of in-vitro blood flow studies performed in order to analyze the fluid-dynamics of a variety of devices used in extra-corporeal circulation (pumps, valves, filters etc.). Even if the flow is transparent, whenever it is confined by walls, the wall must have optical windows transparent to the laser light, so that the laser radiation can reach the measurement volume and the scattered light can reach the photomultiplier. While this is easy in test rigs dedicated to aerodynamic and hydrodynamic flow studies, such as wind-tunnels and water tunnels, the optical access to internal flows often poses serious questions that may lead to modifications in the flow boundaries, therefore sometimes turning LDV into an invasive measurement technique. In case of forward-scatter LDVs, two optical accesses are needed, while only one optical access is necessary for back-scatter configurations; actually this is one of the reasons why most commercial systems propose back-scatter probes.

When realizing an optical access the window should meet several specifications, which are only in part related to optics. In fact, first of all the window should have physical and mechanical characteristics, which allow to withstand all stresses applied by the fluid and the environment.

Typical stresses are caused by pressure, by temperature, by vibrations or by chemical aggression. This is why a variety of options can be found in literature; typically windows are made of: a) Perspex, plexiglass or other polymeric materials; b) Glass; c) Quartz.

As a second aspect, the optical window should be installed in such a way to keep the same geometry of the flow boundaries; this is seldom possible, especially if flat windows are installed on a curved surface. It is up to the experimenter to evaluate the intrusivity effects caused in each case. When possible, a complete part of the test rig may be realized in transparent material; this is often the case in pipes, where a full section of the duct may be realised in transparent material keeping exactly the same geometry of the pipe under study (with exception of surface roughness). There exist also a variety of biomedical devices which are already transparent, such as certain centrifugal blood pumps or blood filters, made in policarbonate or similar transparent materials. The window is an optical interface, and therefore it affects the light beam propagation, depending on its shape, on its refractive index, on its absorption and reflection coefficient. Surface roughness must be as low as possible, so to reduce diffraction and surface scattering effects; a polished surface whose roughness is a fraction of the light wavelength (typically $\lambda/4$), will behave as an optically flat surface. Such values are easily obtained with polished glass, perspex, and quartz. Transmission and absorption depend on window material and light wavelength; reflection may be a concern, especially in case of multiple windows. Stray reflections and scattered light may increase Doppler signal noise level and need to be blocked for safety reasons. Refraction across optical interfaces is the most important question for optical access. Refraction occurs at any optical interface between two media having different refractive indices n_1 and n_2, according to Snell law:

$$n_1 \sin \vartheta_1 = n_2 \sin \vartheta_2 \tag{16}$$

being ϑ_1 and ϑ_2 the angles between the normal to the surface and the light propagation direction. In case of transparent windows two optical interfaces are present; the first one between air and window, the second

In general, measurements close to walls are noisy, because the wall surface scatters light towards the receiving optics. A diffusive surface, even if painted in black to increase its absorption, may still give large noise; therefore it is often better to have a polished reflecting surface instead of increasing surface absorption. In this case, if the optical axis is not aligned perpendicular to the surface, reflected light will never reach the receiving optics, therefore reducing noise caused by the wall.

7. Seeding

As already said, LDV measures the velocity of seed particles which act as light scatters. Therefore it is essential that particles have the same velocity as the fluid and measurement uncertainty depends directly on the ability of seed particles to accurately track fluid motion. Particles immersed in a moving fluid are subject to a number of forces. The main forces are: a) buoyancy and pressure gradient forces; b) inertial forces; c) drag forces. Ideally, a particle whose density ρ_P matches that of the fluid ρ_F would follow the flow. If $\rho_P/\rho_F \neq 1$, then particle size d_p becomes important. The smaller the size, the better the particle will follow the flow; this is why in LDV particle diameter is typically in the $d_p \approx 1\mu m$ range, or less. Seeding of liquid flows appears easier than gaseous flows, because liquids have a much larger density which is closer to that of typical seed materials, rather than gases. Particles should also behave as efficient light scatterers.[27] Due to their small size, Mie theory applies for predicting light scattering. While a small particle will follow accurately the flow, a larger particle will scatter more light, thus for given optical system sensitivity, too small particles cannot be detected. Non-spherical particles tend to align with flow lines and tend to scatter light in preferential directions, therefore spherical particles should be preferred. Particle concentration will affect data rate, therefore particle concentration is adjusted in order to reach the desired data rate. Particular care has to be taken to avoid conglomeration of particles and strong beam attenuation in case of excessive particle concentration. Typical seeding is made of:

- solid particles,
- liquid particles and aerosols.

Solid particles are frequently used in liquids, even though in measurements in water, it may not be necessary to seed the flow, because naturally present particles could be sufficient. Usually a suspension of solid particles is diluted in the liquid, so to achieve the optimal particle concentration. Polystirene latex particles (sometimes fluorescent) and glass microspheres (eventually aluminium coated) are available for liquids in different sizes. Solid particles are also used for combusting flows; typically they are made of metal oxides (TiO_2, Al_2O_3, ZrO etc.), because they withstand the high temperatures and the reacting environment of a flame. When solid particles are employed in gaseous flows, they are introduced into the flow either by fluidised beds or by atomisation of a concentrated solution of particles in water or ethanol. Aerosols and smoke particles are frequently used to seed gas flows. A variety of atomizers exist, which can produce fine aerosols of water or oils which meet the requirements for LDV seeders. Monodisperse seed particle size distributions should be preferred, because they improve signal to noise ratio at the photomultiplier; in fact most of the noise is caused by light scattered from very small particles, which cannot provide detectable Doppler bursts, but whose scattered light adds up to increase the noise floor. Furthermore the different response times of particles having dispersed diameters causes a spread in the velocity distribution which is erroneously interpreted as turbulence. Usually atomized liquids do provide polidisperse size distributions, while monodisperse size distributions are more frequent with solid particles. Polystirene latex particles are excellent for this purpose.

All particles used in LDV represent a potential health hazard, due to their small size. Ventilation of the environment where operators work and confinement of the seeded flow and filtering is recommended and should be applied in laboratory practice.

8. Laser Doppler Velocimeter Signal

When a particle moves across the LDV probe volume, it scatters light and a Doppler burst is detected by the photomultiplier tube. This signal is amplitude modulated by the gaussian beam profile and it has a carrier frequency determined by the frequency shift Δf_B applied by the Bragg cell present in the optical system, and a Doppler frequency depending on particle velocity. Figure 10 shows a typical Doppler burst.[28] It has a pedestal having a gaussian shape and a modulation due to the fringes.

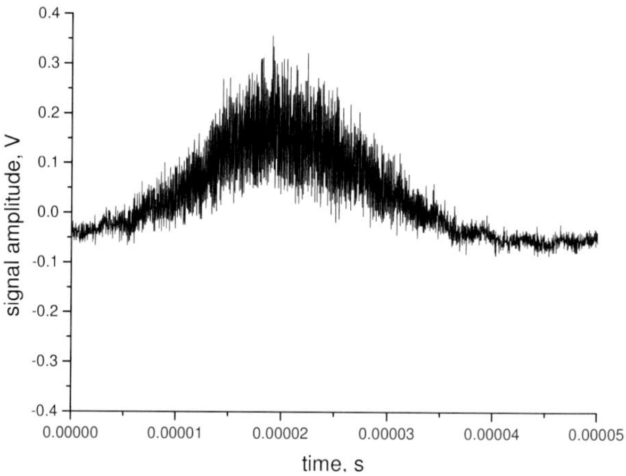

Figure 10. Typical unfiltered Doppler burst.

Pedestal is usually filtered out, because it does not convey information on particle velocity. Figure 11 shows a Doppler burst after high pass filtering for pedestal removal and low pass filtering for noise reduction.

Particle size and particle trajectory across the probe volume determine main characteristics of the Doppler burst. In fact signal peak amplitude depends on the intensity of scattered light, which increases for large particles and for particles crossing the centre of the probe volume. The visibility of the oscillating part of the signal depends on particle size d_p relative to fringe spacing S; if particle size is smaller than fringe spacing the fringe visibility is maximum, while if particle size is larger than fringe spacing the fringe visibility decreases. This is another reason why particle size should be small.

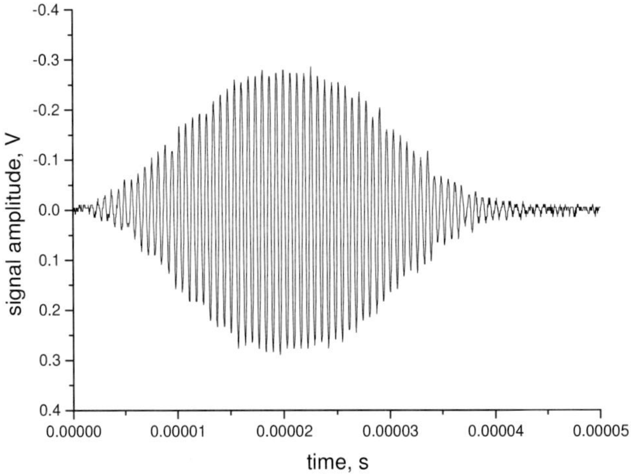

Figure 11. Typical Doppler burst after filtering.

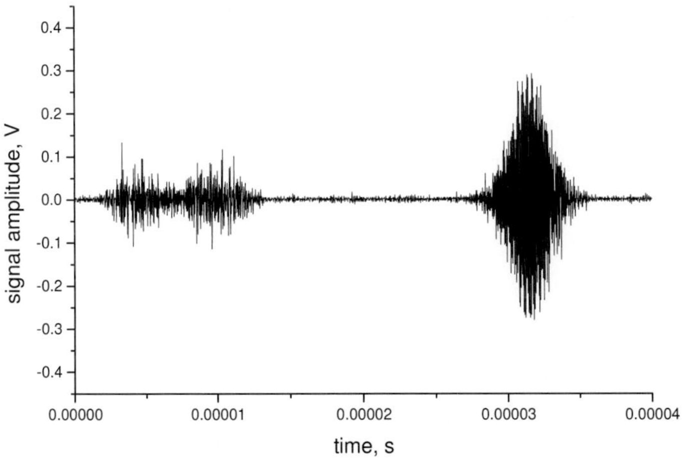

Figure 12. Typical random sequence of Doppler bursts.

Figure 12 shows a typical Doppler signal as observed in a flow. Flow velocity is randomly sampled. Proper corrections are needed to correct for velocity bias in statistical analysis and resampling is necessary to allow spectral analysis of the velocity signal.

9. Doppler Signal Analysis

Each Doppler burst must be detected and then processed in order to measure its frequency, so to compute particle velocity. Each Doppler burst provides one velocity sample. Processing can be done either in time domain or in frequency domain. Signal is always band-pass filtered before processing, so to eliminate pedestal and high frequency noisy components. A variety of analogue and digital signal processing techniques have been historically applied.[14, 28-32] Frequency trackers were amongst the first systems developed; they needed signals with large data rate, to avoid signal drop out in case of absence of particles in the probe volume. Period counting techniques have also been widely employed; their main limitation was the difficulty in processing noisy signals.

Modern processors employ digital algorithms running on DSP processors and provide large data rates even on very noisy signals. In these processors the signal is digitized and then processed. Two main classes of algorithms are used: a) Fast Fourier Transform; b) auto and cross correlation. Doppler burst detection is based on complex triggering procedures, which employ Doppler signal amplitude (envelope or pedestal) or consider the amplitude of correlation peaks emerging from the noise floor.

References

1. Y. Yeh and H. Z. Cummins, *Applied Physics Letters*, **4** (10), 176 (1964).
2. J. W. Foreman, E. W. Jeorge and R. D. Lewis, *App. Phy. Lett.*, **7** (4), 77 (1965).
3. C. M. Penney, *IEEE J. of Quantum Electronic*, **QE-5**, 318, (1969).
4. D. A. Jackson and D. M. Paul, *Phy. Lett.*, **32A** (2), 77 (1970).
5. R.V. Edwards and J. C. Angus, *J. of App. Physics*, **2** (2), 837 (1971).

6. R. J. Adrian and R.J.Goldstein, *J. of Physics E: Scientific Instruments*, **4**, 505 (1971).
7. M. J. Rudd, *J. of Scientific Instruments*, **2(2)** 55 (1969).
8. F. Durst and J. H. Whitelaw, *Proceedings of the Royal Society of London* **A**, 324, 157 (1971).
9. L. Lading, *Opto-electronics*, 5, 175 (1973).
10. L. E. Drain, J. of Physics D: Applied Physics, **5**, 481 (1972).
11. C. Greated and T. S. Durrani, *J. of Physics E: Scientific Instruments*, **4**, 24 (1971).
12. G. R. Grant and K. L. Orloff, *Applied Optics*, **12(12)**, 2913 (1973).
13. W.V. Smith and P. P.Sorokin, in *The laser* New York, McGraw-Hill (1966).
14. L. E. Drain, in *The Laser Doppler Technique*, New York, Wiley (1980).
15. T. S. Durrani and C. A.Greated, *Laser systems in flow measurement*, New York, Plenum Press (1977).
16. D. A. Compton, J. K. Eaton, *Experimental Fluids*, **22**, 111 (1996).
17. A. K. Tieu, M. R. Mackenzie, E. B. Li, *Experimental Fluids*, **19**, 293 (1995).
18. J. D.Broadway and E. Karahan, *DISA Information*, **26**, (1981).
19. A. F. Bicen, *TSI Quarterly*, vol. VIII, issue 2, (1982).
20. E. P. Tomasini, M. Gasparetti, N. Paone, *Measurement Science & Techniques*, **7**, 576 (1996).
21. N. Paone, P. Castellini and M. Gasparetti, *Int. Symp. On Applications of Laser Anemometry to fluid Mechanics*, Ed. D. F. C. Durao (1996).
22. M. Pinotti, N. Paone and E. P. Tomasini, Int. Symp. On Applications of Laser Techniques to Fluid Mechanics, Lisbon, 3.2.1 (1994).
23. F. Durst, T. Keck, and R. Kleine, *Proceedings of the 6th Symp. On Turbulence in Liquids*, Univ. of Missouri-Rolla (1979).
24. I. G. Edwards and A. Dybbs, *Int. Symp. On Applications of Laser Anemometry to Fluid Mechanics*, 171, ed. D. F. C. Durao, London (1984).
25. J. C. F. Pereira, in *Instrumentation for Combustion and Flow in Engines*, 267, Eds. D. F. C. Durao et al., (1989).
26. C. Vafidis, J. H. Whitelaw, 2nd Int. Conf. On Methodology and Innovation in Automotive Testing and Process Control, Florence, 1347 (1988).
27. Van den Hulst, *Light scattering from small particles*, New York, Dover publ., (1981).
28. F. Durst, A. Melling, J. H. Whitelaw, *Principles and practice of laser-Doppler anemometry*, New York, Academic, 1981.
29. R. J. Adrian, *Laser Velocimetry*, p.155, in Fluid Mechanics Measurements, Washington DC, R.J. Goldstein, (1983).
30. C. A. Greated, T. S. Durrani, *Laser Systems in Flow Measurement*, New York, Plenum (1977).

31. R. J. Adrian et al. *Laser Techniques And Applications in fluid Mechanics*, Berlin, Springer-Verlag, (1998 and previous editions).
32. R. J. Adrian, *Laser Doppler Velocimetry*, SPIE Milestone Series, MS78 (1993).

PHOTOACOUSTIC SPECTROSCOPY USING SEMICONDUCTOR LASERS

Pietro Mario Lugarà,[a,b,*] Angela Elia[a,b] and Cinzia Di Franco[a]

[a]*Laboratorio Regionale LIT*[3]*, CNR-INFM, Via Amendola 173, 70126 Bari, Italy*

[b]*Dipartimento Interateneo di Fisica, Via Amendola 173, 70126 Bari, Italy*

[*]*E-mail: lugara@fisica.uniba.it*

The basics of photoacoustic spectroscopy in solid and gaseous samples are summarized. A survey of the applications of near-infrared diode lasers and mid-infrared quantum cascade lasers in photoacoustic spectroscopy is reported.

1. Introduction

The PhotoAcoustic Spectroscopy (PAS) can be considered as a calorimetric method, based on the direct measurement of the heating produced by the non-radiative relaxation of the excited states, populated by optical absorption; therefore PAS is intrinsically different from all the conventional optical absorption methods, such as simple transmission spectroscopy, Fourier-transform infrared spectroscopy (FTIR), cavity ring-down and intracavity spectroscopy. If the excitation light source is modulated, a periodic heating is produced. In gaseous samples, this results in a modulation of local pressure and, thus, in the generation of an acoustic wave. In the case of solid samples or powders, the acoustic wave is produced in the surrounding gas by the transfer of the periodic heating from the absorbing sample to the gas. Whatever the sample is, an acoustic wave must be detected and we need only small sensitive microphones, independent of the light wavelength.

The photoacoustic effect, discovered in 1880 by A.G. Bell,[1] found its first application only more than 50 years later in a PhotoAcoustic (PA) system. Viegerov[2] used the PA technique for the first spectroscopic gas analysis in 1938; he studied blackbody infrared light absorption in gases to detect the gas concentrations in a mixture. Later, Luft[3] enhanced the detection sensitivity of the PA system to the parts per million (ppm) levels. The exploitation of gas and solid state lasers developed from 1960 as sources in photoacoustic spectroscopy increased the advantages in terms of sensitivity and selectivity, as a result of the high power and monochromaticity of the laser sources.[4,5] The PA systems devoted to gas samples reach now sub-ppb (parts per billion) detection limits and have a linear response in a wide range of gas concentration, covering 6 - 8 orders of magnitude.

2. Photoacoustic Spectroscopy in Solid Samples: Theory

The PA signal generation in a gas surrounding a solid sample can be modeled using the heat equation and studying the one-dimensional heat flow through the sample and the gas in the opposite direction to the light beam. The solid sample (*s*) can be considered optically and thermally homogeneous; its rear face is assumed in good thermal contact with a backing material (*b*), whereas the front surface, impinged by light, is exposed to the gas (*g*). Figure 1 sketches this model in the case of a thermally thick, optically absorbing sample.

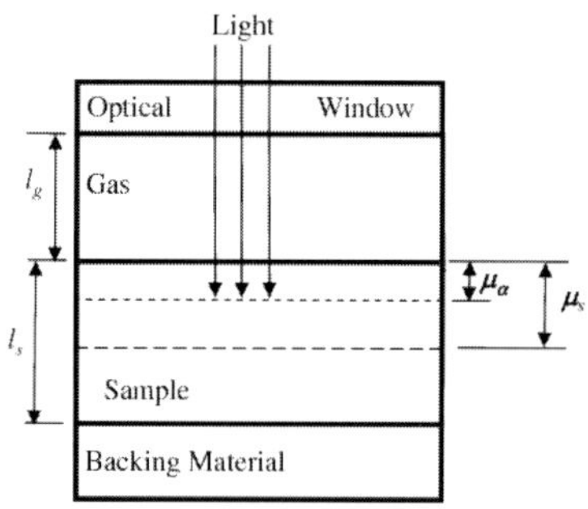

Figure 1. Schematic configuration for photoacoustic signal generation in a solid sample.

According to Rosencwaig and Gersho,[6] the quite complicated expression of the acoustic signal can be studied in six well defined cases, resulting in the simplified expressions reported in Table 1.

Both the gas and the backing material are assumed to be optically transparent. The sample absorption coefficient is α ($\bar{\nu}$), depending on the light wavenumber, $\bar{\nu}$; the sample optical depth is $\mu_\alpha = 1/\alpha$. The thermal diffusion length of the sample is $\mu_s = 1/a_s$, with a_s the thermal diffusion coefficient of the sample, given by $a_s = (\ \omega/2d_s\ )^{1/2}$; ω is the light modulation frequency; the thermal diffusivity is $d_s = k_s/\rho_s C_s$, where k_s is the thermal conductivity, ρ_s and C_s are the density and specific heat of the sample. These thermal parameters are also defined for the gas and the backing material and appear in Table 1 with the subscript *g* and *b*, respectively.

The constant factor Y in Table 1 is defined as:

$$Y = \frac{\gamma P_0 I_0}{2\sqrt{2}\, l_g T_0} \qquad (1)$$

In the case of optically and thermally thin samples (T1, T2), the signal A is proportional to $\alpha\ l_s$ and shows a ω^{-1} dependence ($\mu_b/a_g \propto 1/\omega$); the signal amplitude is managed by the thermal properties of the backing material.

In the case of either optically thin or thick, but thermally thick samples (T3, O3) the signal has a $\omega^{-3/2}$ dependence, but it is worth noting that it is proportional to $\alpha\ \mu_s$ rather than to $\alpha\ l_s$; thus the signal is only due to the light absorbed within the first thermal diffusion length; the sample thermal properties influence the signal amplitude.

In optically thick samples (O1, O2), the photoacoustic signal is independent of α, and then of $\bar{\nu}$, and varies as ω^{-1}. The thermal properties of the backing material affect the signal amplitude of thermally thin samples (O1). This is the case of a strong, spectrally flat absorber such as carbon black, commonly used for the calibration of photoacoustic spectrometers.

The phase lag between the photoacoustic signal and the modulation waveform of the light beam is due to the finite propagation time of thermal waves during signal generation. The phase angle measurements provide information on the depth from which the signal is generated within the sample. Both phase and amplitude spectra, at different modulation frequencies, are measured in samples such as films and/or layered structures.

sample morphology, the combination of PAS and FTIR has been mainly applied to characterize polymers. They are difficult to characterize by conventional IR spectroscopy, because of they are often pellets, chips, membranes or as manufactured articles.

Boccaccio et al.[12], for example, characterized several poly(vinylidene fluoride) membranes with different porous structures using different spectroscopic techniques. They demonstrated the usefulness of FTIR-PAS in studying porous materials which, as a rule, give low quality infrared spectra when other sampling techniques are used.

Fourier transform infrared PAS is also a successful IR technique to analyze catalytic systems. Examples are presented by Ryczkowski.[13]

The PA technique, apart from providing direct optical absorption spectra, can also be used to characterize thermal properties of a large variety of materials, especially semiconductors.

Raveendranath[14] et al. reported an initial attempt to understand the effects of lithium distribution within the $LiMn_2O_4$ system on the thermal properties of the complex itself and of its de-lithiated form. In particular, they investigated the variation of thermal diffusivity measured using an open photoacoustic cell. The thermal diffusivity was evaluated from the phase data, as a function of modulation frequency of excitation light (He–Ne laser, 632.8 nm, 20 mW) in transmission detection configuration. The results of this study reveal that the thermal diffusivity depends on the lithium concentration in $LiMn_2O_4$ related systems and point out the possibility of studying in detail the dependence of thermal diffusivity on lithium concentration in $Li_xMn_2O_4$ system at different values of x.

The photoacoustic technique has proved to be a powerful tool for thermal properties investigations (thermal diffusivity, thermal conductivity) of two-layer systems with minimal experimental arrangement.[15,16]

In particular, Qing Shen and Taro Toyoda[16] studied the thermal properties of porous silicon (PS) films with greatly different porosities and deposited on *p*-type Si substrates by electrochemical anodic etching (20 % – 60 %). The effective thermal diffusivities of the two-layer PS on Si samples were determined by studying the dependence of the PA signal on the light modulation frequency (12 – 400 Hz), at a fixed wavelength (a 300 W xenon arc lamp associated with a monochromator), measured under a transmission configuration. In addition, they evaluated the thermal conductivity of the PS films using a two-layer model and demonstrated that the thermal conductivity of the films depends strongly on its structure.

The PA technique was also used to derive the thermal diffusivity values of intrinsic InP and InP doped with S, Sn and Fe.[17] The authors evaluated the thermal diffusivity from the phase of the PA signal at a frequency range for which the sample thickness matches the thermal diffusion length of the sample. The advantage of using this chopping frequency range consists of the more reliable measurements of the thermal diffusivity, since, in this range, thermal diffusion is the only heat generation mechanism. They used an open PA cell irradiated by optical radiation from an argon ion laser (488 nm) modulated by a mechanical chopper. The periodic pressure fluctuations, produced in the coupling medium as a result of the heat generated due to the non-radiative de-excitations within the sample, are measured using a sensitive microphone. The phase of the PA signal is measured as a function of chopping frequency, using a dual phase lock-in amplifier.

The same experimental set-up has been employed to measure the thermal diffusivity of lanthanum phosphate ceramics prepared by the sol–gel process and sintered at different temperatures.[18] The thermal diffusivity value was obtained via the transition frequency between the thermally thin to thermally thick region from the log–log plot of photoacoustic amplitude versus chopping frequency. The influence of the sintering temperature on the propagation of heat carriers and hence on the thermal diffusivity value was also investigated. The results were interpreted in terms of variations in porosity versus the sintering temperature, as well as changes of grain size.

The PA techniques have also been used for investigating the different physical and chemical properties of polymers, as well as how the processing conditions of these materials affect their physical properties.[19,20] Rodríguez et al.[19] measured the thermal diffusivity and thermal conductivity of amylose, amylopectin and starch using an open photoacoustic cell. The thermal conductivity of these polymers was predicted from the heat diffusion theory and the ratio of the photoacoustic amplitude of the sample and other tested materials. The values obtained were in good agreement with those reported in the literature. Another application is the thermal characterization of doped polyaniline and its composites with cobalt phthalocyanine.[20] The authors, using an open PA cell, demonstrated that the effective thermal diffusivity value can be tuned by varying the relative volume fraction of the components.

3. Photoacoustic Spectroscopy in Gas Samples: Theory

The non-radiative relaxation process in optically excited gaseous samples occurs when the relaxation time can compete with the radiative lifetime of the excited energy levels. Radiative decay has a characteristic 10^{-7} s lifetime at visible wavelengths, as compared with 10^{-2} s at 10 μm. For non-radiative decay these values depend on the pressure (decay time is inversely proportional to the pressure) and can vary strongly at atmospheric pressures (10^{-3}–10^{-8} s). By modulating the radiation source at an acoustic frequency the temperature changes periodically, giving rise to a periodical pressure change that can be detected using a sensitive microphone.

Several theoretical studies of the photoacoustic effect in gases are available.[21,22] The generation and the detection of the PA signal can be divided into three main steps: i) absorption of modulated light of appropriate wavelength by molecules and heat release in the gas sample due to non-radiative relaxation (molecular collisions) of the excited states; ii) generation of an acoustic wave due to the periodic heating of the gas sample; iii) detection of the acoustic signal in the photoacoustic cell using a microphone.

The first step deals with the absorption processes and the subsequent heat production in the gas, due to the energy transfer from vibrational to translational degree of freedom. In particular, heat production in a gaseous sample excited by an intensity modulated laser beam may be described using a simple model in which the absorbing gas can be modelled by a two-level system involving the ground state and the excited state.[23] For typical atmospheric conditions the total relaxation time τ of the excited state can be approximated by the non-radiative decay time τ_{nr}, thus the whole absorbed energy in the sample is released as heat. The optical absorption coefficient of the gas is given by $\alpha(\lambda) = \sigma(\lambda)\, N$ where $\sigma(\lambda)$ is the molecular absorption cross section and N the number density of molecules. In the case of weak absorption and intensities (photon flux ϕ small enough not to saturate the molecular transition) and for modulation frequencies $\omega = 2\pi f$ that fulfill the condition $\omega\tau \ll 1$ (ω in the kilohertz range or below), the heat production rate $H(\vec{r}, t)$, as a function of the spatial coordinate $\vec{r}$ and the time t, is: $H(\vec{r}, t) = \alpha I_0(\vec{r})\, e^{i\omega t}$. Its modulation directly follows the modulation of the incident radiation $I(\vec{r}, t) = I_0(\vec{r})\, e^{i\omega t}$ without any phase lag.

The second step deals with the acoustic wave generation based on the heat production rate $H(\vec{r}, t)$. The physical laws managing the system are the energy, momentum and mass conservation laws and the state equation.

Morse and Ingard[24] have derived the inhomogeneous wave equation relating the acoustic pressure p and the heat source H:

$$\nabla^2 p(\vec{r}, t) - \frac{1}{c^2}\frac{\partial^2 p(\vec{r}, t)}{\partial t^2} = -\frac{\gamma - 1}{c^2}\frac{\partial H(\vec{r}, t)}{\partial t} \tag{2}$$

where c indicates the sound speed in the gas and $\gamma = C_p / C_V$ the ratio of specific heats. In this equation, the dissipative terms due to viscosity and thermal conduction are neglected. This equation is an inhomogeneous wave equation that can be solved by taking the time Fourier transform on both sides and expressing the solution $p(\vec{r}, t)$ as an infinite series expansion of the normal mode solution $p_j(\vec{r})$ of the homogeneous wave equation.[21]

For a sinusoidal modulation of the incident radiation with an angular frequency ω, the pressure amplitude p can be expressed as a superposition of normal acoustic modes:

$$p = p(\vec{r}, \omega) = \sum_j A_j(\omega)\, p_j(\vec{r}) \tag{3}$$

where $A_j(\omega)$ is the complex amplitude of the normal mode j and $p_j(\vec{r})$ are the solutions of the homogeneous wave equation:

$$\left(\nabla^2 + \frac{\omega_j^{\ 2}}{c^2}\right) p_j(\vec{r}) = 0 \tag{4}$$

which satisfy the boundary condition of vanishing normal gradient of $p_j(\vec{r})$ at the cell walls.

The orthonormal modes $p_j(\vec{r})$ for a cylindrical geometry of the PA cell are given by the superposition of longitudinal, azimuthal and radial modes.[24]

Each Fourier coefficient $A_j(\omega)$ is given by:

$$A_j(\omega) = -\frac{i\omega(\gamma - 1)}{\left(\omega_j^{\ 2} - \omega^2 - i\omega\omega_j / Q_j\right)} \frac{1}{V_o} \int_{V_o} p_j^* H dV \tag{5}$$

which takes into account the orthonormal conditions for the eigenfunctions, p_j, and the mode damping and where V_o is the cell volume, Q_j the quality factor of the resonance and the integral describes the geometrical coupling between laser radiation and the acoustic mode.

If the modulation frequency is equal to one of the acoustical eigenfrequencies of the cavity ($\omega = \omega_j$), the energy is accumulated in a standing wave and its amplitude is amplified in comparison to a non resonant cell ($\omega_j = 0$) by a factor equal to the quality factor Q_j. Therefore, the use of specially designed cells, which are acoustically resonant at the modulation frequency, is an effective method for sensitive trace gas detection.

The PA signal measured by a pressure sensor, usually a microphone, is given by:

$$S = C \cdot P(\lambda) \cdot \alpha(\lambda) \tag{6}$$

where C is the cell constant in the unit of V cm/W, P the optical power of the laser source and $\alpha = N\sigma$. The cell constant depends on the cell geometry, the microphone response and the nature of the acoustic mode.

PA signals are proportional to the pump laser power and therefore the maximum detection sensitivity can be realized by means of the PAS technique with high-power laser excitation. However, also diode lasers in overtone region have been extensively used.

3.1. Semiconductor Laser Sources

Laser-based photoacoustic devices have attracted a lot of interest as they exhibit high detection sensitivity and selectivity (including differentiation between isotopomers and isomers), multicomponent capability and large dynamic range (several orders of magnitude in concentration), and generally neither sample-preparation nor pre-concentration are required. However, the laser source characteristics in terms of available wavelengths, tunability, linewidth, power, operation temperature, etc., as well as the combination with appropriate sensitive detection schemes are crucial for the success of laser-based sensing. In this section a short review on the most used semiconductor laser sources, i.e. diode laser and quantum cascade laser, will be given.

Semiconductor diode lasers, made from the III-V group of semiconductor materials, have been used to access the overtone and the combination band transitions in the near infrared region. They are based on band-to-band radiative recombination at the *p-n* junction. The emission wavelength depends on the semiconductor material band gap. The tunability of diode lasers is achieved via the variation of active layer; in particular coarse tuning can be obtained by varying the current flowing through the junction, fine tuning by adjusting the temperature of the laser.

Many lattice matched structures such as AlGaAs, InGaAs, GaAlAs and GaAlAsP have been explored to cover the range 1100–1800 nm where many gases have vibrational transitions. For longer wavelengths, antimonide compounds, grown by molecular beam epitaxy on GaSb substrates and employing compressively strained GaInSbAs quantum wells (QWs) between Ga(Al)Sb(As) barriers in the active region, have been developed. They exhibit continuous wave (cw) room temperature lasing at wavelengths above 2 μm with optical powers up to 20 mW.[25] Narrow ridge Fabry-Perot GaInSbAs/GaSb type II electrically pumped QW lasers emitting at 2.35 μm have been demonstrated.[26] These lasers emit in the fundamental spatial mode and exhibit single mode operation over a wide range of currents and temperatures. They are resonant with the overtones and combination absorption lines of gases such as CO, CH_4, NH_3 and NO_2. Antimonide diode lasers in the range 2-3 μm[27] and InAsSb/InAs lasers in 3-5 μm[28,29] spectral region have been demonstrated.

Lead salt-based diode lasers have been the most diffuse tunable laser source, both for high sensitivity trace gas detection and for the determination of molecular line parameters. Sensors based on lead salt diode laser sources exhibit sensitivity of the orders of ppb.[30,31] For many applications, however, lead-salt diode laser spectrometers are limited by the requirement for cryogenic cooling of the lasers and detectors as well as the low output emission power in the sub-mW range which is often spread over several discrete wavelengths (multimode emission).

For spectroscopic applications, the main requirement is the single mode emission. This has been obtained via two different approaches. The first consists of implementing wavelength-selective elements such as gratings into the laser cavity, either directly into the diode substrate by holographic printing in the active area as in the distributed feedback (DFB) laser, or external to the diode as in the external cavity diode laser (ECDL). In figure 2 the schematic structures of DFB diode laser and ECDL are shown. Typical single mode optical gains for DFB lasers and ECDL are in the milliwatt range with room temperature operation.

We have to note that operating in the near-infrared region decreases the detection sensitivity since vibrational overtones and combination bands fall in this spectral range. On the other hand, diode laser are easy to use, robust, reliable. They can operate in a single mode regime at room temperature with relative high optical powers (mW). In addition, they can be coupled with fiber optical technology and low cost spectrometer components. The main advantages of diode lasers reside in the very narrow wavelength light emission so that interferences from other transitions are negligible and in the tunability of the output wavelength. These benefits open up a wide range of new applications such as on line sensors for manufacturing processes and for the atmospheric environment. The quantum cascade lasers (QCLs) are unipolar semiconductor lasers based on intersubband transitions in a multiple quantum-well heterostructure.

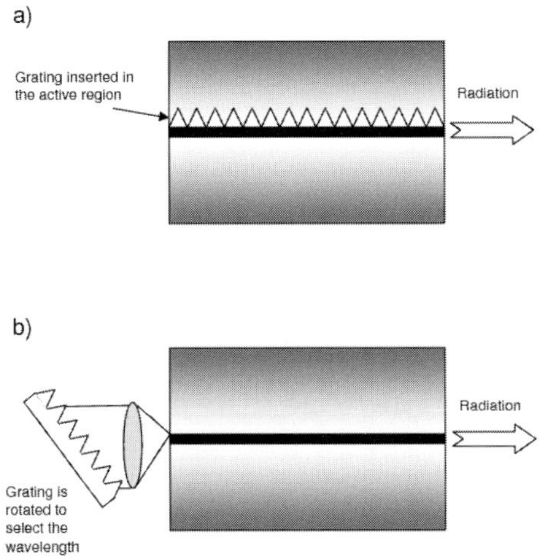

Figure 2. Schematic diagrams of (a) a distributed feedback diode laser, (b) an external cavity diode laser.

They are designed by means of band-structure engineering and grown by molecular beam epitaxy techniques. The benefit of this approach is a widely variable transition energy dictated primarily by the thicknesses of the quantum well and barrier layers of the active region rather than the band gap as in diode lasers. Typical emission wavelengths can be varied in the range 3.4 – 17 μm. Although the QCLs had been fabricated using InP-based or GaAs-based III-V materials, widely used in optoelectronics, and being able to cover a wide frequency range, the upper state lifetime was limited by the optical phonon emission; it was of the order of picosecond at room temperature, thus limiting the QCL operation to cryogenic temperatures. To overcome this effect, novel active region architectures to maintain a population inversion even at high temperatures and under strong optical fields were used, such as the so-called two-phonon or bound-to-continuum approaches.[32,33] A typical band-to-continuum structure under high electric field is sketched in figure 3.

A second approach was to reduce the waveguide losses by using buried heterostructures, where the active region is "buried" by a semi-insulating InP layer, creating a low-loss lateral waveguide. In order to reduce the thermal management, device planarization with Y_2O_3:Si_3N_4 dielectric layers has been reported by Spagnolo et al.[34]

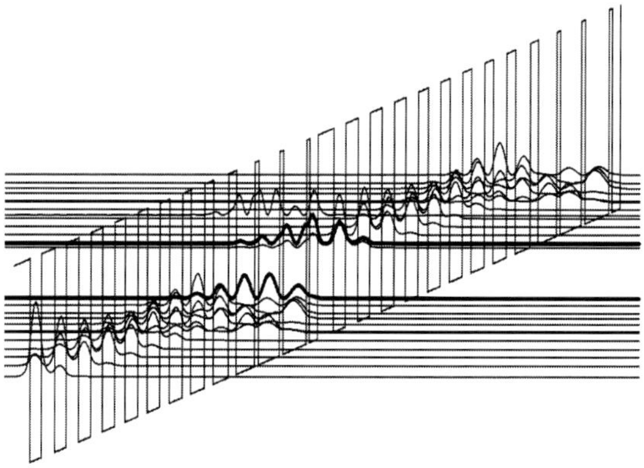

Figure 3. Typical band-to-continuum structure under an electric field of 33.5 kV/cm. The layer sequence in nm from left to right, beginning with the injection barrier is 4.0/ 2.0/ 0.8/ 5.8/ 0.9/ 5.2/ 0.9/ 4.5/ 2.0/ 3.7/ 2.1/ 3.4/2.1/ 3.2/ 2.3/ 3.1/ 2.3/ 3.1/ 2.5/ 3.1/ 2.8/ 3.0/ 3.0/ 2.9/3.25/ 2.85 where $Al_{0.45}Ga_{0.55}As$ barriers are in bold. Underlined layers are doped to a sheet carrier density of 8.3×10^{11} cm^{-2}.

In figure 4, a schematic of the planarized device heterostructure is shown. If this planarization is combined with thick gold electroplating and epilayer-side mounting of the device, the thermal resistance is further reduced. The innovations in QCLs led to the first demonstration of cw at room temperature at the wavelength of 9 μm in 2002.[35]

In 2003, Yu and colleagues[36] achieved a room-temperature cw at shorter wavelengths and very large output powers by reducing the doping in the active region and ridge width. To achieve the single frequency required by chemical sensing applications, a Bragg grating was integrated into the laser waveguide for the first time at Bell Laboratories by Gmachl and coworkers, resulting in a DFB laser, operating at cryogenic temperatures.[37]

The latest generation of QC-DFB lasers is based on a "top-grating" approach that takes advantage of the characteristics of a mid-infrared waveguide. For mid-infrared wavelengths below 15 μm, dielectric waveguides of low-doped semiconductor layers with a proper refractive index modulation are used.[38] At longer wavelengths, the waveguide is overlaid with metal. In this case the radiation is driven not only by the dielectric but also by a surface plasmon mode.[39] To date, room temperature cw DFB QCLs with an optical power up to 50 mW have been reported.

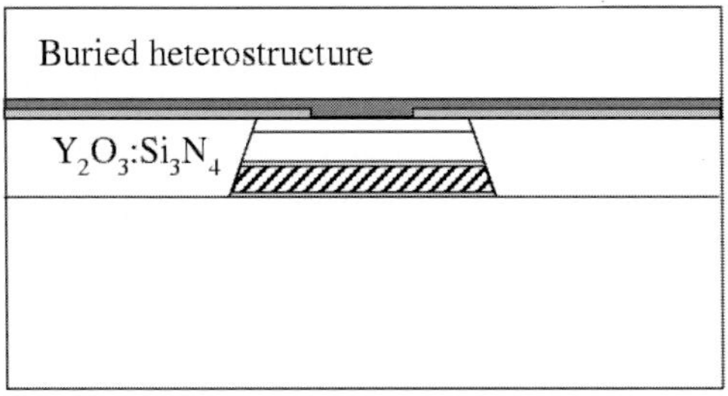

Figure 4. Schematic of QC heterostructure planarized with Y_2O_3:Si_3N_4.

Continuous wavelength tunability without mode hops is achieved through the temperature dependence of the waveguide parameters. The temperature can either be varied by a temperature change of the heat sink on which the device is mounted or more rapidly by changing the direct QC laser excitation current. Characteristic total tuning ranges per current sweep are typically around 0.4% of the emission wavelength. For many spectroscopic applications, the spectral linewidth of the laser emission is as important as continuous tunability. Because of their narrow linewidth (a few megahertz) and large optical powers, cw DFB-QCLs are suitable for gas spectroscopy applications. QCLs generally require several amperes of current in cw operation, and compliance voltages of 5–10 V. The resulting thermal load to the laser is relevant and good thermal management is necessary to reach room-temperature operation. In addition, the tuning range of a DFB-QCL covers one or two absorption lines of a gas. However, some applications are based on the detection of multi-component gas matrix, requiring a large tuning range. Fortunately, the intersubband transitions can be tailored to enable the design of active regions with very large gain bandwidth. One possibility is to have active regions tuned to different transitions coexisting in the same waveguide structure, as demonstrated in 2002 by Gmachl et al.[40] However, the selection of the single frequency by an external cavity is easier if the broad gain is achieved using a bound-to-continuum transition that offers an essentially homogeneous broadening of the gain spectrum. In 2004 Maulini and coworkers[41] demonstrated the first external cavity QCL (EC-QCL) operating in a continuous wave over a record frequency span of 175 cm^{-1}, using a bound-to-continuum QC structure with an optical power up to 10 mW. Coarse tuning can be obtained by rotating the grating, while changing the cavity length and laser chip temperature allows the fine tuning. The main advantage of EC-QCLs with respect to DFB QCLs is a broader tuning range, limited only by the spectral bandwidth of its gain element. The usefulness of these lasers for spectroscopic applications has recently been demonstrated by Wysocki et al.[42] who used a thermoelectrically cooled cw EC-QCL for spectroscopic absorption measurement of nitric oxide and water.

A broader tunability of several hundreds of wave numbers will allow the detection of entire absorption bands and enhance the flexibility of QCLs for trace gas analysis.

3.2. Photoacoustic Spectroscopy in Gas Samples: Applications

This section covers several recent application examples to illustrate the advantages of NIR diode laser and quantum cascade laser–based photoacoustic spectroscopy.

3.2.1. PAS Applications with NIR Diode Lasers

Bozoki et al. developed in 1999[43] a photoacoustic sensor system for automatic detection of low concentrations of water vapor. A Littman-configuration external-cavity diode laser operating at 1125 nm was used as a light source. It was a Fabry–Perot type double-channel InP/InGaAsP buried-heterostructure diode laser, with anti-reflection coating on one side. A gold-coated 1200 lines mm^{-1} holographic grating, illuminated at grazing incidence, was used as a wavelength-selective element and a mirror with a high-reflectivity coating was used for optical feedback. Wavelength tuning was achieved by moving the mirror with the help of a micrometer screw. A high-sensitivity PA cell with a resonance frequency of 3345 Hz, a reference photoacoustic cell and PC-controlled electronics were also integrated in the sensor. The system was calibrated by using synthetic air samples from several cylinders having different water-vapor contents by mixing them with the use of mass-flow controllers. Different water-vapor concentrations were prepared in a randomly selected sequence. The laser-modulation frequency was adjusted to the peak of the resonance. From the slope of the fitted calibration curve and by taking into account the 1.5 mW light power accessible in the measurement, the 50 mV Pa^{-1} sensitivity of the measuring microphone and the optical absorption on the top of the measured line of $9.5x10^{-8}$ cm^{-1}/(μmol per mol), the cell constant of the measuring PA cell was deduced to be 3900 Pa cm W^{-1}. From the standard deviation of the linear fit the minimum detectable concentration of water vapor was determined to be 13 μmol per mol.

A novel approach to wavelength-modulation photoacoustic spectroscopy was reported by M. E. Webber et al.,[44] to detect NH_3. They integrated a diode laser and an optical fiber amplifier into the sensor, to enhance sensitivity. The tunable diode laser operated near 1532 nm with more than 30 dB optical isolation and greater than 20 mW of output power to ensure that the erbium-doped fiber amplifier (EDFA) was fully saturated. The ammonia transition near 1532 nm was selected because of its isolation from H_2O and CO_2 interferences. The diode laser operated in wavelength modulation mode, the output of the EDFA, up to 500 mW, was collimated into free space and aligned through the optoacoustic cell with a double-pass configuration to yield a total path length of 18.4 mm. The EDFA unit also had a built-in uncalibrated photodiode for monitoring the relative output power, which is used for normalizing the optoacoustic signal to account for any power changes. The signals from the 5 kHz bandwidth microphones were conditioned and demodulated at twice the modulation frequency by use of a lock-in amplifier. The gas-handling system was composed of a source bottle with a NIST-traceable mixture containing 10.9 ppm of NH_3 in a balance of N_2, a bottle of N_2 for dilution, and two mass-flow controllers, used to create mixtures with ammonia concentrations of 100–1000 ppb. The signal-to-noise ratio (SNR) indicated that the detection limit for this sensor was better than 6 ppb of ammonia. The normalized minimum detectable fractional optical density, $\alpha_{min}l$, was calculated to be 1.8 x10^{-8}; the minimum detectable absorption coefficient, α_{min}, was 9.5x10^{-10} cm^{-1}; and the minimum detectable absorption coefficient normalized by power and bandwidth was 1.5x10^{-9} W $cm^{-1}/Hz^{1/2}$.

The best detection limits for H_2O, HCl and CH_4 were obtained by Besson et al. in 2006.[45] They used a fiber-coupled multi-gas PA cell, made of three identical stainless steel gold-coated cylinders of 3 mm inner radius and 17 cm long, terminated by two large diameter buffer volumes. The length of the buffer volumes was designed to realize altogether a large impedance difference and an efficient acoustic notch filtering corresponding to a quarter wavelength of the resonator standing wave (L_{buff} = 85 mm). The diameter of these volumes was optimized (d = 50 mm) to enable the implementation on the outer flange of all required elements, such as windows, loudspeaker and gas inlets. Each of the three resonators was excited by a DFB laser, in order to benefit from the full optical power from each laser. The resonators operated in their first longitudinal mode at a resonance frequency close to 1 kHz in air. Three optical fibers terminated by a built-in collimator, were fixed at the outer flange of the buffer volume. A fine mechanical alignment of the collimators allowed the beams to pass through the resonators without touching the walls to avoid any wall noise. A piezo-electric transducer, fixed on the outer flange of the second volume, has been used as a sound emitter for automatic tracking of the resonance frequency. The gas measurements were performed for H_2O at 1368.6 nm, CH_4 at 1651.0 nm and HCl at 1737.9 nm. Detection limits of 80 ppb of CH_4, 24 ppb of H_2O and 30 ppb of HCl, were extrapolated by considering a SNR = 3.

3.2.2. PAS Applications with QCLs

The commercial availability of high performance QCLs has given a great impulse to the development of PA sensors operating in the spectral range 3-24 μm, where most molecular species exhibit strong rotational-vibrational absorptions with characteristic features. These features allow their sensitive and selective detection.

Paldus et al.[46] reported photoacoustic spectra of NH_3 and water vapor using a cw cryogenically cooled QC-DFB laser operating at 8.5 μm with a 16 mW power output. They used a PAS resonant cell (1.66 kHz) which consisted of an acoustic resonator (100 mm long, gold-coated copper) and two buffer volumes (50 mm long). The laser beam intensity was modulated at 1.66 KHz using a mechanical chopper. The QC-DFB was scanned in wavelength over 35 nm by temperature tuning for generation of absorption spectra or temperature stabilized for real-time concentration measurements. The detection limit of ammonia was 100 ppbv at standard temperature and pressure with a measurement time of 10 minutes.

Hofstetter et al.[47] reported PA measurement of CO_2, CH_3OH and ammonia using a pulsed 10.4 μm QC-DFB laser operated at 3-4% duty cycle with 25 ns long current pulses (2 mW average power) and close to room temperature. The QC-DFB was scanned in wavelength over 3 cm^{-1} by temperature tuning with a linewidth of 0.2 cm^{-1}. They used a resonant multi-pass PA cell consisting of an acoustic resonator (120 mm long, gold-coated copper) and two buffer volumes (60 mm long) integrated into a Herriott multipass arrangement (36 passes with an effective pathlength of 15 m). The PA cell was equipped with a radial 16-microphone array to increase sensitivity. The laser beam intensity was mechanically chopped at the first longitudinal resonance (Q=70) of the PA cell (1.25 kHz). The ammonia detection limit was 300 ppb, which corresponds to a minimum measurable absorption coefficient of α_{min}=2.2×10-5 cm^{-1}, with a SNR of 3 and a pressure of 400 mbar. PA absorption spectra of CO_2, CH_3OH and NH_3 were also reported.

Da Silva et al.[48] have reported the PA measurement of O_3 with a commercial DFB-QCL (Alpes Lasers) emitting at 9.5 μm and working in pulsed operation (duty cycle 2 % and 50 ns long current pulses for determination of concentrations, duty cycle 0.8 % and 20 ns long current pulses for determination of spectra) near room temperature (thermoelectrically cooled). The QCL (2 - 4.6 mW average power) was modulated by an external TTL signal at 3.8 KHz to excite the first longitudinal mode of the differential PA cell (with a Q=36). PA spectra were measured by scanning the wavelength of the QCL by temperature tuning. Photoacoustic measurements were performed for concentrations ranging from 4300 ppm to 100 ppb. The detection sensitivity of ~100 ppb corresponds to a minimum measurable absorption coefficient of α_{min}=1.24×10-6 cm^{-1} with a SNR of 1.

Elia et al. developed quantum cascade laser-based photoacoustic sensors for the detection of NO[49] and hexamethyldisilazane (HMDS)[50] with a detection limit on the order of hundred parts in 10^9. They used a photoacoustic spectrometer consisting of amplitude modulated QCL, a photoacoustic cell and a signal acquisition and processing equipment. The resonant PA cell was a cylindrical stainless steel resonator of 120 mm length and 4 mm radius with λ/4 buffer volumes on each side used as acoustic filters. The cell was closed by two antireflection coated ZnSe windows. The resonator operated in the first longitudinal mode at 1380 Hz and was equipped with 4 electret microphones (Knowles EK 3024, 20 mV/Pa, 0.5 $\mu V/Hz^{-1}$), placed at the position of maximum acoustic amplitude to increase the signal-to-noise ratio. The source used to detect NO was a DFB-QCL (Alpes Lasers) operated in pulsed mode (pulse duration of 42 ns and a duty cycle of 1.4%) with an optical average power of 8 mW at a wavelength around 5.3 μm. The laser beam intensity was modulated by a mechanical chopper at the first longitudinal resonance frequency of the photoacoustic cell. The detection limit for NO measurement was 500 ppb with a 10 s integration time constant. The minimum detectable absorption coefficient at SNR=1 was $\alpha_{min}=4.4\times10^{-6}$ cm^{-1}, and the minimum detectable absorption coefficient normalized to power and detection bandwidth was 1.1×10^{-7} $cm^{-1}W/Hz^{1/2}$. They also reported the PA detection of HMDS at ppb level. The laser source was a home made Fabry-Perot QCL with a superlattice active region and a relaxation-stabilized injector,[51] emitting a peak power of 2 W at temperatures below 120 K and several 100 mW at room temperature. The laser was mounted in a helium closed-cycle cryostat, working at 20 K. The optical power was modulated at the acoustic frequency of the cell using a pulse generator. All the photoacoustic measurements were performed at a duty cycle of 0.014 % (100 ns pulse length). A minimum detectable concentration of 200 ppb was obtained, essentially limited by the background signal (~ 500 μV).

Von Lilienfeld-Toal et al.[52] demonstrated the feasibility of a QCL-based PA sensor for non-invasive glucose measurements. They used two laser source, supplied by Alpes Lasers emitting at 1080 and 1066 cm^{-1}. These wavelengths were in the maximum and the minimum of a glucose mode, respectively. Furthermore, the mode at 1080 cm^{-1} showed a high correlation with the glucose concentration in the in vitro studies, and the mode at 1066 cm^{-1} was much less influenced by the presence of glucose. Therefore they expected the ratio between these two to indicate the changes of glucose. They demonstrated the reliability of the method on the basis of the clinical laboratory data.

The Groupe de Spectrometrie Moleculaire et Atmospherique (Reims, France)[53] detected atmospheric methane using a QCL, supplied by Alpes Lasers, emitting near 7.9 μm on the v_4 band of CH_4. It was a cw DFB laser, working at cryogenic temperature and requiring a cooling system, with a power of 80 mW at 80 K. They used a differential Helmholtz resonance PA cell, made with stainless steel cylindrical volumes, 10 cm in length, 5mm radius. They were linked by capillaries with 10 cm length and a radius of 2 mm. On the middle of these capillaries, valves were placed for the inlet and outlet of gas. The BaF_2 cell windows had a thickness of 3 mm. The resonance of this cell was 315 Hz. The acoustic signal was measured with two microphones (Bruel & Kjaer 4179) placed at the centre of the two volumes. The achieved detection limit was of 3 ppb. Moreover they found a CH_4 concentration in air slightly above 2 ppm. Note that the average concentration of methane in ambient air is generally 1.7 ppm.

Lima and co-workers[54] reported the PA detection of NO_2 and N_2O. A pulsed 6.2-μm QCL and a 8-μm QCL were employed, respectively, as versatile tunable light sources. The InGaAs-AlInAs/InP DFB-QCL laser sources were commercially available single-frequency lasers (Alpes Lasers), designed for pulsed operation near room temperature. The QCLs were excited with a pulse duration of 50 ns (duty cycle 2%) and modulated by an external TTL signal at 3.8 kHz to excite the first longitudinal mode of the resonant differential PA cell. A sensitivity of 80 ppb, for NO_2 detection, was derived for SNR = 1. In the case of N_2O measurements, the sensitivity, calculated from the standard deviation of the background signal, was 84 ppb at SNR = 1.

4. Conclusions

The theory of photoacoustic signal generation in solids and gases has been highlighted; we have also reported a survey of the applications of diode and quantum cascade semiconductor lasers in photoacoustic spectroscopy, selecting only few of the most significant experimental results obtained up to now.

References

1. A. G. Bell, *Am. J. Sci.* **20(118)**, 305 (1880).
2. M. L. Viegerov, *Dokl. Akad. Nauk SSSR* 19, 687 (1938).
3. K. F. Luft, *Z. Tech. Phys.* **5**, 97 (1943).
4. E. L. Kerr and J.G. Atwood, *Appl. Opt.* **7**, 915 (1968).
5. L. B. Kreuzer, *J. of Appl. Phys.* **42**, 2934 (1971).
6. A. Rosencwaig and A.Gersho, *J. of Appl. Phys.* **47(1)**, 64 (1976)
7. S. Park, S. K. Lee, J. Y. Leet, J. E. Kim, H. Y. Park, H. L. Park, H. Lim and W.T. Kim, *J. Phys. Condens. Matter.* **4**, 579 (1992).
8. N. L. Pickett, F. G. Riddell, D. F. Foster, D. J. Cole-Hamilton and J. R. Fryerb, *J. Mater. Chem.* **7(9)**, 1855 (1997).
9. G. C. Astrath, A. C. Bento, M. L. Baesso, A. Ferreira da Silva and C. Perssonll, *Thin Solid Films*, **515**, 2821 (2006).
10. J. M. Jáñes-Limón, J. F. Pérez-Robles, J. González-Hernández, Y.V. Vorobiev, J.A. Romano, F. G. C. Gandra and E. C. da Silva, *J. Sol-Gel Sci. Techn.* **18**, 207 (2000).
11. T.A. El-Brolossy, S. Abdallah, T. Abdallah, M. B. Mohamed, S. Negm and H. Talaat, *Eur. Phys. J. Special Topics* **153**, 365 (2008).
12. T. Boccaccio, A. Bottino, G. Capannelli and P. Piaggio, *J. Membr. Sci.* **210**, 315 (2002).
13. J. Ryczkowski, *Catalysis Today*, **124**, 11 (2007).
14. K. Raveendranath, Jyostna Ravi, S. Jayalekshmi, T.M.A. Rasheed and K. P. R. Nair, *Materials Science and Engineering* ***B*** **131**, 210 (2006).
15. A. Cruz Orea, I. Delgadillo, H. Vargas, J. L. Pichardo and J. J. Alvarado, *Solid State Communications*, **100(12)**, 855 (1996).
16. Q. Shen and T. Toyoda, *Rev. of Sci. Instr.* **74(1)**, 601 (2003).
17. S. D. George, P. Radhakrishnan, V. P. N. Nampoori and C. P. G. Vallabhan, *J. Phys. D: Appl. Phys.* **36**, 990 (2003).
18. S. D. George, Rajesh Komban, K. G. K. Warrier, P. Radhakrishnan, V.P.N. Nampoori and C. P. G. Vallabhan, *Intern. J. of Thermophys.* **28(1)**, 123 (2007).
19. P. Rodríguez and G. González de la Cruz, *J. of Food Engineering* **58**, 205 (2003).

20. S. D. George, S. Saravanan, M. R. Anantharaman, S. Venkatachalam, P. Radhakrishnan, V. P. N. Nampoori and C. P.G. Vallabhan, *Phys. Rev.* **B 69**, 235201 (2004).
21. L. B. Kreuzer, *The physics of signal generation and detection*, in Optoacoustic spectroscopy and detection, edited by Y-H. Pao (Academic Press, New York, 1977).
22. A. Rosencwaig, in *Photoacoustics and photoacoustic spectroscopy*, vol. 57 (J. Wiley and sons, New York, 1980).
23. A. C. Tam, in *Ultra sensitive laser spectroscopy* (Academic press, New York, 1983).
24. P. M. Morse and K. U. Ingard, in *Theoretical Acoustics* (Princeton University, Princeton, N. J. 1986).
25. Y. Rouillard, F. Genty, A. Perona, A. Vicet, D. A. Yarekha, G. Boissier, P. Grech, A. N. Baranov and C. Alibert, *Philos. Trans. R. Soc. London, Ser.* **A 359**, 581 (2001).
26. P. Werle, *Spectrochim. Acta* **A 54**, 197 (1998).
27. A. Fried, S. Sewell, B. Henry, B. P. Wert, T. Gilpin and J. R. Drummond, *J. Geophys. Res. Atmospheres* **102**, 6253 (1997).
28. B. Lane, D. Wu, A. Rybaltowski, H. Yi, J. Diaz and M. Razeghi, *Appl. Phys. Lett.* **70**, 443 (1997).
29. P. Werle, A. Popov, *Appl. Opt.* **38**, 1494 (1999).
30. P. Werle, K. Maurer, R. Kormann, R. Mücke, F. D'Amato, T. Lancia and A. Popov, *Spectrochim. Acta* **A 58**, 2361 (2002).
31. A. Fried, Y. Wang, C. Cantrell, B. Wert, J. Walega, B. Ridley, E. Atlas, R. Shetter, B. Lefer, M. T. Coey, J. Hannigan, D. Blake, N. Blake, S. Meinardi, B. Talbot, J. Dibb, E. Scheuer, O. Wingenter, J. Snow, B. Heikes and D. Ehhalt, *J. Geophys. Res.* **108(D4)**, 8365 (2003).
32. J. Heinrich, R. Langhans, M.S. Vitiello, G. Scamarcio, D. Indjin, C. A. Evans, Z. Ikonić, P. Harrison, S. Höfling andA. Forchel, *Appl. Phys. Lett.*, **92**, 141111 (2008).
33. M. S. Vitiello, G. Scamarcio, V. Spagnolo, A. Lops, Q. Yang, C. Manz and J. Wagner, *Appl. Phys. Lett.*, **90**, 121109 (2007).
34. V. Spagnolo, A. Lops, G. Scamarcio, M. S. Vitiello and C. Di Franco, *J.of Applied Physics*, **103**, 043103 (2008).
35. M. Beck, D. Hofstetter, T. Aellen, J. Faist, U. Oesterle, M. Ilegems, E. Gini and H. Melchior, *Science*, **295**, 301 (2002).
36. J. S. Yu, S. Slivken, A. Evans, L. Doris and M. Razeghi, *Appl. Phys. Lett.*, **83** (13), 2503 (2003).
37. C. Gmachl, J. Faist, J. N. Baillargeon, F. Capasso, C. Sirtori, D. L. Sivco and A.Y. Cho, *IEEE Photon. Technol. Lett.*, **9**, 1090 (1997).
38. R. Kohler, C. Gmachl, A. Tredicucci, F. Capasso, D. L. Sivco, S. N. G. Chu and A.Y. Cho, *Appl. Phys. Lett.*, **76**, 1092 (2000).

39. J. Faist, D. Hofstaetter, M. Beck, T. Aellen, M. Rochat and S. Laser, *IEEE J. Quant. Electron.*, **38**, 533 (2002).
40. C. Gmachl, D. L. Sivco, R. Colombelli, A.Y. Cho and F. Capasso, *Nature*, **415**, 883 (2002).
41. R. Maulini, M. Beck, J. Faist and E. Gini, *Appl. Phys. Lett.*, **84**, 1659 (2004).
42. G. Wysocki, R. F. Curl, F. K. Tittel, R. Maulini, J. M. Bulliard and J. Faist, *Appl. Phys.* **B***: Lasers Opt.*, **81**, 769 (2005).
43. Z. Bozoki, J. Sneider, Z. Gingl, A. Mohacsi and M. Szakall, *Meas. Sci. Technol.*, **10**, 999 (1999).
44. M. E. Webber, R. Claps, F. V. Englich, F. K. Tittel, J. B. Jeffries and R. K. Hanson, *Applied Optics*, **40(24)**, 4395 (2001).
45. J. P. Besson, St. Schilt and L. Thevenaz, *Spectrochim. Acta* **A 63**, 899 (2006).
46. B. A. Paldus, T. G. Spence, R. N. Zare, J. Oomens, F. J. M. Harren, D. H. Parker, C. Gmachl, F. Capasso, D. L. Sivco, J. N. Baillargeon, A. L. Hutchinson and A. Y. Cho, *Opt. Lett.*, **24**, 178 (1999).
47. D. Hofstetter, M. Beck, J. Faist, M. Nagele and M. W. Sigrist, *Opt. Lett.*, **26**, 887 (2001).
48. M.G. Da Silva, H. Vargas, A. Miklós and P. Hess, *Appl. Phys.* **B 78**, 677 (2004).
49. A. Elia, P. M. Lugarà and C. Giancaspro, *Opt. Lett.*, **30(9)**, 988 (2005).
50. A. Elia, F. Rizzi, C. Di Franco, P. M. Lugarà and G. Scamarcio, *Spectrochimica Acta* **A 64**, 426 (2006).
51. G. Scamarcio, M. Troccoli, F. Capasso, A. L. Hutchinson, D. L. Sivco and A. Y Cho, *Electronics Lett.*, **37(5)**, 1 (2001).
52. H. von Lilienfeld-Toal, M. Weidenmuller, A. Xhelaj and W. Mantele, *Vibrational Spectroscopy*, **38**, 209 (2005).
53. A. Grossel, V. Zeninari, L. Joly, B. Parvitte, D. Courtois and G. Durry, *Spectrochim. Acta* **A 63**, 1021 (2006).
54. J. P. Lima, H. Vargas, A. Miklos, M. Angelmahr and P. Hess, *Appl. Phys.* **B 85**, 279 (2006).

DIGITAL HOLOGRAPHY: A NON-DESTRUCTIVE TECHNIQUE FOR INSPECTION OF MEMS

Giuseppe Coppola,[a,*] Simonetta Grilli,[b] Pietro Ferraro,[b]

Sergio De Nicola[c] and Andrea Finizio[c]

[a]*Istituto di Microelettronica e Microsistemi, CNR*
Via P. Castellino, 111, Napoli, Italy
[b]*Istituto Nazionale di Ottica Applicata, CNR*
Via Campi Flegrei, 34, Pozzuoli (Na), Italy
[c]*Istituto di Cibernetica, CNR*
Via Campi Flegrei, 34, Pozzuoli (Na), Italy
[*]*E-mail: giuseppe.coppola@cnr.it*

This chapter describes the possibility to use Digital Holography as a tool to carry out a non-contact and non-destructive characterization and inspection of micro-electro-mechanical systems (MEMS). The technique allows to evaluate quantitatively, with high accuracy, different features of a typical MEMS: the profile; the deformations induced by external influences; the behavior when actuated under operation conditions. Digital holography provides two main advantages consisting in the possibilities to perform a dynamic characterization of the MEMS and to reconstruct the in-focus image of the object for 3D structures. The evaluation of the MEMS performance is particularly useful when studying and understanding the effectiveness of the design and of the fabrication process. Several examples of MEMS inspection are illustrated to demonstrate the reliability of the technique.

1. Introduction

In recent years the development of highly innovative microdevices, for different fields of application, is mainly due to the development of Micro-Electro-Mechanical-System (MEMS) and Micro-Optical-Electro-Mechanical-System (MOEMS) structures.[1] The realization of such microstructures is based on highly sophisticated and complex micro-machining methods. These techniques allow to obtain highly integrated microdevices with high level tolerances and performances. However some factors can affect the fabrication of reliable devices.

The even more complex geometries of microstructures, demanded by new designs, impose the development of even different and more complex micromachining techniques. In fact it is possible to say that it does not exist, for such microstructures, a standard fabrication process to be adopted for different devices: each microdevice needs a specific fabrication process. Moreover, the absence of reliable numerical models, that can predict the overall behavior of the microdevices during operation and during the fabrication process, is another critical point. In fact, microstructures behave very differently by macro-structures because of the effects due to atomic and surface science interactions that are usually negligible for macro-structures. Reliable characterization tools are strongly required in order to fully exploit the MEMS and MOEMS technologies on very large scale market. Those tools would provide quantitative information for assessing reliable fabrication processes and would allow non destructive testing and evaluation of microstructures and microdevices.

Interferometric techniques are very powerful tools to inspect and characterize both the fabrication process and the final functional behavior of microdevices. Interferometric analysis at different wavelengths, combined with microscopes can measure with high accuracy and in full-field mode the different important parameters such as three-dimensional profile, refractive index, strain and local displacement induced by external mechanical and thermal loads. Nowadays, interferometers are successfully adopted for such applications but some drawbacks are limiting their potentialities. In fact, in most common interferometers, the optical path difference (OPD) is measured by means of the well known phase-shifting technique. This method requires an optical component able to create precise phase-shifts in one arm of the interferometer. The phase, and consequently the OPD, is retrieved by digitizing at least three shifted interferograms. The phase-sifting requires the acquisition of more than one image, thus preventing a consistent real-time analysis. That constitutes one of the main drawbacks of the method. Recently, a new holographic method has been developed to overcome this limitation, the Digital Holography (DH).[2-6] DH allows the registration of the hologram directly by means of a solid state detector (CCD or CMOS) and the subsequent numerical reconstruction of the amplitude and phase the object beam. DH is characterized by the acquisition of only one image, thus giving the possibility to study behaviors of MEMS and MOEMS in real-time.

2. Principle of Operation of Digital Holography

In holography, an object is illuminated by a collimated, monochromatic, coherent light with a wavelength λ. The object scatters the incoming light forming a complex wavefield (the *object beam):*

$$O(x,y)=|O(x,y)|e^{j\phi(x,y)} \tag{1}$$

where $|O|$ is the amplitude and ϕ the phase, x and y denote the Cartesian coordinates in the plane where the wavefield is recorded (*hologram plane*). The phase $\phi(x,y)$ incorporates information about the topographic profile of the MEMS under investigation because it is related to the optical path difference (OPD):

$$\phi(x,y)=\frac{4\pi}{\lambda}\cdot OPD \tag{2}$$

where a reflection configuration has been considered. The purpose of holography is to capture the complete wavefront, and in particular the phase, ϕ, and reconstructs this wavefront in order to obtain a quantitative information about the topographic profile of the object. Since all light sensitive sensors respond to intensity only, the phase is encoded in the intensity fringe pattern adding another coherent background wave $R(x,y)=|R(x,y)|e^{j\varphi(x,y)}$, called the *reference beam.* Both waves interfere at the surface of the recording device. The intensity of this interference pattern is calculated by:

$$\begin{aligned} I(x,y)&=|O(x,y)+R(x,y)|^2 = \\ &=|R(x,y)|^2+|O(x,y)|^2+O(x,y)R^*(x,y)+R(x,y)O^*(x,y) \end{aligned} \tag{3}$$

where * denotes the conjugate complex. The *hologram* is proportional to this intensity: $H(x,y)=\alpha\cdot I(x,y)$.

In Digital Holography the hologram is acquired by a CCD (or CMOS) camera array, i.e. a two-dimensional rectangular raster of $M \times N$ pixels, with pixel pitches Δx and Δy in the two directions. Thus, hologram patters recorded by a CCD are nothing less than digitized version of the wavefields that impinge on the CCD surface. Therefore, Δx and Δy are the sampling intervals in the observation plane. Mathematically this two-dimensional spatial sampling can be described using the following relation:

$$H(m,n) = H(x,y)rect\left(\frac{x}{M\Delta x}, \frac{y}{N\Delta y}\right)\sum_{m=1}^{M}\sum_{n=1}^{N}\delta(x - m\Delta x, y - n\Delta y) \qquad (4)$$

where δ(*x*, *y*) is two-dimensional Dirac-delta function, *m* and *n* are integer numbers, (*M*Δx)×(*N*Δy) is the area of the digitized hologram, *rect*(*x*,*y*) is a function defined as a constant amplitude value, if the point of coordinated (*x*, *y*) is inner to the digitized hologram, and is zero elsewhere. The sampling process should satisfy the Shannon sampling theorem in order to obtain a perfect reconstruction of the object image. In particular, to satisfy this theorem, the distance between two contiguous fringes of the interference pattern, described by Eq. (3), must be recorded by at least two pixels of the acquisition system. Figure 1(a) shows an hologram relative to a micro-balance realize by micromaching. An optical image of the characterized structures is shown into the inset of the Fig. 1(a).

Hologram reconstruction is achieved by multiplying the recorded intensity distribution of the hologram *H(m, n)* by the reference wave-field in the hologram plane:

$$\begin{aligned} R(m,n)H(m,n) &= R(m,n)\alpha\left|R(m,n)\right|^2 + R(m,n)\alpha\left|O(m,n)\right|^2 + \\ &+ \alpha\left|R(m,n)\right|^2 O(m,n) + \alpha R^2(m,n)O^*(m,n) \end{aligned} \qquad (5)$$

The first term on the right side of this equation is the attenuated reference wave, the second one is a spatially varying “cloud” surrounding the first term. These two terms constitute the zero-th order of diffraction. The third term is, except for a constant factor, an exact replica of the original wavefront and for this reason is called *virtual image*. The last term is another copy, the *conjugate image*, of the original object wave, but focused on the opposite side on the holographic plane (*real image)*. These three terms are overlapped if the reference beam and the object beam are collinear (in-line configuration).

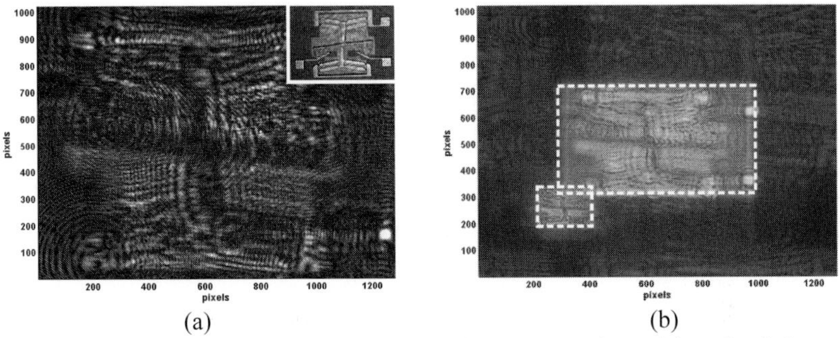

Figure 1. (a) Acquired hologram of a micro-balance; *inset*: view of the microbalance. (b) The three terms obtained by the reconstructing process when an off-axis configuration is adopted. The small white broken frame is relative to the virtual image, the big one is relative to the real image, whereas the zero-th order of diffraction is spread on the entire figure.

Several methods have been proposed to solve the twin-image problem or at least the elimination of the dc term,[7-12] in order to extract only the virtual image. Many of these techniques require the acquisition of more than one hologram. This methodology can be a serious limitation for the real-time characterization of dynamic MEMS. An alternative configuration uses a small angle between the reference beam and the object beam (off-axis configuration). In this way, as reported in Fig. 1(b) the three diffraction orders reported in Eq. 5 propagate along different directions and can be spatially filtered.

In the reconstruction process, the hologram can be seen as an amplitude transmittance that diffracts the reference wave. In other words, the wavefront scattered by the object under investigation is obtained through the propagation of the $R(m, n){\cdot}H(m, n)$ from the holographic plane to the image plane. This back propagation is numerically obtained by using a discretization of Fresnel-Kirchoff paraxial approximation of the Rayleigh-Sommerfeld's diffraction formula:[8]

$$Q(m,n) = -\frac{j}{\lambda d} e^{j\frac{2\pi d}{\lambda}} e^{j\frac{\pi \lambda d}{MN}\left(\frac{m^2}{\Delta x^2}+\frac{n^2}{\Delta y^2}\right)} \cdot$$
$$\cdot \sum_{k=0}^{M-1}\sum_{l=0}^{N-1}\left[H(k\Delta x, l\Delta y) R(k\Delta x, l\Delta y) e^{j\frac{\pi}{\lambda d}\left(k^2\Delta x^2 + l^2 \Delta y^2\right)} e^{-2j\pi\left(\frac{km}{M}+\frac{\ln}{N}\right)}\right] = \tag{6}$$
$$= -\frac{j}{\lambda d} e^{j\frac{2\pi d}{\lambda}} e^{j\frac{\pi \lambda d}{MN}\left(\frac{m^2}{\Delta x^2}+\frac{n^2}{\Delta y^2}\right)} \cdot DFT\left\{ H(k\Delta x, l\Delta y) R(k\Delta x, l\Delta y) e^{j\frac{\pi}{\lambda d}\left(k^2\Delta x^2 + l^2 \Delta y^2\right)}\right\}$$

where DFT{…} denotes the Discrete Fourier Transform and d is the distance between the holographic plane and the image plane. The reconstructed image is an M × N matrix with elements (m, n) and steps:

$$\Delta\xi = \frac{d\lambda}{M\Delta x} \qquad \Delta\eta = \frac{d\lambda}{N\Delta y} \tag{7}$$

along the two transversal directions. Thus, by the back propagation, a discrete version of the complex optical wavefront present on the surface of the object can be reconstructed. The possibility to manage numerically this reconstructed optical field allows to simultaneously determine both its intensity and especially its phase distribution $\phi(m,n)$. By inverting the Eq. 2 and considering an homogeneous material, from the reconstructed phase distribution the height distribution $s(m,n)$ of the object under investigation can be obtained; namely:

$$s(m,n) = \frac{\lambda}{4\pi}\phi(m,n) = \frac{\lambda}{4\pi}\arctan\frac{\mathrm{Im}[Q(m,n)]}{\mathrm{Re}[Q(m,n)]} \tag{8}$$

where *Im* and *Re* are the imaginary and real part of the reconstructed optical field, respectively. As reported in Eq. 8 the phase distribution is obtained by a numerical evaluation of the *arctan* function, so the values of the reconstructed phase are restricted in the interval [-π,π], i.e. the phase distribution is wrapped into this range. In order to resolve eventual ambiguities arising from height differences greater than $\lambda/2$, phase-unwrapping methods have to be applied.[13-18]

The possibility to process a numerically reconstructed profile $s(m,n)$ allows to compare the MEMS topography between two different operational states. In other words, if $Q_1(m,n)$ and $Q_2(m,n)$ are the reconstructed complex wavefields of two holograms recorded at two different states of the object, the corresponding phase change $\Delta\phi_{1\text{-}2}=\phi_1-\phi_2$, and consequently profile variation $\Delta s_{1\text{-}2}$ are given (considering the Eq. 8) by:[19-23]

$$\Delta s_{1-2}(m,n)=\frac{\lambda}{4\pi}\Delta\phi_{1-2}(m,n)=$$
$$=\frac{\lambda}{4\pi}\frac{\mathrm{Re}[Q_2(m,n)]\mathrm{Im}[Q_1(m,n)]-\mathrm{Im}[Q_2(m,n)]\mathrm{Re}[Q_1(m,n)]}{\mathrm{Re}[Q_2(m,n)]\mathrm{Re}[Q_1(m,n)]+\mathrm{Im}[Q_2(m,n)]\mathrm{Im}[Q_1(m,n)]} \tag{9}$$

In the case of deformation measurements, the two states are states of deformation of the object under investigation and the calculated profile variation provides information about the displacement of the surface of the investigated object.

Equations 6 and 7 show that, when the Fresnel-Kirchoff approximation is involved and starting from an hologram with dimension of $M\Delta x \times N\Delta y$ pixels, an image with a dimension of $M\Delta\xi \times N\Delta\eta$ pixels is reconstructed. As consequence, the dimension of the reconstructed images depend (see Eq. 7) on the features of the CCD o CMOS employed camera; (number of pixels N, M and pixels size Δx, Δy), the wavelength of the illumining laser source and the reconstruction distance *d*. Furthermore, the use of the Fresnel-Kirchoff integral limits the resolution, a limitation that turns out to be particularly severe when trying to achieve better resolution, as in microscopy application where aberrations introduced by microscope objective and imaging lenses are unavoidable present and high-spatial frequency components are diffracted at large angles. In this condition, the Fresnel-Kirchoff integral can be still used as a valid tool for numerical reconstruction of the hologram, but the possibility offered by Digital Holography to manage the phase of the reconstructed image has to be exploited in order to remove and/or compensate the unwanted wave front aberrations.[6,24-28] Finally, the Fresnel-Kirchoff approximation greatly reduces the calculation time, through the use of the two-dimensional DFT algorithm.

3. MEMS Inspection

The previous section reported the procedure for retrieving the phase distribution of the object wave field, allowing to obtain quantitative phase contrast imaging in microscopy. Therefore, Digital Holography, by means of the reconstructed phase distribution, can be directly used for metrological applications and, in particular, for inspection and quantitative evaluation of microstructure surface morphology.

Nowadays, other 3D imaging methods based on interferometry are available. These allow the measurement of minute displacements and surface profiles. Methods like holographic interferometry, fringe projection, and speckle metrology can provide full-field non-contact information about coordinates, deformations, strains, stresses and vibrations. However, many of these methods do not allow an easy compensation of the optical aberrations and/or the direct calculation of the full-field map of the object through the calculation of the complex wave front from a single exposure.[19-21,27] As a consequence, both the acquisition time and the sensitivity to external parameters, such as temperature, pressure, mechanical stability, etc are reduced.

Figure 2 shows an experimental recording holographic set-up used to characterize MEMS with several different geometries and shapes, such as cantilever beams, bridges and membranes. This set-up refers to a reflection configuration. However a configuration in transmission mode can be easily arranged for the characterization of transparent objects. The set-up is a Mach-Zehnder interferometer, where the first beam splitter is employed to create both the reference and object beam from a collimated, monochromatic, coherent light source with a wavelength λ. The same polarization direction is imposed to both the beams in order to maximize the fringe contrast of the hologram. The hologram is obtained by a second beam splitter that combines the reference beam with the wavefield reflected by the MEMS. Moreover, this beam splitter allows to impose a small angle between the directions of propagation of the object and reference waves and so create off-axis holograms.

This set-up, for example, has been employed to create and acquire the hologram shown in Fig. 3(a) and relative to a group of silicon cantilevers. Circular fringes can be noted in the hologram. These are due to the interference of the parabolic phase factor (the object beam) superimposed onto the characteristic phase distribution of the object wave front and plane wave front of the reference beam. In particular, the parabolic phase factor accounts for the wave front curvature introduced by the optical lens (generally a microscope objective) positioned between the object and the digital camera and thus it has to be removed in order to obtain an accurate profile reconstruction of the object under inspection.

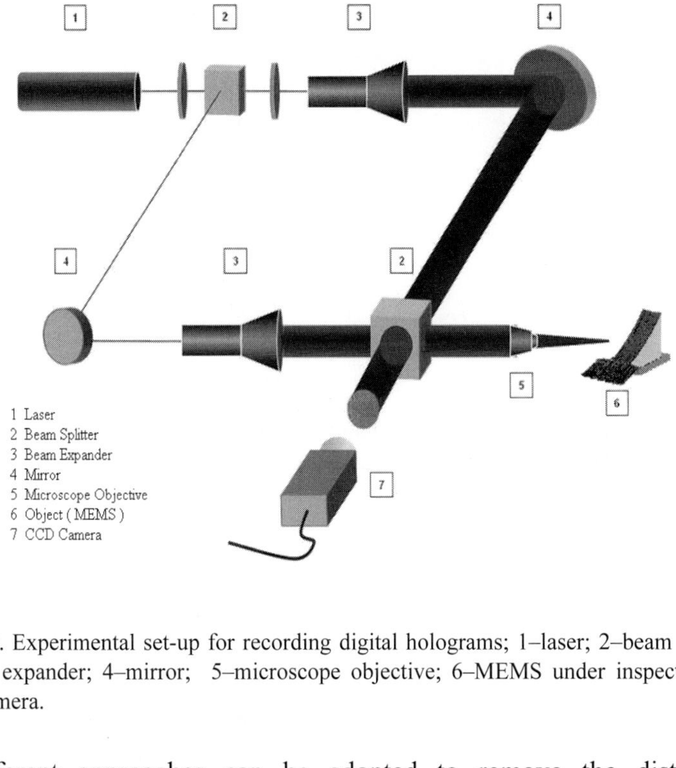

Figure 2. Experimental set-up for recording digital holograms; 1–laser; 2–beam splitter; 3–beam expander; 4–mirror; 5–microscope objective; 6–MEMS under inspection; 7–CCD camera.

Different approaches can be adopted to remove the disturbing parabolic phase factor in the reconstructed image plane.[23-29] However, the presence of wide areas of the MEMS, acting as a plane mirror surface, is often used to easily remove the contribution due to the curvature introduced by the optical system.

Therefore, the quantitative phase contrast imaging of the object under investigation can be retrieved from a hologram, such as that reported in

Fig. 3(a), without any contribution due to the optical aberration. In Fig. 3(b) is reported the density plot of the phase map $\phi(m,n)$ relative to the hologram of Fig. 3(a) and reconstructed at a distance d = 100 m. Figure 3(c) illustrates the reconstructed morphology of the MEMS analyzed and evaluated by Eq. 8. Finally, a qualitative image of the same object obtained by a Scanning Electron Microscope (SEM) is shown in Fig. 3(d).

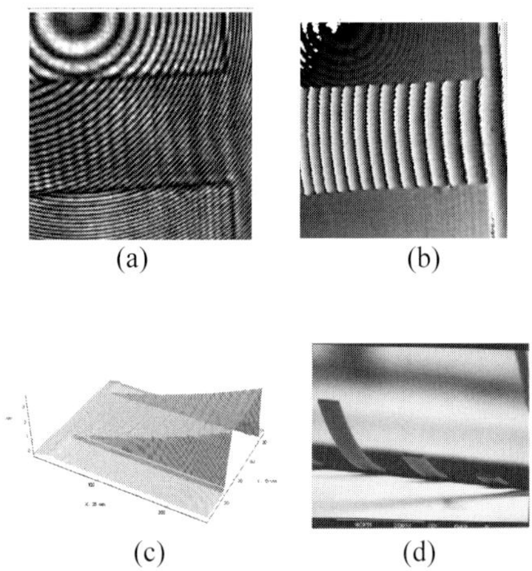

Figure 3. (a) Hologram of a silicon cantilever; (b) retrieved quantitative phase map contrast; (c) quantitative reconstructed morphology of the cantilever; (d) SEM image of the object characterized.

3.1. Tuning of the Size and Focus of the Reconstructed Image

As can be noted in the above-reported images, the reconstructed objects always appear well-focused. Obviously, these images can be obtained by means of a well knowledge of different parameters such as the focal length of the microscope objective, distance between the object and the microscope objective, and distance between the plane of the hologram and the microscope objective. However, it may be difficult to know or measure those parameters. An alternative is given by the unique characteristic of the Digital Holography to perform a numerical reconstruction. In fact, as reported in Section 2, the procedure to carry out a reconstructed image depends on the distance d. Thus, the focus can be sought by performing numerical reconstructions at different distances and visually estimating the in-focus quality of the obtained image, in analogy to mechanical translation of the microscope objective in conventional optical microscopy. However, if for an object a magnification of the size of some regions is required, microscope objectives with different magnifications have to be employed in the experimental set-up illustrated in Fig.2. Higher is the magnification μ of the objective, shorter is the depth of focus, so if the sample experiences even very small displacements δ along the optical axis (for example due to the thermal expansion of the object), a very large change Δd=-$\mu^2 \cdot \delta$ occurs in the distance to the imaging plane and, as a consequence, the image can be recorded out of the focus. The in-focus amplitude and/or phase-contrast image can be obtained again by modifying the distance d used in the numerical reconstruction process according to the afore-mentioned quantity Δd. This updating of the reconstruction distance can be performed automatically allowing the dynamic characterization of MEMS. Bearing in mind that the experimental set-up illustrated in Fig. 2 is a Mach-Zehnder interferometer, is obvious that the axial displacement δ causes a phase shift ($\Delta\varphi=4\pi\delta/\lambda$) in the fringe pattern of the hologram. Thus, it is possible to monitor the eventual phase shift of a small (few pixels) flat portion of the MEMS under investigation to determine the displacement Δd. This value can be used to correct the reconstruction distance d. So, in-focus amplitude and phase-contrast images, for each recorded hologram, can be reconstructed. In other words, if d_0 is the initial distance between the object and the plane of the camera, the corrected reconstruction distance $d'(t_i)$ for each acquired hologram, can be obtained by:

$$d'(t_i) = d_0 + \Delta d(t_i) = d_0 - \mu^2 \delta(t_i) = d_0 - \mu^2 \frac{\lambda}{4\pi} \Delta\varphi(t_i) \quad (10)$$

where t_i, for i=1, 2, 3….n, is the acquisition instant of the hologram and it is related to the frame rate of the CCD or CMOS employed camera.

Fig. 4 shows the hologram relative to a micro-heater realized by integrating an heater resistor on top of micro-machined suspended membranes.[31-32]

In this structure the temperature can be increased by the Joule effect, i.e., through a current flow in a resistor. The suspended structure aids to dissipate the heat only above the resistor. This structure during the working condition can reach temperature values around 700°C, and the temperature change causes a deformation of the structure itself but an expansion of both the object and its mechanical support, too. In order to perform a quasi real time characterization of the micro-heater, the phase shift induced by the thermal expansion has to be evaluated.[33]
Figure 4b reports the signal relative to this phase-shift and acquired on the small white square shown in Fig. 4a. This signal can be acquired with an high frame rate because it is relative to an area of few pixels. By means of a DFT procedure, the phase shift $\Delta\varphi$ can be estimated and applying the Eq. 10 the correct reconstruction distances for each acquired hologram can be obtained and, as a consequence, to reconstruct in-focus amplitude and phase-contrast images.

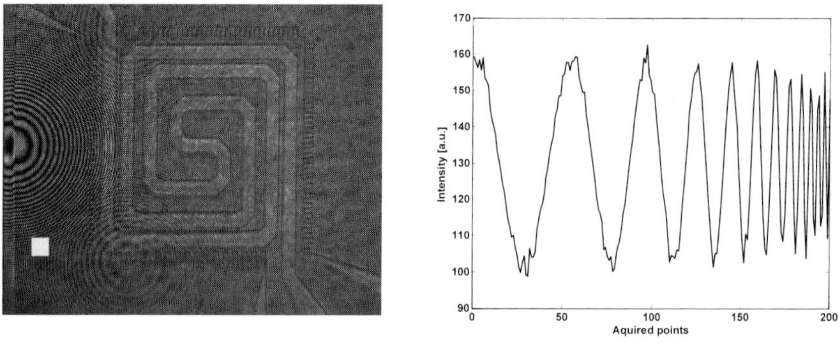

(a) (b)

Figure 4. (a) Digital recorded hologram relating to a micro-heater; (b) Intensity of signal recorded on the small white square shows in Fig. 3a and related to the phase-shift of the fringes.

As described in Section 2, according to Eq. 7, the dimension of the reconstructed images depends on the number of pixels M, N and pixels size Δx, Δy of the digital employed camera, on the wavelength of the illumining laser source and on the reconstruction distance *d*. Thus, according to the above described procedure for the focus tracking, each reconstructed image is characterized by a different dimension because reconstructed at a different distance *d*. In particular, the pixels size of each reconstructed image is:

$$\Delta\xi(t_i) = \frac{\lambda d(t_i)}{M\Delta x} \quad \& \quad \Delta\eta(t_i) = \frac{\lambda d(t_i)}{N\Delta y} \tag{11}$$

So, these images cannot be compared among them to evaluate some profile deformation due to the working conditions,[6,23,34] i.e., Eq. 9 cannot be applied because, for each state, the reconstructed images present different sizes. In order to avoid this limitation, i.e. to obtain always the same reconstruction pixels ($\Delta\xi$ and $\Delta\eta$) independent of the reconstruction distance, it is possible to numerically change the number of pixels M and N for each hologram to be reconstructed. In other words, the number of the pixels M and N is augmented by padding the matrix of the hologram with zeros in both the horizontal and vertical directions such that the reconstruction pixel can be constant for each reconstruction distance $d(t_i)$.[5,6,34]

In particular, the number of zeros to add to the matrix of the hologram along the two axes is calculated according the following relations:

$$\begin{cases} \Delta M = M_j - M_i = \dfrac{\lambda}{\Delta\xi\Delta x}(d(t_j) - d(t_i)) \\ \Delta N = N_j - N_i = \dfrac{\lambda}{\Delta\eta\,\Delta x}(d(t_j) - d(t_i)) \end{cases} \tag{12}$$

When the following relationships are imposed:

$$\begin{cases} \Delta\xi = \Delta\xi(t_j) = \Delta\xi(t_i) \\ \Delta\eta = \Delta\eta(t_j) = \Delta\eta(t_i) \end{cases} \quad \forall i,j \tag{13}$$

Thus, applying simultaneously both the focus tracking procedure and the controlling image size procedure, a sequence of reconstructed images in focus and with the same size can be easily and automatically obtained. By this approach, the images reported in Fig. 5 have been reconstructed and have been used to estimate the quantitative 3D full-field deformation experienced by the above citied micro-heater. The side bar inserted in each image reports the temperature on the heating element. Analyzing the images, the quantitative profile deformation of the membrane, on which the micro-heater is realized, can be estimated.

The operation time and so the temperature variation of the previous structure are very slow because a thermal effect is involved. As a consequence the reconstructed images reported in Fig. 6 can be considered as relative to "stationary states".

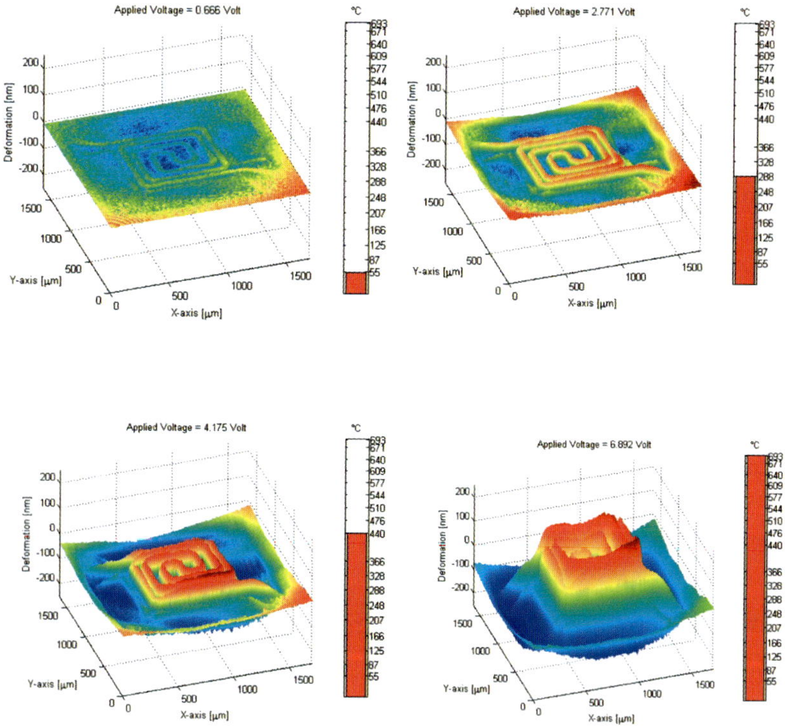

Figure 5. Four images relative to the 3D full-field deformation experienced by the micro-heater.

Anyway, the above described methods have been applied to characterize MEMS based on rapid profile variations, too. As example the characterization of an RF MEMS is reported. In particular, the inspection of a micromechanical shunt switch in coplanar waveguide configuration for microwave application is reported.[35] The structure is shown in Fig. 6 where in the right side of the figure, the movable bridge of the RF MEMS is illustrated. When a suitable DC voltage is applied between the gold bridge and the coplanar waveguide, the bridge experiences an electrostatic force for the actuation and goes down (OFF state).

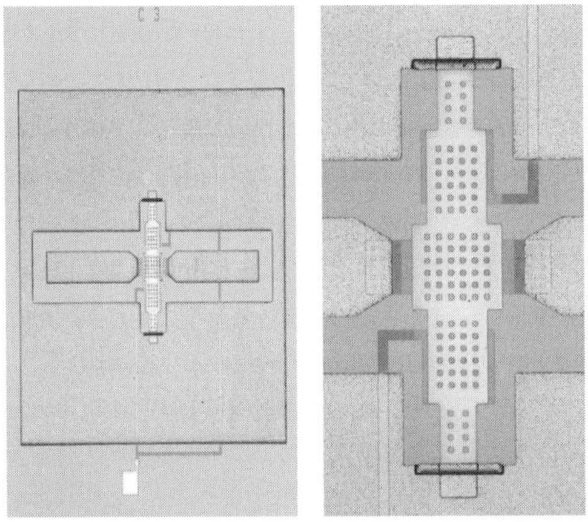

Figure 6. RF MEMS: shunt switch realization in coplanar configuration (left) and detail of the movable bridge (right).

When removing the applied voltage, the electrostatic force disappears and the bridge returns to the initial position (ON state).[36-37] DH has been applied to characterize the actuation of the bridge, analyzing both the vertical movement and the shape of the bridge during its actuation. To this aim, the RF MEMS has been actuated by a voltage ramp signal and a sequence of holograms has been acquired during the working condition. A CMOS camera with acquisition rate of 500fps has been employed for the acquisition. Three reconstructions corresponding to the above sequence are reported in Fig.7, where both the retrieved phase contrast and the correspondent reconstructed profile of the RF MEMS are shown. This kind of characterization has been very interesting because an asymmetry in the actuation of the bridge has been shown. This type of behavior would be difficultly extracted by another conventional method of characterization of RF MEMS; such as electrical equivalent model.

3.2. Extended Focused Image

In all above-illustrated examples the size of the analyzed structures is smaller than the focal depth of the objective microscope employed for

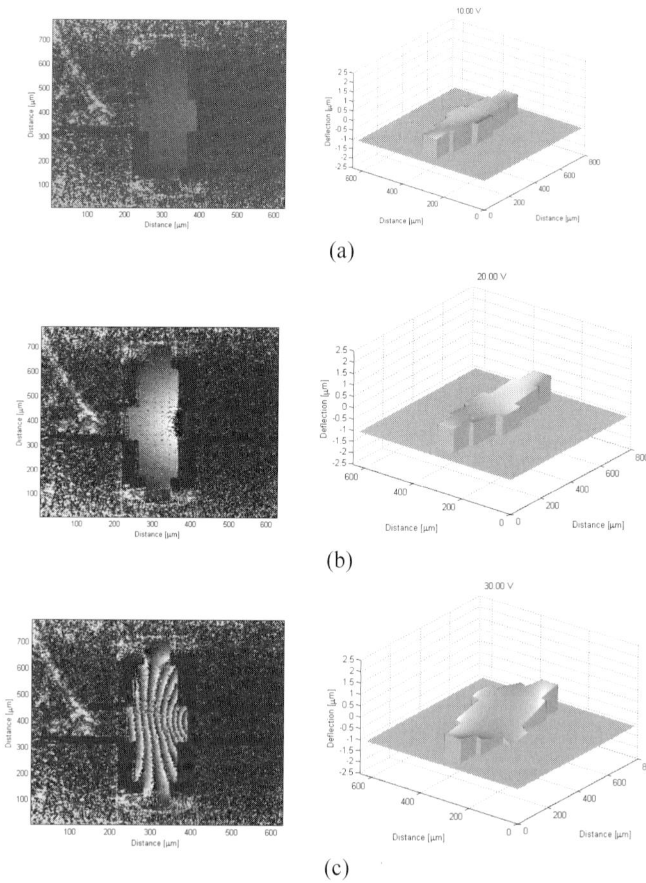

Figure 7. Retrieved phase maps and corresponding 3D full-field variation of the actuated bridge and relative to three different values of voltage extracted from a ramp signal: 10 V (a), 20 V (b), 30 V (c).

the observation. In other words, the object in each analyzing state is all in focus. At most, if the working state changes, the object can experience a translation that can induce a focus lost, but the focus can be recovered by the above-described methods. However, in some applications the size of the MEMS could be larger that the depth of focus of the employed microscope objective and so a single image in which the whole longitudinal volume of the object is in-focus can not be obtained. If an accurate analysis of the whole object has to be performed, it is necessary to have a single sharp image in which all details of the object, even if they are located at different planes along the longitudinal direction, are still in focus.[38] By means of DH and, in particular, by the possibility to numerically manage the reconstructed complex wave front, an extendedfocused image (EFI) of a 3D object can be obtained, without any mechanical scanning or complex experimental set-up. The method has been borrowed from the optical microscopy. In fact, in this case an EFI is composed by moving the microscope objective along the optical axis and acquiring an image for each longitudinal step. Numerical software, based on the contrast analysis, allows to identify for each acquired image the in sharp focus portion. Then, the different identified portions are composed together to give single images in which all details are in-focus, i.e., the EFI. Obviously, the acquired stack of images has to be composed of numerous images in order to obtain a detailed EFI.

In DH the stack of images with different in-focus portions can be obtained without any mechanical translation but by modifying the reconstruction distance. In other words, starting from a single hologram, reconstructed images at different image planes can be obtained for each reconstruction distance *d*. As described in previous sections, when the Fresnel-Kirchoff approximation is used, the size of the reconstructed image depends on the reconstruction distance *d*. Thus, for each image plane, the size of the reconstructed image has to be controlled according the Eq. 12. The range of variation of *d* can be easily evaluated from any retrieved phase map contrast because this phase map incorporates information about the topographic profile of the object under investigation. In fact, from Eq. 8 the longitudinal extension of the object under investigation can be estimated; in particular:

$$\Delta s = s_{\max} - s_{\min} = \frac{\lambda}{4\pi}\left(\tilde{\varphi}_{\max} - \tilde{\varphi}_{\min}\right) \tag{14}$$

where $\tilde{\varphi}$ is the unwrapped phase value. Thus, if the μ is the magnification of the microscope objective, in order to obtain all the volume of the object in focus a stack of images has to be reconstructed varying the reconstruction distance in the following range $\Delta d_{\max}$:

$$\Delta d_{\max} = -\mu^2 \frac{\lambda}{4\pi}\left(\tilde{\varphi}_{\max} - \tilde{\varphi}_{\min}\right) \tag{15}$$

The EFI is obtained by "cutting" the obtained stack of reconstructed amplitude images along the entire volume of the object by the surface of the object obtained by the Eq.(8). In other words, the reconstructed profile gives the coordinate *(x, y)* of the surface along which to "cut" the reconstructed volume to compose the EFI. Actually, the "cutting" operation means that slices of pixels were taken at the intersections between the retrieved surface and the volume of the stack, from each image of the stack. Those slices of pixels were stitched together to form EFI. Of course, how each slice is wide in terms of pixels depends on the resolution required that is related also to the axial resolution (i.e. the distance between two planes of the stack). In the following, step by step, the conceptual flow process to get the EFI from a single digital hologram is summarized:

- Step I: recording the digital hologram;
- Step II: reconstruction of the complex whole wave field from the hologram;
- Step III: extraction of phase-map of the object from the complex wave field;
- Step IV: the range of distances given by Eq.(13) is evaluated though the phase map;
- Step V: amplitude reconstruction of a stack of images of the entire volume from the lowest to the highest point in the profile of the object (adopting size controlling Eq. 12);
- Step VI: extracting the EFI image from the stack of amplitude images on the basis of the phase map obtained by the previous step III.

EFI of MEMS obtained by DH can be very useful to detect, as they appear in different locations of the structure under observation, for example, defects or cracks due to the technological processes or handling operations, or aging process.[39] Figure 8 illustrates the above-described technique applied to a silicon cantilever covered with a thin layer of aluminum. The combination of initial residual stress and the deposition of the aluminum layer have caused a progressive breakage of the structure.

It is clear that in Fig. 8(a) the tip of the cantilever is severely out of focus while the initial part of the crack is visible at the anchor point of the cantilever. Conversely, (Fig. 8(b)), at a different plane of focus, corresponding to the tip location of the cantilever, the left side of the image is in focus (note the black dot close to the tip), while the base is blurred and completely out-of-focus.

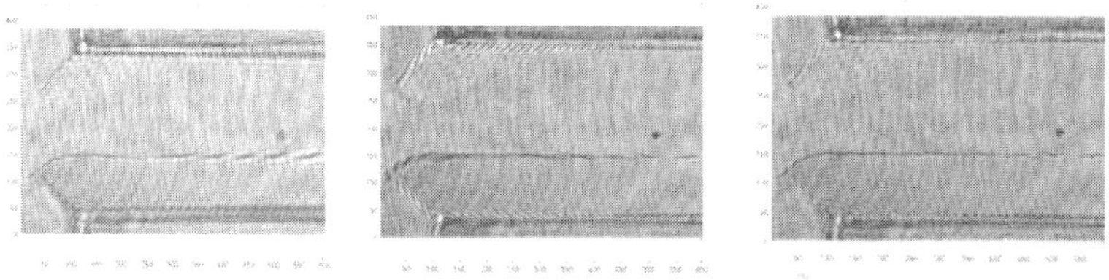

(a) (b) (c)

Figure 8. (a) Amplitude reconstruction of the cantilever where the base appears in focus; (b) Amplitude reconstruction of the cantilever where the tip appears in focus; (c) EFI image of the cantilever where all the details are in focus.

Finally, Fig. 8(c) shows the EFI where the crack is clearly visible and in good focus all along its length. This image is obtained by cutting a stack composed of 35 amplitude images each obtained by reconstructing the hologram at step of 1 mm from 156 mm to 190 mm.

One more EFI reconstruction applied to another cantilever of the same family of that shown in Fig. 8 is reported in Fig. 9. In this case, the cantilever presents abrupt breaks and cracks in different locations on its surface.

In Figure 9, the EFI obtained by DH (Fig. 9(a)) is compared to the EFI obtained by a optical microscope (Fig. 9(b)) where the distance between the microscope objective and the cantilever is changed by a piezoelectric mechanism.

Due to presence of different breaks and cracks a more detailed EFI has reconstructed using a stack of 50 images.

Finally, in Figure 9(c), a combination of 3D plot of phase map and holographic EFI of the cantilever is reported.

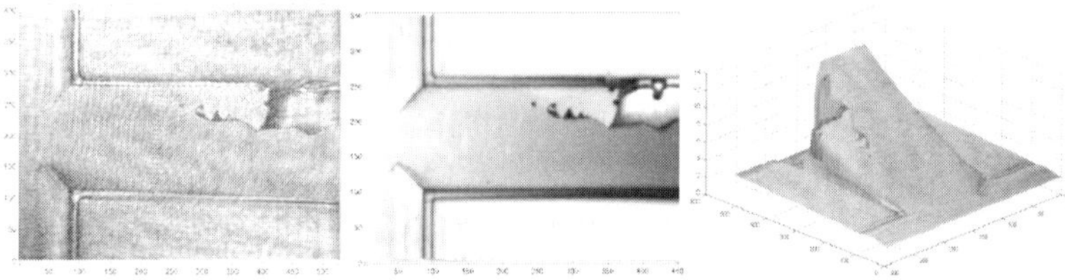

(a) (b) (c)

Figure 9. EFI image comparison between extended focused images obtained by (a) holographic method and by (b) an optical microscope; (c) 3D rendering of the cantilever carried out combining the phase map and the stack of the amplitude.

4. Conclusions and Outlook

This chapter has described how the DH can be efficiently used for analyzing, with high accuracy, Micro-Electrical-Mechanical structures. This analysis is useful both to obtain the topography of the structures and to study how all geometrical imperfections inherent to the technology, stress gradient effects, real boundary conditions and damping mechanisms, affect the real mechanical behavior of MEMS.

Since the hologram is coded numerically as a digitized image, both the intensity and the phase of the reconstructed wavefront can be manipulated. This allows both compensating any type of aberration and tuning, in automatic way, the focus and the size of the reconstructed images. Moreover, the whole 3D information intrinsically contained in the digital hologram is usefully used to construct a single image with all portions of a 3D object in focus.

The very important advantage of the described method is the possibility of obtaining an EFI of a microscopic object without a mechanical scanning operation. This possibility is particularly interesting for all that structures that vary their profile and/or position during the working condition and for all that structures where a fluid or living "object" are involved.

5. Acknowledgments

The authors are grateful to C. Magro and G. E. Spoto of the Silicon-Based Optoelectronics, Bio and Nano System Group of STMicroelectronics, Catania (Italy), P. Maccagnani and R. Marcelli of the Istitute for Microelectronic and Microsystem for allowing utilization of their MEMS.

References

1. S. D. Senturia, Ed., *Microsystem Design* (Kluwer–Academic, London, 2001).
2. C. M. Vest, Ed., *Holographic Interferometry* (John Wiley & Sons, New York, 1979).
3. P. K. Rastogi, *Holographic Interferometry*, (Springer Verlag, Berlin, 1994).
4. T. M. Kreis and W. Jüptner, in Fringe 97 Conference Proceedings, W. Jüptner, and W. Osten, Eds., (Academic Verlag, Berlin, 1997), 253.
5. P. Ferraro, S. De Nicola and G. Coppola, in *Controlling image reconstruction process in digital holography*, Ed. B. Javidi, (Springer, New York, 2006).
6. P. Ferraro, S. De Nicola and G. Coppola, in *Digital Holography and Three-Dimensional Display: Principles and Applications*, Ed. T. C. Poon, (Springer, New York, 2006).
7. T. M. Kreis and W. P. O. Jüptner, *Opt. Eng.,* 36 (1997).
8. Y. Takaki, H. Kawai and H. Ohzu, *Appl. Opt.*, 38 (1999).
9. I. Yamaguchi and T. Zhang, *Opt. Lett.*, 23 (1997).
10. S. Lai, B. King and M.A. Neifeld, *Opt. Comm.*, 173 (2000).
11. E. Cuche, P. Marquet and C. Depeursinge, *Appl. Opt.,* 39 (2000).
12. C. Liu, Y. Li, X. Cheng, Z. Liu, F. Bo and J. Zhu, *Opt. Eng.*, 41 (2002).
13. U. Schanrs and W. Juptner, *Appl. Opt.*, 33 (1994).
14. D. C. Ghiglia and M. D. Pritt, Eds., *Two-Dimensional Phase Unwrapping: Theory, Algorithms, and Software* (Wiley, New York, 1998).
15. K. M. Hung and T. Yamada, *Opt. Eng.*, 37, (1998).
16. A. Baldi, *Appl. Opt.*, 40, (2001).
17. L. Aiello, D. Riccio, P. Ferraro, S. Grilli, L. Sansone, G. Coppola, S. De Nicola and A. Finizio, *Opt. and Lasers in Eng.*, 45 (2007).
18. S. Baltiysky, I. Gurov, S. De Nicola, P. Ferraro, A. Finizio and G. Coppola, *Imag. Science J.*, 54 (2006).
19. X. Lei, P. Xiaoyuan, A.K. Asundi and M Jianmin, *Opt. Eng.*, 40 (2001).

20. X. Lei, P. Xiaoyuan, M. Jianmin and A. K. Asundi, *Appl. Opt.*, 40 (2001).
21. S. Seebacker, W. Osten, T. Baumbach and W. Juptner, *Opt. Las. Eng.*, 36, (2001).
22. B. Nilsson and T. Carlsson, *Opt. Eng.*, 39, (2000).
23. W. Osten, T. Baumbach and W. Juptner, *Opt. Lett.*, 27 (2002).
24. A. Stadelmaier and J. H. Massig, *Opt. Lett.*, 25 (2000).
25. G. Pedrini, S. Schedin and H. J. Tiziani, *J. Mod. Opt.*, 48 (2001).
26. S. De Nicola, P. Ferraro, A. Finizio and G. Pierattini, *Opt. Las. Eng.*, 37 (2002).
27. S. Grilli, P. Ferraro, S. De Nicola, A. Finizio, G. Pierattini and R. Meucci, *Opt. Exp.*, 9 (2001).
28. S. De Nicola, P. Ferraro, A. Finizio and G. Pierattini, *Opt. Lett.*, 26 (2001).
29. E. Cuche, P. Marquet and C. Depeursinge, *Appl. Opt.*, 38 (1999).
30. P. Ferraro, S. DeNicola, A. Finizio, G. Coppola, S. Grilli, C. Magro and G. Pierattini, *Appl. Opt.*, 42 (2003).
31. G. S. Chung, *Sens. Actuators* **A**, *Phys.*, 112 (2004).
32. L. Dori, P. Maccagnani, G. C. Cardinali, M. Fiorini, I. Sarago, S. Guerri, R. Rizzoli and G. Sberveglieri, in XI Eurosensors Conference Proceedings, Vol. I, (Warsaw, Poland, 1997), p.289.
33. G. Coppola, V. Striano, P. Ferraro, S. De Nicola, A. Finizio, G. Pierattini and P. Maccagnani, *J. MEMS*, 16 (2007).
34. P. Ferraro, , G. Coppola, D. Alfieri, S. De Nicola, A. Finizio and G. Pierattini, *IEEE JSTQE*, 10 (2004).
35. V. Striano, G. Coppola, P. Ferraro, D. Alfieri, S. De Nicola, G. Pierattini, A. Finizio, R. Marcelli and P. Maccagnani, in *Fibres and Optical Passive Components*, IEEE/LEOS Conference Proceedings, (2005) p. 97.
36. G. M. Rebeiz, Ed. *RF MEMS: Theory, Design and Technology*, (Wiley Interscience, Hoboken, N.J., 2003).
37. R. Marcelli, G. Bartolucci, G. Minucci, B. Margesin, F. Giacomozzi and F. Vitulli, *Electr. Lett.*, 40 (2004).
38. L. Mertz, Ed. *Transformation in Optics*, (Wiley, New York, 1965).
39. P. Ferraro, S. Grilli, D. Alfieri, S. De Nicola, A. Finizio, G. Pierattini, B. Javidi, G.Coppola and V. Striano, *Opt. Exp.*, 13 (2005)

INFRARED DETECTORS

Carlo Corsi*
Consorzio C.R.E.O. Centro Ricerche Elettro Ottiche
Via Pile, 60, 67100 L'Aquila, Italy
E-mail: corsi@romaricerche.it

Infrared technologies (materials, devices and systems) generally have been confined within a selected scientific community till in the 80's, when the Focal Plane Arrays development was a real breakthrough which could supply smart solutions for new product manufacturing and for opening new wide markets. These developments integrated with the advanced signal processing, thanks to the elimination of cryogenic cooling in the new microbolometers, allow to foresee that the integrated structures of "Smart Sensors" will be strategic components for important areas like transports, environment, territory control and security. Theory and design of the most important IR Sensors, their main historical developments and their trends in future developments are described.

1. Introduction

Infrared (IR) detectors are devices transforming the radiant energy of the IR region (from 0.7 μm to 300 μm) of electro-magnetic (e.m.) spectrum incident on the sensor in another form of energy which can be easily measurable like, generally, in the form of electric signal.
This conversion of energy normally is done by:

Photoelectric effects

a) photoelectric external effects: by generating free electrons from the surface of the sensor that has been hit by a photon with a sufficient energy;
b) photoelectric internal effects: the most relevant phenomena in the sensors development, by generating a couple of hole-electron inside a photoelectric material, generally a photoconductor or a photovoltaic structure.

Bolometric effects

Effects based on variations of vibration energy of crystalline reticle which are generally measured by the variation of electrical resistance. Recently advanced developments of IR detectors have been achieved (the so-called silicon microbolometers and also the pyroelectric detectors, based on the variation of the dielectric constant and therefore on the variation of the electrical charge in a capacitive sensor structure).
The IR detectors using photoelectric effects are *photon detectors* or *quantum detectors* and are measuring the photons with a quantum energy higher than the internal conduction energy gap, while the *bolometric* or *thermal detectors* are measuring the average of incoming radiative energy, independently of the spectral content (assuming a constant absorbing coefficient).

2. Parameters Characterizing the IR Detectors

The main parameters of IR detectors are: spectral response, signal-to-noise ratio per incident unit power, response time and working temperature.[1,2]
- The *spectral response* R_λ (λ) in V/W is generally represented by a curve showing the responsivity vs wavelength, where R_λ is the rms (root mean square) of the electrical output voltage of the sensor per unit of radiating rms power at the wavelength, λ.
- The *response time* that is the time needed for the signal to achieve 70,7% of the equilibrium value.-The signal-to-noise per incident unit power in Watt , the so called detectivity, is given by:

$$D = \frac{S}{N}\frac{1}{EA} \qquad (1)$$

where S, N are respectively the rms of the signal voltage and of the noise voltage, E (irradiance) is the rms of the incident radiation, A is the sensitive area of the IR sensor. The *detectivity* is a very important parameter for characterizing IR detectors and generally is indicated as D (T, f, Δf), where T is the blackbody temperature of the radiating source, f

is the frequency of modulation of the signal and Δf the normalized bandwidth of the signal output amplifier.
A normalized detectivity is given by:

$$D^* = \frac{S}{N} \frac{\sqrt{\Delta f}}{\left(EA^{1/2}\right)} \tag{2}$$

In reality, assumptions made in the above formula are only valid if noise is proportional to $A^{1/2}$ and to Δf. The first hypothesis is only valid for the photon detectors and for variation of the sensor area of one-two orders of magnitude; it is not valid for some bolometric sensors; the proportionality to Δf is also valid in a limited frequency variation.
For a specific value of the sensitive area and of the working temperature and for a certain environmental radiance, maximum theoretical value of the detectivity can be obtained and, in the case of cooled photon detectors, this value can be obtained by the so called BLIP (Background Limited Infrared Photodetectors) curve (Fig.1).

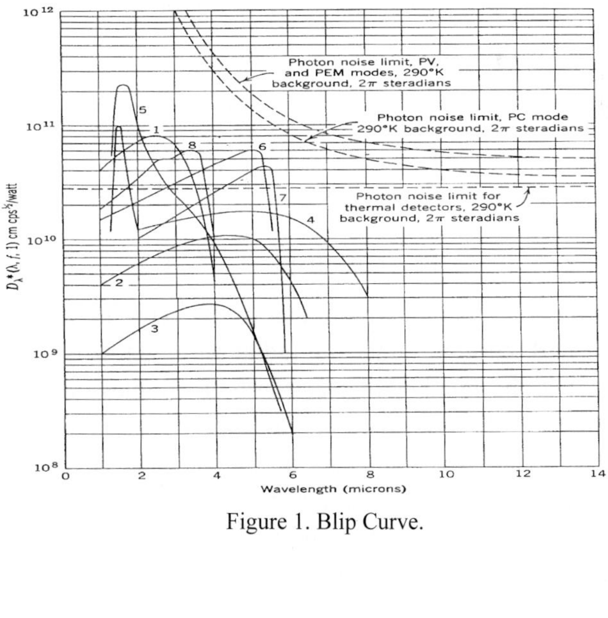

Figure 1. Blip Curve.

3. Theory of Intrinsic Photoconductivity

The intrinsic photoconductive sensor is essentially a photoresistance, that is a resistance changing its value when is radiated by some form of e.m. energy, normally light, like it is shown in Fig. 2.
When the energy of a photon of the radiation light is higher than the forbidden energy gap "Eg" in the phoconductive semiconductor, a couple of electron-hole is generated, therefore creating a voltage variation that is measured as a signal output at the load resistance R_L (Fig.2).

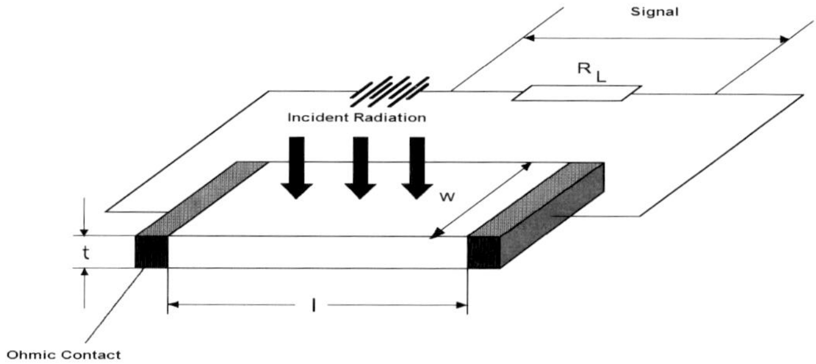

Figure 2. Intrinsic photoconductive sensor.

4. General Theory

The basic expression, which describes the photoconductivity, intrinsic and extrinsic, in a semiconductor in equilibrium state, is given by [2,3]

$$I_{ph} = q\eta A\Phi_s g \tag{3}$$

where I_{ph} is the (dc) photocurrent, that is the current increase, with respect to the dark current, under the radiation effect; η is the quantum efficiency, that is the number of electron-hole couples generated, with respect to the equilibrium state, per absorbed photon unit; q is the electronic charge and Φ_S is the density of the photon flux generated either by the radiation of the signal emitted by the object to be detected, either by the radiation emitted from the background; g is the photoconductivity gain as defined later on.

Generally the photoconductivity is a phenomenon based on the two types of electrical charge, electron and holes, therefore the total current I_{ph} is given by

$$I_{ph} = \frac{qwt\,(b\Delta n + \Delta p)}{L}\mu_h V_b \tag{4}$$

where *w*, *t*, *L* are the dimensions of the photoconductor (see Fig. 2) and where $b = \mu_e/\mu_h$; μ_e is the electronic mobility, μ_h is the hole mobility and V_b is the polarization voltage

$$n = n_0 + \Delta n \; ; \; p = p_0 + \Delta p \tag{5}$$

where n_0 and p_0 are the density of the thermal carriers, at equilibrium, Δn and Δp are the concentration of the excess photogenerated carriers.
In the case that $\Delta n \cong \Delta p$, the lifetime of carries is given by

$$\tau = \frac{\Delta nt}{\eta \Phi_s} \tag{6}$$

where from (3) and (4) we obtain

$$g = \frac{t\mu_h V_b(b\Delta n + \Delta p)}{L^2 \eta \Phi_s} \tag{7}$$

This means that, when $\mu_e >> \mu_h$ using the (6) and the (7) the photoconductivity gain is given by the ratio between the lifetime of the free carriers and the transit time inside the electrodes of the photoresistor

$$g = \frac{\tau}{L^2 / \mu_e V_b} = \frac{\tau}{T_t} \tag{8}$$

The gain of photoconductivity is therefore higher or lower than 1 depending if the length of diffusion is lower or higher than the interelectrode space L. When the load resistance R_L is much higher of R, the voltage at the load resistence is essentially like in an open circuit:

$$V_s = I_{ph} R_d = I_{ph} \frac{1}{qwt(n\mu_e + p\mu_h)} \qquad (9)$$

where R_d is the sensor resistance. Therefore using the (2) and (4) and assuming $\Delta n \cong \Delta p$ the responsivity is given by

$$R_v = \frac{V_s}{P_\lambda} = \frac{\eta}{Lwt} \frac{\lambda\tau}{hc} \frac{V_b(b+1)}{bn+p} \qquad (10)$$

where the monochromatic absorbed radiation power is $P_\lambda = \Phi_s A h\nu$ where $\nu = c/\lambda$.

Assuming that the change in the electrical conductivity due to the irradiance is small in comparison to the dark conductivity, the voltage responsivity is expressed by

$$R_V = \frac{\eta}{Lwt} \frac{\lambda\tau}{hc} \frac{V_b(b+1)}{bn_0 + p_0} \qquad (11)$$

This means that the basic requirement for achieving a high responsivity in photoconductor at a certain wavelength λ is requiring a high quantum efficiency (η) and a long lifetime (τ) of the excess carriers, and also a small distance between the electrodes, a low concentration of thermally generated carries n_o, p_o at equilibrium, with the highest applicable polarization voltage V_b.

In reality this model is ignoring important effects deriving from a too short interelectrodes distance or by other physical-structural characteristics, which can originate "draining" effects or surface recombination.

5. Photovoltaic Detectors

The photovoltaic effect is present in sensors structures with an internal voltage difference which is moving the photogenerated carries (electrons and holes) in opposite directions. The most common example of photovoltaic effect is based on an abrupt p-n junction in a semiconductor and is named "photodiode".

The photons with energy higher than the energy-gap which are incident over the surface of the sensor are creating a couple "electron-hole" in both sides of the p-n junction that ,thanks to the induced field, diffuse within the diffusion path of the junction reaching the region of spatial carrier where the high electric field is separating the electron-holes couples in such a way that the minority carriers are quickly accelerated becoming majority carriers in the other side of the electrical junction. In this way the generated photocurrent, modifying the current-voltage characteristics, is given by the inverse negative current I_{ph} as it is shown in Fig. 3:

$$I_{ph} = \eta q A \Phi \tag{12}$$

where A is the photodiode area, Φ is the flux of incident photons and η is the quantum efficiency.

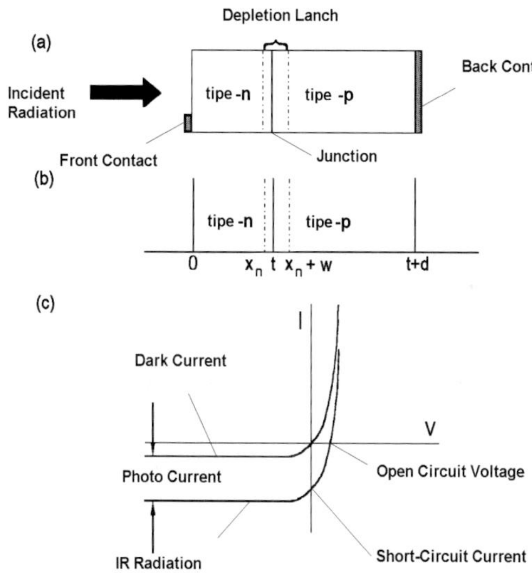

Figure 3. Photodiode.

Normally the gain in current in a photovoltaic detector with a simple structure (e.g. not of avalanche or tunnel type) is equal to 1.

When the p-n junction is working with an open circuit, the accumulation of electrons-holes couples in the junction carriers produces an open circuit voltage which, in the case of a load resistance R_L connected to the diode, is generating a current which achieves its highest value when the diode is short-circuited (short-circuit current I_{sh}).

The open circuit voltage can be obtained by just multiplying I_{sh} by the incremental resistance $R = \left(\frac{\delta v}{\delta I}\right)$ when $V=V_b$ that is

$$V_{ph} = \eta q A \Phi R \tag{13}$$

where V_b is the voltage polarization and I = f(V) is the I-V curve; in many cases the photodiode is operating at polarization voltage equal to zero:

$$R_0 = \left(\frac{\partial I}{\partial V}\right)^{-1}_{Vb=0} \tag{14}$$

A figure of merit normally used is the value of the R_0A product

$$R_0 A = \left(\frac{\partial J}{\partial V}\right)^{-1}_{Vb=0} \tag{15}$$

where J=I/A is the current density.
In the normal detection the photodiodes are working at zero bias, while the inverse polarization is used for high frequency applications for reducing the RC constant of the device.

6. Photodiode Currents

Various mechanisms are involved in the current phenomena in a photodiodes, the most important are:

a) Dark currents mainly due to thermally generated carriers in the crystal and in the depletion layer of the p-n junction, and surface currents due to surface states and surface leakages.
b) Diffusion currents given by

$$J_D = J_S\left[\exp(qV/kT) - 1\right] \tag{16}$$

where J_s is given by

$$J_s = (kT)^{1/2} n_i^2 q^{1/2} \left[\frac{1}{p_{po}} \left(\frac{\mu_e}{\tau_e} \right)^{1/2} + \frac{1}{n_{no}} \left(\frac{\mu_h}{\tau_h} \right)^{1/2} \right]^{-1} \quad (17)$$

where n_i is the concentration of intrinsic carriers, p_o and n_o are the concentrations of the majority carriers and τ_e and τ_h are the lifetime of electrons and holes respectively in the p region and in the n region.
The diffusion current is changing vs temperature as the square of intrinsic density of electrons (n_i) and generally is the current dominating at high temperatures.

c) Generation-recombination current

Such mechanism can be the dominant one at low temperatures and is given by

$$J_{GR} = \frac{qn_i w}{(\tau_{eo}\tau_{ho})^{1/2}} \frac{2sinh(qV/2kT)}{q(V_{bi} - V)/kT} f(b) \quad (18)$$

where V_{bi} is the induced voltage and τ_{eo} and τ_{ho} are the lifetimes of the electrical carriers in the depletion layer and f(b) is a complex function which is normally close to ≈1 The generation-recombination (g-r) current can be simplified by

$$J_{GR} = \frac{qwn_i}{2\tau_o} \quad (19)$$

which, taking into account that the width of the barrier is varying as the square root of the applied voltage ($w \cong V^{1/2}$) in presence of abrupt junctions or as the cubic root ($w \cong V^{1/3}$) for linearly graded junctions ,is showing the proportional dependence for the g-r current, moreover the g-r current is proportional to n_i, while the diffusion current is proportional to n_i^2, therefore there is a temperature T_e where the two currents are comparable, while, below T_e, the g-r current is dominant.
Other current phenomena are related to tunneling effects and more important, to surface leakages, the dark current taking care of all these phenomena is expressed by

allowing to get the value of the lifetime τ just by measuring with a spectrum analyzer the knee of the curve at $\nu=1/\tau$.

- The so called *1/f Noise* characterized by a spectrum where the noise power is inversely proportional to the frequency f, according to

$$I_{1/f} = \left(\frac{KI_n^{\alpha} Af}{f^{\beta}} \right)^{1/2} \tag{27}$$

where K is a proportionality factor, I_b is the current bias, α is a constant almost equal to 2 and β a constant almost equal to 1.

The 1/f noise is normally associated to the presence of potential microbarrier at the boundaries of polycrystalline grains in the semiconductor and the reduction of 1/f noise is almost an art which has to care the electrical contacts realization and the preparation of photosensitive surfaces.

Normally the IR photodectors show a 1/f noise at low frequencies, while at higher frequencies the predominant noise is the g-r noise till the $1/\tau$ frequency where the Johnson Noise is starting to prevail. (Fig.4)

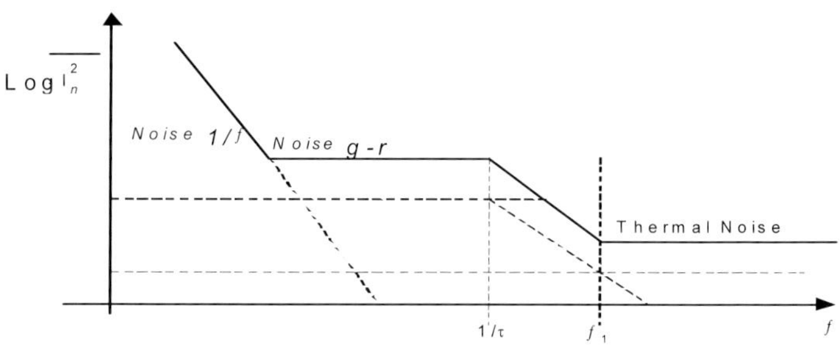

Figure 4. Electronic Noise vs Frequency.

9. Complex Devices: IR Focal Plane Arrays (FPA)

The most important application of IR detectors is thermovision which is allowing to see the thermal emission of the scene. The thermovision systems have strongly evolved in time since the first thermovision systems developed during the Second World War, based on a simple opto-mechanical scanning focusing scene onto a single detector.

Such type of image reconstruction, based on a serial scanning of the image points (pixels) has been improved in the time, mainly by enhancing the signal-to-noise ratio thanks to the increase of the sensors number. This has been achieved firstly, by using linear detectors arrays, with as a serial read-out integrated in time (Time Delay Integration TDI-Fig. 5a) or by using a parallel structure to read simultaneously more rows of the scanned image and then by integrating them in the image reconstruction (Fig. 5b).

It is evident the importance of the development of FPA detectors working in the so called "staring" mode, that is capable of seeing the image by a simultaneous vision of the scene, thanks to a mosaic sensors structure positioned in the focal plane of the image avoiding the use of any opto-mechanical scanning or movement.

Various types of electronic read-out have been developed since the 70's, starting with a pseudo-bidimensional read-out based on the sequential read-out of rows and columns by using multiplexers and shift-register, (Fig. 6a), passing then to the so called X-Y addressing by using integrated read-out devices (Fig. 6b) or external addressing capable of selecting specific areas of the mosaic sensors (Fig. 6c).

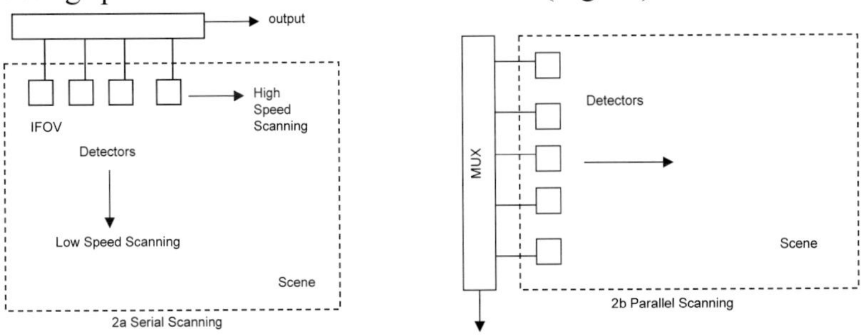

Figure 5. Linear Detector Array: a) serial and b) parallel structure.

The readout have been continuously evolving till 1970 when the new charge-coupled-device CCD read-out based on electronic charge transfer has allowed a completely integrated read-out (Fig. 6d), with the development of FPA sensors with a high number of detectors (more than 1 million pixels).

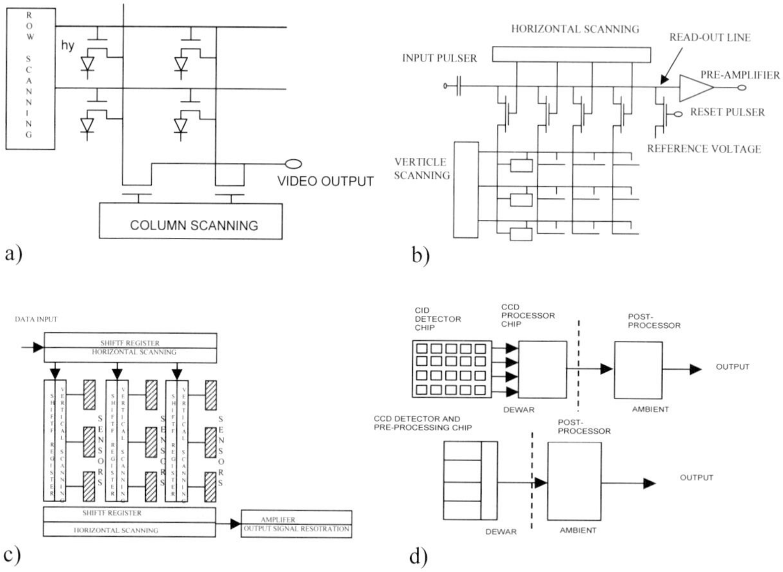

Figure 6. a) (x-y) Addressing by CMOS Switching; b) Rows-Columns (x-y) Scanning; c) Rows-Columns (x-y) Scanning; d) Integrated (x-y) Scanning.

10. Historical Scenario

The history of IR radiation sensors[3] is starting in 1800 when the astronomer, William Herschel, discovered the existence of infrared radiation, a form of energy of the light beyond the "red" (therefore from the Latin "infra" below-"red"), by trying to measure the heat of the separate colors of the rainbow spectrum. After noticing that the temperature of the colors increased from the violet to the red part of the spectrum, Herschel decided to measure the temperature just beyond the red portion of the spectrum in a region apparently devoid of sunlight. To his surprise, he found that this region had the highest temperature of all. In April 1800 he reported to the Royal Society as *Dark Heat*[4] and, making further experiments on what he called the "calorific rays" that existed beyond the red part of the spectrum, he found that they were reflected, refracted, absorbed and transmitted just like visible light. The basic laws of IR radiation (*Kirchhoff's law, Stefan-Boltzmann's law, Planck's law and Wien's displacement law*) have been developed many years after the discovery of IR radiation. In 1859 Gustave Kirchhoff found that a material that is a good absorber of radiation is also a good radiator. Kirchhoff's law states that the ratio of radiated power and the absorption coefficient: 1) is the same for all radiators at that temperature, 2) is dependent on wavelength and temperature, and 3) is independent of the shape or material of the radiator. If a body absorbs all radiation falling upon it, it is said to be "black." For a blackbody the radiated power is equal to the absorbed power and the emissivity (ratio of emitted power to absorbed power) equals one. In 1884, L.E. Boltzmann, starting from the physical principles of thermodynamics, derived the theoretical formula of T4 Black Body Radiation Law, stated empirically in 1879 by J. Stefan's, by developing the *Stefan-Boltzmann's Law*:

$$W=\sigma T^4 \tag{28}$$

where W= Radiation Power, T =Absolute Temperature, σ= Stefan-Boltzmann's Constant. In 1901, Nobel Prize Max Karl Ernst Ludwig Planck developed the *Planck's law* which stated that the radiation from a blackbody at a specific wavelength can be calculated from

$$I(\nu)d\nu = \frac{2h\nu^3}{c^2}\frac{1}{\exp\left(\frac{h\nu}{kT}\right)-1}d\nu \tag{29}$$

(where $I(\nu)\delta\nu$ is the radiation power emitted per unit of surface and solid angle unit, in the frequency interval($\nu\div\nu+\delta\nu$); T=Absolute Temperature, c = speed of light; h = Plank's constant).

Soon after Wilhelm Wien (Nobel Prize 1911) established the *Wien's displacement law* taking the derivative of the Plank's law equation to find the wavelength for maximum spectral radiance at any given temperature:

$$\lambda_M T=2897.8 \ (\mu m\ K) \tag{30}$$

IR detectors development, even after the discovery of Infrared Radiation by Sir H. Herschel, were mainly based on the use of thermometers which dominated IR applications till the 1st World War, although in 1821 J.T. Seebeck had discovered the thermoelectric effect and in 1829 L. Nobili had fabricated the first thermocouple, allowing in 1833 the multi-element thermopile development by Macedonio Melloni with the first detection of a human being at 10 meters. Early thermal detectors, mainly thermocouples and bolometers, were sensitive to all infrared wavelengths and operating at room temperature and normally, until few years ago, they were relatively low sensitivity and slow response time.

The first photon detectors (based on photoconductive effect discovered by Smith in 1873 in Selenium and, later on, by Bose in photovoltaic lead sulfide, but not applied for many years) were developed by Case in 1917 and in 1933 Kutzscher developed IR PbS detectors (using natural galena found in Sardinia): these sensors were widely used during the 2nd World War. These detectors have been extensively developed since the 1940's. Lead sulfide (PbS) was the first practical IR detector, sensitive to infrared wavelengths up to ~ 3 μm. In the mean time Cashman developed TaS, PbSe and PbTe IR detectors with high performances strongly supporting the great developments in England and US. The history of IR detector developments has been therefore strongly conditioned by military applications for many decades driving the main projects of the IR industry and in some way of R&D labs.

After the war, thanks to discovery of transistors in 1948 a wide variety of new materials were developed for IR sensing. Lead selenide (PbSe), lead telluride (PbTe), and indium antimonide (InSb) cooled detectors extended the spectral range beyond that of PbS, providing sensitivity in the 3-5 μm medium wavelengths (MWIR) atmospheric window. Infrared technology during the 50's was enjoying a great growth, especially in applied technologies, thanks to the development of solid state IR sensors (extrinsic photoconductive germanium detectors were allowing to reach long wavelength spectral region, although needing very low temperature through the use of liquid helium). Almost contemporarily the first InSb, detectors at liquid nitrogen temperature with high detectivity in the 3-5 μm spectral wavelength were developed.

Photolithography, available in the early 1960's, was applied to make IR sensor arrays: linear array technology was first demonstrated in photoconductive PbS, PbSe, and InSb detectors and photovoltaic (PV) detector developments began with the availability of single crystal InSb material. But we had to wait till the 60's to see the first advanced developments coming out thanks to direct gap photon materials based on ternary semiconductor compounds (HgCdTe and PbSnTe). This was a real breakthrough, also because in the mean time microelectronics was offering new advanced manufacturing technologies like photomasking and integrated microsoldering and assembly. So, thanks to these advanced technologies, the first linear arrays of tens of elements were developed at the end of 60's with a strong competition between the HgCdTe and PbSnTe (this compound could offer more stability and reliability in performances, moreover at the beginning of 70's Lincoln Labs. MIT researchers were developing the first solid state IR 10 μm lasers applied to environmental control).[5] On the other hand, PbSnTe was showing higher dielectric constant, limiting high frequency performances, and a high thermal expansion coefficient, with a strong limitation for the integration in silicon microelectronics.

This choice, mainly made by US industries, was allowing the development of the" first generation of linear detector arrays", which allowed to obtain BLIP detectors at liquid nitrogen temperature (this first generation of CMT linear arrays was the basis for the "Common Modules" LWIR FLIR systems with a number of pixels from 60 up to 180, in which the read-out of the detectors was done connecting each element of linear arrays with feed-through to the read out electronics).

The invention of Charge Coupled Devices (CCDs) in 1969[6], with functioning devices at the beginning of the 70's, made it possible to start the developing of the "second generation" FPAs detector arrays coupled with on-focal-plane electronic signal readouts. In the middle 70's , in USA , while the 1st Common Module IR Arrays were produced , the first CCD IR bidimensional arrays [7,8] were appearing and, in Italy, the first Smart Sensors based on LTT RF sputtered thin films using X - Y addressing read - out were developed[9].

In 1975 the first CCD TV camera was realized and this was allowing to forecast the "2nd generation FPAs" capable of a staring vision, although the necessity of very high spatial resolution and high reliability even in complex structures, with extremely high number of pixels (up to one million pixels), were pushing towards alternative solutions, with materials less difficult than CMT, in the manufacturing process (e.g. extrinsic silicon detectors). Early assessment of this concept showed that

information. The Smart Sensor technology, based on the use of a smarter sensor architecture, allows to integrate technical design and development from optics, detector materials, electronics, and algorithms into the sensor's functions rather than trying to get the required performance by relying on drastic improvements in just one aspect of the technology, for instance the number of sensor pixels.

One of the most advantageous application area for "Smart Sensors" is the Infrared Field where the information to be extracted is generally based on very small signals buried in highly intensive and diffused background noise and often high intensity “unwanted signals”.

This implies that infrared imaging devices require some processing of detector output signals to correct non-uniformity and remove the background effect.

Without this on-focal-plane processing, most of the data would be useless clutter or unwanted data, because of the whole acquired pattern only a few pixels contain targets information of selected targets. Therefore conventional approaches need to process these complex data through the read out electronics, the analogue to digital converters, and the digital signal processor before finally separating and rejecting the clutter. In contrast, the Smart Sensor rejects this clutter before it is read off the focal plane sensors so that most of the useless data is not processed.

The “Smart Sensor” design concept is based on the processing capabilities, at least at some stage of thresholding, inside the sensors structure itself. This means that the “Smart Sensor” in some way emulate living eye, in the early stage, at least at a primordial level like “insect eye”, and, in future perspective, could reach performances close to the “human eye”, thanks to neuronal network development, which can allow patterns recognition and objects discrimination.[31] In general, background clutters are extended objects and are more slowly varying spatially than target, therefore temporal filtering as well as spatial filtering, further complemented by multi-spectral filtering, are required for target signal detection and extraction.

IR DETECTORS		
1800	IR radiation	Sir W. Herschel
1821	Thermoelectric Effect	Seebeck
1829	Thermocouple	G. Nobili
1833	Thermopile	Macedonio Melloni
1836	Optical pyrometer	Becquerel
1873	Photodection (Selenium)	Smith
1884	IR Radiation Law	Boltzmann
1902	Photonductivity effect	Bose
1917	Lead Sulphide	Case
1933	Lead Sulphide (Galena)	Kutzsher
1940	TI_2S	Cashman
1942	Golay Cell	Golay- Quenn Mary College
1948	Transistor	Bardeen-Brattain- Shockley
50s 1959	PbS, PbSe, PbTe HgCdTe	T. Moss RRSE W. Lawson, J. Putley
60s 1969	Ge:X, InSb CCD	 Boyle-Smith (Bell Labs)
70s 1973 1975 1978	PbSnTe/HgCdTe,Si:X, Common modules IR Smart Sensors Si:X/CCD/PtSi/CCD HgCdTe/CCD	Lincoln Labs, SBRC Hughes, Honeywell, Rockwell, Mullard Night Vision Lab C. Corsi Elettronica SpA RCA Princeton Lab W. F. Kosonocky, F. Shepherd D. Barbee - F. Milton - J. Steckel
80s	HgCdTe SPRITE InGaAs QWIP	T. Elliot RSE F. Capasso, L. Esaki, B. F. Levine M. Razeghi, L. J. Kozlowski.
90s	Pyroelectric FPAs MicroBolometer FPAs Multi-colour FPAs Advanced FPAs	RRSE-BAE R. A. Wood (Honeywell) JlTissot, P. R. Norton A. Rogalski, H. Zogg S. D. Gunapala, D. Z. Ting (Jet Propulsion Labs)
2000	MEMS FPAs –Cantilever IR Nanotubes/Nanowires	B. Coole, S.R. Hunter, X. Zhang J. Xu, S. Huang, Y. Zhao S. Maurer, G. Jiang, D. J. Zook

Table 1. History of IR Detectors.

One of the simplest feature extraction and in the same time, most appealing for the numerous applications, is the discrimination of point sources from extended background emissions and/or of fast events (moving targets or changeable emissions) for static or slow moving scenario just as it is operating the fly-eye. In this case a reticule structured detector, which is electronically modulated to obtain a spatial-temporal correlation of the focused spot target, buried within the diffused background emission, can allow the detection of point source or a well defined shaped target improving the signal to clutter ratio: this correlation associated to appropriate temporal signatures, can allow to discriminate and identify the targets, like it is performed by an insect eye thanks to a spatial-temporal correlation. [32,33] At last, it is important to underline that the important recent developments of neural networks for advanced computing allow foreseeing an impressive growth of the "Smart Sensor" concept especially for those detector technologies which will take advantage of the possibility of integrating processing devices.

12. Future Infrared Detectors

The main efforts in nowadays IR detectors developments are oriented towards Focal Plane Array with the highest number of pixels (more the 10^6 elements), with integrated electronics for signal read-out and elaboration, with working temperature close to room temperature and with high uniformity supported by the fact that, thanks to the increased integration time possible in staring arrays, NETD values close to BLIP limit can be easier achieved. This with the main task of achieving high optoelectronic FPA's performances with smaller and lighter structures, with possibilities of applications in civil areas thanks to cost reduction by eliminating opto-mechanical scanning and cryogenic low-temperature cooling.

There is a large research activity directed towards 2-D staring arrays detectors consisting of more than 10^6 elements.

IR FPAs have a similar growth rate as dynamic random access memory (RAM) integrated circuits (ICs) [35,36] (it is consequence of Moore's Law,[37] which predicts the ability to double transistor integration on each IC about every 18 months) but with a lag behind by about 5-10 years and with some positive break-points (linear arrays and FPAs). Consequently, whereas FPAs with number of pixels up to 10^4 were available in the early 1980s, several companies are now producing monolithic FPAs with 10^6 pixels, although limited to short-middle IR regions .

Figure 7 illustrates the trend of arrays size for various technologies over the past 40 years and some forecasting for the future. Actually the largest HgCdTe FPA is a short wavelength IR (SWIR) hybrid 2048 x 2048 with a unit cell size of 18 μm x18μm for astronomy and low background applications.[32]

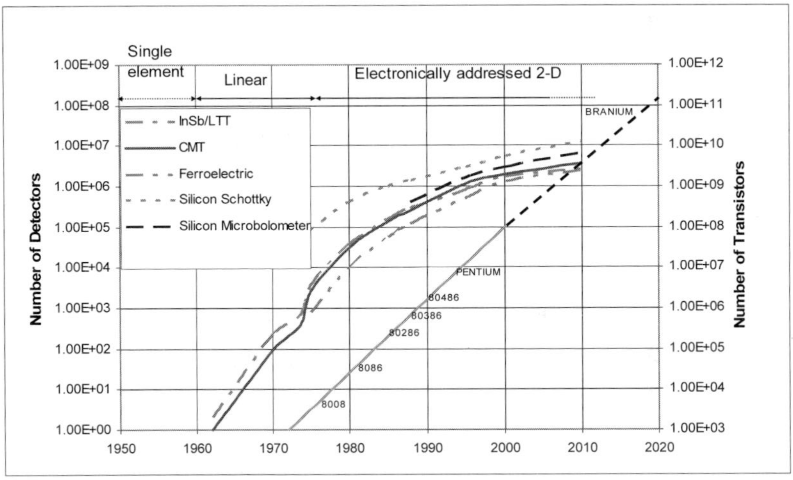

Figure 7. Number of pixels in infrared detectors arrays (Moore Law).

We have to take in consideration that pixel size is conditioning the achievable number of pixels with a feasible FPA chip size, also conditioned by the size and the cost of optics and overall by the fundamental limit on the pixel size determined by diffraction law.[35,36]

In fact the size of the diffraction-limited optical spot, or Airy disk, is given by $d = 2.44\ \lambda\ F$, where d is the diameter of the spot, λ is the wavelength and F is the f number of the focusing lens (for high luminosity F/1 optics at 10-μm wavelength, the diffraction limited spot size is ~ 25 μm). In future, a reduction of the LWIR pixel size, in case

of applications requiring high spatial resolution, could reach the limit of ~ 5 μm, thanks to oversampling (up to a factor of 4).
Key parameters of a single pixel sensor such as the ultimate sensitivity (measured by NETD), response time and working temperature are integrated more and more by key parameters of FPAs as number of pixels, uniformity, reliability and cost. Operational requirements (mainly of maintenance and reliability) were pushing for new advanced sensors avoiding the cryogenic needs. All these requirements will be strongly conditioned by the complete integrability with silicon microelectronics technologies. The competition among various technologies has been strong with emerging new actors in the last years (overall, Room Temperature Microbolometers).
The new microbolometers technology, completely integrable with silicon technology and therefore often named silicon microbolometers, have been emerging in the last years with a high promising for future IR sensors market growth.[38] So, for the first time, thanks to the elimination of cryogenic cooling, IR Smart Sensors are emerging on the international market, becoming strategic components for the most important civil areas, like transports, (especially cars, aircrafts and helicopters), environment and territory control, biomedicine and helps to "human beings better life" (intelligent building, energetic control, thermo-mechanical structuring, auxiliaries to handicapped people etc).

References

1. R.A. Smith, F. E. Jones and R. P. Chasmar, *Detection and Measurement of Infrared Radiation*, Oxford, (1958).
2. P.W. Kruse, Mc GlauchLin and R. B. McQuistan, *Elements of Infrared Technology*, Wiley, New York, (1962).
3. E S .Barr, *Amer. J Phys.*, **28**, 42 (1960).
4. William Herschel *Experiments on the Refrangibility of the visible Rays of the Sun" Phil.Transactions of the Royal Society of London*, **90**, 284 (1800).
5. I. Melngailis and T. C. Harman, *Semiconductors and Semimetals*, **5**, 11 Ed. Willardson and Beer (Academic Press, New York, 1970).
6. W. S. Boyle and G. E. Smith, *Bell Syst. Tech. J.*, **49**, 587 (1970).
7. D. F. Barbe, *Proc. IEEE,* **63**, 38 (1975).
8. A. J. Steckl, R.D.Nelson et al., *IEEE Proceedings,* **63** (1975).

9. C. Corsi, *IEEE Proceedings,* **63**, 62 (1975).
10. C. T. Elliot, D. Day and D. Wilson, *Infrared Physics*, **22**, 31 (1982).
11. F. Shepherd, *Proceedings of SPIE*, **443**, 42 (1983).
12. W. Kosonocky, *Proceedings of SPIE* , **443**,167 (1983).
13. C.T.Elliott, *IEEE Proceedings*, Conf. Publ. n. 3211, 61 (June 1990).
14. C. Corsi, "*Rivelatori infrarosso: stato dell'arte e trends di sviluppo futuro*", *Atti Fondazione Giorgio Ronchi* , Firenze XLVI, **5**, 801 (1991).
15. C. Corsi, *Proceedings of 2nd Joint Conference IRIS-NATO*, London 25 - 28 June 1996
16. P. R. Norton, *Proceedings of SPIE*, **3379**, 102 (1998).
17. M. Razeghi, *Opto-Electr.*, **6**, 155 (1998).
18. L. J. Kozlowski and W. F. Kosonocky, in *Handbook of Optics*, Ed. M. Bass, Williams, and W. L. Wolfe, (McGraw-Hill 1995).
19. P. Norton et al., *Proceedings of SPIE*, **4130**, 226 (2000).
20. A. Rogalski *Proceedings of SPIE*, **4413**, (2001)
21. A. Rogalski, in *Handbook of Infrared Detection Technologies*, **27**, 59 Ed. Henini and Razeghi (Oxford, 2003).
22. F. Bertrand, J. T. Tissot and G. Destefanis, in *Physics Sem. or Devices*, **II**, 713, Ed. V. Kumar - Agarwal, (Narosa Pub. House, New Delhi 1998).
23. B. F. Levine, *J. Appl. Phys.*, **74**, 1 (1993).
24. S. D. Gunapala and K. M. S. V. Bandara, *Thin Films*, **21**, 113 (Academic Press, 1995).
25. S. D. Gunapala and S. V. Bandara, in *Handbook Thin Devices*, **2**, 63, Francombe, Ed. (Academic Press 2000).
26. R. Watton, *Ferroelectrics*, **91**, 87 (1989).
27. R. A. Wood et al., *IEEE Proceedings*, Solid State Sensors & Actuators Workshop, June 1992, USA.
28. C.Corsi, *Int. NATO Electronics Warfare Conference*, Washington DC (1978).
29. C. Corsi et al., Nat. Patent n.47722°/80, Tech. Rep. PT-79 Elettronica S.pA.(1979).
30. T. F. Tao *Proceedings of SPIE*, **178**, 2 (1979).
31. C.Corsi, *Microsystem Technologies*, 149 (1995).
32. A. Moini, *Techn. Report,* The University of Adelaide, SA 5005, Australia 8 (1998).
33. C. Corsi, *8th AITA -IR Physics*, **49**, n.3, 192 (2007).
34. K. Vural, L. J. Kozlowski et al., *Proc. SPIE*, **3698**, 24 (1999).
35. C. Corsi, *4th AITA 1997*, Atti Fondazione Giorgio Ronchi, LIII, **1-3**, 11 (1998).
36. P. R. Norton *Proceedings of SPIE*, **3698**, 652 (1999).
37. P. E. Ross "*Moore's Second Law*", Forbes, 116, March25, 1995. "Int. Roadmap for Semiconductors ", ITRS, 2000 (http://public.itrs.net).
38. C. Corsi, *6th AITA Conf.*, Fondazione Giorgio Ronchi LVII, **3**, 363 (2002).

TERAHERTZ: THE FAR-IR CHALLENGE

Massimiliano Dispenza,[a,*] Annamaria Fiorello,[a] Alberto Secchi[a] and Mauro Varasi[b]

[a]*Selex-Sistemi Integrati*
Via Tiburtina km 12.400, Rome, Italy
[b]*Finmeccanica*
Piazza Monte Grappa, 4, 00195, Rome, Italy
[*]*E-mail: mdispenza@selex-si.com*

This chapter is an overview on terahertz technologies and applications for sensing. The most advanced imaging and spectroscopy techniques are described, considering current opportunities and limitations in comparison to probes in the adjacent regions of the e.m. spectrum. Potential applications are highlighted, with a specific focus on security for detection of illicit substances and revealing of hidden objects. The technological status and current bottlenecks on sources and detectors are reviewed and future trends discussed.

1. Introduction

Interest in the THz spectral region has its deep-rooted origins in several different areas of science. In the 1920s, astrophysicists began to take an interest in this field,[1] followed shortly afterwards by spectroscopists.[2]
Lastly, in 1974, the electronics community[3,4] started to use the term THz when referring to the range of frequencies between 300 GHz and 10 THz (or 10-300 cm^{-1} in wave numbers, 1 mm-30 μm in wavelengths, 1.25 - 37.5 meV in energy, 14 - 480 K in temperature units).
Today, the increasing need to counter the growth of the asymmetrical threats, posed by international terrorism has generated an urgent necessity for technology capable of quickly identifying aggressive acts, using explosives, chemical and bacteriological agents, in addition to those using conventional weapons.
THz technology is a potentially interesting area for the development of solutions in terms of sensors, for at least three reasons:

-Terahertz radiation is easily transmitted through many common barrier materials such as packaging, clothing, shoes, book bags, etc. in order to search for potentially dangerous materials contained in them.[5]

-The transmitted and reflected spectra of many materials of interest to security applications, including explosives drugs and other chemical and biological agents, contain THz absorption fingerprint useful to identify these hidden compounds.[6]

-Terahertz radiation is considered biologically safe to the subject being scanned or to the system operator due to the low photon energies unable to cause harmful photoionization.[7-9]

The possibility of combining the capability of imaging an object to that of spectroscopic analysis opens up a new perspective for a complete system ranging from imaging the threat, identifying its nature and therefore its danger, thus promising solution for Concealed Weapons Detection (CWD) and Concealed Explosives Detection (CED).[10] This possibility could be better implemented in the stand-off configuration, with distances in the range of tens of meters. Unfortunately, in this context, the attenuation of the THz waves caused by the atmosphere, especially by interaction with water molecules (see Fig. 1), poses severe requirements on the system.

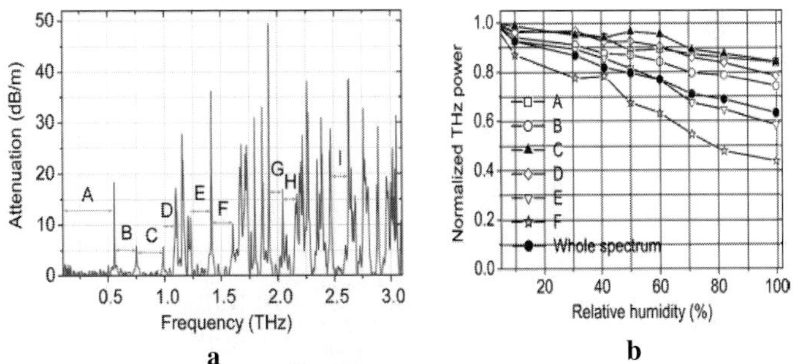

Figure 1. Atmospheric attenuation.[10] (a) Nine major THz transmission bands in the range 0.1-3 THz. (T=23°C, R.H.=26%): A: 0.1-0.55THz; B: 0.56-0.75THz; C:0.76-0.98THz; D: 0.99-1.09THz; E:1.21-1.41 THz: F:1.42-1.59THz; 1.92-2.04 THz; 2.05-2.15THz; 2.47-2.62 THz – (b) THz transmitted power vs relative humidity in six transmission bands.

This chapter intends to review the potential of THz technologies both for spectroscopic analysis (section 2) and imaging functions (section 3), also with a focus on the current state of the art for THz sources (section 4).

Particular emphasis will be placed on possible applications in the security sector.

2. THz Spectroscopy

A large set of molecular and crystalline modes can be excited due to interaction of THz waves with matter: rotational modes of gas molecules, crystal phonons, intermolecular bonds such as H-bonds are clearly identifiable in the absorption plots across the THz range (Fig. 2).
Polar liquids, such as water, are highly absorptive.[11] Crystals formed from polar liquids are significantly more transparent because dipolar rotations are frozen out, but these crystals may display phonon resonances. Non-polar, non-metallic solids such as plastics and ceramics, and non-polar liquids, are transparent at least partially.
Vibration features are also associated with intermolecular hydrogen bond relative motions.[6] Other types of intermolecular vibration modes are also excited. These characteristics make THz radiation particularly suited to identifying a wide range of substances, from organic macromolecules to solid crystals, and it is thus used for spectroscopic analysis and chemical detection of several compounds of interest, such as explosives and illicit drugs.

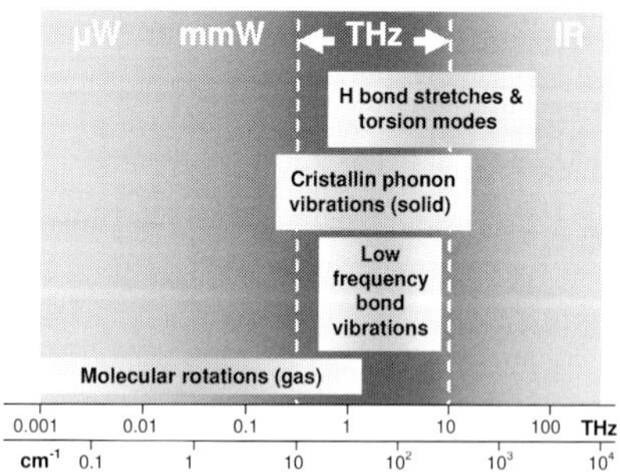

Figure 2. Electromagnetic field with matter interaction modes.

2.1. *THz vs Raman*

THz absorption analysis can also be considered as complementary to

Raman spectroscopy.[12-13] In fact, the Raman approach relies on a non linear type of radiation-to-matter interaction, thus providing different cross-section values, involving different roto-vibrational transitions, in the same energy region, due to the different selection rules involved in the two processes.

Figure 3 provides an example of comparison between Raman and THz absorption spectra for two chemically similar drugs: cocaine free base and cocaine hydrochloride. This is one case in which it is very obvious that THz spectra are much more suitable for distinguishing molecules of similar structure.

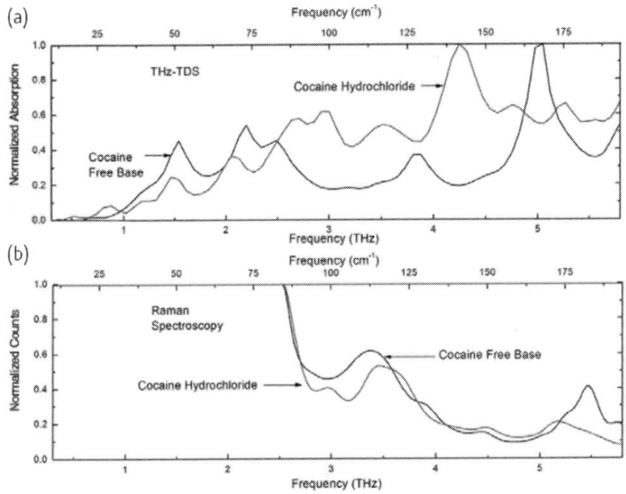

Figure 3. The THz spectra of polycrystalline cocaine free base and cocaine hydrochloride[13] obtained using (a) THz-TDS (Time Domain Spectroscopy) and (b) Raman spectroscopy.

2.2. Spectroscopic Techniques

Several methods have been employed in THz spectroscopy, the most important of which are based on:

- Fourier Transform Spectroscopy (FTS),
- Terahertz Time Domain Spectroscopy (TTDS),
- Frequency Scanning Spectroscopy.

FTS is a classical method for spectral analysis in the IR portion of the

spectrum. Its use has been extended towards the THz range, which is also known as Far Infra Red.
It uses blackbody-like broadband sources, typically consisting of a high pressure Hg arc lamp combined with an interferometer (see Fig. 4).

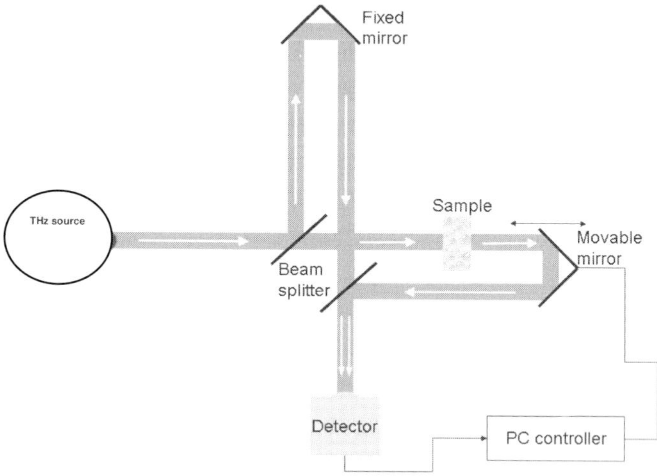

Figure 4. Fourier Transform Spectroscopy apparatus.

The beam with a power spectrum **s(f)** coming from the source is split into two equal parts and propagates along two axes. The sample with a transmission coefficient **t(f)** and a phase shift **Φ(f)** is put in one axis. The path length of the other arm can be varied to give a phase change of **2πfΔ**, where **Δ** is a variable delay. The detector measures the frequency integral of the absolute square amplitudes and the interferogram is given by [14]:

$$I(\Delta) = 2\int_0^{\infty} t(f) \cdot s^2(f) \cdot e^{i(2\pi f \Delta + \phi(f))} df \quad (1)$$

This measurement technique is derived from common approaches in IR spectroscopy, and is therefore affected by the low brightness of available sources of this type in THz domain and requires cryogenically cooled detectors.

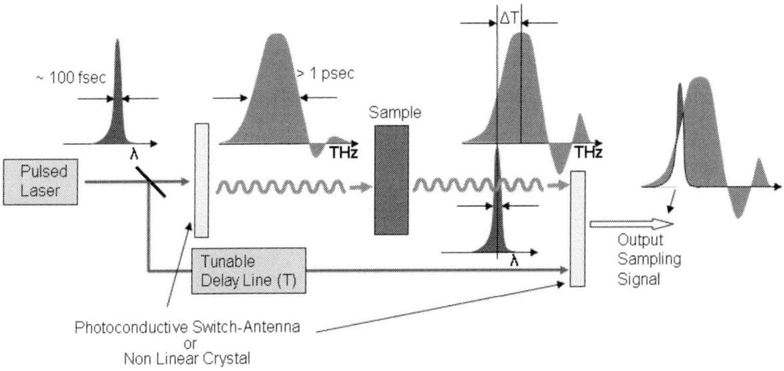

Figure 5. TTDS apparatus.

In TTDS (see Fig. 5), on the other hand, a pulsed laser source is used to generate a THz pulse, and then to sample the THz signal resulting from the transmission/reflection from the specimen.

First, a train of optical pulses, the duration of which is in the order of hundreds of fs and repetition rates up to tens of MHz, is generated by a suitable laser source, usually Ti:sapphire.

The optical pulse train is directed onto a THz emitter commonly implemented using two possible techniques. The first possibility involves using a photoconductive substrate, such as Low Temperature Grown (LTG) GaAs, upon which a dc biased (> 40V) antenna is printed. A current pulse is thus originated due to the incoming optical pulse, and this causes the emission of a picosecond pulse from the antenna (Fig. 6a).[15] The second option, known as optical rectification, relies on the use of non linear optical crystals (e.g. ZnTe) which cause the mixing (i.e. Difference Frequency Generation) of the spectral components in the broadband (>1THz) optical pulses, and thus the emission of THz signals.[16,17] The first method increases efficiency, but a wider bandwidth can be achieved using the second one.

After transmission through or reflection off the sample, the THz wave collides with the detector together with part of the optical pulse beam (Fig. 6b) which had been split before emission and delayed as required.

In a similar way as for generation, two approaches can also be used for detection: the photoconductive effect and so-called electro-optical sampling (again based on non linear optical crystals).

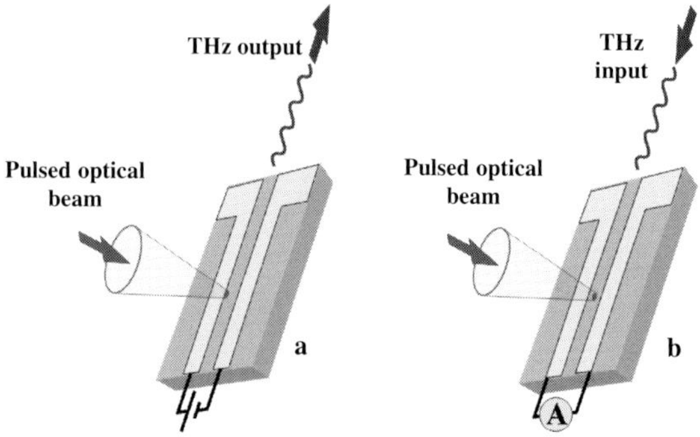

Figure 6. Photoconductive antennas for THz emitters (a) and receivers (b).

In both cases, the delay of the optical pulse train is scanned within a range equal to the duration of the THz pulse, the ratio of the two being of the order of 1:20 or higher. The optical pulse can thus act as a sampler of the shape of the THz wave, providing information on its amplitude and phase.[18-20]

Since multiple pulses are used to map out different points of the THz signal, its pulse to pulse stability in amplitude and shape, and also that of the optical source, is of vital importance.

TTDS does not require the use of a coolant for the detector[21] and this is one advantage with respect to FTS. Another positive characteristic is that, in TTDS, the time varying electric field is measured rather than just the intensity. The Fourier transformation of the THz pulse gives the amplitude and phase of the signal filtered by the sample, and then directly the real and imaginary parts of its dielectric function, without the need to resort to the Kramers-Kronig formula.

TTDS is a widely used technique generally used in commercially available products for Terahertz spectroscopy.[22,23]

While the transmission setup has become a more or less routine method,[24] reflection measurements are desired for practical applications, since most bulky targets are impossible to test in a transmission mode.[25,26] Furthermore, reflection spectroscopy, especially diffuse reflection spectroscopy of irregular surfaces, is the only one applicable for stand-off detection.[27]

As regards the stand-off detection, it is worth noting that in this case the actual target distance is an unknown (and not even constant) variable of the system, and it is therefore not easy to ensure that both the THz and optical beams travel paths of equal delays, to collide with the detector at the same time and perform optical sampling of the THz signal.

In one proposed implementation of the TTDS scheme (see Fig. 7), which may overcome this difficulty, both beams are focused onto and reflected off the sample. The path of both beams can thus be easily adjusted to be the same.[19] This approach has also the benefit of being insensitive to any variations in distance due to target motion.

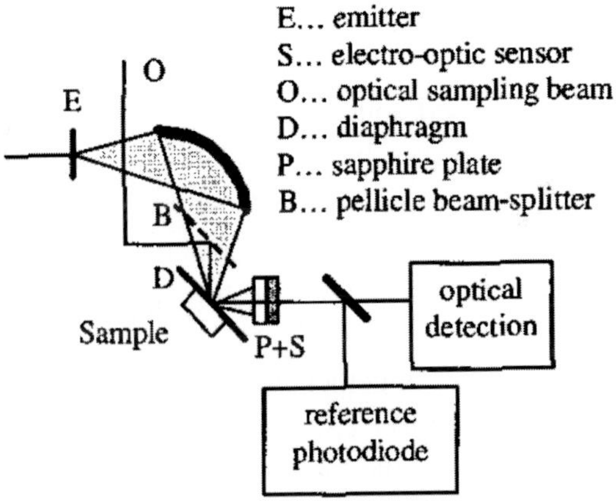

Figure 7. Reflection setup with collinear THz and optical beams.[19]

Lastly, high resolution spectroscopy can also be achieved by Frequency Scanning, i.e. using coherent, narrow band tunable sources based either on electronic (Gunn diodes, Backward Wave Oscillator), non-linear optics (parametric oscillators, photo-mixers), or lasers (Quantum Cascade Lasers, Free Electron Lasers, gas lasers) directly.

Unlike broadband sources which lose a high percentage of their radiated power because most of the signal spectrum falls beyond atmospheric transmission windows, narrow band emitters can be accurately tuned within the windows themselves. A technological limitation of this is due to the limited tuning range of these sources, which makes it difficult to obtain measurements over a wide frequency range. A narrow tuning range can be overcome by employing an array of sources centered at different wavelengths.

FTS	(+) higher available power at f < 4THZ (+) high SNR at f > 4THZ (-) cooled detectors
TTDS	(+) higher power at f < 4THZ (+) high SNR at f < 4THZ (+) no cooled detectors (-) tail of signal bandwidth out of transmission windows
Frequency Scanning	(-) low power (-) low tuning range (+) possibility to match transmission windows

Table 1. Characteristics of the three THz spectroscopic techniques.

The bandwidth of the pulsed sources used in TTDS is defined by laser pulse duration, which may not be reduced below hundreds of fs. This implies a cut-off frequency of not more than a few THz.

The black body source employed in FTS is, on the other hand, peaked in the IR range due to its thermal origin, thus providing little available power at THz frequencies.

Such spectral behavior, together with coherent characteristics of detection in TTDS compared to FTS, is enough to justify the higher SNR performances of the former in the lower part of the THz range, in excess of 10^8 in power from 10 GHz to 4 THz, which is much higher than the SNR of 300 obtained with FTS.

On the other hand, FTS has a higher SNR at higher frequencies.[14]

Lastly, in the case of tunable narrow band sources, the peak frequency can be centered in the desired position within the THz range, but may be affected by various limitations (tuning, output power) depending on the intrinsic technology used (Quantum Cascade Lasers, Backward Wave Oscillators, Optical Parametric Oscillators, etc.), as will be discussed later on (see section 4). Table 1 summarizes the characteristics, pros and cons of the three spectroscopic techniques.

2.3. Database of THz Spectra and Identified Signatures

A basic requirement for the use of THz radiation in the detection of chemical and biological threats is the existence of a reliable database of acquired spectra for the substances in question. A lot of effort has been expended in this direction (see Fig. 8).[5, 6, 10, 28]

It must be taken into consideration initially that some features of the spectral responses are highly dependent on sample characteristics and preparation techniques (grain size, presence of impurities, powder or pellet form), which can cause spurious peaks, resonance effects etc.

Such features must be discarded, as they are obviously not useable as a fingerprint for identification. Table 2 shows some typical absorption peaks, which have been identified as characteristics for a set of explosives and drugs. It must be taken into account that many of the above explosives have low vapor pressure, and THz spectroscopy thus represents an alternative to techniques that require direct interaction with vapor to detect the substance of interest.

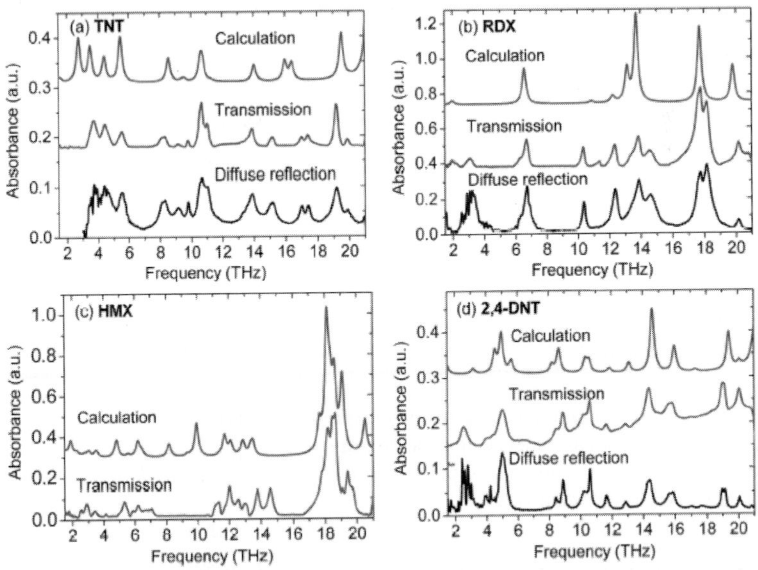

Figure 8. Compared spectra of different explosives (TNT, RDX, HMX, 2, 4-DNT) measured by transmission and reflection TTDS.[10]

Furthermore, it should be noted that, in the preparation of the most widely used explosives, the active compound is usually dispersed in a plastic matrix or mixed with other binders which may change its spectral

Material	Peak Absorption frequencies (THz)	Ref.
Explosives		
Semtex-H	0.72, 1.29, 1.73, 1.88, 2.15, 2.45, 2.57	29
PE4	0.72, 1.29, 1.73, 1.94, 2.21, 2.48, 2.69	29
RDX/ C4	0.72, 1.26, 1.73	29,30,31
PETN[a]	1.73, 2.51	29
PETN[b]	2.01	32
HMX[a]	1.58, 1.91, 2.21, 2.57	29
HMX[b]	1.84	32
TNT[a]	1.44, 1.91	29
TNT[b]	1.7	32
TNT	5.6, 8.2, 9.1, 9.9	31,33
NH_4NO_3	4, 7	32
Drugs		
Methamphetamine	1.2, 1.7–1.8	34
MDMA	1.4, 1.8	34
Lactose α-monohydrate	0.54, 1.20, 1.38, 1.82, 2.54, 2.87, 3.29	29
Icing sugar	1.44, 1.61, 1.82, 2.24, 2.57, 2.84, 3.44	29
Co-codamol	1.85, 2.09, 2.93	29
Aspirin, soluble	1.38, 3.26	29
Aspirin, caplets	1.38, 3.26	29, 34
Achetaminophen	6.5	35
Terfenadine	3.2	35
Naproxen sodium	5.2, 6.5	35

Table 2. Peak absorption frequencies of some explosives and drugs.[6]
[a] Samples are prepared as pellets using spectrographic-grade polyethylene.
[b]Samples are ordered as compressed pellets from Accurate Energetics LLC. All materials are in sensitized form (water-free).

response. As an example, figure 9 displays the spectral behavior of RDX and PETN in pure crystalline form (lower box) compared to those of SX2 (a military explosive containing RDX), Metabel (the active ingredient of which is PETN) and Semtex (a mixture of RDX and

PETN). It is obvious that complex explosives share some absorption peaks with their constituent elements, and these peaks must be identified in order to be used as fingerprint.

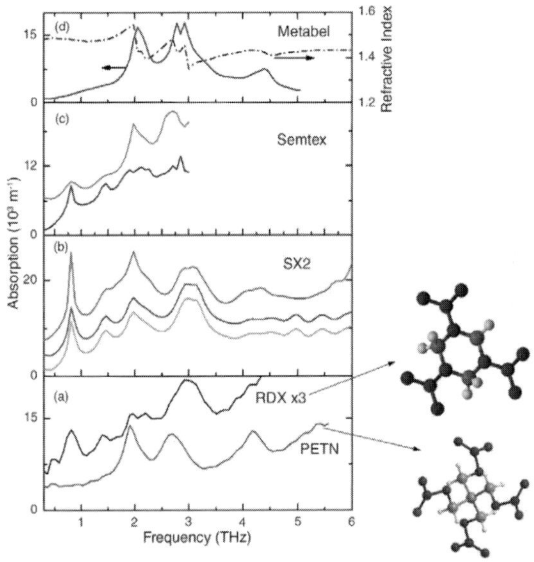

Figure 9. Spectral response of RDX, PETN, SX2, Metabel, Semtex.[13]

Lastly, THz radiation was also considered for the detection of so-called improvised explosives, a category comprising highly oxidizing compounds, such as hydrogen peroxides, nitrogen oxide, or ammonium nitrate, usually employed in mixtures with fuel oil for truck/car bombs.

In particular, a spectrum was acquired for ammonium nitrate,[6] as shown in Fig.10. In this case, due to it being an amorphous material, no sharp peak can be identified, apart from two small oscillations in the 3 -7 THz range and the broad absorption band centered at 3 THz.

This means that THz spectroscopy is of limited use in this case, apart from employing particular techniques, e.g. focusing the slope of absorption increase or checking any large difference in transmission from 0.2 THz to 3 THz, nevertheless merely obtaining an exclusion test rather than a clear identification.

Databases of THz spectra for many compounds have been created so far and in some cases are also available on the internet.[36-37]

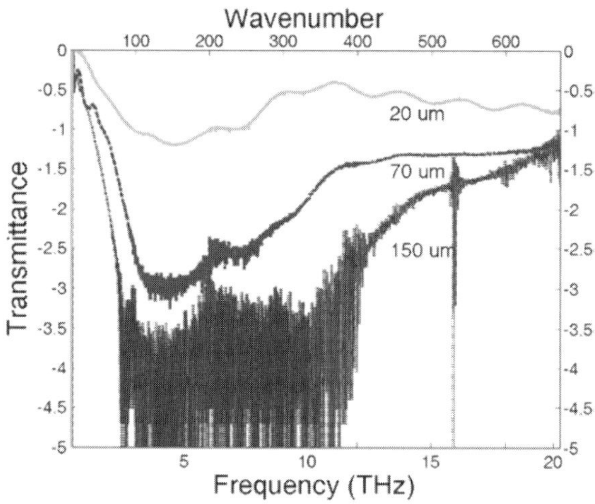

Figure 10. Ammonium nitrate absorption THz spectrum.[6]

3. THz Imaging

THz radiation is a valuable probe for performing imaging with comparative advantages compared to other portions of the e.m. spectrum. The transparency of clothing and other cover materials, together with significant differences in reflectivity between materials such as metal, water, polar and non-polar dielectrics, enable information about concealed objects to be obtained with better contrast and definition than by using IR light.

On the other side of the spectrum, if compared to mm-wave radiation, for which well developed technology is used, a definite improvement can be obtained in terms of spatial resolution, due to the shorter wavelength of the radiation involved.

Furthermore, the presence of spectral fingerprints (already discussed) can be used by performing multi-spectral imaging to identify nature of specific objects in the image.

Obviously, it should be considered that in the THz range, atmospheric absorption is definitely higher than that occurring at mm-wave frequencies. A correct choice of the operating frequencies, which should lie within well defined transmission windows, is required in this case.

3.1. Imaging Techniques

Either passive or active imaging can be performed, both having their pros and cons.

The passive approach relies on the detection of radiation generated by objects due to either:

- e.m. emission of the object according to its own temperature T and emission factor ε;
- reflection of e.m. waves radiated from surrounding objects at temperature T_S according to its reflection factor *r*;
- transmission of e.m. waves radiated from background objects at temperature T_B behind the object according to its transmission factor *t*.

Apparent temperature T_0 at each point of the image is thus given by

$$T_0 = \varepsilon T + r T_S + t T_B \qquad (2)$$

and contrast in the scene is related to the difference in apparent temperature.

In the case of screening of people, the main source of illumination is body heat. Concealed objects are displayed if they are colder than the human body or absorb part of radiation emitted by the body.

Passive imaging allows simplification in technology and design since one has not to tackle the lack of efficient sources in THz spectrum.

Equipments based on this approach and dedicated to security applications have already been developed and are now available on the market.[38-39]

These equipments also has commercial advantages: passive imaging, being unregulated, enables users to skip long timelines for the regulatory requirements necessary for an active system, and privacy issues are of less concern, as concealed objects are unambiguously revealed but anatomic details are not displayed in this case (see figure 11).

Active imaging on the other hand provides clear advantages in terms of sensitivity, but at the expense of higher system complexity.

The active approach proposes a set of alternative options as regards the nature of the THz signal (pulsed vs CW) or the detection scheme (direct vs heterodyne).

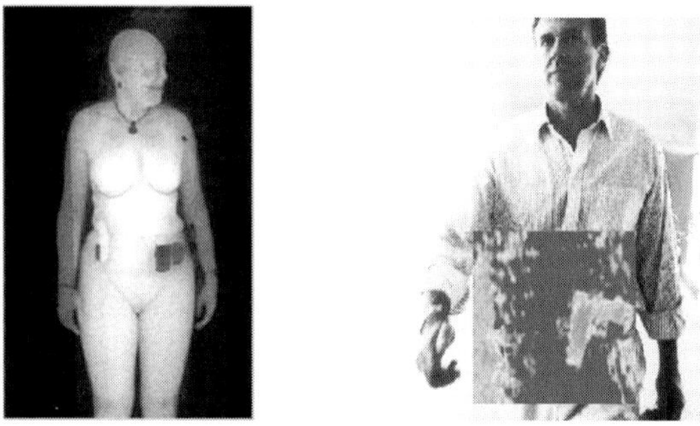

Figure 11. Active (a) vs passive (b) imaging: privacy concerns.

Pulsed signals may be successfully used to perform a 3D tomography of the object in a similar way to a short-range radar. This can also improve system sensitivity by techniques such as range gating, to select signals at a given depth while discarding surface brightness.

Active illumination also enables the use of a set of radar techniques such as signal chirping, phased array and SAR (Synthetic Aperture Radar).[40-43]

Moreover, the broadband nature of the signal also enables the spectral analysis of the pixels in the picture (see Fig. 12).[44]

By using heterodyne detection, sensitivity (and SNR) can be increased by up to 8 orders of magnitude compared to direct detection.[45]

Another important aspect for any imaging system, besides spatial resolution and SNR, is its acquisition speed (or frame rate).

This is also a historical limitation of THz systems due to being unfeasible and practically unaffordable to build large detector arrays, as they would be required for conventional picture resolutions (1 Mega pixel or higher).[46] Detectors are quite expensive and they cannot easily be integrated in small areas. When real time image acquisition, one image at a time, is not practical, faster scan techniques (i.e. mechanical scan of a single detector across a scene, using mirrors) must be used. This degrades the frame rate to 50 pixels/s.[47] One proposed method of dealing with this problem is electro-optical imaging: the beam received is converted from THz to optical, so that acquisition through a visible CCD camera with a high enough resolution is possible at a frame rate up to 5000 pixel/s.[48, 49]

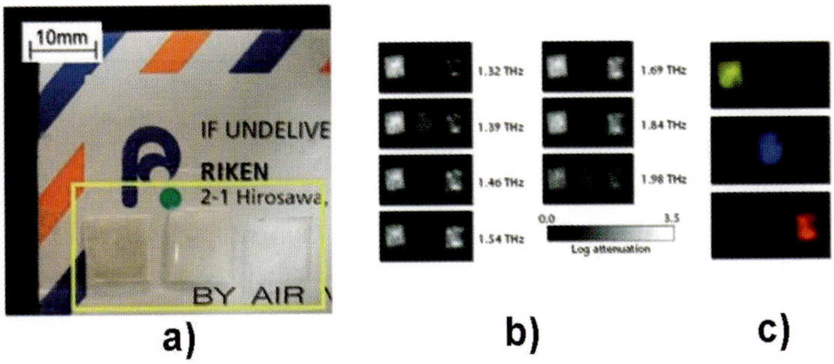

Figure 12. Example of multispectral imaging.[44] (a) View of the samples. The small polyethylene bags contain (left to right): MDMA, aspirin and methamphetamine. The bags were placed inside the envelope and the area indicated by the yellow line was scanned; (b) Multispectral image of the target, recorded at seven frequencies between 1.32 and 1.98 THz; (c) Spatial patterns of MDMA (yellow), aspirin (blue) and methamphetamine (red) extracted from the multispectral image by use of fingerprint spectra .

Conversion is achieved by modulating an optical beam with the THz signal by using a non linear crystal. In this way speed is improved at the expense of SNR.

Another technique to avoid scanning by increasing the effective number of pixels for a given array of N detectors is interferometric imaging: when THz radiation reaches the array, the relative phase and amplitudes of the signal in the complex u-v plane is measured for each pair of detectors.

Thus, each pair of detectors gives rise to a couple of u-v numbers, which means N(N-1)/2 couples for the whole array.

A rather complicated analysis shows that N real pixels generate a resolution equivalent to N(N-1)/2 effective pixels, thus providing a quadratic increase in picture definition.[46-50]

3.2. Close Range vs Stand-Off

Usually, two imaging modes, stand-off and close range, are distinguished according to whether the target distance is greater or shorter than 1-2 meters respectively. For security applications, it should also be considered that the target could be moving. The main critical issues that arise when passing from the analysis of close range static targets in the lab to the stand-off detection of moving targets in real application scenarios are related to:

1. Atmospheric absorption

Propagation losses increase with increasing frequency and may be quite high outside of transmission windows. Tails of broadband signals fall well beyond the limits of these windows, thus causing signal loss and range reduction. This is the case when spectral signatures and bare imaging are required, since broadband TTDS signals need to be used.

2. Atmospheric dispersion

Pulse spread due to dispersion in atmospheric moisture has also to be considered, since it spoils signal shape and reduces its strength. For example, a 1 ps pulse is broadened to a time duration in excess of 100 ps through 100 m of a humid atmosphere.[51]

3. Resolution

The diffraction-limited spatial resolution at a distance **d** of an imaging system of aperture size **D** is given approximately by **λd/D**.
This implies that for 1 cm resolution at 1THz, the required aperture size is 10 cm at 3m distance, increasing to more than 30 cm at 10m.

4. Proper timing in pulse sampling for TTDS.

The critical aspects already discussed in section 2 also arise when spectroscopic functions need to be used together with imaging to identify specific objects or compounds in the image.
Stand-off spectroscopy is affected by all the limitations inherent in the need to operate in reflect mode, thus sensitive to any target motion from pulse to pulse in a TTDS approach.
A simple analysis of detection range is useful to give an idea of the problems that occur in real-life applications. This can be done by comparing the power received after one round trip (transmission, target reflection and travel back) with detector sensitivity.
A basic formula relates the received THz power ($\mathbf{P_r}$) to the transmitted one ($\mathbf{P_O}$) according to

$$P_r = A_r \left[\frac{P_0}{\Theta(2d)^2} \right] e^{-2\alpha_b L_b} e^{-2\alpha_a d} e^{-2\alpha_e L_e} \tag{3}$$

where:
Θ is the solid angle in which the transmitted power is directed,
$\mathbf{A_r}$ is the effective area of the receiver,
$\mathbf{L_b}$ and $\mathbf{L_e}$ are the thickness of the layers of the barrier and explosive respectively,
$\boldsymbol{\alpha_b}$, $\boldsymbol{\alpha_b}$, $\boldsymbol{\alpha_b}$ are the attenuation coefficients of the barrier material, of the explosive and of the atmosphere,
$\boldsymbol{d}$ is the distance from the target.

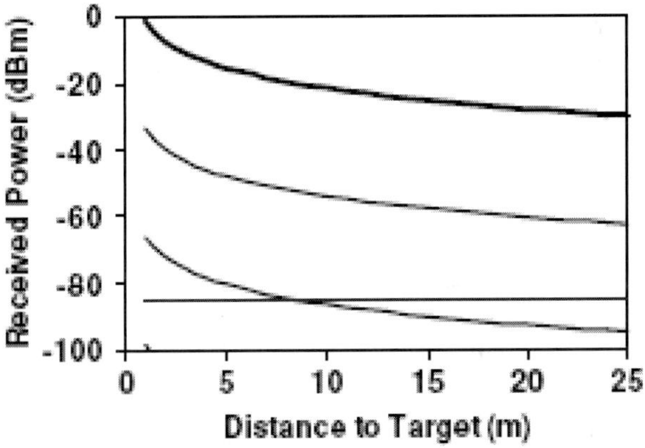

Figure 13. Calculation of detected THz power through a wool sweater of thickness 0, 3, 6 and 9 mm.[6]

Using this formula, Federici et al.[6] showed (see Fig. 13) that detection range may be heavily dependent on barrier layers and even very common covers, such as a wool sweater, may have significant effect in reducing the received power below detection levels (10^{-11}W in this example).
Combined spectroscopic and imaging operation in stand-off conditions is the most challenging objective in THz sensing for defense and security purposes, as it lies at the intersection of most of the critical issues discussed in this chapter.
To summarize all the issues discussed above, Table 3 aims to include all different aspects related to the THz sensing domain outlining the pros and cons of the different options for each category identified.

Detection	**Direct detection** (+) simplicity (-) lower sensitivity
	Heterodyne detection (+) up to 8 orders of magnitude dynamic increase (-) complexity (commercially not affordable)
Image acquisition	**Raster Scan** (+) simple (-) slow frame rate
	Electro-optic Imaging (+) fast (-) reduced dynamics
	Interferometric imaging (+) increasing resolution and frame rate with fewer detectors (-) hardware and computational complexity
Passive	(+) no THz source required (+) unregulated (commercially preferred) (+) no anatomic details (privacy) (-) lower sensitivity
Active	**CW** (+) possibility to match transmission windows
	Pulsed (+) imaging & spectroscopy possible (-) broadband means absorption out of atmospheric windows (-) depth information can be acquired (3D tomography) with even sensitivity improvement by range-gating techniques

Table 3. Summary of THz imaging detection techniques.

4. THz Sources

Many different approaches for THz signal generation have been employed over the years.[52] They can roughly be divided in the following categories:

- Electronic sources based on multiplied output, either from solid state devices or vacuum tube components.

- "Optical style" sources based either on laser effect, on down conversion from optical frequencies, or on black body thermal emission.

Each of these THz generators has its limitations as regards current technological capabilities and the performance levels achievable, which still justify talking about the so-called "THz gap" in terms of available sources operating in the range between IR and mm-wave.

Electronic solid-state sources based on semiconductors, i.e. oscillators and amplifiers, rely either on frequency multiplication from mm-wave sources or on direct generation.

Sources for direct generation are traditionally Gunn,[53,54] IMPATT or TUNNETT diodes, based on GaAs, InP or wide band gap semiconductors such as GaN. They are affected by high-frequency roll-off due to reactive parasitic or high transit times.

Directly generated CW sources exist at 100 GHz, with a narrow relative line width ($< 10^{-6}$), with output power up to 100 mW. Output power fades as $1/f^2$ and then as $1/f^3$ as the frequency increases.

Frequencies above 1 THz are obtained by multiplication[55] exploiting up to 3rd or 4th order chains of Schottky barrier diodes. They are limited by low frequency conversion coefficients.[56,57]

Electronic vacuum tube sources were developed in various configurations such as Klystrons, Travelling Wave Tubes (TWT), Backward Wave Oscillators (BWO) and Gyrotrons.[58, 59]

The most noteworthy of these is BWO, as it provides the highest power and tuning range at THz frequency.

A BWO is a tube in which a free electron beam, flowing along an electric field (1 to 10 kV) spiraling around a magnetic field (about 1 T), transfers its energy to a slow wave counter-propagating electromagnetic structure.

Power up to a few mW at frequency > 1THz can be obtained, with a bench-top device much smaller than other free electron based sources. The tuning range may reach 0.4 THz at 1.3 THz centre frequency for commercially available devices.[60]

In spite of its positive features, the need for extremely high fields (both magnetic and electric), as well as high current densities still limit its field of application.

THz radiation can be also generated directly by laser effect, as in gas lasers and Quantum Cascade Lasers (QCL), or indirectly by the down-conversion of an optical carrier.

Gas lasers are optically pumped by CO_2 laser to excite the roto-vibrational levels of gas molecules. Methanol is the most widely used active gas operating at 2.5 THz and achieving a power of few tens of mW. Usable frequencies are limited to the available molecular levels, which can be excited in the range between 0.3 to 5 THz. They are also still bulky, and expensive devices, but also widely distributed and available from several companies such as Coherent Inc.[61] and Edinburgh Inst. [62]

QCLs are semiconductor lasers in which stimulated emission is obtained by transition between two states lying within the conduction band (so called inter-sub-bands or intra-band transitions). In fact, standard inter-band transitions used in photonic laser devices are not capable of providing radiation at frequencies lower than 10 THz even using narrow band gap lead-salt materials.

Intra-band discrete levels are obtained by suitably engineered hetero-structures giving rise to a so-called super-lattice. Moreover the lasing effect in these devices is only due to electron transition rather than electron-to-hole recombination (uni-polar devices). Super-lattice structure is periodical, and thus, after recombination, electrons tunnel to the next period, where a new transition can occur, allowing a quantum efficiency greater than one. The latter effect is at the origin of the cascade attribute in the name.[63,64]

The first QCL in the THz domain was obtained by a joint cooperation between NEST-Pisa and the Cavendish Lab. It operated at 4.4 THz with about 2mW output power at 50K.[65] Recently, the demonstration of terahertz QCLs emitting 248 mW peak power in pulsed operations at 4.4 THz, and up to 138 mW of power in CW operations, have been reported.[66]

Resonant-phonon terahertz QCLs have also been demonstrated up to temperatures of 164 K in pulsed mode and 117 K in CW mode.[67]

The main difficulties involve achieving population inversion between narrowly separated sub-bands and mode confinement at long wavelengths. Nevertheless QCL operation at low THz frequencies ranging from 2.0 THz down to 0.84 THz was also demonstrated.[68]

External cavity lasers have been produced to increase tunability. In the simplest implementation a mirror has been inserted in the cryostat head, in close proximity to the laser facet, without any coupling optics in between. By changing the mirror position, the cavity length is varied and the emission is tuned. Tuning ranges up to 3 cm^{-1} (~ 0.1 THz) have been achieved in this way.[69]

Optical down-converters for THz sources are usually of two types and are used both for narrow-band and broad-band generation: photocurrent-based photo-mixers and photo-mixers based on non-linear optical crystals.[70]

In the first case,[71-73] two phase-locked CW lasers are focused onto a photo-conductive substrate with very short (<1ps) carrier lifetime, e.g. low temperature grown GaAs or Uni-Travelling Carrier photodiode.[74,75] This generates charged carriers between closely spaced electrodes printed onto the substrates and connected to an antenna which radiates out the signal.[76-78]

A similar effect can be obtained by Difference Frequency Generation, which can also be implemented in an OPO cavity, using Non Linear optical substrates such as ZnTe, GaP or $LiNbO_3$.[44, 79]

As the two incoming optical waves are of CW type, narrow band THz signals are emitted.

The use of broadband optical sources, on the other hand, enables the generation of conversely broadband THz waves,[80] in the same way as illustrated in the above paragraph on TTDS.

Black body sources provide a broad band (wavelength spectral range from 5 mm to 5 μm), low power spectral density and inconsistent radiation.

These sources are available commercially and typically consist in low pressure, water-cooled mercury arc lamps. They are widely used in Fourier transform spectroscopy. Table 4 provides a summary of achieved performances for some of the sources for which commercial products are available, together with a short recall of advantages and limitations of each single approach.

5. Conclusions and Future Outlook

To conclude this brief overview, we can observe that we are now coming to the end of two decades in which much effort has been expended on technological developments for emitters (PAs, non-linear optic approaches, semiconductor lasers), receivers (super-conducting bolometers, electro-optical sampling etc, Schottky mixers) and detection techniques (TTDS, heterodyne techniques etc.) in the THz field.

Consequently a wide range of technological solutions are now at the proof-of-concept stage, i.e. TRL (Technology Readiness Level) 3 or higher. Some of these solutions (TTDS, PAs, GaAs Schottky mixers) have already been integrated in real systems operating in relevant environment (TRL 6 or higher).

Few products are already available commercially, in many cases as bench top components for lab analysis or close-range monitoring,[22, 23] but the first systems attempting stand-off imaging[39] were recently introduced onto the market.

A recent analysis[87] based on trends in scientific literature in the THz field during the last 30 years showed that this field is now in a period in which well established and tested technology is available, the potential applications are clearer and this, rather than the autonomous efforts of the THz community, could soon start to become the drivers for technological development.

Immediate applications are expected in the non-destructive, non-invasive inspection of industrial applications. Sensing for homeland security and protection is also commonly taken as a mid-term application.

As regards these two domains, two events were respectively considered to have boosted scientific and technological growth: the Shuttle Columbia disaster, which highlighted problems for which structural inspection and monitoring were recognized as a definitive countermeasure, and 9/11 which increased the fear of global terrorist threats. In three subsequent documents[88] focused on the technological assessment of the THz field, the US National Research Council recently recommended studies and investment with a view to CWD and CED.

Laboratory experiments indicate that, with more powerful and largely tunable sources and more sensitive detectors, THz techniques have massive potential for the detection and identification of concealed explosives even with combined spectroscopic and imaging functions or consolidated stand-off detection.

THz source	Details
Direct Multiplied mm-waves	Available commercially (e.g. Virginia Diodes, Inc.:[81] 12μW, 1.2 - 1.3 THz) (+): small, solid state devices (-): high-frequency roll-off due to reactive parasitics or high transit times
BWO	Available commercially (e.g MICROTECH® instruments inc.:[60] 0.5 mW max output power, operating frequency up to 1.42 THz, 0.4THz tuning range; ISTOK Research and Production Company), (+): relatively high power, tunability (-): still bulky, expensive
Optically pumped gas lasers	Available commercially (e.g. Coherent, Inc. SIFIR-50:[61] power > 50 mW, 0.3-7 THz, CW or pulse); Edinburgh Inst[62] (+): high power (-): bulky, expensive, not tuneable (discrete set of frequencies)
QCL	Commercially available (e.g. Laser Components:[82] oper. freq. 2-5 THz, 10 KHz pulsed, peak power < 0.2 mW) (+): integrated solid state sources (-): requires cooling, still low tunability
Photoconductive antennas (PCAs)	Available commercially (e.g. BATOP GmbH:[83] wavelength range 0.1 -3THz, power < 1mW pulsed) (+): established technology (integrated in several TTDS systems) (-): cannot withstand large power; only pulsed, only broadband
Photomixing	Available commercially (e.g. TOPTICA[84]) (+): Tunable, wideband, even CW operation by mixing of frequency locked CW lasers (-): Efficiency lower than PCAs
Optical parametric oscillator	Available commercially (e.g. M-Squared Firefly-THz:[85]). Room temperature operation, continuous tuning (1.2 - 3 THz), 10 μW pulsed (10 ns pulse at 400 Hz) (+): Room temperature operation, Tunable (-): Low power
Mercury Lamps	Available commercially (e.g. in Sciencetech SPS-300 THz FTIR[86]). Water cooled mercury arc lamp, broad band (from 5 mm to 5 μm), low operational pressure (10^{-3} Torr). (+): Simple broadband sources, room temperature operation, (-): Low power in THz region (tails from IR), bulky.

Table 4. Available performances and typical limitations in THz sources.

Nevertheless, efforts are still required before technology will be mature enough to provide the levels of sensitivity required for wide-ranging use in security applications. More advanced techniques for image acquisition

are also expected before an acceptable frame rate can be achieved and real time imaging will be definitely possible.
Furthermore, the algorithms for image processing and data fusion approaches with complimentary sensors (CCD visible cameras, IR viewers) need further research.
While new technological solutions will become available, the still open issue regarding any hazardous biological effects will be repeatedly arisen, and more definitive and conclusive answers, than those currently available, will be required.
On the other hand, the risk of over-estimating the potential capabilities of this sensing technology must be avoided: e.g. promised applications for the detection of improvised explosives is so far not confirmed by identifiable fingerprints, which are generally missing or confusing compared to those of more common explosive compounds.
Lastly, it can be concluded that, compared to tested technologies and widespread applications in the surrounding regions of the EM spectrum in THz domain there is still a need for more basic research and technological development, but this field is definitely no longer in its infancy and promises to provide its own solutions for practical problems in the next decade.

References

1. E. J. Nichols and J. D. Tear, *Astrophysics J.*, vol. 61, 17 (1925).
2. J. W. Fleming, *IEEE Trans. MTT*, vol. MTT-22, 1023 (1974).
3. J. R. Ashley and F. M. Palka, *IEEE MTT-S Int. Symp. Dig.*, vol. 73, 180 (1973).
4. J. Kerecman, *IEEE MTT-S Int. Symp. Dig.*, 1973, vol. 73, 30 (1973).
5. M. C Kemp in *Millimeter Wave and Terahertz Technology for the Detection of Concealed Threats – A Review, Infrared and Millimeter Waves*, 647 (2007).
6. J. F. Federici, B. Schulkin, F. Huang, D. Gary, R. Barat, F. Oliveira and D. Zimdars, *Semicond. Sci. Technol.*, vol.20, 266 (2005).
7. R. H. Clothier and N. Bourne, *J. Biol. Phys.*, vol. 29, 179 (2003).
8. M. R. Scarfi et al., *J. Biol. Phys.*, 2003, vol. 29, 171(2003).
9. E. Berry et al., *J. Biol. Phys.*, 2003, vol. 29, 263(2003).
10. H. B. Liu, H. Zhong, N. Karpowicz, Y. Chen and X. C. Zhang, in *Terahertz Spectroscopy and Imaging for Defense and Security Applications,* IEEE Proceedings, vol.95 n°8, 1514 (2007).

11. K. Yamamoto, M. Tani and M. Hangyo, in *Terahertz time-domain spectroscopy of ionic liquids and organic liquids*, Joint 30th Intl. Conf. on Infrared and Millimeter Waves & 13th Intl. Conf. on Terahertz Electronics, vol. 2, 413 (2005).
12. A. D. Burnett, W. H. Fan, P. C.Upadhya, in *Broadband terahertz time-domain and Raman spectroscopy of explosives*, Proceedings of SPIE, Vol. 6549, 654905 (2007).
13. G. Davies and A. Burnett, Materials Today, Elsevier, vol. 11, 18 (2008).
14. P. H. Y Han, M. Tani, M. Usami, S. Kono, R. Kersting and X.-C. Zhang, *J. Appl. Phys.*, vol. 89, 2357 (2001).
15. P. R. Smith and D. H. Auston, IEEE *Journal Of Quantum Electronics*, vol. 24. No. 2, 255 (1988).
16. A. S. Nikoghosvan and E. M. Laziev in *Terahertz generation at optical rectification in free space and in a waveguide*, Lasers and Electro-Optics Europe, CLEO/Europe, 428 (2003).
17. M. Tani, S. Matsuura, K. Sakai and S. Nakashima, *Applied Optics*, vol. 36, issue 30, 7853 (1997).
18. L. Duvillaret, F. Garet, J. F. Roux and J. L. Coutaz, *IEEE J. On Elected Topics In Quantum Electronics*, vol.7 n. 4, 615 (2001).
19. F. Pashkin, H.Kadlec, Nemec and P. Kuzel, in *Phase-sensitive time-domain terahertz reflectometry*, Joint 29th Int. Conf. on Infrared and Millimeter Waves and 12th Int. Conf. on Terahertz Electronics, 373 (2004).
20. Y. Cai, I. Brener, J. Lopata, J. Wynn, L. Pfeiffer, J. B. Stark, Q. Wu, X. C. Zhang, and J. F. Federici, *Applied Physics Letters*, vol. 73, issue 4, 444 (1998).
21. Hosako, N. Sekine, M. Patrashin, S. Saito, K. Fukunaga, Y. Kasai, P. Baron, T. Seta and J. Mendrok, in *At the Dawn of a New Era in Terahertz Technology*, IEEE Proceedings, vol. 95, issue 8, 1611 (2007).
22. http://www.teraview.com.
23. http://www.picometrix.com.
24. D. Zimdars, J. White, G. Stuk, A. Chernovsky, G. Fichter and S. L. Williamson, in *Time Domain Terahertz Detection of Concealed Threats in Luggage and Personnel*, Proceedings of SPIE, vol. 6212 (2006).
25. C. Baker, T. Lo, W. R. Tribe, B. E. Cole, M. R. Hogbin and M. C. Kemp, in *Detection of Concealed Explosives at a Distance Using Terahertz Technology*, IEEE Proceedings, vol. 95, issue 8, 1559 (2007).
26. S. G. Kong and D.H. Wu in *Terahertz Time-Domain Spectroscopy for Explosive Trace Detection*, Proceedings of the IEEE International Conference on Computational Intelligence for Homeland Security and Personal Safety, 47 (2006).
27. H. B. Liu, Y. Chen, G. J. Bastiaans and X. C. Zhang, *Optics Express*, vol. 14, 415 (2006).
28. K. Yamamoto, M. Yamaguchi, F. Miyamaru, M. Tani and M. Hangyo, *Jpn. J. Appl. Phys.*, vol. 43, 414 (2004).

29. W. R. Tribe, D. A. Newnham, P. F. Taday and M. C. Kemp, in *Hidden object detection: security applications of terahertz technology*, Proceedings of SPIE, vol. 5354, 168 (2004).
30. F. Huang, B. Schulkin, H. Altan, J. Federici, D. Gary, R. Barat, D. Zimdars, M. Chen and D. B Tanner, *Appl. Phys. Lett.*, vol. 85, issue 23, 2477 (2004).
31. R. Osiander, J. Miragliotta, Z. Jiang, J. Xu and X. Zhang, in *Mine field detection and identification using THz spectroscopic imaging*, Proceedings of SPIE, vol. 5070, issue 1 (2003).
32. D. J. Cook, B. K. Decker, G. Maislin and M. G. Allen *Through container THz sensing: applications for explosive screening*", Proceedings of SPIE, vol. 5354, (2004).
33. M. J. Fitch, D. Schauki, C. Kelly and R. Osiander, *Terahertz imaging and spectroscopy for landmine detection*, Proceedings of SPIE, vol. 5354, issue 45 (2004).
34. K. Kawase, Y. Ogawa and Y. Watanabe, *Opt. Express*, vol. 11, issue 20, 2549 (2003).
35. M. B. Campbell and E. J. Heilweil in *Non invasive detection of weapons of mass destruction using THz radiation*, Proceedings of SPIE, vol. 5070, 38 (2003).
36. http://webbook.nist.gov/chemistry/thz-ir.
37. http://www.frascati.enea.it/THz-BRIDGE/database/spectra/searchdb.htm.
38. C. Mann in *Practical Challenges for the Commercialization of Terahertz Electronics*, Microwave Symposium IEEE/MTT-S International, 1705 (2007).
39. http://www.thruvision.com.
40. J. O'Hara and D. Grischkowsky, *Optics Letters*, vol. 27, issue 12, 1070 (2002).
41. J. O'Hara and D. Grischkowsky, *JOSA* **B**, vol. 21, issue 6, 1178 (2004).
42. B. B Hu and M. C. Nuss, *Optics Letters*, vol. 20, issue 16, 1716 (1995).
43. K. McKlatchy, M. T. Reiten and R. A. Cheville, *Appl. Phys. Lett.*, vol. 79, 4485 (2001).
44. K. Kawase, M. Sato, T. Taniuchi and H. Ito, *Appl. Phys. Lett.*, vol. 68, no. 18, 2483 (1996).
45. K. J. Linden and W. R. Neal, in *Terahertz Laser Based Standoff Imaging System*, IEEE Proceedings of the 34th Applied Imagery and Pattern Recognition Workshop, 8 (2005).
46. J. F. Federici, D. Gary, B. Schulkin, F. Huang, H. Altan, R. Barat and D. Zimdars, *Appl. Phys. Lett.*, vol. 83, no. 12 2477 (2003).
47. G. Zhao, R. N. Schouten, N. Van Der Valk, W. T. Wenckebach and P. C. M. Planken, in *Design and performance of a THz emission and detection setup based on a semi-insulating GaAs emitter*, Review Of Scientific Instruments, vol. 73, no. 4 (2002).
48. Q. Wu, T.D. Hewitt and X.-C. Zhang, *Appl. Phys. Lett.*, vol. 69, 102 (1996).
49. M. Usami, Appl. *Phys. Lett.*, vol. 86, issue 14 (2005).

50. A. Bandyopadhyay, B. Schulkin, M. D. Federici, A. Sengupta, D. Gary and J. F. Federici in *Terahertz near-field interferometric and synthetic aperture imaging,* JOSA A, vol. 23, issue 5, 1168 (2006).
51. T. Yuan, H. Liu, J. Xu, F. Al-Douseri, Y. Hu and X.-C. Zhang, in *Terahertz time-domain spectroscopy of atmosphere with different humidity*, Proceedings of SPIE in Terahertz for Military and Security Applications, vol. 5070 (2003).
52. G. P. Gallerano, in *Overview of Terahertz Radiation Sources,* Proceedings of the FEL Conference, 216 (2004).
53. E. Alekseev, A. Eisenbach, D. Pavlidis, S.M. Hubbard, and W. Sutton, in *Development of GaN-based Gunn-Effect Millimeter-Wave Sources*, Work supported by ONR and DARPA/ONR.
54. T. W. Crowe, *IEEE J. of Solid-State Circuits*, Vol. 40, no. 10 (October 2005).
55. H. Eisele, A. Rydberg and G.I. Haddad, in *Recent advances in the performance of InP Gunn devices and GaAs TUNNETT diodes for the 100-300-GHz frequency range and above*, IEEE Transactions on Microwave Theory and Techniques, vol. 48, issue 4, Part 2, 626 (Apr. 2000).
56. P. H. Siegel, in *Terahertz Technology – Review*, IEEE Transactions On Microwave Theory and Techniques, vol. 50 no. 3 (2002).
57. V. Raisanen, in *Frequency multipliers for millimeter and submillimeter wavelengths*, Proceedings of IEEE, vol. 80, 1842 (Nov. 1992).
58. J. Rodgers, R. Chang, V. L. Granatstein, T. M. Jr. Antonsen, G. S. Nusinovich and Y. Carmelin *Miniature Plasma Cathode for High-Power Terahertz Sources*, Joint 30th Int. Conf. on Infrared and Millimeter Waves and 13th Int. Conf. on Terahertz Electronics, IRMMW-THz 2005, vol. 1, 323 (2005).
59. L. Ives, D. Marsden, M. Caplan, C. Kory, J. Neilson, R Wilcox and T. Robinson, in *Terahertz Backward Wave Oscillators,* Infrared and Millimeter Waves and 12t^{h} International Conference on Terahertz Electronics Conference Digest, 677 (2004).
60. http://www.mtinstruments.com/thzsources/index.htm.
61. http://www.coherent.com/Lasers/.
62. http://www.edinst.com.
63. F. K. Barkan, D. M. Tittel, Mittleman, *Optics letters*, vol. 29, no. 6 (2004).
64. J. Faist, A. Tredicucci and F. Capasso, *IEEE J. of Quantum Electronics,* vol. 34, issue 2 (1998).
65. R. Köhler and A. Tredicucci, *Nature*, 2002, vol. 417, 156 (2002).
66. B. S. Williams, S. Kumar, Q. Hu and J. L. Reno, *Electronics Letters* vol. 42, Issue 2, 89 (2006).
67. B. S. Williams, *Opt. Express*, vol. 13, 3331 (2005).
68. G. Scalari, et al. in *Recent progress on long wavelength quantum cascade lasers between 1-2 THz*, Lasers and Electro-Optics Society, 20th Annual Meeting, 755 (2007).
69. A. Tredicucci, in *Frequency tuning of THz quantum cascade lasers*, Conference on Lasers and Electro-Optics - Pacific Rim, 2007. CLEO/Pacific Rim (2007).

70. R. Stohr, in *Photonic Millimeter-wave and Terahertz Source Technologies*, International Topical Meeting on Microwave Photonics, 1 (2006).
71. S. Verghese, K. A. McIntosh and E. R. Brown, *IEEE Trans. Microwave Theory Tech.*, vol. 45, 1301 (1997).
72. A.W. Kadow, A. C.Jackson, S. Gossard, Matsuura and G. A. Blake, *Appl. Phys. Lett.*, vol. 76, no. 24, 3510 (2000).
73. J. L. Doualan, J. F. Lampin, R. Czarny, M. Alouini, X. Marcadet, S. Bansropun, R. Moncorge, M. Krakowski and D. Dolfi, in *Continuous wave THz generation based on a dual-frequency laser and a LTG - InGaAs photomixer*, International Topical Meeting on Microwave Photonics, 1 (2006).
74. H. Ito and T. Nagatsuma, in *Ultrafast uni-traveling-carrier photodiodes for measurement and sensing systems,* Proc. SPIE - Int. Soc. Opt. Eng. (USA), vol. 4999, 156 (2003).
75. H. Ito, F. Nakajima, T. Furuta, K. Yoshino, Y. Hirota and T. Ishibashi, *Electron. Lett. (UK)*, vol. 39, no. 25, 1828 (2003).
76. R. Stohr, Heinzelmann, K. Hagedorn, R.- Guisten, F. Schafer, H. Sttier, F. Siebe, P. van der Wal, V. Krozer, M. Feiginov and D. Jager, *Electron. Lett.*, vol. 37, no. 22, 1347 (2001).
77. F. Nakajima, T. Furuta and H. Ito, *Electronics Lett. (UK)*, vol. 40, no. 20, 1297 (2004).
78. A.J. Seeds, C. C. Renaud, M. Pantouvaki, M. Robertson and I. Lealman, in *Photonic synthesis of THz signals*, Proceedings of the 36th European Microwave Conference, 1107 (2006).
79. K. Kawase, J. Shikata, T. Taniuchi, and H. Ito, in *Widely tunable THz wave generation using LiNbO optical parametric oscillator and its application to differential imaging*, 4th Int. Millimeter Submillimeter Waves Applicat. Conf., vol. 3465, 20 (1998).
80. J. Hebling, *IEEE J. of Selected Topics In Quantum* Electronics, vol. 14, no. 2 (2008).
81. http://www.virginiadiodes.com.
82. http://lasercomponents.com/jp/fileadmin/user_upload/home/Datasheets/qcl/quanta_tera.pdf.
83. http://www.batop.de.
84. http://www.toptica.com/page/applications_terahertz_thz_cw_dfb_diode.php.
85. http://www.m2lasers.com.
86. http://www.sciencetech-inc.com.
87. R. Sanchez, *IEEE J. of Selected Topics In Quantum Electronics*, vol. 14, no. 2 (2008).
88. Reports by National Research Council (The National Academies Press, http://www.nap.edu):
    - Existing and potential standoff explosives detection techniques (2004).

- Assessment of millimeter wave and terahertz technology for detection and identification of concealed explosives and weapons (2007).
- Countering the threat of improvised explosive devices: Basic research opportunities (2007).

SENSING BY SQUEEZED STATES OF LIGHT

Virginia D'Auria, Alberto Porzio* and Salvatore Solimeno

CNISM – Napoli, CNR-INFM and
Dipartimento di Fisica, Università "Federico II"
Complesso Universitario di Monte Sant'Angelo
Via Cintia, 80126 Napoli, Italy
E-mail: alberto.porzio@na.infn.it

Squeezed states of light represent the most famous type of non-classical radiation states. They are characterized by a reduction of the quantum noise in one of the field observables, with respect to the noise affecting a coherent beam of the same amplitude (the standard quantum limit). This peculiarity in the noise property has suggested the use of these states in particular measurement schemes to beat the limit imposed by standard quantum noise. This contribution aims to briefly review three applications of squeezed states: (i) quantum interferometry; (ii) absorption measurements; (iii) high resolution imaging.

1. Introduction

Squeezed states of light have a long history in quantum optics. They have been introduced more than 30 years ago[i] and since then have found different experimental realizations.[ii] Apart from their interest for investigating fundamental aspect of quantum mechanics, they have been intended as a novel tool for breaking the standard quantum limit is some optical measurements schemes.[5-11]

The name squeezed states is used for indicating a particular class of states pertaining the quantum harmonic oscillator. Together with

---

[i] There are several review papers on the argument. A complete historical survey has been published by V.V: Dodonov[1] where it is possible to find an exhaustive list of theoretical References.

[ii] An exhaustive list of experimental papers will be very long. Some reviews have been published but they are not very recent. Among them, the paper by M. C. Teich and B. E. A. Saleh[2] and the one by V. Buzek and P.L. Knight[3] are probably the most complete. Also recent textbooks like the one by H. A. Bachor and T. C. Ralph[4] may give a general overview.

coherent states they form the Minimum Uncertainty States ensemble. For these states the uncertainty product, written for any couple of orthogonal quadratures[iii] is at the Heisenberg limit. However, contrarily to coherent states, where the uncertainty is uniformly distributed between any couple of orthogonal quadratures, squeezed states present a redistribution of the noise among a particular pair of quadratures, so that on one quadrature the noise is reduced at the expenses of an enhancement on the orthogonal one. This fundamental issue opens a window on the possible applications of squeezed light to experimental techniques where the sensitivity is limited by the field's uncertainties.

It has to be noted that together with squeezed states other kind of "non-classical states" *i.e.* states whose properties cannot be derived from classical Maxwell equations, are today endowed with application to high sensitive measurements.

After a glance to the theory behind the squeezed states and on their properties we will sketch the quantum Langevin equations governing an Optical Parametric Oscillator in order to show how these devices actually generates non-classical radiation states. Then, we will introduce some concepts relative to the detection and the characterization of these states. The second part of the chapter will focus on three different fields of applications: i) precision interferometric measurements; ii) spectroscopy and absorption measurements; iii) high resolution imaging. For each of them a brief description will give a sketch on the method together with some experimental results.

2. The Squeezed States of Light

The study of the quantum harmonic oscillator, since the seminal Schrödinger work in 1926,[12] has been concentrated on "non spreading wavepackets", *i.e.*, the search for solutions whose properties are time independent. Approaching the problem from this view point has taken researcher to the definition of a particular class of states that, since the 1967 Glauber paper,[13] where named "coherent states".[iv]

[iii] For a definition of the quadrature operator see Eq. (6) below.

[iv] We remind that all these concepts apply to single modes of the *e.m.* field.

The simplest way to define them is to look for the eigenstates of the non-Hermitian annihilation operator a:

$$a|\alpha\rangle = \alpha|\alpha\rangle . \tag{1}$$

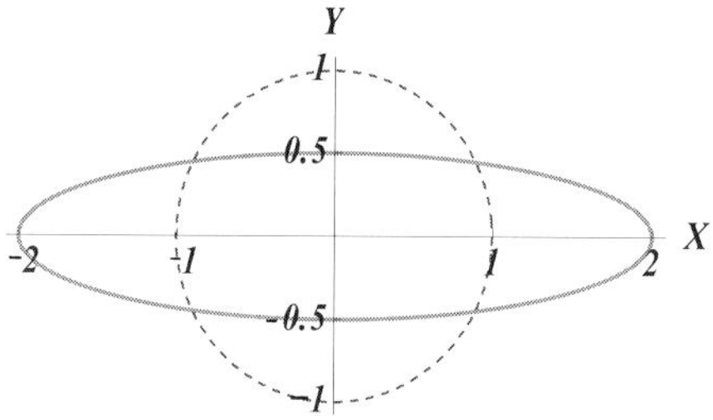

Figure 1. Phase space representation of the normalized uncertainties relative to a coherent (red dashed) and a quadrature phase squeezed (green thick) state obtained for a squeezing parameter ζ=0.7 (see Eq. (2)). For the squeezed state there is a redistribution of the noise, with the noise along Y reduced to 0.5 and the noise on X enhanced to 2 in order to comply with the Heisenberg limit.

They can be obtained from the vacuum state by the action of the displacement operator D(α) defined by:

$$D(\alpha) = \exp(\alpha a^{+} - \alpha^{*} a) \qquad \text{so that } |\alpha\rangle = D(\alpha)|0\rangle .$$

These states have been recognized as the class of states lying at the border line between the classical and quantum realms. Consequently, it is enough to slightly modify them to arrive at various families of states, sometimes indicated as "generalized coherent states", which will be non-classical. Non-classicality implies that the time dependent e.m. field cannot be fully described by the Maxwell equations being the corpuscular nature of the photon somehow preponderant.
We will focus here on the class of squeezed states obtained from the coherent ones by the action of the so-called squeezing operator:[14]

$$S(\varsigma) = \exp\frac{1}{2}(\varsigma a^{2} - \varsigma^{*} a^{+2}) \tag{2}$$

where $\varsigma = r e^{i\theta}$ is the complex squeezing parameter. When applied to a coherent state, the operator (2) *squeezes* the uncertainty circle representing the coherent state in the quadrature phase space, into an ellipse with a minor axes e^{-2r} and oriented along (θ+π)/2 (see Fig. 1). As

we will see, particular relevance has assumed the so called vacuum squeezed state obtained by $|\varsigma,0\rangle = S(\varsigma)|0\rangle$.
The definition (2), although extremely formal, gives an intuitive view on the physical processes that can be used for generating such states. $S(\varsigma)$ is made of a quadratic combination of annihilation and creation boson operators so mimicking physical situations in which a simultaneous creation (or annihilation) of a pair of photons happens. Such a process is the quantum version of the generation of signal and idler beams in non-linear optic. As a matter of fact, all the experimental realization of squeezed states has been obtained in particularly designed non-linear optical devices.[15-22]
The experimental realization of states characterized by squeezing on one quadrature, has opened a new scenario for all the contest where the noise in the field quadrature limit the performance of a device.
We conclude this section by noting that the concept of squeezed or in general of "*non-classical*" light apply to any situation where it is possible to obtain a noise reduction on one physical observable with respect to the noise affecting the same observable measured on a coherent state.[v]

3. Generating and Detecting a Squeezed State

While coherent states are characterized by a Poissonian photon number distribution, a squeezed field presents a sub-Poissonian distribution. In theory, being the result of the action of a two photon operator on the vacuum state, a squeezed vacuum field presents zero probabilities of having an odd number of photons while photons show-up into pairs.[vi]
The first experimental realization of a non-classical state dates back to 1977 when the photon anti-bunching effect, a signature of a sub-Poissonian field, was observed at MIT. Almost a decade later, in 1986, the first experimental observation of quadrature noise reduction in a squeezed vacuum state generated by a degenerate parametric amplifier was reported.[15]

---

[v] The most famous example is a photon pair generated via Spontaneous Parametric Down Conversion (SPDC) showing EPR correlation.[23]
[vi] For this reason the first name proposed for such states were two-photon coherent states.[14]

3.1. An Actual Squeezer: the Degenerate OPO Below Threshold

As explained above, the squeezing operator $S(\varsigma)$ can be mimicked by a non-linear optic interaction, where single pairs of photons are created/annihilated simultaneously. This situation is realized in an Optical Parametric Oscillator (OPO) where the interaction of three field, indicated as pump (*p*), signal (s) and idler (*i*), gives a non linear interaction Hamiltonian that assumes in the degenerate case, *i.e.*, when signal and idler collapse into a single field mode, exactly the form of $S(\varsigma)$:[25,26]

$$H_{NL} = -i\hbar\chi^{(2)}\left(a_p\left(a^+\right)^2 - a_p^+\left(a\right)^2\right) \tag{3}$$

with $a_s = a_i = a$. Equation (3) describes the process of annihilation of a pump photon into a pair of photons in the down-converted mode and vice versa. This purely quantum sketch gives an intuitive comprehension of the nature of a squeezed state.

Looking at the OPO dynamics it is possible to distinguish two distinct working regimes. For pump power exceeding a given threshold value, the OPO generates intense beams. Below the threshold, the device acts as a phase dependent amplifier able to amplify/de-amplify the vacuum noise entering the system through loss mechanisms. In this regime it is possible to analytically evaluate the noise properties of the down-converted beam by writing the Langevin equation for this field generated inside an optical cavity:[vii]

$$\dot{a} = -\gamma a + \chi_{eff}\mathrm{E}a^+ + \sqrt{2\gamma}a_0 \tag{4}$$

with γ the cavity damping, χ_{eff} the effective non-linear strength, E a measure of the pump amplitude and a_0 the vacuum field coupling into the cavity through loss mechanism.[viii] Linearizing Eq. (4), with respect to the field fluctuations, and moving to the frequency domain, one obtains

---

[vii] A simple model can be found in the Collett and Gardiner seminal paper.[27]

[viii] It is important to stress that approaching the same phenomenon from a classical view-point would have excluded any contribution from an external vacuum field (classical vacuum fields are truly empty!) whose role, even if it can appear as an artifact, is essential in any quantum treatment. As a matter of fact, the presence of a vacuum field preserves the commutation relation between a and a^+ so guaranteeing the self-consistency of the quantum theory.

an algebraic equation. The solution, combined with the solution obtained for the variable a^+, is used for calculating the quadrature[ix] noise spectra:[27]

$$S(X(0)) = 1 + \frac{4E}{(1+E)^2 + \omega^2}$$
$$S(X(\pi/2)) = 1 - \frac{4E}{(1+E)^2 + \omega^2} \qquad (5)$$

where the frequency ω is normalized to the cavity bandwidth and E is the pump power normalized to the threshold value. The two spectra above represent two Lorentzian shaped curves.

The second of them, (plotted in Fig. 2 for an ideal device working close to the threshold) is relative to the squeezed quadrature.

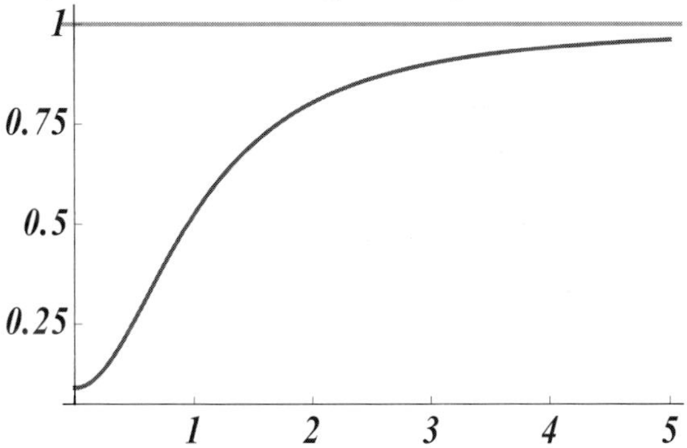

Figure 2. Typical squeezing frequency spectrum for a below threshold degenerate OPO. The frequency ω is normalized to the cavity linewidth. The spectrum is an up-side down Lorentzian with a width equal to the OPO cavity linewidth. Maximum squeezing occurs at low frequency while it becomes negligible for frequency above the cavity linewidth. The standard quantum limit, indicated by a straight green line, is equal to 1.

Noise reduction occurs mainly in a bandwidth equal to the cavity linewidth. Maximum squeezing occurs at zero frequency while it becomes negligible outside the cavity bandwidth. In the ideal case, as the one herein discussed, the product of the two spectra, vs. the frequency, is

ix The quantum quadrature operator, defined in Eq. (6), can be viewed as the quantum analogue of the quadrature of the classical *e.m.* field.

always equal to the minimum uncertainty limit so assessing the MUS property for the relative squeezed state.

3.2. Detecting Squeezed Light

Squeezing marks itself onto the noise properties of the quadrature. Luckily, the quadrature quantum operator, defined by:

$$X(\theta)=\frac{1}{\sqrt{2}}\left(a\,e^{-i\theta}+a^{+}\,e^{i\theta}\right) \tag{6}$$

is a physical observable and can be accessed by a quantum homodyne detector. A schematic of an homodyne is given in Fig. 3. The quantum homodyne is the optical replica of the electrical homodyne. It is based on the interference, on a 50÷50 beam splitter, between a strong coherent state, indicated as local oscillator (LO) and treated as a classical field (i.e. with a given phase and amplitude), and the signal to be analyzed. Balanced detection of the BS outputs, by a pair of high quantum efficiency photodiodes (PD), is eventually performed. The difference of the photocurrents of the two PDs gives a signal directly proportional to the signal field quadrature. The phase θ of the quadrature corresponds to the difference in phase between the LO and the signal fields.[x]

The homodyne performances critically depend on the overall detection efficiency η. This parameter accounts for essentially three different mechanisms that affect the detection. The first is the so called *collection efficiency* and takes into account the probability that a photon of the signal mode, generated inside the OPO cavity, reaches the homodyne beam splitter.[xi]

The second is the *visibility* of the interference between the LO and the signal modes and it is related to the geometrical mode matching at the

[x] A complete treatment on the homodyne detector can be found in the PhD thesis of Virginia D'Auria[29].

[xi] This efficiency can be further specified as made of two contributions, the first is the cavity coupling efficiency, accounting for the fact that real cavities are not single ended devices, the second the transmission efficiency, accounting for possible transmission losses between the cavity output mirror and the homodyne beam-splitter.

beam splitter. The last one is the intrinsic *quantum efficiency* of the detectors, *i.e.*, the probability that a single photon give rise to a photo-electron. Although the latter represents an intrinsic limit of the homodyne detection, the first two parameters can be experimentally improved. There are, essentially, two different ways of treating homodyne data.

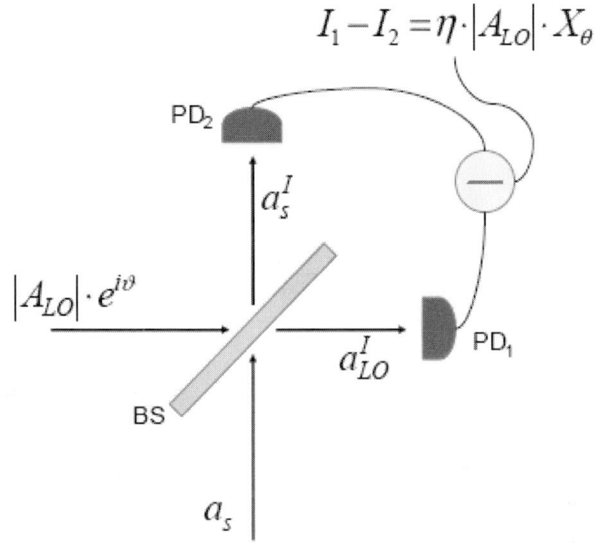

Figure 3. Schematic of a homodyne detector. A classical intense beam, the local oscillator (LO) interferes, on a balanced beam splitter, with the signal field a_s. Two high quantum efficiency photodiodes detect the two beams outing the beam splitter. The difference of the two photocurrents is proportional to the quadrature X(θ), where θ is fixed by the relative phase between the LO and the signal field.

The most common is the direct acquisition of the quadrature spectrum by measuring, by a spectrum analyzer, the difference photocurrent power spectrum. In this way it is possible to retrieve the typical shape plotted in Fig. 2 and to measure the effective noise reduction.

A more complex approach, named quantum homodyne tomography[30-32] (QHT), allows a complete reconstruction of the field state.[33-36] QHT relies on the observation that the marginal distribution $p(X,\theta)$ of the possible value assumed by the quadrature X_θ is the Radon transform of

the state Wigner function[xii] so that is can be used, by applying suitable numerical routines, to reconstruct all the state properties.
In the last decade QHT has evolved from this simple picture to more complicate analysis that allows recovering not only the Wigner function but the expectation value on the state of a given operator provided that it is possible to find a suitable pattern function to be averaged over the homodyne data.[38] Doing so, the complete density matrix of the state can be obtained.

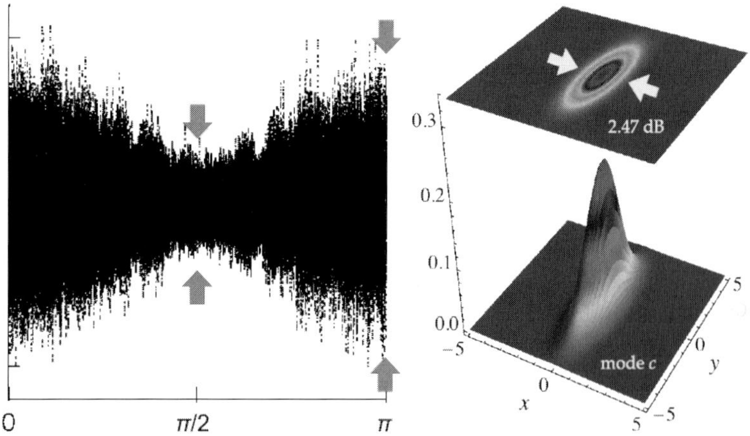

Figure 4. Experimental homodyne trace (left) and reconstructed Wigner function (right) for a 2.47 dB quadrature squeezed state generated by a below threshold OPO.[28] The Wigner function has been obtained by pattern function quantum tomography applied to 10^6 homodyne data.

4. The Squeezed Interferometer

One of the major breakthroughs on the possible application of squeezed light in measurement apparatus was published in 1981 by C.M. Caves[39] showing how the use of squeezed light can enhance the performance of a Michelson Interferometer aiming at the detection of Gravitational Waves (GW). In such a device, aiming at measuring very small perturbation in

[xii] The Wigner function is a quasi probability distribution often used in quantum optics for describing field state in the phase space. It was introduced by E. P. Wigner in 1932.[37]

the mirror positions, there are two competing noises due to the quantum nature of light.

The first is the so called *shot-noise* namely the intrinsic noise due to the corpuscular nature of light,[4,25] the second is the radiation pressure that affects the mirrors dynamic. The two quantum noise dominates in different frequency bands. Radiation pressure noise exerts a force on the interferometer mirrors, which responds to the force only at low frequencies due to the mechanical susceptibility. Shot noise dominates at higher frequencies where the mirror response to radiation pressure noise becomes smaller. Moreover, the *shot-noise* decreases as the circulating optical power increases while the radiation pressure noise increases. Both shot noise and radiation pressure noise are caused by quantum fluctuations of a vacuum electromagnetic field that enters the unused port of the interferometers.[39] The original Caves' idea was to inject a vacuum squeezed state though the unused port of the interferometer so to reduce the influence of the quantum noises in the interferometric signal.

Above few hundreds of Hertz only the *shot-noise* contribution survives with a term of the form:

$$\Delta x_{shot} = \sqrt{\frac{\hbar c \lambda}{\pi P}} \tag{7}$$

where c is the speed of light in vacuum, λ the laser wavelength, and P the circulating power.

The ultimate limit, achievable with a squeezed light interferometer is give by:

$$\Delta x_{sq} = e^{-r} \sqrt{\frac{\hbar c \lambda}{\pi P}} \tag{8}$$

where r is the squeezing parameter (see Eq. (2)).

This simple idea has been studied in details by looking at the modifications induce by the squeezed light in the noise spectrum of the device.[40] The proposed scheme has also been the object of two proof of principle experiments performed on small interferometers. The first was made on a Mach-Zehnder interferometer and reported a signal-to-noise-ratio increasing of 3 dB,[5] while the second one was obtained in a polarization interferometer showing an improvement of 2 dB.[6]

The limit of this approach was recognized to be more technical than physical. To move from optical benches toward long baseline GW detectors, as the EGO[xiii] (European Gravitational Observatory) or the LIGO[xiv] (Large Interferometer Gravitational Observatory) systems, a very stable squeezed light source, with outstanding noise properties, would have been needed.

Very recently, for the first time, 10 dB of stable squeezing have been obtained.[41] Moreover, a group at the MIT, working inside the LIGO collaboration, successfully proved an enhancement of the performances, of a squeezed light 40 m long baseline interferometer.[42]

The role of squeezing is to lower the noise floor of the interferometer in the frequency region where the noise is dominated by the *shot-noise*.[xv] In Fig. 5 the main achievement of this experiment[xvi] is shown. The GW detector noise-floor, measured in normal operation, is compared to the noise floor obtained by injecting the squeezed light into the interferometer through its unused port. The plots also report a simulated GW signal centered @ 50 kHz. The GW signal absolute height is similar in both cases while the signal-to-noise-ratio (SNR), retrieved as the peak heights with respect to the detector noise floor, is increased by 44%. This result opens a new scenario for the future generation of GW detectors.

[xiii] http://www.ego-gw.it/.

[xiv] http://www.ligo.caltech.edu/.

[xv] The performances of a long baseline interferometer are influence by many noise sources. A complete model is rather complex and goes beyond the scope of this contribution. An up-to-date review is given by the PhD thesis of dr. Keisuke Goda.[43]

[xvi] By courtesy of Dr Keisuke Goda.

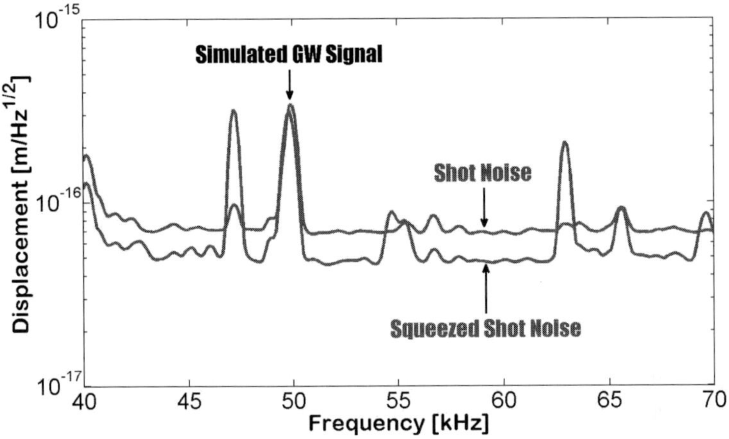

Figure 5. The measured noise floor of the GW detector with a simulated GW signal at 50 kHz with and without the injection of squeezed vacuum. The shot noise floor is reduced, along all the band, by the injection of squeezing while the strength of the simulated GW signal is retained. This corresponds to a 40% increase in SNR or detector sensitivity (*this picture appears by courtesy of Dr Keisuke Goda*).

5. Absorption Measurement with Squeezed Light

Beside the interest for application to GW interferometer, squeezed states, and other class of states showing non-classical features, have been applied to different measurement schemes. Among them, some relevance has gained the use of non-classical light for enhancing spectroscopic apparatus or improving absorption measurement.

The use of a vacuum squeezed field in a spectroscopic measurement was introduced in 1992 at Caltech[7] and consisted in illuminating the atomic vapor by a strong coherent field and a squeezed one. In this case the squeezed beam acted as a noise eater beam effectively reducing the influence of the shot noise floor onto the SNR. They obtained a sensitivity enhancement of 3.1 dB.

Later on, non-classical intensity correlations between bright beams from an above threshold OPO have been exploited for enhancing the SNR in modulated absorption set-ups.[44-46]

These measurements were based on the idea that the noise in one of the intensity correlated beams can be used for reduce the role of the noise in the other beam.[47]

Recently,[48] we have proved that squeezed light can be used not as an auxiliary noise eater beam but as a direct probe of the absorption coefficient for a partially transmitting sample (see Fig. 6 for a sketch of the experimental set-up). Standard methods, for measuring the transmittivity of a sample, rely on direct measurement of the radiation intensity entering and leaving the sample. Sufficient accuracy can be achieved by using beams so intense to contrast the shot-noise, although, in some circumstances, using high input intensity is either not useful (in the case of very low absorption) or unwise (strongly nonlinear materials or samples whose structure may be altered by intense photon fluxes).

In the proposed method, the sample is directly irradiated with a squeezed vacuum field and only downstream the sample, it is combined with a coherent LO in a balanced homodyne detector. The interaction of the squeezed vacuum with the sample modifies the spectrum of the homodyne current by changing its variance; as a consequence, the transmittivity is determined by measuring the variance changes. The main advantage of this method is a number of photons interacting with the sample as low as 10^7 photons per second (few pW of optical power).

The experiment was set-up at the output of degenerate OPO below threshold.[36] The actual state outing such a device is a *squeezed-thermal-state*[49] whose total number of photon is given by:

$$N_{tot} = n_{sq} + n_{th} + 2 n_{sq} n_{th} \tag{9}$$

where n_{sq} and n_{th} are the squeezed and the thermal photon number respectively. Physically they measure the deviation of the actual state from the minimum uncertainty one and its effective squeezing. Being the squeezed beam very weak n_{sq} and n_{th}, and hence N_{tot}, have to be measured by QHT; through the measurement of the squeezed (ΔY^2) and anti-squeezed (ΔX^2) quadratures' variances:[xvii]

[xvii] See the paper by D'Auria *et al.*[48] for details.

$$
\begin{aligned}
n_{th} &= 2\left(\sqrt{\Delta X^2\,\Delta Y^2} - \frac{1}{4}\right) \\
n_{sq} &= \frac{1}{4}\left(\sqrt{\frac{\Delta X^2}{\Delta Y^2}} + \sqrt{\frac{\Delta Y^2}{\Delta X^2}} - 2\right)
\end{aligned}
\tag{10}
$$

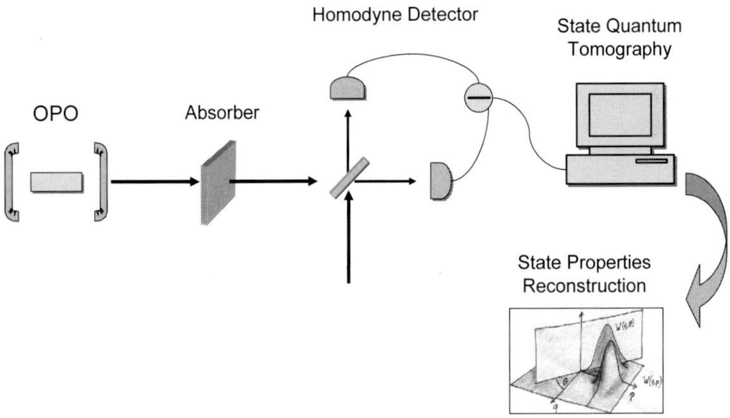

Figure 6. Schematic of the set-up used for absorption measurement with squeezed light. The OPO output is sent through a variable absorber. The transmittivity T is obtained by comparing the state parameters, retrieved by quantum homodyne tomography, up- and down-stream the sample.

In this way the transmittivity T can be retrieved from the measurement of the variances up- and down-stream the sample even for very low photon fluxes. Being N_{tot} equivalent to the beam intensity, it linearly depends on the sample transmittivity T.

On the contrary, n_{sq} and n_{th}, slightly deviate from the linear behavior; their effective behavior can be approximated by a straight line only for a reasonable range of T.

The method efficiency has been experimentally tested by simulating the effect of the sample interaction by a variable attenuator. In Fig. 7 we report T as obtained by QHT compared to the value measured by standard intensity measurements. The accuracy of the method depends in turn on the degree of squeezing available at the OPO output and the number of data processed by quantum tomography. A greater degree of squeezing makes the state more sensible to losses so that the limit accuracy increases for higher degree of squeezing.

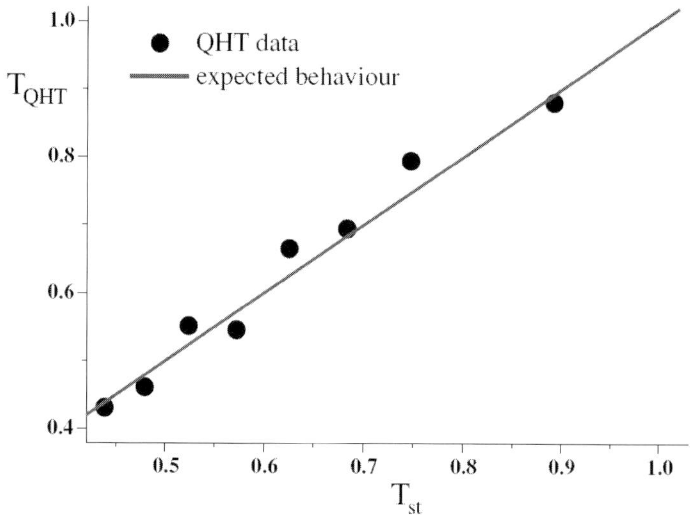

Figure 7. Transmittivity T_{QHT} as retrieved by squeezed light method (via QHT) versus the value obtained by standard intensity measurements (T_{st}). Experimental points are plotted together with the expected behavior. A linear regression of the data with $T_{QHT} = A + B\, T_{st}$ gives $A = -0.05 \pm 0.07$ and $B = 1.1 \pm 0.1$ in good agreement with the expected values of $A = 0$ and $B = 1$ respectively.

Moreover, a request on the accuracy fixes the total number of photon hitting the sample during the measurement time: the so-called *photon dose*.[48] We proved that for getting a relative accuracy of $\delta T / T = 0.01$ requires, in the case of a squeezed probe field, a total photon dose of two orders of magnitude lower than for standard intensity measurements is requested.

The method results in being a convenient approach in the case a low photon dose is required. In this case the accuracy obtainable with a squeezed probe surpasses the one relative to standard intensity measurements.

6. High Resolution Imaging

So far we have discussed situations in which the time correlation properties of photons, *i.e.*, their simultaneous birth, have been exploited for breaking the classical limit in some particular set-up. The simple model of OPO we have discussed does not say anything about other type of correlations that can be realized. It has been proved that a multi-mode OPO model leads to spatial correlations.[50]

High precision imaging using CCD cameras or photodetector arrays encompass many areas of science. The ultimate limit in classical imaging resolution is given by the *shot-noise* associated to the intensity of the beam. Of particular importance is the measurement of image displacements, for example, the position of a laser beam (often indicated as the *laser pointer* problem). As a matter of fact, information on the tilt or the shift of a probe laser beam are required in (i) atomic force microscopy;[51] (ii) very small absorption coefficient via the mirage effect;[52] (iii) observation of the motion of single molecules.[53]

Shifts from initial direction/position, are usually observed in differential measurements. More specifically, the beam is sent to a two-quadrant photodiode, split into two halves $x>0$ and $x<0$, and the difference photocurrent between the two halves is measured. For a beam centered on the detector, the mean photocurrent is zero; any displacement d, along x, will cause a variation of the photocurrent (positive or negative, according with the direction of the displacement).

In classical optics, the beam used as laser pointer can be assumed to be in a coherent state. In this condition, it can be shown that, the minimum d that can be measured (*i.e.*, the d corresponding to a SNR=1) is:[54]

$$d_{SQL} = \sqrt{\frac{\pi}{8}} \frac{w_0}{\sqrt{N}} \tag{11}$$

where N is the total number of photons recorded by the two halves of the detector during the measurement and w_0 is the spot radius of the beam, assumed to be in a Gaussian TEM_{00} spatial mode. This limit can be surpassed by using multimode non classical light.

It is known that each generic field $E(x)$ can be decomposed in an ortho-normal set of transverse modes. In particular, we will consider the basis

$\{u_i\}$ is build by defining $u_0 = E(x)/|E(x)|$, and u_1 (flipped vector) so that $u_1 = -u_0$ for $x<0$ and $u_1 = u_0$ for $x>0$[xviii] and by constructing the other vectors $u_{i\geq 2}$ by using standard orthonormalization procedures. Although, a priori, all of the modes u_i contribute to the quantum noise, it can be shown[55] that the noise in the displacement d reduces to the difference among the interference between u_0 and u_1 at the two halves of the detector:

$$I_d = \int_{-\infty}^{0} u_0^*(x')u_1(x')dx' - \int_{0}^{\pm\infty} u_0^*(x')u_1(x')dx' \tag{12}$$

We note that the definition of u_0 and u_1 provides $I_d = 1$.
The measurement is analogous to a homodyne measurement. The two modes represent the two input beams while the two halves of the multimode beam are equivalent to the two outputs of the homodyne beam-splitter. Therefore, similar to a homodyne measurement, the noise on the differential measurement is completely canceled when one of the mode is occupied by a perfect squeezed vacuum, with the squeezed quadrature in phase with the coherent field of the other mode.[xix]
These principles have been at the base of experimental proofs in one and two dimensions.[54,56] In both cases the squeezed beams have been generated by sub-threshold degenerate OPOs. In order to simply illustrate the experimental implementation of the method, we focus on the one-dimension experiment.[54] In this case a TEM_{00} squeezed beam provides the u_0 mode while mode u_1 is obtained transmitting a TEM_{00} coherent mode through a specially designed phase-plate introducing a π phase shift between two halves of the beam profile. Modes u_1 and u_0 are then combined into $E(x)$ by using a partially transmitting beam splitter. The output of the BS is sent to a two-quadrant photodiode. With a squeezed u_0, the noise in the intensity difference between the two halves is reduced, thus indicating the setting of a spatial correlation between the right ($x>0$) and left ($x<0$) region of the transverse profile of the field.

[xviii] The explicit form of the rest of the basis vectors is unessential in this contest.
[xix] The same idea can be extended to 2D by including a third *twofold-flipped* mode.

This spatial noise correlation can now be used to improve the precision of displacement measurements in the image plane.
Introducing, by an electro-optic modulator (EOM), a controlled lateral shift of the coherent *flipped* mode allows to evaluate the benefit of the squeezed beam. Applying to the EOM a sinusoidal signal, at a given frequency Ω, the beam shifts and a modulation peak appears on the spectrum of the photocurrent I_d. In the case of mode u_0 in a coherent state, the noise floor around the peak is at the shot noise level. On the contrary, when the mode u_0 is in a squeezed state the noise floor reduces so improving the SNR. Starting from 3.5 dB of noise reduction in the squeezed u_0 an improvement by a factor of 1.7 has been obtained.[54]

7. Conclusions

Squeezed states of light have been presented as a reliable candidate for surpassing the standard quantum limit in three different classes of measurements. The non-classical features of such states allow to: (i) reduce the influence of the *shot-noise* in interferometric GW antennas; (ii) measure the absorption of a sample in the very low photon flux regime; (iii) increase the resolution in imaging set-ups.
In all the reported cases the use of non-classical light allows to surpass the classical limit.

Acknowledgments

The authors wish to thank Dr Keisuke Goda and Prof. Nicolas Treps for their kindness and Dr M.G.A. Paris for useful discussions during the preparation of the manuscript.

References

1. V. V. Dodonov, *J. Opt.* ***B****: Quantum Semiclass. Opt.*, **4**, R1 (2002).
2. M. C. Teich and B. E. A. Saleh, *Quantum Opt.*, **1**, 153 (1989).
3. V. Bužek and P. L. Knight, in *Quantum interference, superposition states of light, and non-classical effects*, Progress in Optics vol. **34**, Ed. E Wolf (North-Holland, Amsterdam, 1995) pp.1–158.

4. H. A Bachor and T.C. Ralph in *A guide to Experiments in Quantum Optics* (Wiley-VCH; 2nd ed., 2004).
5. M. Xiao, L.-A. Wu and H. J. Kimble, *Phys. Rev. Lett.*, **59**, 278 (1987).
6. P. Grangier, R. E. Slusher, B. Yurke and A. La Porta, *Phys. Rev. Lett.*, **59**, 2153 (1987).
7. E. S. Polzik, J. Carri and H. J Kimble, *Phys. Rev. Lett.* , **68**, 3020 (1992).
8. P. R. Tapster, S.F. Seward and J.G. Rarity, *Phys. Rev.* **A 44**, 3266 (1991).
9. C. D. Nabors and R. M. Shelby, *Phys. Rev.* **A 42**, 556 (1990).
10. J. J. Snyder, E. Giacobino, C. Fabre, A. Heidmann and M. Ducloy, *J. Opt. Soc. Am.* **B 7**, 2132 (1990).
11. V. D'Auria, C. de Lisio, A. Porzio, S. Solimeno and M. G. A. Paris, *J. Phys.* **B**: *At. Mol. Opt. Phys.* **39**, 1187 (2006).
12. E. Schrödinger, *Naturwissenschaften*, **14,** 664 (1926).
13. R. J. Glauber, *Phys. Rev. Lett.* , **10**, 84 (1963).
14. D. Stoler *Phys. Rev. D* **1**, 3217 (1970); *ibid.* **4**, 1925 and **4**, 2309 (1971).
15. L.-A. Wu, H. J. Kimble, J. L. Hall, and H. Wu, *Phys. Rev. Lett.*, **57**, 2520 (1986).
16. B. Yurke, P. Grangier, R. E. Slusher and M. J. Potasek *Phys. Rev.* **A 35,** 3586 (1987).
17. B. L. Schumaker, S. H. Perlmutter, R. M. Shelby and M. D. Levenson, *Phys. Rev. Lett.* , **58,** 357 (1987).
18. S. F. Pereira, M. Xiao, H. J. Kimble and J. L. Hall, *Phys. Rev.* **A 38,** 4931 (1988).
19. M. Vallet, M. Pinard and G. Grynberg, *Europhys. Lett.* , **11,** 739 (1990).
20. A. Sizmann, R. Schack and A. Schenzle, *Europhys. Lett.* , **13,** 109 (1990).
21. P. Kürz, R. Paschotta, K. Fiedler and J Mlynek, *Europhys. Lett.*, **24,** 449 (1993).
22. M. Fox, J. J. Baumberg, M. Dabbicco, B. Huttner and J. F. Ryan, *Phys. Rev. Lett.*,**74,** 1728 (1995).
23. T. E. Kiess, Y. H. Shih, A. V. Sergienko and C. O. Alley, *Phys. Rev. Lett.*, **71,** 3893 (1993).
24. H. J. Kimble, M. Dagenais and L. Mandel, *Phys. Rev. Lett.*, **39**, 691 (1977).
25. L. Mandel and E. Wolf in *Optical Coherence and Quantum Optics* (Cambridge University Press, 1995).
26. D. F. Walls and G. J. Milburn in *Quantum Optics* (Springer, 2nd ed., 2008).
27. M. J. Collett and C.W. Gardiner, *Phys. Rev.* **A 30,** 1386 (1984).
28. V. D'Auria, S. Fornaro, A. Porzio, S. Solimeno, S. Olivares and M. G. A. Paris, "Full characterization of Gaussian bipartite entangled states by a single homodyne detector", submitted to *Phys. Rev. Lett.* (http://arxiv.org/abs/0805.1993).
29. V. D'Auria, "Dynamics and Behavior of Triply Resonant OPOs below the threshold", PhD thesis, University "Federico II" of Naples, (2005) available on-line at: http://www.fedoa.unina.it/view/people/D=27Auria,_Virginia.html.
30. U. Leonhardt, in *Measuring the Quantum State of Light* (Cambridge University Press, Cambridge, 1997).

31. M. G. A. Paris and J. Řeháček, *Quantum States Estimation* (Lecture Notes in Physics vol. 649) (Springer, Heidelberg 2004).
32. G. M. D'Ariano, M. G. A. Paris and M. F. Sacchi, *Adv. Imaging Electron Phys.*, **128**, 205 (2003).
33. G. Breitenbach, S. Schiller and J. Mlynek, *Nature*, **387** 471 (1997).
34. Zavatta, F. Marin and G. Giacomelli, *Phys. Rev.* **A 66**, 043805 (2002).
35. G. Mauro D'Ariano, M. De Laurentis, M. G. A. Paris, A. Porzio and S. Solimeno, *J. Opt.* **B 4**, S127 (2002).
36. V. D'Auria, A. Chiummo, M. De Laurentis, A. Porzio, S. Solimeno and M. G. A. Paris, *Opt. Express*, **13**, 948 (2005).
37. E. P. Wigner, *Phys. Rev.*, **40**, 749 (1932).
38. G. M. D'Ariano, C. Macchiavello and M. G. A. Paris, *Phys. Rev.* **A 50**, 4298 (1994).
39. C.M. Caves, *Phys. Rev. Lett.*, 45, 75 (1980), *Phys. Rev.* **D 23**, 1693 (1981).
40. F. Pace, M. J. Collett and D. F. Walls, *Phys. Rev.* **A 47**, 3173 (1993).
41. H. Vahlbruch, M. Mehmet, S. Chelkowski, B. Hage, A. Franzen, N. Lastzka, S. Goßler, K. Danzmann and R. Schnabel, *Phys. Rev. Lett.*, **100**, 033602 (2008).
42. K. Goda, O. Miyakawa, E. E. Mikhailov, S. Saraf, R. Adhikari, K. McKenzie, R. Ward, S. Vass, A. J. Weinstein and N. Mavalvala, *Nature Physics*, **4**, 472 (2008).
43. K. Goda, "Development of Techniques for Quantum-Enhanced Laser-Interferometric Gravitational-Wave Detectors", PhD thesis, MIT, Boston (2007) available on line at: http://www.ee.ucla.edu/~goda/thesis/main.pdf.
44. C. D. Nabors and R. M. Shelby, *Phys. Rev.* **A 42**, 556 (1990).
45. J. J. Snyder, E. Giacobino, C. Fabre, A. Heidmann and M. Ducloy, *J. Opt. Soc. Am.* ***B*** **7**, 2132 (1990).
46. P. R. Tapster, S. F. Seward, and J. G. Rarity, *Phys. Rev.* **A 44**, 3266 (1991).
47. A. S. Lane, M. D. Reid, and D. F. Walls, *Phys. Rev.* **A 38**, 788 (1988).
48. V. D'Auria, C. de Lisio, A. Porzio, S. Solimeno and M. G. A. Paris, *J. Phys. B*, **39**, 1187 (2006).
49. P. Marian, *Phys. Rev.* **A 45,** 2044 (1992).
50. M. I. Kolobov, *Rev. Mod. Phys.*,**71**, 1539 (1999).
51. C. A. J. Putman, B. G. de Grooth, N. F. van Hulst and J. J. Greve *Appl. Phys.* , **72**, 6 (1992).
52. C. Boccara, D. Fournier and J. Badoz, *Appl. Phys. Lett.*, **36**, 130 (1980).
53. H. Kojima E. Muto, H. Higuchi and T. Yanagida, *Biophys. J.,* **73**, 2012 (1997).
54. N. Treps, U. Andersen, B. Buchler, P.K.Lam, A. Maitre, H-A. Bachor, C. Fabre, *Phys. Rev. Lett.*, **88**, 203601 (2002).
55. C. Fabre, J. B. Fouet and A. Maître, *Opt. Lett.*, **25**, 76 (1999).
56. N. Treps, N. Grosse, W. P. Bowen, C. Fabre, H.-A. Bachor and P. K. Lam, *Science*, **301**, 940 (2003).

FIBER OPTIC SENSORS IN STRUCTURAL HEALTH MONITORING

Maurizio Giordano,[a] Jehad Sharawi Nasser,[a] Mauro Zarrelli,[a] Andrea Cusano[b,*] and Antonello Cutolo[b]

[a]*Istituto per i Materiali Compositi e Biomedici, CNR Piazzale Enrico Fermi 1, 80055, Portici (Napoli), Italy*

[b] *Dipartimento di Ingegneria, Università del Sannio, Corso Garibaldi 107, 82100 Benevento, Italy*

[*]*E-mail: a.cusano@unisannio.it*

A great demand exists nowadays to special systems that permit the monitoring and controlling in real time, and under diverse range of operating conditions, the performance of structural and mechanical elements or structures with minimum cost and effort. The objective of this chapter is to build a base of knowledge containing a brief demonstration of a candidate system and highlighting the most important fields of its applications. The practical experience of famous authors, researchers, companies and great industries, rather in presenting or/and in utilizing this technology is going to be discussed through the following lines.

1. Structural Health Monitoring - An Overview[1]

Structural Health Monitoring (SHM) and Damage Detection is considered one of the most promising issues for the intrinsic characteristics of safety and the cost reduction it incorporates. The expected rise of traffic would bring to a need for their insertion, in spite of current safety levels being high. Cost reduction is a more direct manner, affecting several issues as the downtime, cost of repair, substitution of vehicles. The implementation of these systems on real commercial structures, demands an effort to establish the requirements these systems should accomplish. Generic specifications can be addressed, because the specific applications could significantly vary one from the other.

There exist two types of SHM systems, called PASSIVE and ACTIVE systems. In the first kind, only sensors are installed in the structures to measure their response due to unknown external load. These measurements are taken in real time, while the structures are in service, then the collected information are compared with a set of reference (healthy) data. The sensor-based system estimates the conditions of the structures based on data comparison. Hence, the techniques of data comparison for interpretation of structural conditions are crucial for a reliable system. The system would require either a pre-stored data bank, or a structural simulator to generate reference data. While, the ACTIVE systems consists of the same facilities of PASSIVE system, plus the built-in devices called actuators used to apply external loads either mechanical or non-mechanical. Here, it is easy to predict the response of the structure under a variety of possible situations and load combinations because when the inputs are known, the difference in local sensor measurements, based on the same input, is strongly related to a physical change in the structural condition, such as the introduction of damage.[2]

1.1. What Is Structural Health Monitoring?

SHM is the "Knowledge" of the "Integrity" of "In Service" structure on a "Continuous Real Time" basis. This knowledge is the ultimate objectives for the end users, maintenance crews, as well as manufacturers. With such knowledge, the users can count with confidence on the optimal use of the structures and minimize the downtime and avoid catastrophic failures, while the manufactures can improve their products (safety and reliability), reduce inventory and minimize the cost. Currently, only limited knowledge can be accumulated in real-time through scheduled maintenance or periodic inspections with extensive labor and causing downtime, and are expensive.

Recent advances in sensing technologies and material/structural damage characterization combined with current developments in computations and communications, have resulted in a significant interest in developing new technologies for monitoring the integrity, and for the detection of damage of both existing and new structures in real time with minimum human involvement.

Using distributed sensors to monitor the "Health" condition of "In Service" structures, became feasible as these sensors systems are able to reflect the conditions of the structures through a "Continuous Real Time" data processing, which can be integrated and automated to perform real time inspection and damage detection.

2. The SHM System

The structural health monitoring system consists of two major entries, hardware and software, adapted to achieve the reliability of the monitoring activity. The system includes five major parts:

(a) Sensing Technologies

Sensors are considered to be the core of the SHM system. They are the devices for measuring and feeling the state of the structure. All sensors have the same working principle, that is, to give a signal when they undergo a change in state, either due to a thermal or a mechanical load.

(b) Diagnostic Signal Generation

In the ACTIVE sensing systems, the signals are used to excite the sensor measurements and simulate the local abnormal behavior of the structure. This means that, the determination of the diagnostic signals and generation critically affects the measurements and the identification of the event. Consequently, high importance should be given to the size and power of signal, as the actuators must be small enough to allow their embedding into the structure but powerful in generating the diagnostic signal for neighboring sensors.

(c) Signal Processing

Collected signals contain much information, in addition to that, the environment and noise might corrupt them. This makes signal processing crucial before they can be used for interpretation.

(d) Identification and Interpretation

It was found from different experiments that, the determination of the physical conditions of the structure based on sensor measurements is a non-linear inverse problem. Thus, several numerical and analytical techniques have been used for the purpose of analyzing the responses (Modal Analysis, Genetic Algorithm, Neural Network and Optimization Algorithm). In this field, it floats up the need of new algorithms to relate sensor measurements to the physical conditions of the structure, an accurate and fast computational technique for large structural components, and a clarification of the relation between damage or defects and the measurable physical quantities of the structure.

(e) Integration

The final system must be reliable and be able to work as a single unit. The structural integrity should be maintained as the installation of the actuators and sensors into the structures, may affects its performance. Moreover, the user interface must be well studied to make the system easy to be operated. It is desirable that the system could display through its initial interrogation the conditions of the structure in an easy way. Finally, the entire system implemented should be as small as possible.

3. Comparison between SHM and NDE

It is highly important at this point to clear the differences between the well known currently used Non Destructive Evaluation (NDE) techniques and the Structural Health Monitoring (SHM).

The importance of this comparison comes from the fact that, some conventional NDE techniques can be considered within the framework of the SHM. The traditional NDE techniques tend to use direct measurements to determine the physical condition of the structures where no history data is needed.

NDE	SHM
No Need For History Data (One measurement at each place)	History Data Is Crucial (Many measurements at the same place in different times)
Accuracy relays heavily on the resolution of the measurements and upon the used equipment	Accuracy depends on the sensitivity of sensors and on the interpretation algorithm (software)
Human error is effective	Least human involvement (error)

Table 1. Comparison between SHM and NDE.

The accuracy of the diagnosis strongly depends upon the resolution of the measurements, which relay heavily on the equipment. On the other hand, the SHM techniques would use the change in the measurements at the same location at two different times to identify the condition of the structure. Hence, the history data is crucial for the technique.

The accuracy of the identification strongly depends upon the sensitivity of the sensors and the interpretation algorithm, which means that the NDE relies more on the instruments, whereas; the SHM is more dependent upon the interpretation software.

Finally, concerning the human involvement, the NDE techniques are more dependent upon the human effort, in installation and measuring work, which might be of highly harmful effect on the results, compared with the SHM systems. Table 1 above, summarizes these differences. Another important comparison can also be shown at a different level, the Maintenance. The two exiting types of maintenance philosophies are described in the following sections.

3.1. The Current Maintenance Philosophy

It is the traditional type that uses data from the initial design and manufacturing process to create a service manual (see Fig.1).

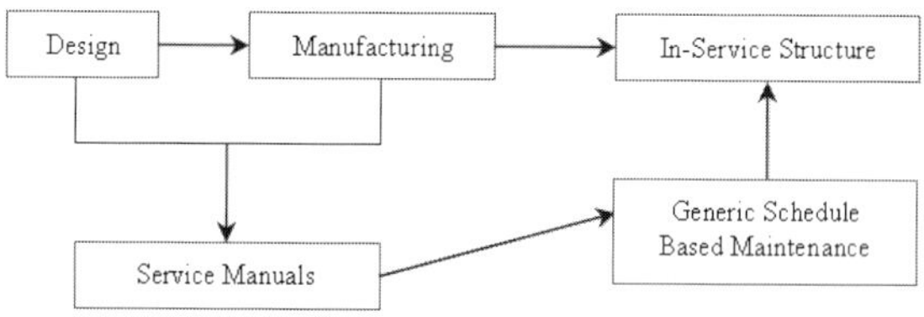

Figure 1. Traditional Maintenance Philosophy.

The maintenance manual, which was derived from laboratory coupon tests and analytical modeling, informs the schedule-based maintenance over the designated service life of each structure produced.

The level of maintenance is generic for the structure and, with the exception of the knowledge that comes with experience, there is no information from the in-service structures feeding back into the design and manufacturing stages, and thus into the service manual. This means that to increase confidence in the reliability of an in-service structure, operator needs to increase the frequency of inspection, and doing so increases the cost of maintenance. In fact, such a mechanism can only ever delay the gradual reduction in reliability as the structure ages.

3.2. The New Maintenance Philosophy

In this type of maintenance strategy using the structural health monitoring helps in creating feedback loops within the design, manufacturing and maintenance procedures by providing additional knowledge about a specific design performance, material quality and structure condition (Fig. 2).

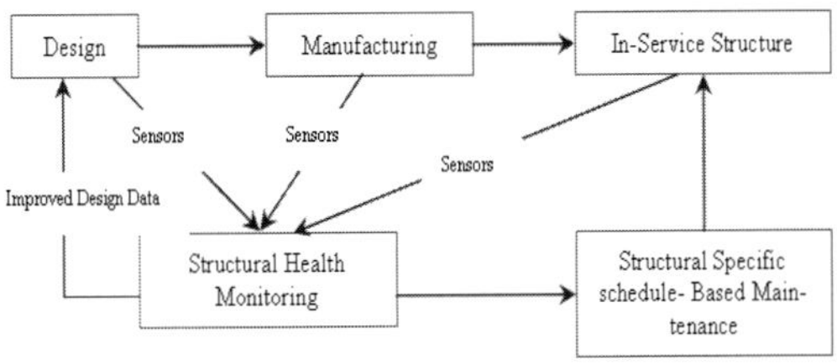

Figure 2. SHM Maintenance Philosophy.

Here, a continuous data flow back is guaranteed towards the initial design function by the sensors, which provide information at each stage of the structure life (design, manufacturing and in-service).
Studying this data enables the operator to create a maintenance schedule based on the condition, history and performance of each structure. This means that targeted maintenance can be done effectively and the inspection activities are only done when they are needed. Such a system helps in maintaining the reliability of the structure at a constant level throughout the service life.

4. Applications of SHM Technology

Nearly all in-service structures require some form of maintenance for monitoring their integrity and health conditions in order to prolong their life span or to prevent a situation of catastrophic failure.

Fiber Optic Sensors are being candidate because of their special characteristics and merits such as: self-referencing capability, ability to be easily multiplexed, ability to be embedded into materials, high temperature capability & operation, small size & light weight, immunity to Electro-Magnetic & Radio-Frequency interference, excellent fatigue properties, relatively low cost, large sensing strain range, ability of be interrogated in both reflection & transmission modes, resistant to environmental attack and finally, their stability over time.[3-8]
Several different optical sensing techniques have found their way into the market place but Fiber Bragg Gratings (FBGs) are commercially one of the most successful. Recently, the dramatically falling costs of electronic units have driven an upsurge in the number of fiber Bragg gratings being used commercially for sensing. In the following few lines, some of these applications are briefly discussed and the great role this technology plays will be underlined.
The potential applications of the SHM technology are very broad, ranging from Aerospace Structures and Military Applications to Civil Infrastructures; from Marine and Offshore Structures to Medical applications; from Oil Production to Automotive and Railways transportations.
The following part of this chapter will give a brief idea about the SHM in these fields and the participation of fiber optic sensors in many applications.

4.1. Aeronautics Applications

It is a really hard mission to show in few lines how important SHM technology is in such a strategic field like aeronautics.
This field falls in three main sectors: Civil, Military and Space. Although these sectors are different in terms of goals, usage and materials, they are similar in: having elevated costs, needing to be regularly maintained and to be extremely safe.

4.1.1. Civil Sector

Aircrafts are widely used for civil applications. In this field very large fleets exist in the world for both human and goods transportation.
Size, capacity, and traveling speed are the major basis in classifying these structures. It is being noticed that the major parts of the aircraft failure happened due to fatigue loading and aging, and the strange thing is that most of the airplanes being destroyed in these accidents were passing a regular inspection schedule, instead; the real problem was hard to be diagnosed as small undetected cracks were accumulating with time. These incidents have led to more stringent inspection procedures, done at fixed intervals and represent the major part of the aircraft health monitoring program.
This program (Fig. 3) divides the airplane into four systems: aircraft structure, hydraulic system, electronics/avionics, and propulsion systems. The monitoring system is highly advanced, providing in-flight data on revolution per minute, vibrations, temperature, pressure, and rate of fuel consumption. Collected data inters an immediate analysis and provides warnings of any potential before it can become serious.

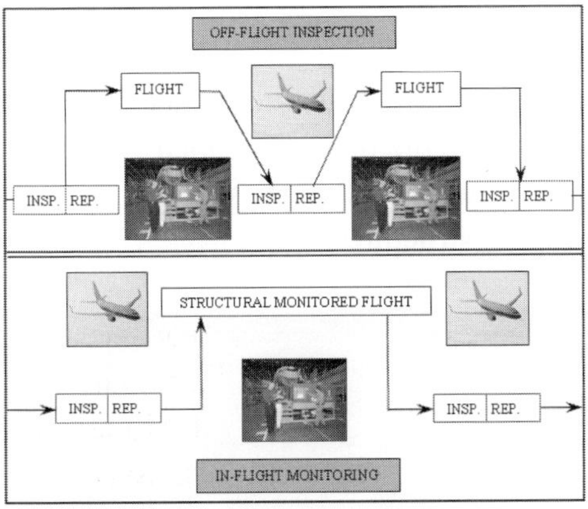

Figure 3. In-Flight SHM.[9]

The system is also a very good time saver, as it reduces the down time.[9]
Some of the most important applications in this field are here described:

Structural Test

Fiber optic sensors are used to detect strain in specific points of complex structures in order to have useful information for the design and stress analysis of critical details.

The small fiber dimensions guarantees the ease of embedding the sensor in composite materials, fiber metal laminates and bonded structures, while in service, their minimized dimensions do not change the local structural properties of the part neither create local strain variation. Moreover, the high temperature endurance of the sensor's material permits the ultimate curing process of materials in the autoclave, here; monitoring of curing process and residual stresses can also be done.

Figure 4 shows how a fiber optic Bragg sensor has been embedded in the composite "J" spar.

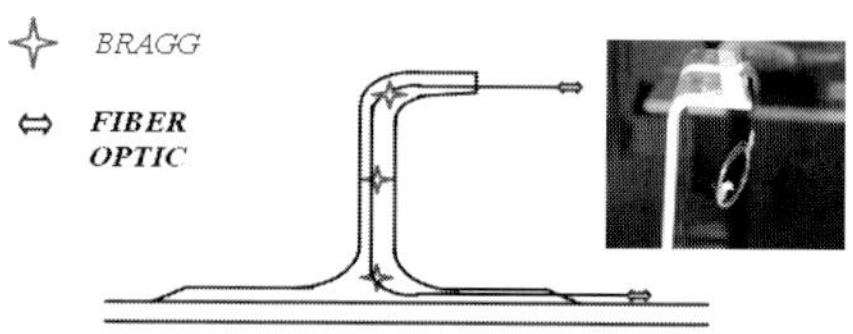

Figure 4. Co-bonded " J " Spar with Embedded FBG Sensor.[10]

In Figure 5 a more complex structural element is shown. This element, used by ALENIA within the program, called AHMOS, is a test bench to demonstrate capability of different sensing techniques of detecting damages on aerospace structures.

Figure 5. EFA TYPHOON.[10]

Loads Monitoring

Measurements of actual aerodynamic loads like wing, empennages, control surfaces are very crucial in controlling the performance on aerospace structures. The structures are instrumented with several strain gages positioned at suitable points i.e. the structure itself is used as a load gage and loads monitoring loads are correlated to measure strain through a linear combination of strain measured at same time in different points.

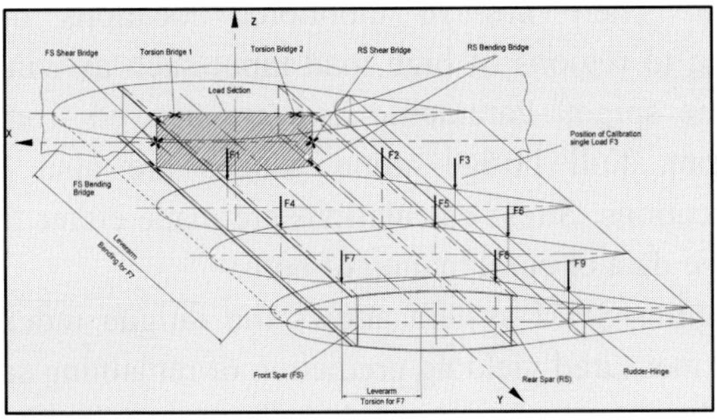

Figure 6. Strain Points for Wing.[10]

4.1.2. Military Sector

The concept of self-monitoring structures is clearly very appealing from both an economic and safety point of view. The airframe of a military airplane is extremely complex and it can involve hundreds of thousands of parts, a sizable number of them are critical and highly loaded that need to be always under control to prevent failure.[9]

Military air-fighters producers and users, spend millions of millions of dollars every year in inspecting and controlling their fleet. The examples are quite many; some of them are briefly mentioned:[1]

- scheduled inspection and repair of each EF-11A aircraft, have risen from 2200 to 8000 hr from 1985 to 2004;
- 18 fighter: $ 35 million per year, assuming 33 hr of flight per aircraft per month and 1000 aircraft fleet;
- U$ 9 million per year for the inspection of the fighter T-38, assuming 420 hr of flight per aircraft per year and 720 aircraft fleet.

EURO-Fighter: Needs of Structural Health Management System[11]

When the Euro-Fighter enters service, it will have the first fleet-wide fit of a structural health monitoring system integrated with a ground support system. A final goal will be to have a diagnostic as well as prognostic health monitoring technologies on future military aircrafts.

It has a dual parametric and strain gauge based structural health monitoring system to enable the fatigue consumption of each aircraft to be measured. There are 16 monitored locations on the aircraft corresponding to regions of high load input such as wing attach, lungs and fore-plane spigot. Parametric data are derived from the aircraft's control system. Full bridge strain gauges are also located at the monitored locations. Stress calculations are made either from parametric or strain gauge data or combination of both.

The Euro-Fighter SHM system allows the fatigue index of individual aircraft to be measured yielding prediction of remaining safe life. For the future aircraft, introducing a parallel diagnostic technology that can detect actual damage, this prognostic capability indicating the withdrawn of the component from service before these damages reach critical limits could further enhance.

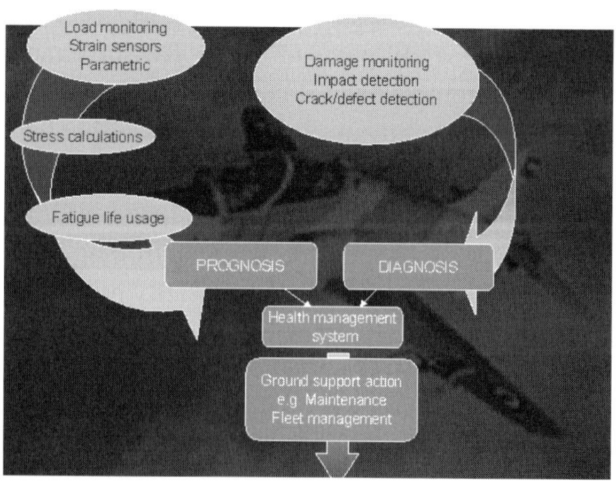

Figure 7. Prognostic & Diagnostic SH Management.[11]

TORNADO-Fighter: Maintaining Aging Military Aircraft[12]

The in-service loads monitoring task for the Tornado aircraft can be described as an integrated concept, consisting of three inter-relater actions: Individual Aircraft Tracking (IAT), Temporary Aircraft Tracking (TAT), and Selected Aircraft Tracking (SAT).

Each aircraft in the fleet is subjected to the IAT using a Pilot Parameter Set (PPS). Selected aircraft are additionally subjected to TAT or SAT. The principle idea behind this concept is to validate the fatigue consumption calculation with the limited amount of flight data available through the calculation performed by the Full Parameter Set (FPS).

An efficient program is used to resolve the problem of estimating the fatigue consumption, which is a very complicated issue that involves large number of components and locations of the structure.

This program consists of the following: firstly, The Qualification Action, where a Full Scale Fatigue Testing (FSFT) is being done to detect the critical components and locations such as wings and front fuselage. Then comes The Strain Gauge Positioning activity, which is being defined according to the guideline derived from the FSFT. These in-service strain gauges permit the direct comparison between the Qualification and the real usage of the structure. Finally, the IAT and the SAT activities are then being performed.

4.1.3. Space Sector

Structures in this category have no limits in terms of costs, but have to operate at an ultimate level of accuracy and efficiency.

The structural health monitoring in general is still a new technology for space, and must undergo the necessary tests to be qualified, such as: withstanding of shock, vibration, temperature, humidity, and long term storage conditions often encountered by space systems. Even with the recent advances in automated ground-based non-destructive evaluation methods, the vast majority of inspections are visual.

There are three key motivations to pursue sensor-based SHM capabilities.

First, given the inspection and maintenance techniques currently available, there is a potential that indications of structural degradation could be missed.
Second, SHM capability could enable on-condition maintenance of airframe structures. This type of maintenance would simplify periodic checks, improve productivity by minimizing structure downtime, and allow the maintenance program to be tailored to the individual airframe.
Finally, SHM is an integral part of a comprehensive condition analysis capability.
Based on the previous information, ARINC, in collaboration with NASA, Pennsylvania Univ., and Luna Inn., has developed and demonstrated a prototype multiplexed sensor system for airframe structure and compatible real-time damage models for board characterization of multiple and synergistic failure modes in current and future airframes.[13]
Another application in this field is reported by the Space Center of Liége in Belgium,[14] where a reusable launch vehicle, intended for future space transportation systems, requires more complex design and much technology performance levels in comparison with actual expendable launchers. According to them, a health monitoring system can be defined as the process of monitoring and assessing the state or conditions of a system by on-board or not, operating in real-time or off-line, continuously or in burst mode acquisition mode.
This system is used to collect data to diagnose the conditions of each subsystem of the overall vehicle and to ensure that preventive maintenance is performed in the most cost efficient manner.
The major objectives foreseen for the structural health system are:
- *Enhancing the maintenance operations*

In continuous operating phase, maintenance is made just when needed while during the OFF service phase, a very fast, automatic and timesaving inspection is achieved;

- *Reducing vehicle/platform turnaround time and cost*

This goal can be achieved by operating the structural health monitoring system in flight on a condition-based maintenance way and operating it on ground to perform an automated assessment of the vehicle's structure, to ensure that no critical damage occurred during the previous flight.

- *Monitoring real -time platform health status*

The starting point for any diagnostic or prognostic capability is to have accurate and appropriate data that detail status, environment, and performance of the component of interest.

4.2. Civil Engineering Applications

Civil engineering structures like bridges, tunnels, highways, railways, dams, seaports, and airports, represent an enormous financial investment. These structures should work well-inside the limits of safety for all their lifetime, or at least when they are in use.

That's because a failure may cause terrible casualties, not only in terms of capital, but more important, in terms of human lives.

In the USA, assets are estimated at U$ 20 trillion. They are huge in size and are exposed to harsh environments at all times, maintenance and damage inspection for them can be costly and time consuming.

Moreover, an interior damage after an earthquake is very critical.

These reasons lead to the idea of continuously controlling the state of these structures while they are in service.

4.2.1. Buildings and Historical Structures

The necessity of condition assessment or structural health monitoring of building arise for contingent reasons associated with natural event or also, as in the case of historical buildings, to monitor the integrity degradation of the structure.

Assessment of civil structures or structural elements are generally performed by conventional techniques such as: ultrasonic scanning, transient pulse and infrared thermography and ground radar.[15-18] Although these established technique are adequate and reliable they are not suitable at all for real time acquisition and *in situ* operation.

These limitations are generally overcome by using different types of sensors as electrical resistivity gauges,[19] acoustic emission[20] or optical fiber.[21-24]

4.2.2. Bridges

The Federal Highway Administration has classified 42 % of the United States' 578.000 bridges as structurally or functionally deficient—if not obsolete. The estimated cost for correcting all these bridges exceeds U$ 90 billion. Similar problems are also found in other developed countries. Bridges defects may arise not only because of environmental attacks, but also due to horrible car accidents and earthquakes that may yield cracks in the concrete slabs.[9]

In all these applications, both configurations bonded and embedded fiber have showed excellent results either in term of signal resolution and accuracy; drastically cutting installation time and handling problem.

The first stay-cable bridge in the world, the Storck's Bridge[9] (Fig. 8) was instrumented with a combination of fiber optic strain sensors and resistive foil strain gauges. This monitoring system has been in operation for several years and is providing a useful comparison between the steel and composite cables. In 2006, the Hong Kong's Tsing Ma Bridge, the world longest suspension bridge, with both railway and regular road traffic was considered as test item for the installation and testing of FBG sensors. Results were compared with existing conventional health monitoring system, previously installed on the bridge.

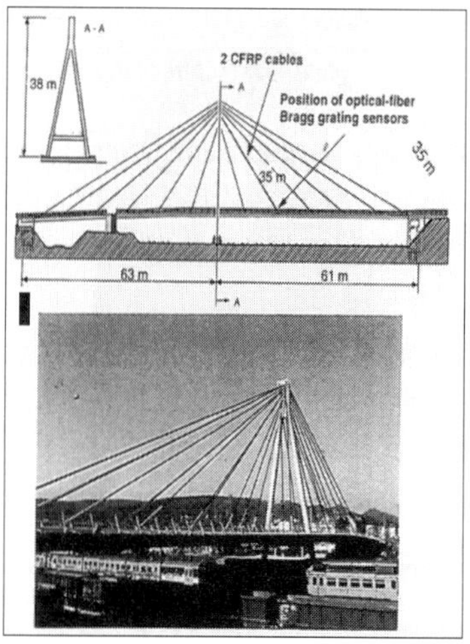

Figure 8. Storck's bridge.

4.2.3. Tunnel[23]

A good example of the application of structural health monitoring in tunnels controlling is within the STABILOS project frame.

This project concerns the monitoring of the Mont-Terri tunnel which was constructed in a stratified rocky zone.

The stratified rock, composed of anhydrite materials, after excavation started to absorb water, and therefore high non-symmetric stresses induced changes of the tunnel section (Fig. 9).

In turn, maintenance should take place for preventive purpose, and in tunnels where similar phenomena occur; this causes the tunnel to timely close. For this reason, repair based on continuous monitoring would have considerable economical impacts.

In the Mont-Terri tunnel, special openings have been left in the concrete in order to integrate the sensors at the end of the building of the tunnel. The spectrum acquired by the sensors has been analyzed continuously to predict any changes in the tunnel's state.

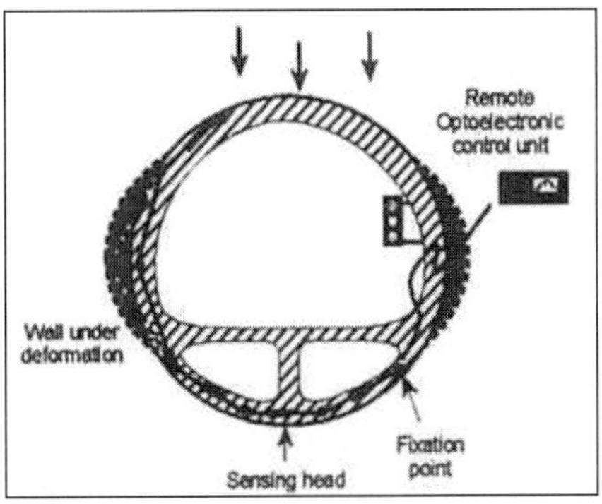

Figure 9. FBG Instrumentation.[23]

4.2.4. Nuclear Industry

The potential hazardous of nuclear plants and the severe degradation conditions of many elements have lead nuclear company and governments to focus on leading edge project of structural health monitoring.

In 1995 an industrial prototype of an eight channel fiber-optic temperature sensor network, based on spectral modulation encoding techniques, has been settled in the nuclear plant of TRICASTIN (France).[24]
The system was built and installed on the stator of 900 MW turbo-generators, operating in the plant to monitor the thermal condition of the element during continuous operation.
In 2004, variations in structural behavior due to aging were monitored, within the frame of a national collaboration between the OXAND company and the Electrical company (France).
The project was focused on the installation and testing of a non-destructive monitoring system of structural performance.
FBG sensors to monitor crack displacement and to measure deformation on a half scale model of concrete confinement wall of a nuclear power plant were also installed in 1997, within the MAEVA project (France).

4.3. *Geotechnical Applications*

After the 1995 Hyogo-Ken Nanbu earthquake, in Japan, the necessity to strength expertise within the area of structural health monitoring for geotechnical application has strongly arisen (5500 people injured and 40.000 buildings destroyed).
Due to this event, many steel buildings reported sever damages to supporting elements and beam –column joints. During subsequent engineering assessments, it was found that many damages were not clearly visible unless fire-protection removed and for many structures, the damage level was not undoubtedly evaluated as the earthquake affected the overall structures.
In the last 15 years, the potential applications of non destructive techniques for structural integrity monitoring and the possibility to use simpler and more economical seismograph to prevent this kind of events has come out.
Accelerators based on Bragg fibers have boosted the technical applications in this area. Optosmart[25] has successfully applied these sensors to build an accurate and reliable seismic wave monitoring system (Fig. 10) and direction detection by an array of three FBGs.

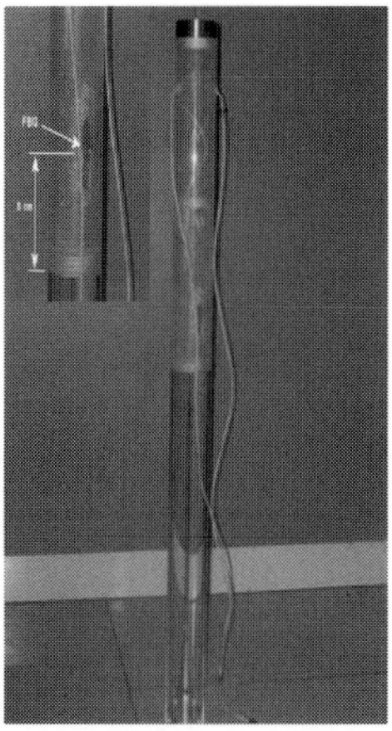

Figure 10. Seismic wave monitoring system (Optosmart[25]).

The basic idea is to measure the axial deformations due to bending stress of an oscillator subjected to a dynamic acceleration at the clamped basis by integrating three FBGs dynamical strain sensors in a mechanical structure acting as a harmonic oscillator.

Fiber optic sensor have also been used to develop seismic station for sub-sea oil well field. The Optowave™ ocean bottom cable system,[27] uses an optical sensor technology comprising 4C seismic stations that contain 3 orthogonal accelerometers and a hydrophone, also developed using optical sensor technology. The system is also made up of optical fibers lead-in cables and an interrogation laser that is placed at the surface, either on a platform or other surface facility, which allows many thousands of sensors to be interrogated by the laser instrumentation.

4.4. *Automotive Applications*

Health monitoring systems in automotive field is quite new, and being done at low scale. A reason for this might be the cost of installing the system in an asset that is quite cheap with respect to the other fields of industry.

But, on the other hand, when considering the goals of having efficient, safe, and economic machines it is also to take it into account that is a powerful technique which could surely enhance the functioning and reduce the consumption of automotive engines. An example is discussed briefly in the following sections.

4.4.1. Combustion-Pressure Sensors for Automotive Engines[8]

A reliable, long-life, high-temperature, miniature and low-cost cylinder pressure sensor is the enabling element of advanced control systems having potential for significant fuel economy improvements, reduced levels of combustion pollutants and improved engine reliability and performance.

The fiber optic pressure sensors have been developed to be suitable for integration into higher-functionality devices such as "smart" ignition systems, fuel injectors, or glow plugs.

The design and performance of the integrated sensors are reported for car engines demonstrating a total error due to non-linearity, hysteresis, and thermal shock. The sensor head consists of a metal housing with a welded sensing diaphragm, a fiber holding ferrule, and two fibers bonded inside the ferrule (Fig. 11).

The sensor utilizes the principle of light reflection from a flexing metal diaphragm to respond to pressure originating from the displacement of a diaphragm by changing the optical signal transmitted.

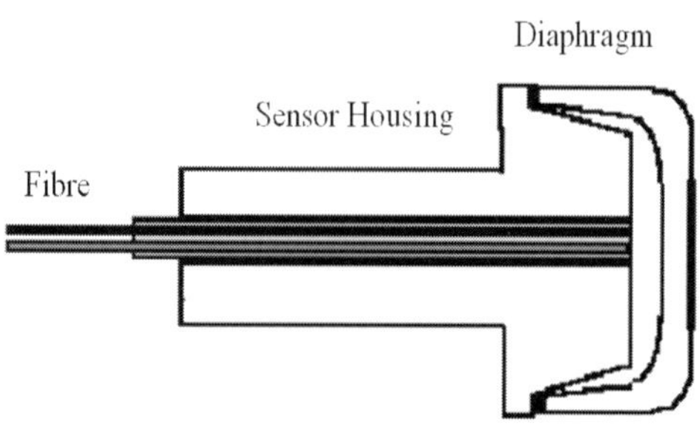

Figure 11. Combustion-Pressure Sensor.

4.4.2. *Fuel Tanks for Natural Vehicles*

The development and the promotion of new fuel systems for the whole transportation industry, which will lead to lower the hazardous emission and to reduce the Europe's dependency on oil, is considered of great potential.
Despite these potential benefits, the natural car market, which is considered to be the cleanest internal combustion vehicles available today, is still inhibited by important factors.
The major problem for this kind of cars is related to safety, maintenance and, not least, the high manufacturing cost compared with the standard installed system.
The ZEM[28] project, namely "Zero-hard gas storage by multi-sensing optical monitoring system", funded under the 5th Framework Programme, aimed to develop a monitoring system based on optical sensors which simplify the periodical control and evaluation of the structural integrity of composite high-pressure tanks for natural gas or hydrogen.
Stationary and mobile applications were considered during the three year work and a demonstrator related to vehicle propulsion was developed. The project approach, based on fiber optic sensors, facilitated a simple but detailed evaluation of the structural integrity during tank re-fuelling.
Fiber optic embedment into the material, sensorized composite tank manufacturing, signature recognition algorithms and interrogation methods were the project main goals.

4.5. Railways Transportation

Railways are one of the most used ways of passengers and goods transportation.
It means that the entire railroads systems (rails, electrical lines, poles, attics, railway exchanges and so on) must be continuously monitored to optimize maintenance, prevent troubles and forced shuts down of the service, reducing operating costs.
Up to now special and very expensive traveling laboratories are used to perform periodically rails and electrical lines screening determining disturbance to the normal service.

Since the technology of fiber optics has caught up an elevated reliability and affordable costs, optoelectronic systems have been developed to perform different kinds of sensing based on different principles of operation.
A recent application has been described by Bosselmann et al.[29-30] who proposed FBG sensors applied to the electrical lines of a railroad in order to monitor their temperature and to ensure that no temperature overhead could cause mechanical strength deterioration of the catenary construction.
The work, performed by Optosmart[26] at the Tel Station Bolzano or at "S. Giovanni" railway station[30] (Naples) by using multiple FBG sensors along a single fibre,[31] can be taken as example to show the potentiality of this new sensing technology for both distributed and localized temperature and static strain sensing.
Engineers bonded FBG on actuator arms and directly on the rails to allow measuring both the strain induced on the floating arms and rail deformations during train service operations (Fig. 12).

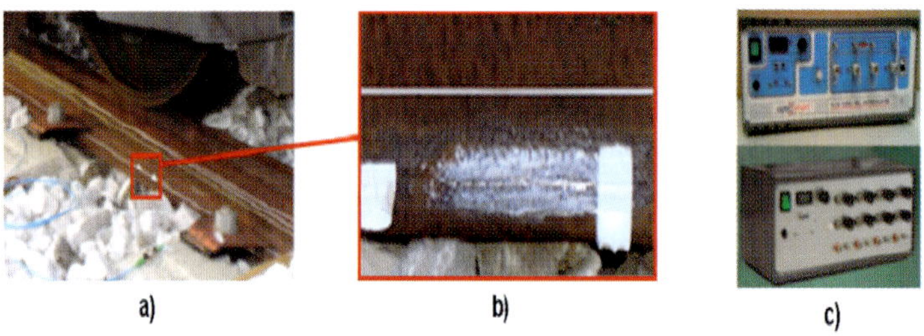

Figure 12. FBG sensors on the rail (a); bonding region (b) and interrogation system (c).

4.6. Wind energy

The world market for wind energy grew by 30 % in 2007 and 32 % in 2006. Industrially emerging country, as China, had experienced three-fold growth, while the U.S. market had doubled in size.
In order to generate multi-megawatt power output, turbine rotor-blade diameters of over 100 m and nacelle heights of over 120 m are becoming standard systems.

Huge turbines utilize the newest composite material technology, stretching the designer's knowledge and the material's capability to the limit in order to balance both costs and performances.
For these reasons, it is becoming increasingly important to put in place systems to monitor their condition in real time.
Optical fiber sensors can be of great benefit as they provide structural performance feedback to the design function, ratifying FEA models and allowing for the development of lighter blades[34-36] with lower production cost and improved performance.

4.7. Oil Production and Pipeline Industry

Oil production and pipeline industry have seen a large development of optoelectronic sensor based systems over the last 5-10 years.
As the easily accessed oilfields are depleted, oil and gas exploration and production is moving to increasingly remote, hostile and environmentally sensitive parts of the planet.
Also pipelines and distribution stations will experience the same problems due to the extremely "hash" environmental conditions during their life. The potential economical benefits of SHM, for the oil production, pipeline industry and related infrastructure could be of huge importance. This is mainly due to the possibility to monitor hundreds of kilometers of pipelines,[25] or to monitor, in real time, the structural integrity of oil tanker in service. This can be done by simply acquiring a light signal from a single instrument or also by intermediate checking of the health structure at localized diagnostic controlling point.
The use of fiber optics in oil production took a giant step forward in 2003 when the Norwegian company Optoplan,[37] installed the first ever permanent seismic surveillance system down-hole in a well in Izaute in southern France.
Down hole monitoring is probably the best known application for fiber optic sensors in the oil and gas sector, and an area in which several of the large service companies are making a concentrated effort to leverage the unique properties of the technology.

A study of hull stress monitoring called HullMon+ was also funded by the 5th Framework Programme,[38] to establish how this technology can aid in the tightened safety-at-sea regime.
In offshore oil exploration some companies have replaced heavy electric seismic arrays with lighter fiber optic arrays that ease handling during operation. In 2007 Light Structures[39] finalizes commissioning on the two first SENSFIB fiber optic hull stress monitoring systems for the QatarGas LNGC at Hyundai HI and Samsung HI.

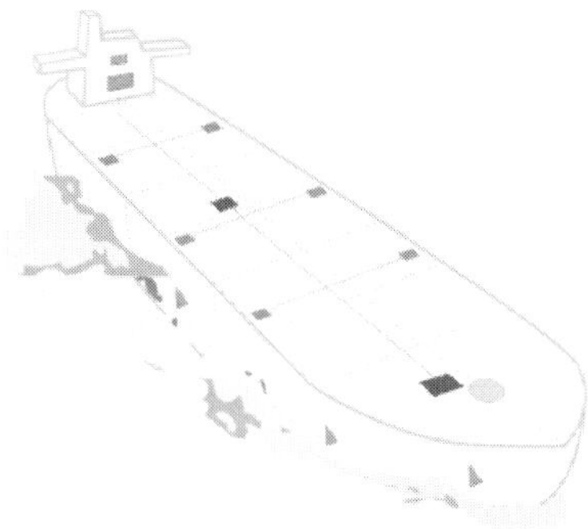

Figure 13. Layout of sensors for Hull stress monitoring (Ref. 37).

4.8. Medical Applications

One of the very last applications of SHM systems based on optoelectronic sensors is related with biomechanics and medical devices. The human musculoskeletal system represents a fundamental information for orthopedists and experts to push forward the frontier of medical possibilities to operate on human natural locomotion.[40-42]
Therefore, it is very important to be able to monitor the deformation occurring on each bone element during normal day human activities.
FBGs can represent the ideal solution to acquire the necessary information on deformation during human locomotion, not only because they are easier to adhere, insensitive to electromagnetic interference and very resistant to aggressive environment agents, but also because they can be used in series onto the same fiber, leading to multiply

measurements with a single fiber. Moreover, due to their reduced dimension, FBGs cause a very small intrusion within the human body.

Final results reported on human femur diaphasys and compared with acrylic cylinder, have shown that there is no much difference between strain gauge and optical fibers when comparing results.

These sensors are very suitable to monitor dynamic and static deformation in bones in vitro and should be suitable as well for use in vivo.

References

1. F. Chang, Summary report on *The First Stanford Workshop on the Structural Health Monitoring*, USA (1997).
2. F. Chang, *Proceedings of the 1st European Workshop on Structural Health Monitoring 2002* – France, p.3, USA DEStech Publications, Inc. (2002).
3. A. Kersey et al., *J. Light Wave Techn.*, Vol. 15, No. 8, 1442 (1997).
4. P. Foote, *Opt. Fiber Bragg Grating Sensor for Aerospace Smart Str.* IEEE (1995).
5. F. Chang, *The Sec. Int. Workshop on SHM*, USA, Stanford University (1999).
6. A. Morin et al., *National Optic Institute, USA*, Vol. 2718, 427 (1996).
7. K. Voss and K. Wanser, *SPIE the Intern. Society for Optical Eng.* (1994).
8. M. Wlodarczyk, T. Poorman, L. Xia, J. Rnold and T. Coleman, *"Embedded Fiber Optic Combustion Pressure Sensor For Automotive Engines"*, California, SPIE (1997).
9. R. Measures, *Structural Monitoring with Fiber Optic Technology*, USA, Academic Press, chapter II 2001.
10. C. Voto, S. Inserra and F. Camerlengo, *MUSEAS First Workshop, Capua, Italy,* (2001).
11. P. Foote, J. McFeat and I. Hebden, *MUSEAS First Workshop, Capua, Italy,* (2001).
12. M. Buderath, *Proceedings of the 1st European Workshop on Structural Health Monitoring* 2002 – France, p. 76, USA DEStech Publications, Inc. (2002).
13. T. Munns, et al., Langley Research Center, Virginia, (2002).
14. L. Renson and Y. Stockman, *MUSEAS First Workshop, Capua, Italy,* (2001).
15. F. Bastianini, A. Di Tommaso, G. Pascale, *Compos. Struct.* 563 (2001).
16. M.R. Clark, D.M. McCann, M.C. Forde, *NDT and E Int.*, **36**, 265 (2003).
17. H.Wiggenhauser, *Infrared Phys. Technol.*,**43**, 233 (2002).
18. M. Scott, A. Rezaizadeh, A. Delahaza, C.G. Santos, *NDT and E Int. NDT and E Int.*, **36**, 245 (2003).
19. C. Cremona and J. Carracilli, *Key Eng. Mater.*, 204/205,47(2001).
20. K. Kageyama et al. *Smart Mater. and Struct.*,**14**, S52 (2005).

21. G. Kister et al. *Eng. Stru.*, 29 (2007)
22. J. Vehi, N. Luo and R. Villamizar, *Proceedings of the 1st European Workshop on Structural Health Monitoring 2002* – France, USA: DEStech Publications, Inc. p.965 (2002).
23. P. Ferdinand et al., *Virginia, Intern. Con. on Optical Fiber Sensors* OFS'97, Oct. 28-31, 1997.
24. www.smartec.ch.
25. www.optosmart.it.
26. www.wavefield-inseis.com.
27. Research EU, the *Magazine of the European Area*, no.1, January 2008.
28. N. M. Theune et al. *EWOFS 2004*, vol. 5502, 536 (2004).
29. T. Bosselmann, *OFS 17*, vol. 5855 (I), 188 (2005).
30. Laudati, F. Mennella, M. Esposito, *Convegno Nazionale AEIT*, (2006)
31. G. Breglio et al. *J. Sens. and Act.* **B**, 110 (1-2) 147 (2004).
32. K. Schroeder et al. *Meas. Sci. Technol.* **17** 1167 (2006).
33. M. Jones, *Nature Photonics*, 2, 153 (2008).
34. L. Rademakers et al., *European Wind Energy Conference*, London, 2004.
35. The Oil & Gas Review 2003 V.2 at www.touchoilandgas.com.
36. www.roctest.com and www.cordis.lu.
37. www.lightstructures.biz and www.ship-technology.com.
38. T. Fresvig, P. Ludvigsen, H. Steen and O. Reikeras, *Medical Eng. & Phy.* 30, (2008).
39. G. Wang, K. Pran and G. Sagvolden, *Smart Mat. and Struct.*, **10**, 472 (2001).
40. T. Finni et al. *Eur. J. Appl. Physiol. Occup. Physiol.*, **77**, 289 (1998).

ELECTRO-OPTIC AND MICRO-MACHINED GYROSCOPES

Valerio Annovazzi-Lodi,[a,*] Sabina Merlo,[a] Michele Norgia,[a]
Guido Spinola,[b] Benedetto Vigna[b] and Sarah Zerbini[b]

[a]*Dipartimento di Elettronica, Università degli Studi di Pavia*
Via Ferrata 1, 2700 Pavia, Italy
[b]*ST Microelectronics s.r.l.*
Via Tolomeo 1, 20010 Cornaredo, Italy
[*]*E-mail: valerio.annovazzi@unipv.it*

The purpose of this chapter is to present the gyroscope technologies available on the market or still under development. Besides mechanical and electro-optical systems, a new class of miniaturized devices based on the vibrating structure is described. In particular, the technological solution investigated at ST Microelectronics to realize a silicon micro-machined device is described. A new interferometric optical method is then applied for the characterization of the vibration modes of the fabricated prototypes.

1. Introduction

Inertial navigation units use accelerometers and gyroscopes (inertial sensors) to measure the state of motion of the vehicle by noting changes in that state caused by accelerations. By knowing the vehicle's starting position and noting the changes in its direction and speed, one can keep track of the vehicle's current position. Accelerometers are devices measuring physical accelerations and by mathematical integration they allow to get the velocity and the distance traveled by the moving object.
In this chapter we will not talk about the accelerometers, but we will focus our attention on gyroscopes. Gyroscope derives its origin from two Greek words: *gyros* which means ''rotation'' and *skopein,* ''to view''.
So gyroscope is a device able to measure angular rate of a moving object with respect to a fixed reference frame. And it is extremely useful to understand if the moving object is experiencing an angular rotation for dead reckoning applications!

Until the fall of the Berlin wall, navigation and guidance instruments have evolved to meet operational requirements without any economical constraints, but following only scientific and technical progresses. Cost of gyros has become an important issue only in the last two decades. Thus alternative gyroscope technologies exploiting new aspects of physics supported by sophisticated signal processing techniques and innovative manufacturing methods have made their appearance. Miniaturized, less power hungry and cheaper devices are now available for standard market applications, like aeronautics and military, and relatively young market applications like the vehicle dynamic control in the automotive sector.

The purpose of this chapter is two folds:

To present the basic technologies of the gyroscopes available on the market so that the reader can understand the advantages and the drawbacks of the different types;

To point out where the research on this field is headed and which market applications are foreseen in the medium term.

Nonetheless the first gyroscopes available to the humankind were mechanical and they were based on the angular momentum conservation of spinning wheels in inertial space (around 1860) it was only the modern technology of guided wave optics, fiber optics and integrated optics to enable cheaper, more reliable and high performances gyroscopes (~1980). In fact these optical devices replaced classical spinning wheels and floating gimbals in civil aircraft such as the new Boeing 777. Anyway, mechanical and optical gyroscopes are used wherever the stability and the degree of performances of the devices are critical for their use. This means that all the military applications requiring low drift over long mission time (typically one-mile-a-day), like for example nuclear submarines and intercontinental missiles, need mechanical gyroscopes while for all the transoceanic civil flights and short range missiles optical gyroscopes are widely used.

But apart from mechanical and electro-optical devices a new class of gyroscopes would be even smaller, more reliable and less expensive:

the Vibrating Gyroscopes. They are based on the coupling of two orthogonal vibration modes of resonating solid bodies induced by Coriolis force. Engineers, looking for alternatives to the wheel, tried to use vibrating rather than rotating bodies to provide gyroscopic torques from the Coriolis acceleration. Nature has provided flying insects, diptera, with a tuning fork for flight control!

Most gyro engineers started to work in this new field in 1965, but since most of them discovered technical problems they abandoned the field until 15 years ago. Now, there is a lot of research on this kind of gyroscopes since this technology fits quite well with the typical manufacturing methods of the existing silicon industry. Moreover, the Coriolis gyroscopes, as they are called, start to make their appearance in application fields where the technical issues are not so demanding: vehicle dynamic control in automotive market and image stabilization in video- cameras. Still a lot of research must be done about the silicon-integrated gyroscope to improve its performances and achieve higher stability and resolution levels, but the field is quite promising.

In this chapter after a description of integrated electro-optical gyroscopes and mechanical classical gyros we will review the implementation of vibrating gyroscope in a silicon micro-machining process running at STMicroelectronics.

2. Electro-Optic Gyroscopes

In the inertial navigation systems used on aircrafts and spacecrafts, the measurement of angular rotation must be performed with high accuracy.

Rotation measurements by means of electro-optical methods are based on the Sagnac effect.[1-8] When a ring optical cavity is rotating with respect to an inertial reference frame, two counter-propagating lightwaves experience in the cavity itself a different optical path. Although the Sagnac effect dates back to 1903, only with laser sources it has been finally possible to make practical optical sensors for detecting inertial rotation. These devices are characterized by the absence of moving parts, which may yield improved reliability, reduced cost, reduced warm-up delay, and insensitivity to acceleration. In the classical view, the optical path difference, measured along the two propagation directions in a rotating cavity, can be considered as a Doppler frequency shift. The time delay between the clock-wise (CW) and counter-clockwise (CCW) propagating optical waves results in an optical phase difference, Φ_S given by

$$\Phi_S = 8\pi S\,\Omega\,/\lambda c \qquad (1)$$

where S is the area enclosed by the optical path, Ω is the component of the angular velocity orthogonal to the cavity plane, λ is the wavelength in vacuum, and c is the speed of light.

Three configurations of the electro-optic gyroscope should be mentioned here: (1) the Ring Laser Gyro (RLG); (2) the Fiber-Optic Gyro (FOG); (3) the Ring Resonator Fiber-Optic Gyro (RFOG).

The Ring Laser Gyro employs a triangular gas-laser cavity to realize an active interferometer (Fig. 1a).[1] As a consequence of the phase difference given by (1), the two counter-propagating cavity modes oscillate at two different optical frequencies that must satisfy the resonance condition. The frequency difference Δf, given by $\Delta f = 4S\Omega/p\lambda$, where p is the perimeter of the cavity, can be detected if a fraction of the two modes is recombined on a photodetector, thus achieving an output current signal $I = I_o[1+\cos\Phi_S(c/p)t]$, where I_o is the photodetected current at rest. The most popular configurations are the dither RLG (DLAG) and the Zeeman four-frequency RLG (ZLAG).[9] Well-developed devices are now produced in the U.S. and currently employed on civilian airplanes (e.g., Boeing 757, 767).

The Fiber-Optic Gyroscope in its basic version is illustrated in Fig. 1b. In the FOG, the phase difference between the two counter-propagating waves is cumulated over a fiber coil (100 to 1000 meters long) to increase responsivity and compactness.[2,3,5-8]

In the basic set-up, a beam-splitter or a fiber-optic coupler is used to split the laser source radiation into counter-propagating waves in the coil. The optical phase difference is given by (1) by substituting $N{\cdot}S$ for S, where N is the number of coil turns. The output photodetected current, after recombination of the two traveling waves by the beam-splitter, is thus given by $I = I_o(1 - \cos\Phi_S)$.

For typical values of geometrical dimensions, it can be concluded that the measurement of the earth angular rotation ($\Omega = 15°/h$) requires to appreciate about 30 μr in FOG (0.03 μr in RLG), while for inertial

guidance, the accuracy requirements increase by a factor $10^2 - 10^3$, which corresponds to the detection of an optical path difference of the order of 10^{-15} m. It should be here emphasized that this requisite must be satisfied in a dc coupled device.
The phase noise, Φ_n, at the quantum limit is equal to the inverse of the amplitude signal-to-noise ratio, evaluated at $\Phi_S = \pi/2$, that is $\Phi_n^2 = 2qB/I_o = 2hfB/\eta P$, where q is the electron charge, B is the measuring bandwidth, η is the photodetector quantum efficiency and P is the received power. For the RLG a similar result can be found, that is $\Phi_n^2=[2h\nu B(1-r)^2]/P$, where r is the reflectivity of the output mirror.
A more detailed comparison reveals that the standard (HeNe) RLG features a better performance (reaching the quantum limit) than the FOG. A specific advantage of the RLG configuration is the frequency output, suitable for digital processing; on the other side, the FOG configuration features modularity, scalability, reduced requirements on optical and mechanical machining, and potentially low cost.
In a brief analysis of the main problems of electro-optic gyros, we should mention as a first disadvantage the high output non-linearity for small rotation angles ($\Phi_S \approx 0$). This limitation can be overcome by converting the output dependence on $\cos\Phi_S$ into a sine dependence and a phase modulation scheme is generally employed to obtain an all-fiber, compact, rugged and less ambient-sensitive sensor.
Closed loop structures, where a feedback effect is exploited to cancel the Sagnac phase shift by means of an analog or digital phase modulation, have been investigated for improving linearity and dynamic range of the FOG.[6] Birefringent and photonic crystal fibers have been proposed for better noise performances.[10-12]
With regards to the effective noise levels, sensitivity at the quantum limit cannot be achieved in the FOG without the elimination of all the sources of non-reciprocity, other than the Sagnac effect, in the propagation of the CW and CCW waves. While the basic set-up (Fig. 1b)
is not fully reciprocal because the two beams do not travel symmetrically through the optical coupler, reciprocal propagation may be obtained by introducing a supplementary coupler and selecting a different output port, thus realizing the so called "minimum configuration".[6] In addition, accurate modal filtering and polarization controlling techniques are

commonly used to improve the actual FOG performance. Rayleigh back-scattering in the fiber and reflections, from discontinuities in the coil, such as splices as well as from the all-fiber components, represent other sources of error in the FOG and should therefore be carefully minimized. The Ring Resonator Gyroscope includes a ring optical cavity, as in the case of the RLG but without active medium; the resonance-frequency shift of the counter-propagating modes, induced by rotation, is measured by launching the radiation of an external laser into the empty cavity through a coupler.[2,6]

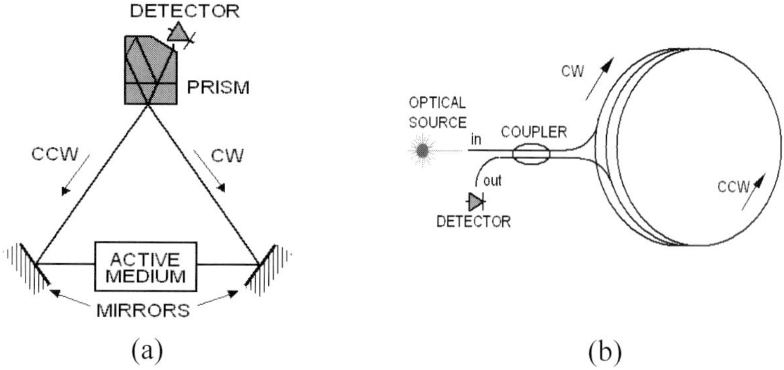

Figure 1. The basic schemes: (a) Ring Laser Gyro, (b) Fiber-Optic Gyro.

3. Mechanical Gyroscopes

The classical mechanical gyroscope that all of us had the chance to see in all the science museums all over the world is based on a simple and clear physical law: the Gyroscopic Law.[4] The essence of the spinning wheel gyroscope is stated by the following equation:

$$M = L\,\Omega \tag{2}$$

That means, if we have a spinning wheel with an angular momentum, L, directed along X-axis and if we apply an external mechanical torque, M, along Y-axis, the wheel will try to align its spin axis with the input torque one having a precession Ω along the Z-axis. The orthogonal external torque is not able to change the amplitude of the angular momentum, but only its direction, as in the figure below.

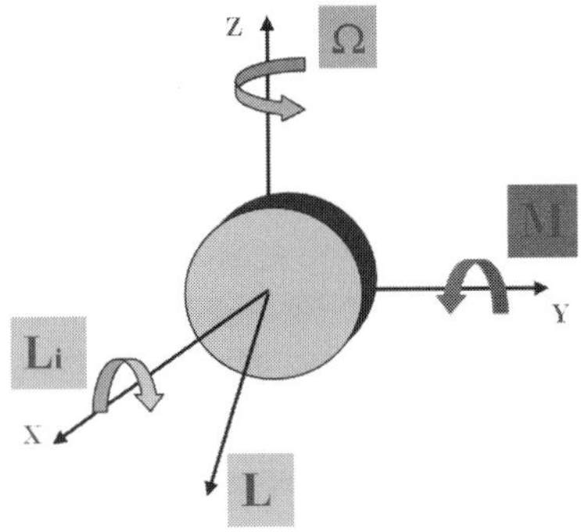

Figure 2. The gyroscopic law.

If we want to use the gyroscopic equation for a mechanical gyroscope it's quite simple. If the X-axis spinning wheel is mounted on a frame experiencing a Z-axis rotation Ω then a mechanical torque (gyroscopic torque) along Y-axis is observed. And by measuring the gyroscopic torque and by knowing the initial angular momentum it is possible to measure the unknown angular rate Ω.
A straightforward implementation of a single axis degree of freedom gyroscope is depicted in Fig. 3. The gyro dynamics is represented by a second order differential equation with the following parameters:

$$I_0(d^2\theta/dt^2)+\gamma(d\theta/dt)+K_{tb}\theta=H\Omega \qquad (3)$$

where
θ = gimbal angle or pick off angle
I_0 = gimbal moment of inertia about the output axis
γ = damping constant about the output axis
K_{tb} = torsion bar stiffness
H = initial angular momentum

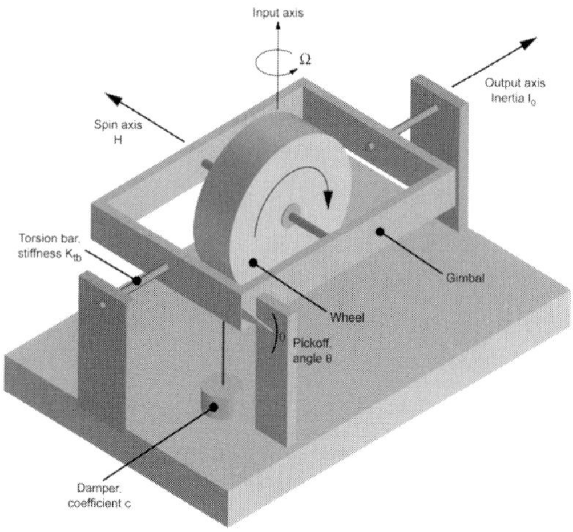

Figure 3. The one-axis gyro.

By looking at the parameters of the gyroscope equation it is useful to make the following comments:

- The angular momentum *H* should be as large as possible to enhance device sensitivity. Since the spinning frequency is limited to the frequency range 200 Hz – 1,000 Hz the main design parameter is the diameter of the spinning wheel (usually in the range of some inches)
- The ball bearings of the spinning wheel must be stiff enough, stable over the desired range of temperature and properly lubricated to prevent destructive metal-to-metal contact.
- Damping fluid is very important for the correct dynamic response and it serves to cushion the gimbal against shock and vibration. Low temperature dependence viscosity fluids as well as smart design to compensate the thermal variations must be chosen.
- The mechanical spring must be stiff enough to have low cross axis sensitivity, higher resonant frequency for high bandwidth and better withstanding mechanical shocks and vibrations. Moreover, the torsion bar has to be proportioned so that the stresses set up in it do not exceed its material's elastic limit; indeed they must be low enough that hysteresis is negligible.

4. Vibrating Gyroscopes

The physical principles of the vibrating gyroscopes are known since almost 200 years. To study the earth's rotation, the French scientist Leon Foucalt used a large pendulum (67 m long), which he built in the Pantheon in Paris. Its iron bob weighted 28 Kg and it swung with a period of about 15 sec. As the earth rotated under the swinging pendulum, the plane of the swing appeared to rotate clockwise at vertical earth's rate because of Coriolis force. In fact since the pendulum has a relative movement with respect to the Earth, i.e. a rotating reference frame, an apparent force must be inserted in the dynamic equation of the pendulum to explain properly its behavior.
All the vibrating gyroscopes exploit the Coriolis induced coupling of two orthogonal acoustic modes of a vibrating solid body. For sensitive devices a high quality resonator with two orthogonal vibration modes is mandatory: excitation mode and Sensing mode must be degenerate, i.e. they have the same frequency.
Moreover, the resonator must be well isolated form the external environment to reject better all the spurious mechanical noise. Different excitation mechanisms are exploited in vibrating gyroscopes available on the market: magnetic, electrostatic and piezoelectric. The following section will describe the principles behind a silicon micro-machined gyroscope, electrostatically actuated and capacitively sensed, that we are realizing at STMicroelectronics.[13-15] Noise limits of the sensing element will be carefully evaluated taking into account the different market applications and measurement results will be shown in the relative section.

4.1. *Micro-machined Silicon Resonant Gyroscope: a Test Case Study*

STMicroelectronics has developed a technological process for the realization of all the inertial sensors (angular, linear accelerometers and gyroscopes), that is a kind of epithaxial micromachining process (Fig. 4).[16] A thermal oxide is grown on standard silicon substrate. On the top of the oxide a thin polysilicon layer is LPCVD deposed in order to realize buried interconnections. Afterwards, a sacrificial oxide layer is deposed on the top of which a thick polysilicon epithaxy is grown (15 μm). This epithaxial polysilicon is the structural material that composes the sensor. The structure of the sensor is defined by deep silicon RIE. At the end, the sacrificial oxide is removed by using a vapor phase HF etching.

With the process described above the sensor results to be thicker than sensors obtained with thin film deposition techniques and therefore shows a higher sensitivity; moreover, with respect to the structures obtained with surface micromachining technique, it has greater stiffness to bending along z-axis and therefore a lower probability to stick on the substrate during sacrificial oxide removal. This process has also a lower cost than processes based on the use of SOI or silicon fusion bonded substrates.

Due to the necessity to protect the devices during wafer sawing process and to guarantee a controlled atmosphere during their lifetime, a process for encapsulating the sensors at wafer level has been developed. The process consists of using another silicon wafer to create the caps for the sensing structure with holes trough it in correspondence to pad region and of using a hermetic wafer to wafer bonding technique to seal the device as shown in Fig. 5.

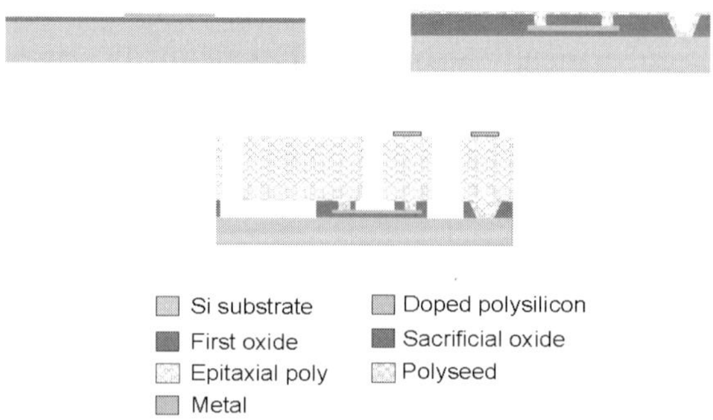

Figure 4. ST THELMA micro-machining process description.

Thelma process, developed at STMicroelectronics, is suitable for manufacturing a wide range of inertial sensors such as accelerometers and gyroscopes. In this section we are going to describe a vibrating gyroscope, in which the angular rate is transduced into the vibration amplitude of a small and complex suspended structure.

Figure 5. Encapsulated inertial sensor: (left) the sensing element and (right) the electronic circuitry.

The design of a micro-machined gyroscope is especially challenging because the Coriolis force is very week, hence resonant structures with high figure of merit and stability must be achieved to cope with market gyro sensitivity and bias-drift requirements.[17]
To evaluate correctly the design trade-offs, a good experimental set-up is needed to examine the behavior of the real structure.
Measurements of gyro mass displacement at submicrometer resolution are reported, using an 800 nm 20 mW laser diode. The resonance curves of the device have been determined at atmospheric pressure and for different vacuum pressure levels. An optimal pressure has been identified around 200 mtorr, for which performance level of the sensor is guaranteed even in the case of low-power and low-voltage supply for the actuation of the gyroscope.
It is intended that all the design trade-offs are decided on a market requirement basis: in our case the driving market is the consumer market. These new micro-machined gyroscopes are based on the Coriolis force acting on a vibrating mass upon rotation. The sensor layout involves two masses of around 10^{-8} Kg suspended in the x-y plane by adequate supporting springs. The masses are forced to vibrate in anti-phase motion by electrostatic actuation, which is generated by applying a periodic voltage to the capacitor comb structure highlighted in Fig. 6.

When the gyro rotates at angular velocity Ω around the vertical z-axis, the resulting force, $F_c=2m(dx/dt)$, causes a vibration along the y-axis (sensing axis), whose amplitude is proportional to Ω. The two suspended masses are forced to vibrate in anti-phase motion in order to have anti-phase Coriolis vibration. This is fundamental to reject linear accelerations because a single mass suspended by springs is also sensitive to acceleration in the sensing direction (y-axis).
To measure the vibration induced by the angular speed along y-axis, a capacitive read-out is implemented. This will give information on displacements and consequently on the angular rate Ω applied to the system.

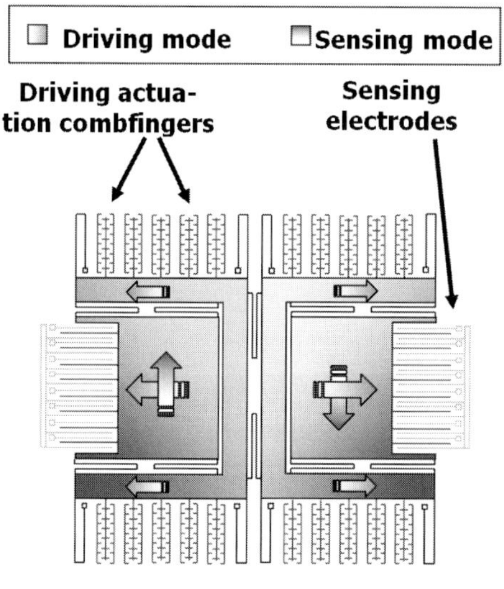

Figure 6. The inertial sensor.

If we apply a harmonic force $F=F_o\sin(\omega t)$ to the driving axis, the displacements on the x-axis and y-axis are also harmonic at the same frequency $\omega/2\pi$. Designing the system with equal resonant frequency for driving and sensing mode, $f_n=\omega/2\pi = (K/m)^{1/2}$, where K is the equivalent spring constant and m the system mass, the vibration amplitude is maximized and amounts to $X_d = Q_xF_o/K$ on the driving axis and to $Y_s = Q_yF_c/K$ on the sensing axis.

For design optimization, it is important to measure the resonant frequency values and also the related figures of merit Q. This is needed to establish the correct trade-off between gyro sensitivity and low-voltage power supply.

4.2. Mechanical-Thermal Noise

Mechanical-Thermal Noise (MTN) represents a major sensitivity limitation for a micro-machined gyro, where one must detect nanometer displacements on a mass m of a few micrograms. This noise source is due to molecular agitation inside mechanical parts, and is the equivalent of Johnson's noise in resistors. In the following, we consider the basic gyro structure of Fig. 7, where damping mechanical resistances $R_{z,y}$ have been introduced to represent air viscosity and other dissipation effects.

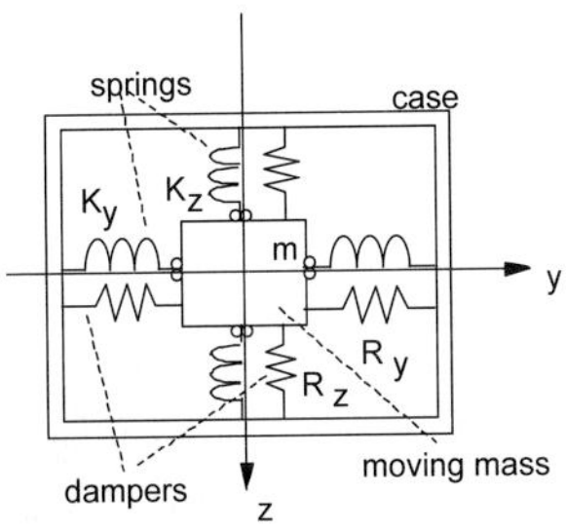

Figure 7. Schematics of the vibrating mass gyroscope. The y and z axes are the reference frame of the gyro case.

The MTN can be modeled by assigning to each damping resistance a r.m.s. force:

$$F_{nz,ny} = \sqrt{4K_B T R_{z,y} B} \tag{4}$$

where K_B is the Boltzman constant, T the absolute temperature and B is the noise bandwidth.[18] The force F_{ny} applied along the y-axis (sensing axis) gives, at the resonance frequency, a displacement $F_{ny}Q_y/K$ (where Q_y is the resonance quality factor, $K=K_y=K_z$ is the spring constant), and a displacement fluctuation along the y-axis:

$$Y_{ny} = \sqrt{2}\sqrt{4K_B TR_y B}\, Q_y / K \tag{5}$$

The signal-to-noise ratio *S/N* can be calculated from (4) and from the gyro responsivity, obtaining:

$$S/N = \frac{2mF_0}{R_y^{3/2}\sqrt{4K_B TB}}\Omega \tag{6}$$

where F_0 is the amplitude of the harmonic force applied to the driving axis and Ω is the angular speed.[19] Introducing the noise equivalent angular speed *NE*Ω, that is defined as the value of *Ω* for which the *S/N* ratio is unity, from eq.(6) we find:

$$NE\Omega = \frac{R_y^{3/2}\sqrt{4K_B TB}}{2mF_0} \tag{7}$$

From this equation, it is evident the importance of designing a gyro with a large mass *m*, and of operating a given device at low damping (i.e., at low pressure inside the package). *NE*Ω has been plotted in the diagram of Fig. 8 by using typical parameter values.[19]

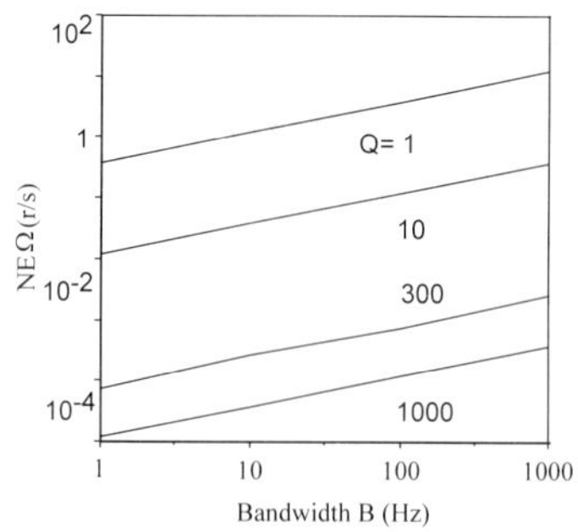

Figure 8. Noise equivalent angular speed *NE*Ω as a function of bandwidth *B*, for *m*=2 10^{-9} Kg and for different values of Q_y, in a typical gyro.

From our analysis, we find a theoretical sensitivity limit due to MTN of the order of 10^{-3}-10^{-4} r/s, for Q_y=100-300, which, for a driving frequency ω/2π=5-20 kHz results in a bandwidth *B* in the range 1-100 Hz. In practice, such performances can be somewhat improved by increasing R_y (by acting on Q_y) and *m*, so that the whole automotive range, down to

10^{-4} r/s (B=10-100 Hz), is in the reach of this technology. Nevertheless, the MTN represents a major sensitivity limitation for vibrating mass gyros, and extending this approach to more demanding applications, such as avionics, would require a major technological breakthrough.

Two other sources of noise are usually considered in suspended-mass micro-machined sensors. They are squeeze damping and electronics noise. The first effect can be safely neglected in gyros, which work at low pressure to get high Q. As for electronics noise, it can be easily shown that standard monolithic integration can provide a noise limitation which is about an order of magnitude smaller than the previously estimated values of $NE\Omega$.[19]

4.3. Optical Characterization of Micro-machined Gyroscopes

At the design stage of a new MEMS, it is important to have a direct measurement of the effective behavior of the prototypes. Indeed, indirect methods, such as the standard electrical capacitive technique, may underestimate important parameters, especially with bare devices, because of parasitic phenomena and cross-talk interference. On the other hand, optical interferometry represents the most powerful tool for detecting small amplitude vibration with high precision. However, classical interferometric schemes are not easy to apply for the detection of displacements in the horizontal plane. Fist of all, it is impossible to monitor the hidden vertical faces of the masses with a laser beam. Moreover, the mass surface does not represent a good optical surface, since it is rough and usually holed.

An interesting optical technique for characterizing vibrating micro-machined structures is feedback interferometry. It is based on the amplitude and frequency modulation arising when a small fraction of the power emitted by a laser diode is allowed to re-enter the laser cavity.[20] The amplitude modulation can be easily measured by means of the monitor photodiode, incorporated inside the package of commercial semiconductor lasers. For very low back injected power (lower than 10^{-6} times the emitted power) the photodiode current, I, exhibits a sinusoidal dependence on the target distance D,

$$I = I_0 + \Delta I \cos\left(\frac{4\pi}{\lambda} D\right) \tag{8}$$

where I_0 is the photodiode current without feedback, ΔI is the modulation depth, depending on the feedback level, and λ is the laser wavelength. This signal resembles the well known, standard interferometric signal with fringes.

This technique is especially suitable for the characterization of MEMS, because it can work very well on a diffusive surface (such as the holed mass), without requiring accurate alignment and wave-front matching.[21-23] Moreover, the interferometer can be miniaturized (few centimeters), thanks to the absence of external optics, other than a focusing lens. It can be easily placed inside a vacuum chamber for characterizations at different pressures, or for measurements of the Coriolis force on rotating devices.[13] Good lasers for this kind of interferometer are single mode, near infrared diodes ($\lambda \approx 800$ nm).

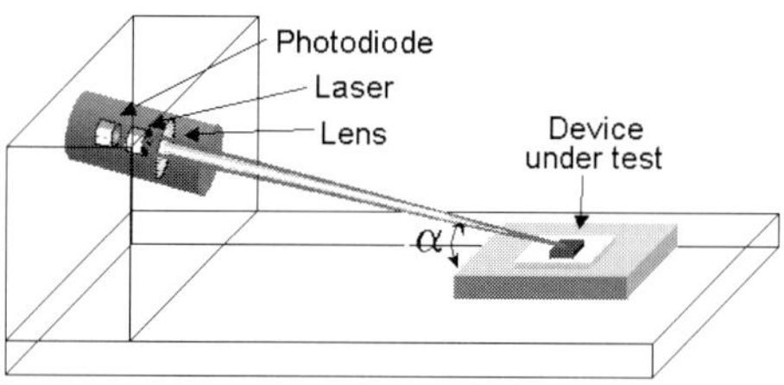

Figure 9. Experimental set-up for the interferometric measurement of MEMS.

The set-up is shown in Fig. 9: the laser beam is focused on the device, at an angle α=20° with respect to the vibration plane, thus detecting the component of the displacement along the beam direction. The spot can be focused to a diameter of few micrometers, in order to measure the movement of single details on the MEMS (such as the springs).

When the mass is electrically actuated and vibrates with an amplitude of more than about 500 nm, we can measure the effective mass displacement, Δs, by counting fringes on the interferometric signal, yielding $\Delta s = M\lambda/(2\cos\alpha)$, where M is the number of fringes.

Some examples of photodetected interferometric signals, obtained by shining the laser on a polysilicon gyroscope electrically driven at different frequencies, are shown in Fig. 10a. By means of this set of data, it is possible to plot the resonance curve of the device (see Fig. 10b), with a resolution of about ±100 nm (a quarter of fringe), and thus to obtain the resonance frequency and the quality factor of the system. This method gives also information about the device response phase-shift.

For example the curve D in Fig. 10a exhibits a 90° phase delay, as expected at the resonance frequency of a mass-spring system.

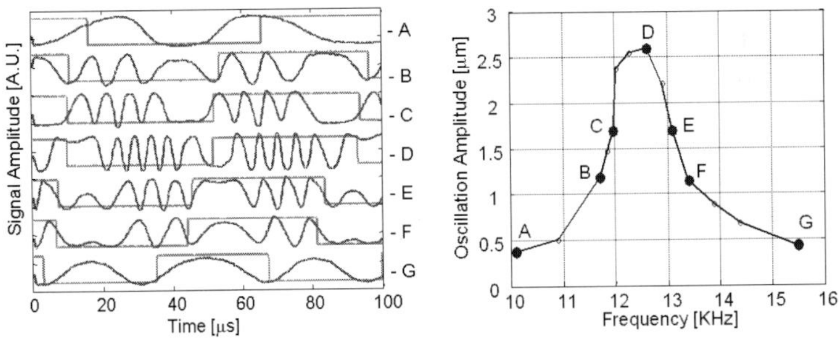

Figure 10. a) Left: Interferometric signals acquired on the mass of a gyroscope, for different driving frequencies (square waves). b) Right: Resonance curve obtained from fringe counting.

For very low vibration amplitude (i.e. <100 nm), it is possible to use the interferometer as an almost linear transducer (in quadrature), thus obtaining a signal directly proportional to the target movement.

The absolute measurement of the signal amplitude is not meaningful, because it depends on the feedback level, which may change randomly over a diffusive surface. Instead, if the system is excited by white noise, the spectrum of the interferometric signal gives a direct measurement of the whole resonance curve of the device.[13] The averaged acquisition of the interferometric signal on a spectrum analyzer is shown in Fig. 11. The device under test is a gyroscope, measured at a pressure of 200 mTorr, exhibiting a quality factor of 5000. A theoretical fitting is also plotted, showing a very good agreement with the acquired data.

The white-noise technique is particularly powerful for a direct real time characterization of new, bare prototypes, since it is able to highlight the frequency location of even unexpected vibration modes.

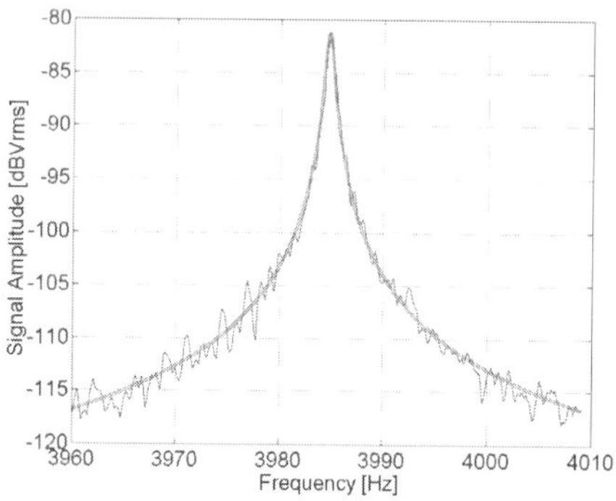

Figure 11. Thin line: gyroscope resonance curve, measured by a spectrum analyzer, after white noise excitation. Thick line: theoretical fitting.

5. Conclusions

In this chapter we have attempted to highlight the progress and the limitations in the field of gyroscope integration techniques and their role in the miniaturization of the sensing element.

Micro-machined gyroscopes for measuring angular rate have drawn attention during the past few years for several applications.[13,14,24-36]

Micromachining can shrink sensor size by orders of magnitude and reduce the fabrication costs. They can be used together with accelerometers to provide information for inertial navigation or for ride stabilization and rollover detection; they can be addressed in some consumer applications such as video-camera stabilization, inertial mouse, robotics, but they still do not have the degree of perfection required by a wide range of aerospace or military applications.

Conventional rotating wheel as well as precision fiber-optic and ring laser gyroscopes are both too expensive and too large to be used in most emerging applications, but they are the only ones to fit the specs of the most demanding markets, i.e. aeronautics and military.

Whether we see a drastic improvement in the performances of the micro-machined gyros so that they will find their path to military and aeronautics market strongly depends on the research and development efforts of micro-machining technologies of the next decade.

Acknowledgements

V. Annovazzi-Lodi, S. Merlo and M. Norgia would like to acknowledge Prof. S. Donati for his continuous scientific guidance.

G. Spinola, B. Vigna and S. Zerbini would like to acknowledge the entire ST technological development group led by Dr P. Ferrari for the tremendous amount of work they spent to fix the micromachining THELMA process.

References

1. F. Aronowitz, *The Laser Gyro*, in *Laser Applications*, New York: Academic Press (1978).
2. G. Cancellieri, Editor, *Single-Mode Optical Fiber Measurement*, Boston: Artech House (1993).
3. S. Donati, Gyroscopes, Chapter 7 in Electro-Optical Instrumentation: Sensing and Measuring with Lasers, Prentice Hall PTR, USA (2004).
4. A. Lawrence, Modern Inertial Technology, Second Edition, Springer (1998).
5. H. C. Lefevre, *Fiber-optic gyroscope*, in *Optical Fiber Sensors*, Vol. 2, Norwood, MA: Artech House (1989).
6. J. M. Lopez-Higuera, Editor, *Handbook of Optical Fibre Sensing Technology*, New York: Wiley (2002).
7. S. Ezekiel and H. J. Arditty, *Fiber-optic rotation sensors and related technologies,* Berlin: Springer Verlag (1982).
8. E. Udd, Editor, *Fiber Optic Sensors*, New York: Wiley (1990).
9. W. W. Chow, J. B. Hambene, T. J. Hutchings, Sanders, V. E. Sargent III and M. O. Scully, *IEEE J. Quantum Electron.*, Vol. QE-16, 918 (1980).
10. S. Blin, H. K. Kim, M. J. F. Digonnet and G. S. Kino, *J. of Lightwave Technology*, Vol. 25, No. 3, 861 (2007).
11. S. Blin, M. J. F. Digonnet, G. S. Kino, *IEEE Photonics Technology Letters*, Vol. 19, No. 19, 1520 (2007).
12. J. Zheng, *IEEE Photonics Technology Letters*, Vol. 17, No. 7, 1498 (2005).
13. V. Annovazzi-Lodi, S. Merlo, M. Norgia, G. Spinola, B. Vigna and S. Zerbini, *IEEE/ASME J. of Microelectromechanical systems*, Vol. 12, No. 5, 540 (2003).
14. R. Oboe, R. Antonello, E. Lasalandra, G. Spinola Durante and L. Prandi, *IEEE/ASME Transactions on Mechatronics*, Vol. 10, No. 4, 364 (2005).

15. O. Colavin, F. Lorenzelli, B. Vigna and A. Hue, *ST J. of Research*, Vol. 3, No. 1, 63 (2006).
16. A. Corigliano, B. De Masi, A. Frangi, C. Comi, A. Villa and M. Marchi, *IEEE/ASME J. of Microelectromechanical Systems*, Vol. 13, No. 2, 200 (2004).
17. N. Yazdi, F. Ayazi and K. Najafi, *IEEE Proceedings*, Vol. 86, No. 8, 1640 (1998).
18. T. B. Gabrielson, *IEEE Transactions on Electron Devices*, Vol. 40, No. 5, 903 (1993).
19. V. Annovazzi-Lodi and S. Merlo, *Microelectronics J.*, Vol. 30, 1227 (1999).
20. S. Donati, G. Giuliani and S. Merlo, *IEEE J. Quant. Electr.*, Vol. 31, 113 (1995).
21. V. Annovazzi-Lodi, S. Merlo and M. Norgia, *IEEE/ASME Trans. on Mechatronics*, Vol. 6, 1 (2001).
22. V. Annovazzi-Lodi, S. Merlo and M. Norgia, *IEEE J. of Microelectromechanical Systems*, Vol. 10, 327 (2001).
23. V. Annovazzi-Lodi, S. Merlo and M. Norgia, *J. of Optics* **A**: *Pure Appl. Opt.*, Vol. 4, S311 (2002).
24. C. Acar and A. M. Shkel, *IEEE /ASME J. of Microelectromechanical Systems*, Vol. 14, No. 3, 520 (2005).
25. S. E. Alper and T. Akin, *IEEE /ASME J. of Microelectromechanical Systems*, Vol. 14, No. 4, 707 (2005).
26. S. An, Y. S. Oh, K.Y. Park, S. S. Lee and C. M. Song, *Sensors and Actuators*, Vol. 73, 1 (1999).
27. F. Ayazi and K. Najafi, *IEEE /ASME J. of Microelectromechanical Systems*, Vol. 10, No. 2, 169 (2001).
28. Y. Dong, M. Kraft and W. Redman-White, *IEEE Sensors J.*, Vol. 7, No. 1, 59 (2007).
29. W. Geiger, W. U. Butt, A. Gaißer, J. Frech, M. Braxmaier, T. Link, A. Kohne, P. Nommensen, W. Lang and H. Sandmaier, *Sensors and Actuators* **A**: *Physical*, Vol. 95, Nos. 2-3, 239 (2002).
30. Huikai Xie and G. K. Fedder, *IEEE Sensors J.*, Vol. 3, No. 5, 622 (2003).
31. R. P. Leland, *IEEE Transactions on Control Systems Technology*, Vol. 14, No. 2, 278 (2006).
32. H. Moussa and R. Bourquin, *IEEE Sensors J.*, Vol. 6, No. 2, 310 (2006).
33. A.S. Phani, A.A. Seshia, M. Palaniapan, R. T. Howe and J. Yasaitis, *IEEE Sensors J.*, Vol. 6, No. 5, 1144 (2006).
34. D. Piyabongkarn, R. Rajamani and M. Greminger, "*IEEE Transactions on Control Systems Technology*, Vol. 13, 185 (2005).
35. S. Lee, S. Park, J. Kim, S. Lee and D.-Il Cho, *IEEE/ASME J. of Microelectromechanical Systems*, Vol. 9, No. 4, 557 (2000).
36. M. Saukoski, L. Aaltonen, and K. A. I. Halonen, *IEEE Sensors J.*, Vol. 7, No. 12, 1639 (2007).

OPTICAL SENSORS IN MEDICINE

Francesco Baldini*

Istituto di Fisica Applicata "Nello Carrara", CNR
Via Madonna del Piano 10, I-50019 Sesto Fiorentino, Firenze, Italy
**E-mail: f.baldini@ifac.cnr.it*

Recent years have witnessed remarkable interest in the study of optical sensors applied in medicine, mainly for the detection of chemical and biochemical parameters. Health-care is surely the application field which seems to have the best future development perspectives for optical sensors, not only considering invasive applications (the high degree of miniaturization of optical fiber sensors, their considerable geometrical versatility, and extreme handiness make it possible to perform a continuous monitoring of numerous parameters, thus enabling performances which are often unique) but also taking into account the developments of optical multiarray biochips for the analysis of multiple parameters. The role of optical sensors in the European Integrated projects CLINICIP - Closed Loop Insulin Infusion in Critically Ill Patients - and CARE-MAN - HealthCARE by Biosensor Measurements And Networking - is also described.

1. Introduction

In biomedicine, an important field of research is the one associated with the development of sensors for the detection of physical and chemical parameters in the human body.

Two classes of sensors can be distinguished:

non-invasive sensors, in which the probe remains outside of the human body and is placed at a distance from or in contact with the skin;

invasive sensors, in which the probe must enter the human body through the natural cavities (nostrils, throat, ears).

The general attitude of physicians is to undertake actions which can be easily tolerated by the patient and which introduce a minimum risk for the safety of the patient. From this point of view, non-invasive sensors are definitely preferable to invasive sensors. On the other hand, for several applications, the insertion of optical sensors inside the human body cannot be avoided. In this case, very severe regulations must be satisfied in order to guarantee the safety of patients.

Optics can be an actual solution in many applications. The potential of optical sensors for continuously monitoring physical and chemical parameters is enormous. The absence of electrical contacts or of the passage of electrical current in the transduction mechanism is of paramount importance in an area in which the safety of the patient is a fundamental aspect. In the case of invasive applications, physicians can also take advantage of the fact that optical fibers can guarantee unique performances, thanks to their geometrical versatility, easy handling, and high degree of miniaturization. Optical catheters with a diameter of the order of dozens of micron and probe heads miniaturized up to 1 μm enable physicians to arrive at places inside the human body that would be unthinkable with other sensor technologies.

This aspect makes it clear why optical fiber sensors for medical applications have been characterized by a remarkable development after the introduction of optical fibers. The first invasive optical fiber oximeter which was described in 1964,[1] can be considered one of the first optical fiber sensors. Since then, many sensors have been proposed which find application in different biomedical areas, ranging from cardiovascular and intensive care to angiology, gastroenterology and ophthalmology. Some of these are still at the prototype level, whereas others are already available on the market.

In a survey published at the beginning of the 1990s[2], a list of chemical parameters was given for which physicians were requiring continuous monitoring (Table 1).

Thanks to optics, some of the parameters mentioned above and others that are not included in the table are currently being measured in hospital. Examples of consolidated or widely diffused optical sensors in physician community are:

- optical oximetry, for the detection of oxygen saturation, based on the on the measurement of the absorption fluctuations which are synchronous with systolic heart contractions (pulse oximetry); [3-5]

Clinical problem	Analyte
Diabetes mellitus	Glucose, potassium, ketomnes, insulin, lactate, pH
Vital function monitoring in intensive care/ anaesthetics/ prolonged surgery	Oxygen, carbon dioxide, pH, haemoglobin, potassium, glucose, sodium, osmolality, lactate
Renal failuring/monitoring dyalisis	Urea, creatinine, potassium, arial natriuretic peptide, pH

Table 1. List of chemical parameters for which physicians were requiring continuous monitoring at the beginning of the 1990s.[2]

- sensors for blood oxygen saturation, pH and blood gases (CO_2 and O_2) in cardiovascular and intensive care, to be used in extracorporeal blood circuits[6] or in benchtop blood gas analysis;[7]
- pressure sensors for the measurement of pressure gradients in the heart, in the circulatory system and in other cardiovascular applications;[8-10]
- sensors for blood flow rate in angiology;[11]
- sensor for bile-containing reflux in gastroenterology.[12-13]

These products are in constant evolution and improved systems are continuously developed. Just at the beginning of 2008, FISO introduced on the market a miniaturized version of the optical fiber sensor for in-vivo pressure. The substitution of the silicon diaphragm with smaller and thinner flexible diaphragm in the Fabry-Perot cavity at the end of a multimode optical fiber, allowed to reduce the diameter of the probe from 500 μm down 125 μm. This size reduction makes this sensor the smallest pressure sensor commercially available, resulting perfectly suitable for intracranial, intravascular, and intrauterine pressure monitoring.

In recent review papers, the state of the art of biomedical optical sensors was investigated in detail.[14-16]

Our attention will be focused, therefore, on the recent developments and on the new sensors that have appeared on the market.

2. Bilirubin Detection in Infant Jaundice

Frequent monitoring of the bilirubin content in blood is an important analysis to be carried out in neonates. Bilirubin is a product of heme

catabolism, and is eliminated through the liver. After the birth, the liver needs to undergo through a maturation process, which generally takes from three to five days, before its excretory action is functioning.
Therefore, the level of bilirubin in the blood may be high during the first days of life: part of the bilirubin can be deposited under the tissue, and the baby's skin can assume a characteristic yellowish color.
All this implies that jaundice in a neonate is very common: nearly 60% of babies develop jaundice during the first week of life. The level of jaundice is to be considered overly high if the measured concentration of total serum bilirubin (TSB) is greater than 17 mg/dl (significant hyperbilirubinemia). If this level persists too long, neurological problems can be manifested, ranging from hearing function abnormalities to severe brain damage.[17]
The most diffuse procedure followed in hospitals is the measurement of TSB by means of a blood sample from the heel of the baby and the subsequent determination of TSB by means of a spectrophotometric analysis of the serum sample. It is apparent that frequent blood sampling is painful for the babies, and also constitutes an expensive procedure. Attempts at a non-invasive determination of TSB have been made by measuring the reflectance of the skin.[18]
An instrument based on this approach has been produced (Minolta jaundice Meter device), but it has been shown that in many cases the effect of skin pigmentation can alter the measurement.[19-20] Therefore, the instrument cannot be considered to be completely satisfactory.
A new method is based on a transcutaneous multiwavelength spectral analysis capable of determining the amount of bilirubin in the skin. A portable device is available on the market: BiliCheck (Fig. 1).[21]
The light from a light bulb is sent by an optical fiber to the infant's forehead and back-reflected. The back reflected light is processed by a microspectrophotometer which makes it possible to obtain the reflectance spectra in the 380-780 nm range. A suitable algorithm enables the information on the bilirubin to be extracted, thus avoiding the interferences coming mainly from oxyhemoglobin, deoxyhemobglobin and melanin. Clinical evaluation has been carried out on racially diverse neonates. Besides the very good correlation between TSB and the bilirubin content in the skin, which testifies to the correctness of the

theoretical approach, the clinical study showed the accuracy and reproducibility of the new device, regardless of the different types of skin pigmentation.[22-23]

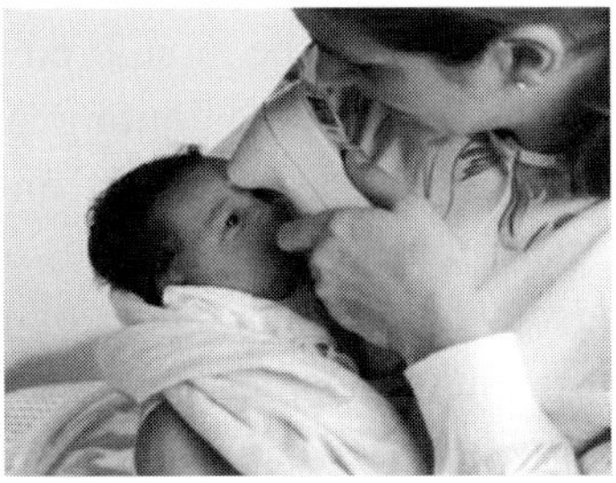

Figure 1. Photo of the Bilicheck sensor applied to the forehead of a baby (Ref. 21).

3. Optical Sensing by Microdialysis

As already mentioned, the general tendency of the physicians is to avoid the use of invasive sensors. On the other hand continuous monitoring of clinical parameters is often unavoidable.

Therefore noticeable efforts are made in order to identify minimally invasive techniques capable to consent continuous measurement. From this point of view, microdialysis seems to meet this need. In microdialysis, a small semipermeable hollow-fiber dialysis catheter is inserted in the tissue and is part of a microfluidic channel, in which a solution flows at very slow flow rate (of the order of μl/min). Within the microdialysis catheter, the equilibrium with the external environment, i.e. extracellular space of tissue, is reached with the diffusion of the analytes present in the interstitial fluid through the dialysis membrane in the flowing solution. The out-coming solution, the perfusate, contains the information related to the analytes within the tissue.[24]

Researchers into traumatic or acute brain injury use microdialysis to learn about how concentrations of ions change in the brain after injury[25] or to provide on-line analysis of brain tissue biochemistry.[26] Other uses include primarily fat and muscle tissue. Recently microdialysis was proposed to collect cytokines at their biological site of action,[27] being cytokines important biomarkers to monitor the effect of drugs influencing the immune system or inflammation.[28]

There are technical and clinical aspects which should be carefully considered in order to make microdialysis technique a reliable technique. First of all, not all the analytes can be monitored with this technique. The cut-off of the dialysis membrane is the limiting factor in the choice of the analytes. Chemical compounds with high molecular weight cannot diffuse through the membrane or their diffusion is very limited. The higher cut-off of the dialysis membrane is of the order of 100 kDal. Clearly the diffusion of the molecules below the cut-off occurs but the diffusion rate is clearly not the same. The recovery rate is an important parameter which gives the ratio between the concentration of the investigated analyte in the perfusate and its real concentration in the tissue. A factor which affects this parameter is the flow rate: the lower the flow rate, the higher the recovery rate, since there is sufficient time for the diffusing material through the membrane to equilibrate with the solution that flows inside the microdialysis catheter. High recovery rates are desirable, since they assure better accuracy in the determination of the analyte, requiring less sensitive sensors. On the other hand, the slower the flow rate the longer the response time, since longer is the required time of the perfusate to reach the sensing point or to be collected in a vial and examined with an external instrumentation.

Microdialysis can also be used to achieve information on the blood concentration of the analytes, becoming an alternative to systematic blood sampling. Surely it guarantees less invasivity of an intravascular approach, and it can become the primary choice in intensive care units where continuous blood loss from diagnostic samples can give rise to critical situations, such as in hospitalized infants or in intensive care patients.[29-30] In this case, a careful analysis and a clinical validation is necessary in order to determine the correspondence between the measured value of the analyte in the tissue and its concentration in the blood, which is often considered by physicians the clinically relevant parameter.

In the first application of microdialysis the fluid coming out form the catheter was collected and analyzed with a bench top instrument.

The necessity of continuous measurements forced the scientist in developing compact, even wearable, instrumentation on line with the microfluidic system (Figure 2).

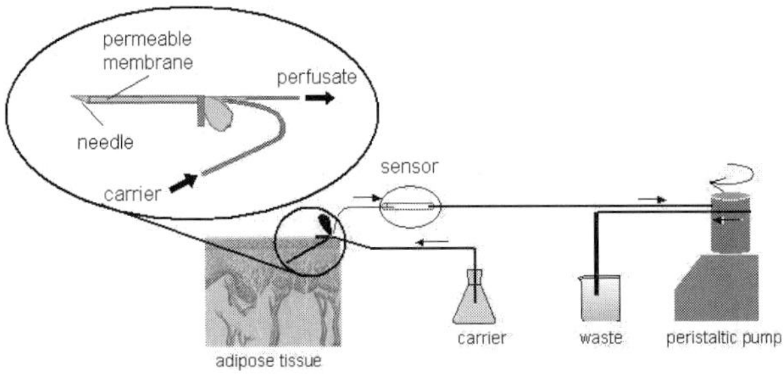

Figure 2. A schematic illustration of the fluidic circuit used in microdialysis for continuous measurement.

Clearly this technique can be coupled with any sensing approach, not necessarily the optical one, but the potential high miniaturization and compactness reachable with optical fibers and the ease of optical transduction of the concentration of the analyte under investigation make the microdialysis-based optical sensors very promising.
In the framework of a recent European project, CLINICIP (Closed Loop INsulin Infusion in Critically Ill Patients, 2004-2007), which had the final objective of developing an intelligent system for improved health status monitoring of critically ill patients, microdialysis was combined with optical sensors for the detection of glucose, pH, partial pressure of carbon dioxide (pCO_2) and partial pressure of oxygen (pO_2).[31]

3.1. Glucose Sensors

World-wide, about 130 million people are believed to suffer from diabetes, a disease which occurs when the body does not adequately produce the insulin needed to maintain a normal circulating blood glucose (80-120 mg/dl). It is estimated that the disease is in rapid expansion (300 million in 2025). Frequent monitoring of blood glucose is crucial for effective treatment and to reduce the morbidity and mortality of diabetes.

Blindness, kidney and heart failure, peripheral neuropathy, pure circulation, gangrene are the severe complications which, over time, are related to diabetes.

It is apparent that monitoring glucose levels in blood is an important element in the treatment of diabetic patients. In intensive care patients, the maintenance of the blood glucose level within normal values, also in non-diabetic patients, has been recently shown an essential issue, since the tight control of glucose levels by intensive insulin infusion improved survival and reduced morbidity of critically ill patients.[32]

The development of a sensor capable of continuously monitoring glucose in diabetic patients is a big challenge which scientists have not still won.[16, 33-35] In connection with many glucose sensors at laboratory level and also with many industrial prototypes, there is no completely reliable sensor available on the market capable to satisfy the strict requirements of a continuous and long-term measurement in diabetic patients.

The enzymatic reaction of glucose with oxygen in the presence of glucose oxidase (GOD):

$$D-glucose+O_2 \xrightarrow{GOD} D-gluconolactone+H_2O_2 \qquad (1)$$

is the base of the development of the first glucose sensors.

In the sixties the first amperometric sensor based on the oxygen consumption was proposed[36] and since then many improvements were carried out. The first glucose optical sensors were based on the measurement of the oxygen consumption[37-38] or on the change in pH caused by the formation of gluconic acid,[39] since D-gluconolactone in (1) is rapidly converted in gluconic acid. The finger prick technique, based on the sampling of a small blood drop from the finger and the subsequent optical analysis by means of colorimetric strips coated with glucose oxidase and peroxidase enzymes, is the present approach followed by diabetic patients at home.

A reagent-less and totally non-invasive optical approach is based on near-infrared spectroscopy.[40]

Light in the near-infrared region (700-2500 nm) illuminates a small area of skin, penetrates through the skin up to a depth of 1-10 μm, and reaches the capillary system below the epidermis. The light scattered back out through the skin is collected, and a near-infrared spectrum is detected. This spectrum contains information on the complex mixture of constituents' tissues and blood, including glucose. It is apparent that special algorithms, based on multivariate mathematical analysis, are necessary in order to extract the information on glucose; it is equally apparent that special calibration is necessary, since the physiological and spectral characteristics of the skin change from subject to subject.

In recent years, subcutaneous adipose tissue has been proposed as a promising site for the continuous measurement of glucose in diabetic patients.[41-43] Within the above mentioned European project CLINICIP, two different sensors were developed to measure continuously glucose levels in interstitial fluid drawn from adipose tissue, one based on the enzymatic reaction with glucose oxidase and one based on direct spectroscopy. In both cases the interstitial fluid is drawn by means of a microdialysis catheter (CMA-60) inserted in the adipose tissue.

In the first sensor, the sensing coating is fixed at the top of a 140 μm optical fiber which is inserted within the microfluidic circuit.[44-45] The sensing coating is a three-layer sensor as shown in Figure 3. On the top of the fiber, there is the oxygen sensitive layer containing a ruthenium complex, the fluorescence of which is quenched by oxygen. The second

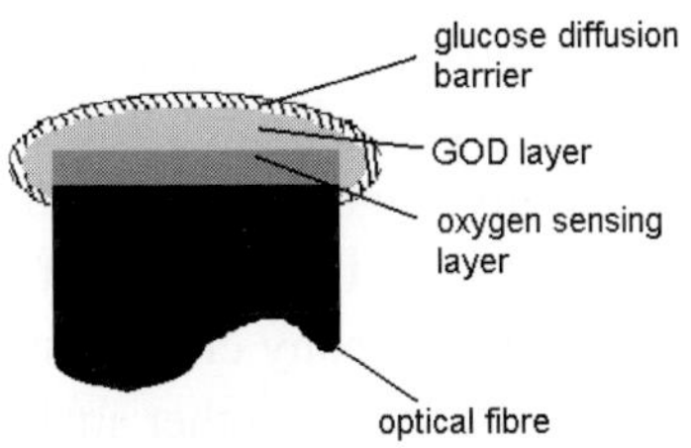

Figure 3. Schematic view of the glucose sensing layer deposited at the tip of the 140 μm optical fiber.

layer is the enzyme layer containing the glucose oxidase and the top layer, formed by a mixture of ethyl cellulose and polyurethane, acts as diffusion barrier for the glucose molecules. An optical fiber oxygen

sensor, formed only by two layers, the oxygen sensor layer and the polymeric protective layer shown in Fig. 3, is used as reference.
The two sensors, thanks to their miniaturized dimensions, can be inserted inside the microfluidic line, just after the microdialysis catheter. Fluorescence life-time is the quantity which is measured with an oxygen meter (Presens GmbH, Regensburg, Germany). The first results obtained on volunteers were very promising, showing the reliability and the selectivity of continuous measurement for the duration of 24 hours.[45]
The second sensor is a reagent-less sensor, which is based on the glucose determination by mid-infrared spectroscopy.[46] In this case a micro-cell is included within the fluidic system and housed inside an infrared mini-spectrometer. The infrared spectrum of the perfusate in the region 1600-900 cm^{-1} is recorded and proper data processing by partial least squares or classical least squares methods allow the glucose quantification. The acquisition of the infrared spectrum allows also the simultaneous measurement of other metabolites, such as urea and lactate. The infrared sensor was tested on volunteers and on intensive care patients with satisfactory results.[47] A vascular body interface was also developed which allows to send a dialyzed sample from diluted whole blood to the measurement micro-cell. The combination of this body interface with the infrared sensor was tested successfully on intensive care patients.[47]
The use of a vascular body interface capable to extract the dialyzed sample from whole blood represents a certain improvement for microdialysis-based measurements. One of the disadvantages of the microdialysis applied to adipose tissue is determined by the fact that it is not always possible to find a clear and reliable correspondence between the measured value of the investigated analyte in the adipose tissue and its concentration in the blood. If this aspect is quite accepted by physicians for glucose, thanks to many clinical studies carried out in the past years, the same can not be said for other analytes.

3.2. pH, pO_2 and pCO_2 Sensors

The demand of very frequent/continuous monitoring of pH, pO_2 and pCO_2, informative indicators of the conditions of a living system, led in the 80's to the development of the first optical fiber sensors for blood pH,[48] oxygen[49] and carbon dioxide.[50] Intravascular catheters for the simultaneous measurement of these three parameters have been described[51] and some of them reached the market in the past years, but presently none is available on the market. Limited stability on long-term and the wall effect, caused by a diffusion gradient of oxygen and carbon dioxide from the blood vessel wall towards its centre are the main drawbacks which undermine the reliability of these sensors.

Therefore, these parameters are generally measured drawing intermittently blood samples and analyzing them in a central laboratory or by a point-of-care blood gas analyzer. On the other hand continuous measurement can be essential in determining the physiologic status of critically ill patients or in monitoring the conditions of operated patients. Optical sensors for pH, pO_2 and pCO_2 have been recently combined with microdialysis applied to adipose tissue within the previously mentioned European Clinicip project.[52-55] The chemical transducers are immobilized inside the internal wall of glass capillaries through which the perfusate is flowing. Absorption changes for pH sensor and modulation of the fluorescence lifetime for pO_2 and pCO_2 are the working principle.

Phenol red covalently bound into the internal wall of a glass capillary by means of the Mannich reaction and platinum(II) tetrakis-pentafluorophenyl-porphyrine entrapped within a polymerized polystyrene layer are the chemical transducers used for pH and oxygen detection, respectively. The ion pair 8-hydroxypyrene-1,3,6-trisulfonic acid trisodium salt/ tetraoctylammonium hydroxide, dissolved in a silicon-based polymeric matrix, is used for the carbon dioxide detection.

If, in the case of glucose the correspondence between the interstitial and blood glucose is widely accepted, the correlation in the case of pH, pO_2 and pCO_2 has to be demonstrated.

Therefore, a clinical validation on animals was carried out to prove the capability of the optical sensor to measure correctly the three parameters in the adipose tissue and to verify the correspondence between the blood values and interstitial values.

Figure 4 shows the pH, pO_2 and pCO_2 monitoring carried on a pig in the presence of 50% blood drawing and blood re-infusion after one hour. The blood drawing induces a decrease of the blood supply at the level of adipose tissue, which implies that the cell metabolism becomes critical. This is demonstrated by the decrease of both pH and oxygen and by the increase of carbon dioxide. In the correspondence of the re-infusion of the blood, the pH and the oxygen start immediately to increase and the carbon dioxide decreases, indicating a return to a healthier situation. The clinical tests showed the feasibility of the continuous measurements of pH, pO_2 and pCO_2 in adipose tissue by means of a microdialysis approach. Moreover, the results obtained with the animal model shows that the adipose tissue can be a valid alternative site for in-vivo continuous monitoring of stress conditions.

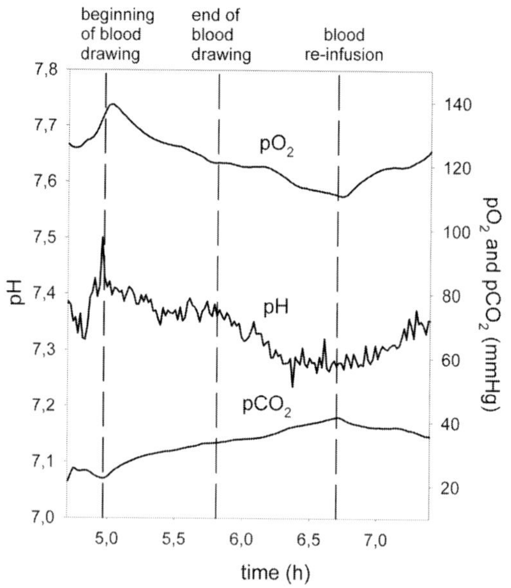

Figure 4. Clinical test on animal: response curves for pH, pO_2 and pCO_2 in the presence of 50% of blood drawing and subsequent blood re-infusion. The beginning and the end of blood drawing and the blood re-infusion are indicated by the vertical lines.

4. Point of Care Testing

In recent years, the request by physicians of compact devices capable to measure bioanalytes at the bedside of the patient has become more and more persistent and, nowadays, point of care testing (POCT), defined as diagnostic testing at or near the site of patient, is a very important application area in biomedicine.[56-57]

This demand is dictated by the need of the physicians to have a fast and reliable response at their disposal, avoiding to delivery of the samples to the central laboratories and to wait a period of time, generally several hours long, to achieve the results of the analysis.

Some portable instrumentations are already available on the market such as the devices from Roche[58] (Cardiac Reader and TROP T) or from Biosite59 (Triage Meters) for the measurement of cardiac markers (troponin, myoglobin, D-dimer, BNP, CK-MB) in fifteen-twenty minutes. It is apparent that in the case of possible infarction, saving time, also a few minutes, in making the correct diagnosis can result crucial for the patient survival. But the timeliness in a correct diagnosis can be essential in many other cases, such as the discrimination of viral and bacterial sepsis in intensive care patient, or the fast identification of the origin of infections.

In any case, from a clinical point of view, the analysis of more than one analyte is essential, but the devices currently available on the market for POCT applications are generally able to measure the different analytes one by one sequentially, and are unable to carry out simultaneous multiassays.

A recently approved European project, CARE-MAN (HealthCARE by Biosensor Measurements And Networking),[60] brings together twenty-six European partners, with the final objective of developing compact and fully automated devices capable of multianalyte detection in cardiovascular disease, coagulation disorders, chronic/acute inflammation, cancer and thyroid disorders. Besides the development of specific biological recognition elements for the detection of the analytes established by the physicians (eg., C-reactive protein,[61] interleukines, neopterin, mieloperoxidase, etc.), optical platforms are under development with the capability to interrogate in-house biochips on which the assays are performed.

Fluorescence[62-63] or chemiluminescence[64] are the optical transduction principles which are used, with the proper labeling of the biological recognition elements.

Figure 5 shows the sketch of one of the fluorescence-based optical platform for the interrogation of a multichannel plastic biochip. The core of this platform is a miniaturized polymethylmetacrylate (PMMA) chip, which consists of two self-made pieces of PMMA opportunely shaped in order to obtain flow channels, 500 μm in width and 400 μm in height.[62] Thanks to the fluorescence anisotropy exhibited by any dipole emitting at a distance from a medium interface of the order of the emitted wavelength,[65] a large fraction of the fluorescence emitted by the sensing layer immobilized on the bottom of the PMMA cover travels along the thickness of the PMMA cover itself up to its end-face where it is collected by a plastic optical fiber connected an optical spectrum analyzer. Figure 5 shows a longitudinal section of the chip with the transversal excitation of the sensing layer with a laser diode and the collection of the emitted fluorescence by means of a 1 mm plastic fiber.

The same chip contains four identical microchannels for the simultaneous detection of more than one analyte.

The potentiality of the optical platform was investigated by a sandwich assay for the C-reactive protein (CRP). C5-clone and DY647-labeled C7-clone were used as capture antibody, immobilized on the

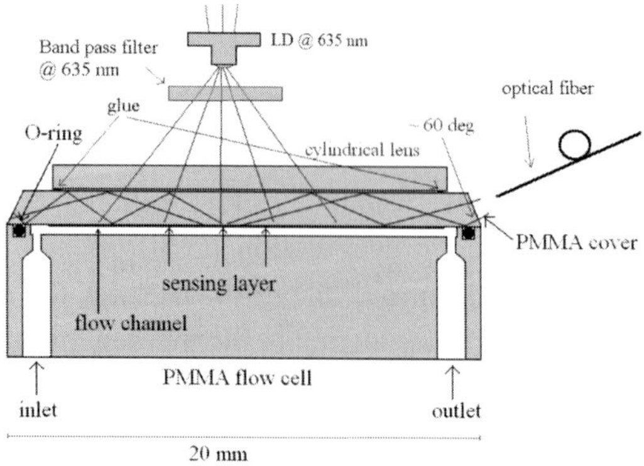

Figure 5. Longitudinal section of the PMMA biochip for multianalyte detection.

PMMA surface, and target antibody, respectively. Different concentrations of CRP (1 ng/mL - 10 μg/mL) were tested by following the developed protocol, with 30 min of incubation time of CRP, 3 min for washing, 30 min of incubation time of 0.1 μg/mL labeled C7-clone and 3 min for washing. The used buffer in all the solutions was HEPES 10 mM, $CaCl_2$ 2 mM, Tween 20 0.005%, pH 6.5. Figure 6 shows the fluorescence emission spectra coming from the PMMA chip for different concentration of CRP and the related calibration curve.

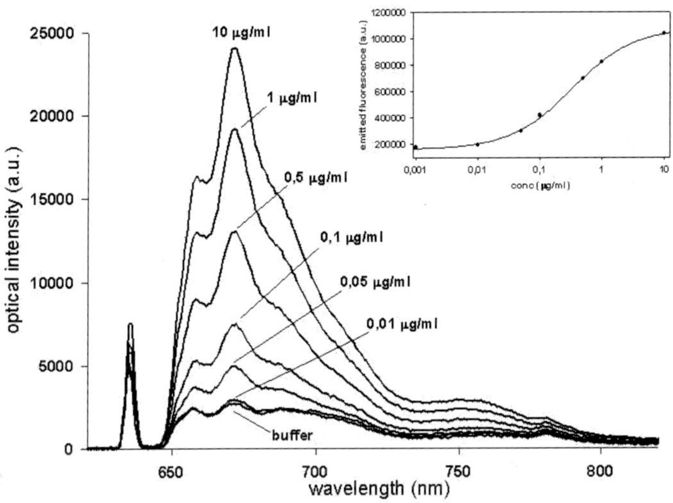

Figure 6. Fluorescence emission spectra coming from the PMMA chip for the sandwich assay for different concentration of CRP. The calibration curve is also shown on the top right, evaluated considering the integral between 650 and 750 nm of each spectrum.

5. Conclusions and Outlook

Optical sensors are in continuous development and are offering to the physicians reliable and efficient tools for the formulation of correct diagnoses and the control of the administrated therapies. The several instruments already available on the market and the more numerous systems developed at an advanced stage in the labs testifies that optical sensors are one of the primary choice for scientists and for physicians when an in-vivo continuous monitoring should be carried out.

In microdialysis, the development of a new intravascular body interface (MicroEye®, Probe Scientific),[66] available on the market since the beginning of 2008, opens new perspectives in the field of microdialysis-based sensing. If the adipose tissue appears the only one practicable site for biological uptake in the case of patients out of the hospital due to safety reason, the intravascular body interface appears the logic one in the intensive care unit, where one or more blood accesses are always available for different purposes.

Optics is also playing a fundamental role in the development of POCT instrumentation, compact and portable close to the patient bedside. In many clinical applications the measurement of a limited number of parameters (5-15) is sufficient to help physicians in the diagnosis of pathologies or in the choice of the appropriate therapy and the present effort is devoted to the design and realization of devices capable to perform multi-analyte assays.

References

1. N. S. Kapany and N. Silbestrust, *Nature*, **208**, 138 (1964).
2. J. C. Pickup and S. Alcock, *Biosens. Bioelectr.*, **6**, 639 (1991).
3. Y. Mendelson and B. D. Ochs, *IEEE Trans. Biomed. Eng.*, **BME-35**, 798 (1988).
4. H. Ugnell and P. A. Oberg, in *Medical Sensors II and Fiber Optic* Sensors, SPIE Conference Proceedings Vol. 2331 (Society of Photo-Optical Instrumentation Engineers, Bellingham, 1994), p. 89.
5. Mallinckrodt Corporate Communications, Corporate Headquarters, 675 McDonnell Blvd., Hazelwood, 63042 MO, USA, http://www.mallinckrodt.com.
6. Terumo Cardiovascular Systems, 6200 Jackson Road, Ann Arbor, 48103-9300 Michigan, http://www.terumo-us.com.
7. Roche Holding Ltd, Grezacherstrasse 124, CH-4070 Basel Switzerland, http://www.roche.com.
8. Y. Lin, T. Sawatari, C. J. Hartley, in *Advanced Characterization, Therapeutics and Systems III*, SPIE Conference Proceedings Vol. 3245 (Society of Photo-Optical Instrumentation Engineers, Bellingham, 1998), p. 51.
9. Samba Sensors AB, Första Långgatan 26, 413, 28, Göteborg, Sweden, http://www.samba.se.

10. FISO Technologies, 2014 Jean-Talon Nord, suite 125, Sainte-Foy (Quebec), G1N-4N6, CANADA, http://www.fiso.com.
11. Perimed AB, Box 5607, S-11486 Stockholm, Sweden, http://www.Perimed.se (Perimed Literature Reference List n. 14, 1998).
12. F. Baldini, in *Fiber Optic Sensors Technology and Applications*, SPIE Conference Proceedings Vol. 3860 (Society of Photo-Optical Instrumentation Engineers, Bellingham, 2000), p. 144.
13. Medtronic, Medtronic World Headquarters, 710 Medtronic Parkway, Minneapolis, MN 55432-5604, http://www.medtronic.co.uk.
14. R. B. Thompson, in *Optical Fiber Sensor Technology Vol. 4*, Eds. K. T. V. Grattan and B. T. (Kluwer Academic Publishers, UK, 1998), p. 67.
15. F. Baldini and A. G. Mignani, in *Handbook of Optical Fiber Sensing Technology*, Ed. J. M. Lopez-Higuera J.M. (John Wiley & Sons, New York, 2002), p. 705.
16. F. Baldini, in *Optical Chemical Sensors*, Eds. F. Baldini, J. Homola, S. Martellucci and A. Chester, NATO ASI Ser. II Math., Phys. & Chem., Vol. 224 (Springer, the Netherlands, 2006), p. 417.
17. G. R. Gourley, *Adv. Pediatr.*, **44**, 173 (1997).
18. D. W. Smith, D. Inguillo and D. Martin, *Pediatrics*, **75**, 278 (1985).
19. R. Grande, E. Gutierrez, E. Latorre and F. Arguelles, *Hum. Biol.*, **66**, 495 (1994).
20. A. Knudsen, *Acta Paediatr.*, **85**, 393 (1996).
21. Respironics, Inc., 1010 Murry Ridge Lane, Murrysville, PA 15668-8525, http://www.respironics.com.
22. V. K. Bhutani, G. R. Gourley, S. Adler, B. Kreamer, C. Dalin and L. H. Johnson, *Pediatrics*, **106**, e17 (2000).
23. F. F. Rubaltelli, G. R. Gourley, N. Loskamp, N. Modi, M. Roth-Kleiner, A. Sender and P. Vert, *Pediatrics*, **107**, 1264 (2001).
24. M. Muller, *Br. Med. J.*, **324**, 588 (2002).
25. D. A. Richards, C. M. Tolias, S. Sgouros and N. G. Bowery, *Pharmacol. Res.*, **48**, 101 (2003).
26. M. M. Tisdall and M. Smith, *Br. J. Anaesth.*, **97**, 18 (2005).
27. A. Xiaoping and J. A. Stenken, *Methods*, **38**, 331 (2006).
28. T. L. Whiteside, *Clin. Diagn. Lab. Immunol.,* **1**, 257 (1994).
29. A. M. Baumeister, B. Rolinski, R. Busch and P. Emmrich, *Pediatrics*, **108**, 1187 (2001).
30. S. Klaus, M. Heringlake and L. Bahlmann, *Crit. Care*, **8**, 363 (2004).
31. www.clinicip.org.
32. G. Van den Berghe, P. Wouters, F. Weekers, C. Verwaest, F. Bruyninckx, M. Schetz, D. Vlasselaers, P. Ferdinande, P. Lauwers and R. Bouillon, *N. Engl. J. Med.*, **345**, 1359 (2001).
33. E. Wilkins and P. Atamasov, *Med. Eng. Phys.*, **18**, 273 (1996).
34. D. C. Klonoff, *Diab. Care*, **20**, 433 (1997).
35. V. R. Kondepati and H. M. Heise, *Anal. Bioanal. Chem.*, **388**, 545 (2007).

36. L. C. Clark, *Ann. NY Acad. Sci.*, **102**, 29 (1962).
37. N. Uwira, N. Opitz and D. W. Lubbers, *Adv. Exp. Med. Biol.*, **169**, 915 (1984).
38. W. Trettnak, M. J. P. Leiner and O. S. Wolfbeis, *Analyst*, **113**, 1519 (1988).
39. W. Trettnak, M. J. P. Leiner and O. S. Wolfbeis, *Biosensors*, **4**, 15 (1988).
40. H. M. Heise, *Horm. Met. Research*, **28**, 527 (1996).
41. F. J. Service, P. C. O'Brien, S. D. Wise, S. Ness and S.M. LeBlanc, *Diab. Care*, **20**, 1426 (1997).
42. J. P. Bantle and W. Thornas, *J. Lab. Clin. Med.*, **130**, 436 (1997).
43. M. Ellmerer, M. Haluzik, J. Blaha, J. Kremen, S. Svacina, W. Toller, J. Mader, L. Schaupp, J. Plank and T. R. Pieber, *Diab. Care*, **29**, 1275 (2006).
44. A. Pasic, H. Koehler , L. Schaupp, T. R. Pieber and I. Klimant, *Anal. Bioanal. Chem.*, **386**, 1293 (2006).
45. A. Pasic, H. Koehler , I. Klimant and L. Schaupp, *Sens. Actuat.* **B**, **122**, 60 (2007).
46. H. M. Heise, U. Damm, M. Bodenlenz, V.R. Kondepati, G. Köhler and M. Ellmerer, *J. Biomed. Opt.*, **12**, 024004-1 (2007).
47. H. Heise, V. Kondepati, U. Damm, M. Licht, F. Feichtner, J. Mader and M. Ellmerer, *Optical Diagnostics and Sensing VIII*, SPIE Conference Proceedings Vol. 6863 (Society of Photo-Optical Instrumentation Engineers, Bellingham, 2008), p. 686308-1.
48. J. I. Peterson, S. R. Goldstein and R. V. Fitzgerald, *Anal. Chem.*, **52**, 864 (1980).
49. J. I. Peterson, R. V. Fitzgerald and D. K. Buckhold, *Anal. Chem.*, **56**, 62 (1984).
50. G. G. Vurek, P. J. Feustel and J. W. Severinghaus, *Ann. Biomed. Eng.*, **11**, 499 (1983).
51. J. L. Gehrich, D. W. Lubbers, N. Opitz, D. R. Hansmann, W. W. Miller, J. K. Tusa and M. Yafuso, *IEEE Trans. BME*, **2**, 117 (1986).
52. A.Bizzarri, H. Koehler, M. Cajlakovic, A. Pasic, L. Schaupp, I. Klimant and V. Ribitsch, *Anal. Chim. Acta*, **573-574**, 48 (2006).
53. A.Bizzarri, M. Cajlakovic and V. Ribitsch, *Anal. Chim. Acta*, **573–574**, 57 (2006).
54. F. Baldini, A. Giannetti and A. A. Mencaglia, *J. Biom. Opt.*, **12**, 24024 (2007).
55. F. Baldini, A. Bizzarri, M. Cajlakovic, F. Feichtner, L. Gianesello, A. Giannetti, G. Gori, C. Konrad, A. A. Mencaglia, E. Mori, V. Pavoni, A. M. Perna and C. Trono, *Optical Sensors*, SPIE Conference Proceedings Vol. 6585, (Society of Photo-Optical Instrumentation Engineers, Bellingham, 2007), p. 658510.
56. C. P. Price, *Br. Med. J.*; **322**; 1285 (2001).
57. P. St-Louis, *Clin. Biochem.*, **33**, 427 (2000).
58. F. Hoffmann-La Roche Ltd, Grenzacherstrasse 124, CH-4070 Basel, Switzerland, http://www.poc.roche.com.
59. Biosite Incorporated, 9975 Summers Ridge Road, San Diego, CA 92121, USA http://www.biosite.com.
60. http://www.care-man.eu.
61. A. Bini, S. Centi, S. Tombelli, M. Minunni, M. Mascini, *Anal. Bioanal. Chem.*, **390**, 1077 (2008).

62. C. Albrecht, N. Kaeppel and G. Gauglitz, *Anal. Bioanal. Chem.*, in press.
63. F. Baldini, A. Carloni, A. Giannetti, G. Porro and C. Trono, *Anal. Bioanal. Chem.*, **391**, 1837 (2008).
64. B. P. Corgier, F. Li, L. J. Blum and C. A. Marquette, *Langmuir*, **23**, 8619 (2007).
65. L. Polerecky, J. Hamrle and B. D. MacCraith, *Appl Opt.*, **39**, 3968 (2000).
66. Probe Scientific Ltd, Bedford i-lab, Stannard Way, Priory Business Park, Bedford, Bedfordshire, MK44 3RZ, United Kingdom, http://www.probescientific.com.

ENVIRONMENTAL AND ATMOSPHERIC MONITORING BY LIDAR SYSTEMS

Antonio Palucci*

ENEA Department of Advanced Physics Technology and New Materials, Laser Applications Section
Via Fermi 45, 00044 Frascati, Italy
**E-mail: palucci@frascati.enea.it*

A scenario of capabilities of laser systems devoted to environmental monitoring control is given with particular emphasis to atmospheric and marine surveillance applications. Theoretical background of the different laser based techniques has been reported together with their limitations. Applications developed at the ENEA Laser Remote Sensing laboratory have been described and compared with complementary measurements obtained from passive remote information. Such systems are far from being considered mature and new perspectives are expected in the future with the application of new laser sources.

1. Introduction

Lidar is the acronym of "Light Detection And Ranging" or commonly referred as "Laser Radar" for the similarity of the radar principles applied in the optical domain. This laser remote sensing technique takes the advantages from the implementation of laser sources characteristics, in terms of powerful, low divergent and monochromatic optical pulses, that made it practical for environmental monitoring.

The Italian lidar community, developed in the eighties of last century, is proud to have a noble father as prof. Giorgio Fiocco, who recorded the first lidar echo reflected from the lunar surface[1], opening the route to a new physical branch of laser activity.

Main centers of excellence in these fields grew in the Milan area (Prof. O. Svelto,[2] CISE,[3] JRC-Ispra[4]), Florence (CNR[5,6]), Rome (Prof.

Fiocco & coworkers, CNR,[7] ENEA[8,9]), L'Aquila (Prof. G. Visconti[10]), Naples and Potenza (Prof. V. Cuomo[11] and N. Spinelli[12]), Lecce (and Prof. M. R. Perrone[13]), Bari (Centro Laser[14]), Cosenza (Prof. C. Bellecci[15]) and Cagliari (Prof. R. Habel).[16]

From the first application, the lidar systems became more and more reliable tools for identification, 3D mapping and evaluation of natural or pollutant components either in atmosphere or in water surveillance and therefore applied in ecological and global changes studies. The methodology has been fruitfully applied to follow tropospheric and stratospheric dynamic evolution due to the high sensitivity (ppb-range) over several kilometers detection range.

Lidar technique offers the advantage to obtain information on the target investigated without physical contact with the object. To the detection of constituents, spectroscopy methods assume high relevance, i.e., frequency and wavelength information are needed to identify target species. Suitable laser frequency emissions are required to perform remote detection of such substances.

Furthermore, lidar technique offers a prompt response, real time analysis, large area scanning, relevant to monitor atmospheric or marine ecosystems.

For marine surveillance, the Laser Induced Fluorescence (LIF) technique can be very effective in significant measurements of the bio-optical parameters in natural waters due to its exceptional sensitivity and low mass detection limits.[17] Qualitative (flow visualization) and analytical applications include measurements of gas-phase concentrations in the atmosphere,[18] plant health status,[19] in monitoring water bodies[20] and crude oil pollutants releases,[21] in experimental fluid mechanics to measure the concentration of a scalar species within a fluid,[22] for fuel visualization in engine environments,[23] in determining the density of a certain atomic level in the plasma directly from the absorption coefficient.[24]

Nowadays, their implementation is moving in stand-off application, security area of interest, for remote sensing of hazardous components released from terroristic attacks in large or in a confined areas. This request arise from the need to enhance the citizens safety & security level in particular form the September 11th 2001 events.

2. Basic Lidar Architecture

A lidar is essentially composed of a transmitter (laser and beam shaping optics) and a receiver (telescope and signal detection electronics). Its principle of operation is illustrated in Figure 1: the target at the distance R from the system sends back part of the laser pulse toward the telescope surface.

A transmitter is utilized to provide high density number of photons (laser pulses) that meet certain requirements depending on application needs (e.g., wavelength, frequency accuracy, bandwidth, pulse duration time, pulse energy, repetition rate, divergence angle, etc). Usually, the transmitter includes lasers, collimating optics, diagnostic equipment, and wavelength control system.

A receiver is demanded to collect and detect returned photon signals while compressing background noise. Usually, it consists of telescopes, filters, collimating optics, photon detectors, discriminators, etc. The bandwidth of the filters determines whether the receiver can spectrally distinguish the returned photons.

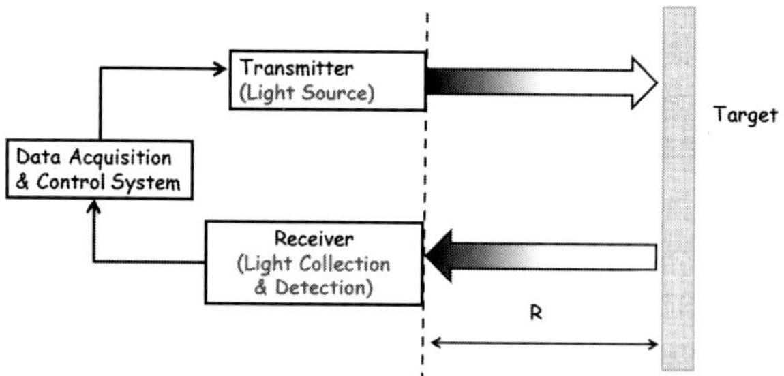

Figure 1. Basic Lidar architecture.

Data acquisition and control system are employed to record returned data and corresponding time-of flight, provides system control and coordination to transmitter and receiver. Usually, it consists of fast transient digitizer, discriminator, computer and software. This part has become more and more important in modern lidars.

The adopted configuration of the transmitter vs receiver can affect the final performances of a lidar apparatus (Fig. 2). Bistatic configuration involves a considerable separation of the transmitter and receiver to achieve spatial resolution in optical probing study. Monostatic configuration has the transmitter and receiver locating at the same site, so that in effect one has a single-ended system. The precise determination of range is enabled by the nanosecond pulsed lasers.

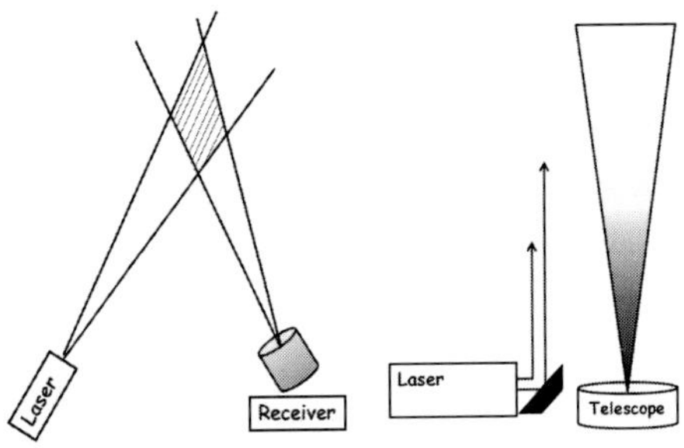

Figure 2. Bistatic (left) and monostatic (right) Lidar configurations.

A monostatic lidar can have either coaxial or biaxial arrangement. In a coaxial system, the axis of the laser beam is coincident with the axis of the receiver optics. This is currently the most frequent configuration adopted in short-mid range probing lidar apparatus.

In the biaxial arrangement, the laser beam only enters the field of view of the receiver optics beyond some predetermined range. Biaxial arrangement helps avoiding near-field backscattered radiation saturating photo-detector, therefore is suitable for mid-long range monitoring. The near-field backscattering problem, in a coaxial system, can be overcome by either gating of the photo-detector or use of a fast shutter or chopper.

In case of atmospheric monitoring, the resolution of lidar measurement is directly related to the laser pulse length, and can be determined by the time-of-flight through equation $R = C \cdot t/2$, where C is the light speed in the medium, t is the time-of-flight, and 2 for the round-trip of the photons traveled (Fig. 3) The ultimate resolution of range determination is limited by the pulse duration time. For example, a 10 ns pulse gives 150 cm as the highest resolution for an atmospheric lidar.

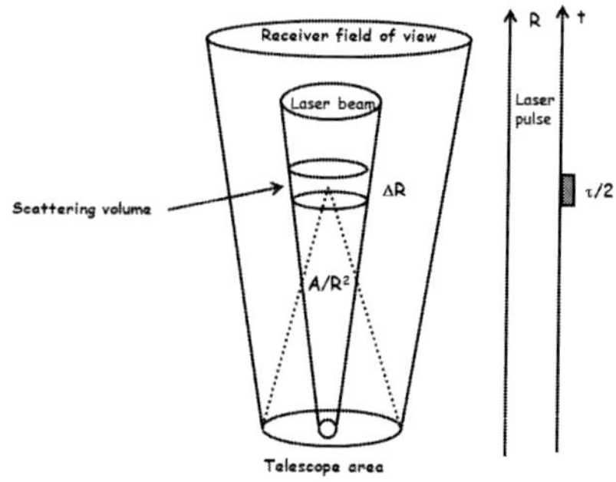

Figure 3. Schematic view of the receiver and illuminated laser scattering volume with indication of the lidar range resolution.

3. Basic Lidar Principle

The lidar equation is the basic equation in the field of lidar remote sensing, which relates the received photon number (power) coming from a scattering region or object to the emitted laser photon number (power), the concentration of the scatterer, the interaction between the light radiation and the scatterer, and the lidar system efficiency.

The lidar equation is developed under two assumptions: the scattering processes are independent, and only single scattering occurs. Independent scattering means that particles are separated adequately and undergo random motion so that the contribution to the total scattered energy by many particles has no phase relation. Thus, the total intensity is simply a sum of the intensity scattered from each particle. Single scattering implies that a photon is scattered only once. Multiple scatters are not considered.

Therefore, in case of a laser pulse sent in the atmosphere, the received power is given by:[25]

$$P = P_0 \frac{A\,\eta_t\,\eta_r\,\rho}{\pi\,R^2} \exp(-2\,\alpha\,R) \qquad (1)$$

where P_0 is the transmitted power, A is the receiver area, η_t and η_r are the transmitter and receiver efficiency, respectively, ρ is the target reflectivity and α is the laser beam extinction coefficient of the involved air volume. In this case, α is regarded as constant simply because R is small. Lidar community prefers to adopt the range corrected signal (RCS) behavior, described as

$$\mathrm{RCS} = \ln\,[P(R)R^2] \tag{2}$$

to display backscattered signal and observe differences in atmospheric propagation.

4. Atmospheric Lidar

When a laser beam is sent into the atmosphere, it interacts with suspended aerosol particles, molecules and atoms present in the air. This scattering is essentially caused by N_2 and O_2 molecules (Rayleigh and Raman scattering) and by the suspended aerosol particles (Mie scattering) present in the atmosphere as dust, water droplets, black carbon, etc.

The relation linking the photon number received, after pulse laser sounding the atmosphere, at the thickness $c\tau_d/2$, is given by:

$$n(R,\lambda) = n_0(\lambda)\eta\beta(R,\lambda)\frac{A}{R^2}\frac{c\tau_d}{2}\exp(-2\int_0^R \alpha(R',\lambda)dR') \tag{3}$$

where n_0 is number of transmitted photons at laser wavelength λ□ and η the overall optical efficiency (mirrors, lens, filters, detectors, etc).

The second term describes the angular scattering probability that a transmitted photon is backscattered by scatters into a unit angle. The volume backscatter coefficient β is the probability per unit distance travel that a photon is scattered into wavelength λ in unit solid angle at angle $\theta = \pi$. The third term is the probability that a scatter photon is collected by the receiving telescope, i.e., the solid angle subtended by the receiver aperture to the scatterer (see Fig. 3); the last term describes the attenuation due to atmospheric transmittance (Fig. 4) and scattering processes.

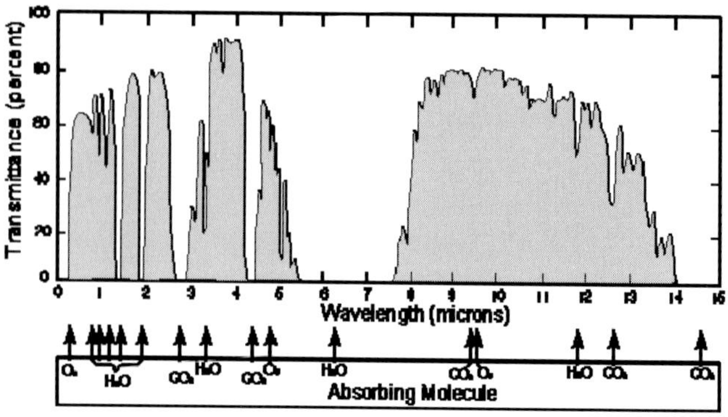

Figure 4. Atmospheric transmittance from ultraviolet (UV) to the mid infrared (MIR) spectral range.

4.1. Differential Absorption Lidar (DIAL)

The remote sensing determination of gaseous compounds dispersed in the atmosphere is an appealing application of the lidar technique. The DIAL (Differential Absorption Lidar) method is based on the use of a pair of wavelengths close to each other, with a large absorption coefficient difference (denoted λ_{on} and λ_{off}, for on-resonance and off-resonance wavelength, respectively). Such a pair of wavelengths, chosen for the detection of a specific pollutant, is sent into the atmosphere and backscattered signals at both wavelengths are compared. If the pollutant is present in the air at a certain location, it will produce a decrease of signal on the λ_{on}-channel but not on the λ_{off}-one. The ratio between the two collected signals is related to the concentration of the investigated gaseous natural or pollutant compound. By applying equation (3) for the DIAL case, we obtain:

$$C(R) = \frac{1}{2\Delta\sigma}\left\{\frac{d}{dR}\ln\left[\frac{n(R,\lambda_{OFF})}{n(R,\lambda_{ON})}\right] - \frac{d}{dR}\ln\left[\frac{\beta(R,\lambda_{OFF})}{\beta(R,\lambda_{ON})}\right] - 2\Delta\alpha'(R)\right\} \quad (4)$$

where C and σ are, respectively, the concentration and the absorption cross-section of the gaseous molecule and

$$\Delta\sigma = \sigma(\lambda_{ON}) - \sigma(\lambda_{OFF}) \quad (5)$$

$$\Delta\alpha'(R) = \alpha'(R,\lambda_{ON}) - \alpha'(R,\lambda_{OFF}) \quad (6)$$

$$\alpha'(R,\lambda) = \alpha(R,\lambda) - C(R)\sigma(\lambda) \tag{7}$$

Equation (4) is valid only in case that λ_{on} and λ_{off} are spectrally very close, in order to avoid changes in the scattering contribution.

Thus, in the lidar equation for DIAL, the presence of molecular species is in the extinction (atmosphere transmission) part, not in the backscatter part. In other words, the molecular absorption contributes to the extinction of light when incident light and scattered light propagate through atmosphere, while the return signals are from the scattering of laser light by air molecules and aerosols.

Therefore if λ_{on} and λ_{off} are properly selected, equation (4) can be simpler expressed as:

$$C(R) = \frac{1}{2[\sigma(\lambda_{ON}) - \sigma(\lambda_{OFF})]}\left\{\frac{d}{dR}\ln\left[\frac{n(R,\lambda_{OFF})}{n(R,\lambda_{ON})}\right]\right\} \tag{8}$$

In Fig. 5, the lidar/DIAL approach is pictorially described, with example of the two laser backscattering echoes at the two on and off wavelengths, their logarithmic ratio and the retrieved range resolved concentration.

In most case, the results from equation (8) are released in molecule/cm^3, while passing to relative concentrations in ppb the following relation can be applied:

$$C[ppb] = \frac{C[molecule/cm^3]}{2.46\times 10^{10}} \tag{9}$$

In lidar/DIAL technique, the minimum range resolved concentration is complex task to be accomplished[25] and it is related to the differential absorption and to the signal to noise ratio (SNR) of the measurement, as:

$$C_{\min} = \frac{1}{2\Delta R\Delta\sigma}\ln\left(1+\frac{1}{SNR}\right) \tag{10}$$

In case of short ranges, the major contribution to equation (10) is given to the statistical errors, while for long ranges, the noise power of the detector dominates the final determination.

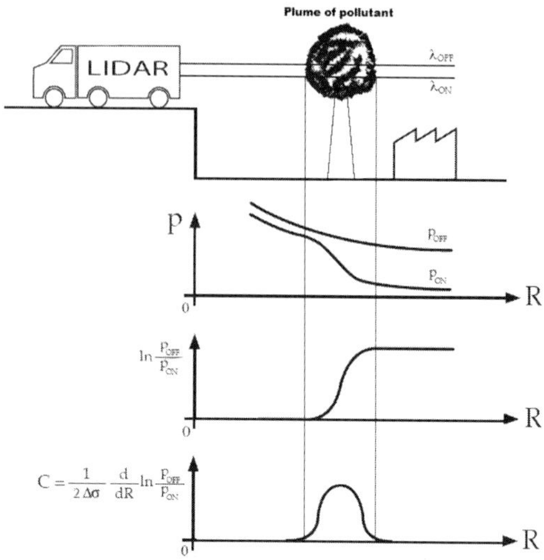

Figure 5. Lidar/DIAL schematic view.

5. Lidar Fluorosensor Technique

LIF is the optical emission from atoms or molecules that have been excited to higher energy levels by absorption of electromagnetic radiation. The main advantage of fluorescence detection compared to absorption measurements is the greater sensitivity achievable because the fluorescence signal has a very low background. In case of resonant excited molecules, LIF provides selective excitation of the analyte to avoid interferences. LIF can be profitable applied to study the electronic structure of molecules and to perform quantitative measurements of analyte concentrations.

LIF technique takes the advantage to employ electro-optics components and therefore to gain the advantages of their peculiarities. These skills allow to settling-up diagnostics tools to be local/remote, real time, to operate with low radiation exposure so that the sampled object is not disturbed or damaged at all. In detail LIF offers the advantage to be:

- Fast (the detection of a substance can be performed in a fraction of a second);

- Remote (the system and the target can be meters far apart);
- Sensitive (better than parts-per-million);
- Specific (substances can be recognized by their spectroscopic fingerprint);
- User-friendly (the system can be deployed in few minutes and does not require a specifically trained user).

From the above considerations, it's straightforward to include LIF in the laser application techniques for environmental monitoring, commonly mentioned as Lidar Fluorosensor. Different platforms were employed to install the monitoring apparatus, ranging from fixed installations,[26] mobile platforms,[27,28] airborne,[29-31] ship vessels[32-34] and submarine payloads,[35,36] both for plant or water remote sensing. We will restrict our interest to marine application of this technique.

5.1. Fluorescence Optical Emission

As in the general schematic representation of a lidar apparatus, a fluorosensor apparatus employs a laser source to transmit high density monochromatic photons towards the target and the backscattered radiation is collected and analyzed by an appropriate optical system.[37]

While in the elastic scattering the main contribution can be ascribed to Raleigh and Mie scattering, inelastic scattering occurs when a light photon is absorbed and re-emitted in a different wavelength range from the excitation. To this respect, Raman and Brillouin scattering, as well as fluorescence by specific chromophores are forms of inelastic scattering. A chromophore is part (or moiety) of a molecule responsible for its color and generally embedded in dissolved or suspended molecules as CDOM (Chromophoric Dissolved Organic Matter) or phytoplankton pigments (chlophyll-a, carotenoids, phycocyanine, phycoerythrin).

Combined measurements of elastic and inelastic scattering, as well as fluorescence and reflection, may allow simultaneous definition of location of the target, its size and chemical composition, as in laser cytometry application of phytoplankton analysis.[38]

In case of lidar fluorosensor, main contributions result from laser backscattered radiation, in case of specular surface reflection (i.e. water), and from frequency-shifted emissions when the target materials are

determined by vibrational and rotational transitions in the molecule. In case of water, 3400 cm^{-1} is the frequency shift for the OH stretch vibrational mode.[39]

Moreover, in order to evaluate concentrations of different chromophores dispersed in water, LIF signals can be calibrated against the concurrent water Raman signal, regarded as an internal standard reference.[40] By rationing the chromophore fluorescence signal F to the water Raman intensity R, we have:[20]

$$\frac{F}{R} = \frac{\sigma_F}{\sigma_R} \frac{k_{ex} + k_R}{k_{ex} + k_F} \tag{12}$$

where the indexes are self explaining and the dependence on system parameters and on the refraction index of water has disappeared.

The ratio of extinction coefficients in Eq. 11 can approximately be regarded as a constant and thus neglected provided a careful choice of excitation and emission wavelengths is performed, in order to avoid errors due to differential absorption. In conclusion, the different chromophore concentrations, expressed in Raman units, are seen to be independent of system parameters. This procedure is usually followed by proper calibration with the matter investigated.

In general, the experimental system is located above the sea water (Fig. 8), at a range R_w from its surface, and the laser beam, after propagation in air, probes a water layer characterized by the extinction coefficient α_w.

The choice of the laser excitation wavelength and the operational height of the lidar fluorosensor apparatus is a trade-off between the extinction coefficients, in air and water, and the excitation efficiency. Even if many of the substances that are present in the waters are more efficiently excited in the UV wavelength range, a lower limit can be observed downwards 300 nm due to the strong atmospheric ozone absorption. For this reason, and also for eye-safety considerations, the most common excitation wavelength lying in the near UV, is an excimer laser @ 308 nm or Nd:YAG III harmonic @ 355 nm.

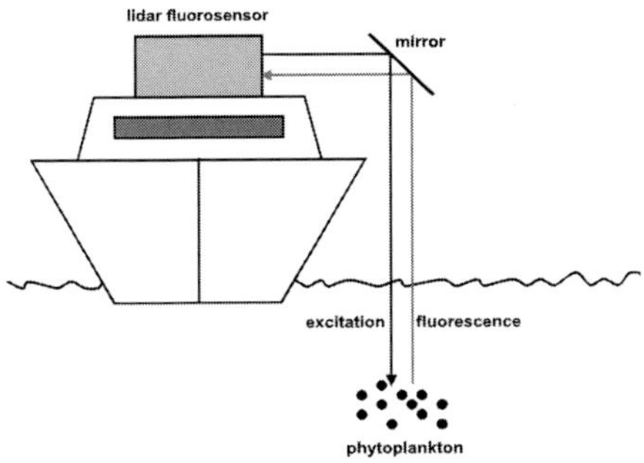

Figure 8. Lidar fluorosensor principle of operation.

6. Lidar Applications

The present section deals with details and significant results of two different lidar systems for atmospheric and marine applications developed at the ENEA Research Centre of Frascati, as examples of capabilities of the two lidar techniques in environmental monitoring.

6.1. Atmospheric Monitoring

Lidar/DIAL instruments based on CO_2 lasers have been developed at the ENEA Research Center of Frascati since the 80's,[41] when a ground based facility was assembled and operated. Different are the advantages for selecting the Mid-Infrared line-by-line wavelength emissions of the CO_2 laser, mainly in the optimal atmospheric window transmission and in the physical interaction of the wavelength transmitted with suspended aerosol particles (Mie scattering) without interference with smaller components strongly affecting the Rayleigh region. Furthermore, many natural occurring and pollutant atmospheric components show strong absorption band in this Mid-IR wavelength region. Therefore, a CO_2 lidar system is more tailored for tropospheric applications than a UV-VIS apparatus, where other physical parameters are predominant.

More recently, a mobile apparatus, called ATLAS (Agile Tuner Lidar for Atmospheric Sensing) was designed, (Fig. 9), and installed on the mobile laboratory ENVILAB (ENVIronmental LABoratory), shown in Figure 10.

ATLAS is based on a TEA (Transverse Excited Atmospheric) CO_2 laser equipped with a fast and accurate scanning mirror and a diffraction grating, allowing one to rapidly tune the transmitter on λ_{OFF} and λ_{ON}, with the evident advantage to reduce the complexity of the overall apparatus. The water vapor concentration has been measured firing alternatively at the 10R18 line (λ=10.260 μm, σ=1.1×10^{-4} cm^{-1} atm^{-1}) and at the 10R20 line (λ=10.247 μm, σ=8.8×10^{-4} cm^{-1} atm^{-1}). The aerosols optical thickness is usually retrieved at the 10P20 line (λ=10.591 μm) because the laser output is maximum at that wavelength.

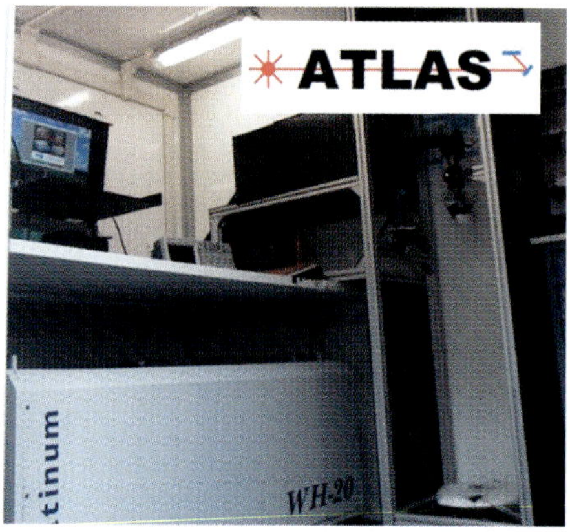

Figure 9. ATLAS (Agile Tuner Lidar for Atmospheric Sensing). Bottom-left: CO_2 laser. Right: Newton telescope. Top-left: control/acquisition computer.

Figure 10. The ENEA ENVILAB (ENVIronmental LABoratory).

Transmitter	Pulse energy	600 mJ (@ 10P20 line)
	Repetition rate	20 Hz
	Pulse width	60 ns FWHM
Receiver	Diameter	310 mm
	Focal length	500 mm
	Coating	Protected Al
Detector	Type	Liquid Nitrogen Cooled MCT
	Diameter	1 mm
	Normalized detectivity	4×10^{10} cm Hz1/2 W^{-1}
Analog-to-digital converter	Type	PCI
	Dynamic range	14 bit
	Sampling rate	50 Ms s-1

Table 1. Main ATLAS specifications.

A typical lidar echo recorded by means of the ATLAS apparatus, during the first experimental trials in atmosphere, is shown in Figure 11.a in slant path configuration (about 30° deg. elevation). The RCS (Eq. 2) shows a useful range of more than 5 Km with the evidence of a thin aerosols layer close to the receiver at about 1 km (Figure 11.b). Soon, the signal decays according to the $1/R^2$ law. The atmospheric extinction coefficient (Figure 11.c) is retrieved by inverting Eq. 1, showing more dense aerosols layers that decrease versus the altitude.

6.2. Marine Survey

A typical fluorosensor instrument is quite similar to a more general lidar lay-out, with a near UV or visible laser transmitter, a sending receiving optics and control electronics. The difference rely on the signal discrimination, generally performed with the help of a dispersive elements (grating, interference filters) behind the optical detectors (photodiode array, photomultiplier tubes).

As reference, the main characteristics of the ENEA Lidar Fluorosensor (ELF)[33] are reviewed and some recent results are described. ELF has been operational aboard the research vessel (RV) *Italica* during four Italian expeditions in Antarctica (13th, 15th, 16th and 18th in 1997-98, 2000, 2001 and 2003, respectively)[42] and the MIPOT (Mediterranean Sea, Indian and Pacific Oceans Transect) oceanographic campaign (2001-02).[43]

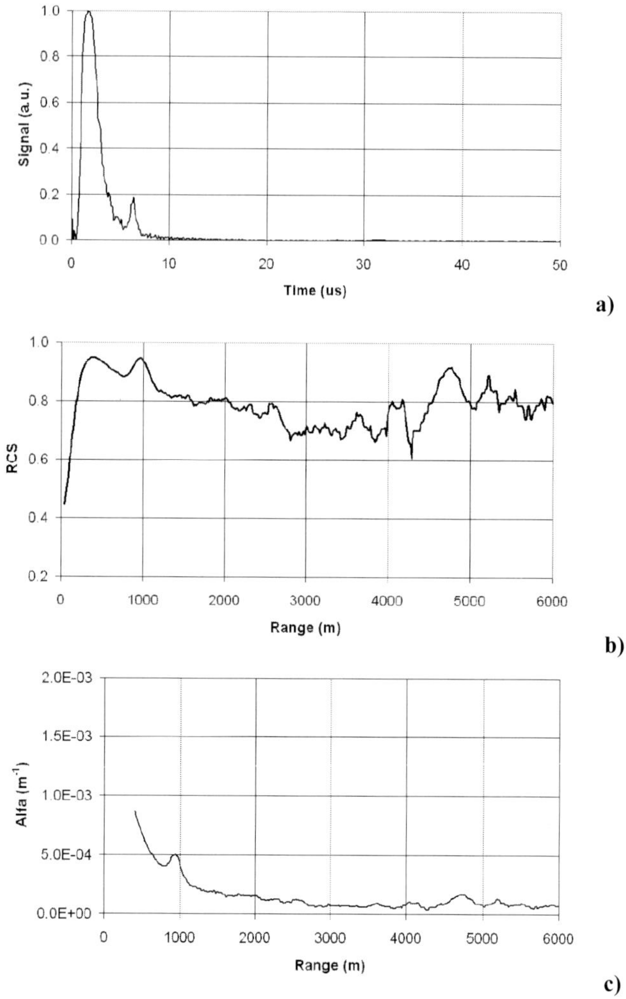

Figure 11. a) Example of lidar signal obtained at 10P20 CO_2 laser line; b) the range corrected signal (RCS) and c) range resolved extinction coefficient.

As we will see in the next section, its data has been used for the validation and/or the calibration of space borne radiometers,[44,45] up to the calculation of new estimations of satellite sensed primary production (PP)[46,47] and CDOM.

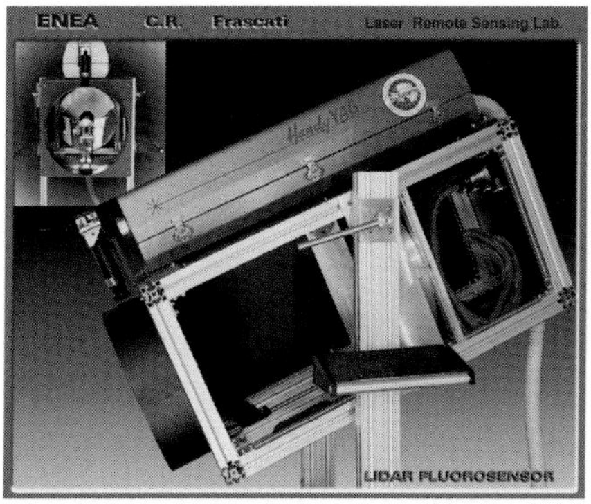

Figure 12. Side view of ELF (a front view is given in the insert). Transmitter: frequency-tripled Nd:YAG laser. Receiver: Cassegrain telescope. Detection: Optical Fiber, Interference Filters and Photomultipliers.

ELF (Fig. 12) is part of a complete laboratory, including local and remote instruments for continuous monitoring and in situ sampling, lodged into an ISO 20' container. It is assisted by ancillary instruments: a lamp spectrofluorometer, a pulsed amplitude fluorometer (PAM), a solar radiance detector, measuring the photosynthetically available radiation (PAR), and a global positioning system (GPS).

The light source is a frequency-tripled Nd:YAG laser (355 nm) followed by a beam expander (BE). Transmitter and receiver are mounted on a common chassis, so as to minimize vibrationally induced optical mismatches, and their axes are coincident. The laser beam is expanded by a factor of three before reaching the sea, both for eye-safety restrictions and in order to increase the laser footprint. The telescope collects and focuses the return optical radiation onto a fiber optic front tip placed behind an high-pass optical filter which cuts the laser backscattered radiation off. This fiber optic bundle splits in four branches routing the signals to four different photomultipliers, respectively. Their electronic output is digitized by analog-to-digital converters (ADCs). A personal computer (PC), embedded in a versa module eurocard (VME) bus, controls all the experimental settings, including the normal or pump-and-probe excitation, the laser transmitter energy, the PMs high voltage (HV) and gating time and the data acquisition parameters. The lidar fluorosensor arrangement on-board of the ship is schematized in Fig. 13, while detailed characteristics of the main components are described in Ref. 33.

Figure 13. Main picture: RV Italica; ELF and the ancillary instruments are housed in a container (inside the circle). Left insert: external view of the container; note on the left of the container the box accommodating the mirror for water surface observation. Right insert: internal view of the container; ELF and the spectrofluorometer are visible behind and on the left of the operator, respectively.

By placing suitable narrow band interference filters in front of the each photomultiplier, four spectral channels were selected, corresponding to the water Raman backscattering (402 nm), the DOM fluorescence maximum (450 nm), the DOM red tail emission (650 nm) and the Chlorophyll peak (690 nm), respectively. The choice of the latter two channels relies on the former knowledge of the main phytoplankton composition. The spectral and electrical responses of the photomultipliers and the related electronics are tested before and after the campaign by using standard lamps and fluorescing targets during the calibration procedure.

6.2.1. Oceanographic Monitoring

Earth oceans play a major role in the climate equilibrium. Photosynthesis in aquatic systems is considered to be responsible for more than 40% of the global carbon fixation on an annual basis, by converting light radiation into organic compounds. At present considerable uncertainties still exist in the understanding of the processes that control artificial and natural CO_2 uptake.

In particular, the key question concerning the Southern Ocean is whether it is able, like the North Atlantic Ocean, to take up atmospheric carbon dioxide, is still open to scientific discussions.

As already described, the ENEA Lidar Fluorosensor (ELF) has joined different Italian expeditions in Antarctica. In particular, the XVI oceanographic campaign, 2000-2001 austral summer, supplied a unique opportunity to compare, for the first time, LIF Chl-a and pCO_2 surface distributions simultaneously operated in this harsh environment.[48]

In Figure 14, the anticorrelation between spot pCO_2 determinations and the continuous stream of LIF data, is clearly evident ($R = -0.8$). The LIF survey revealed the presence of high productivity in polynya regions and close to the ice shelf edges, showing the high variability of the Antarctic Ross Sea.

The overall good agreement between two different experimental approaches confirms the reliability of the lidar fluorosensor apparatus as a sea-truth instrument for large area and real time analysis. Following the success of the present analysis, all data collected were jointly used as input to tune bio-optical algorithms for improving the precision of estimations of primary productivity (PP) in the Southern Ocean.

A discussion of the application of standard models in Antarctica and more details on the model implemented can be found elsewhere:[47] here an example of the results of LIF based and standard PP models is given in Figure 15 based on the ELF measurements carried out during the 16th Italian expedition in Antarctica (January 5th 2001 - February 26th 2001). The present comparison indicates that usual PP models applied to standard chl-a concentrations can underestimate PP up to 50%.

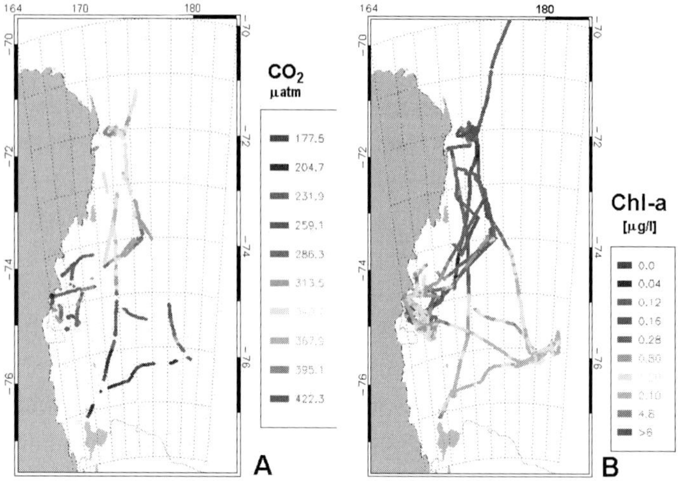

Figure 14. Maps of the distributions measured during the XVI Antarctic expedition; color scale indicates: a) pCO_2 [μatm] and b) Chl-a concentration [μg/l], average values are given for each color.

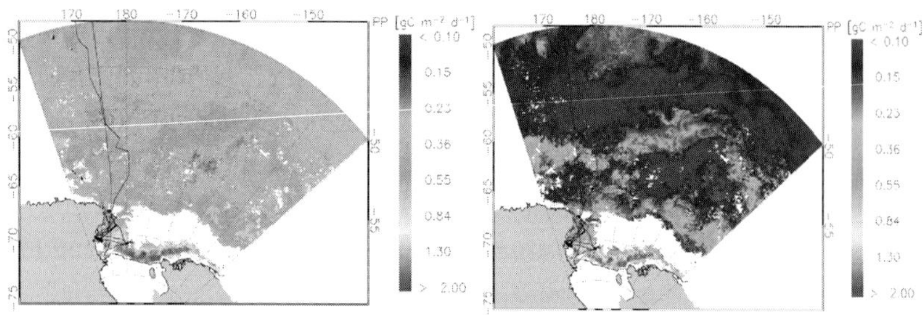

a) **b)**

Figure 15. Average PP, based on the monthly products of January and February 2001, calculated with a) the ELF-calibrated SeaWiFS chl-a bio-optical algorithm and the new PP model, and b) the standard SeaWiFS chl-a bio-optical algorithm and the VPGM model by Behrenfeld and Falkowski. Continuous line: ship track.

6.2.2. Oil Slicks Releases

The Venice lagoon and the nearby open sea area form an extremely peculiar ecosystem with high a risk of both industrial and anthropogenic pollution. Industrial activities carried out at Porto Marghera directly affect the composition of waters and bottom sediments with their liquid exhausts and gaseous emissions, while anthropogenic releases are expected to be of importance in the most populated areas of the lagoon (Venice center and Chioggia). The lagoon water flow system, characterized by limited exchanges with the open sea, makes difficult to eliminate industrial wastes together with all the anthropogenic organic substances and unburned oils used for combustion and local transport. Some of these substances, such as chlorinated poly-aromatic compounds, are characterized by a high persistence in water and a low probability of natural degradation under environmental agents (sun radiance, salinity and temperature). Due to the complexity of biochemical processes involved in the lagoon equilibrium, it is extremely important to monitor the local situation, regards to the presence of specific pollutants which might generate high risk situations.

The shape of most common crude oils is a strong and broad fluorescence emission peaked at blue wavelengths, upon UV laser excitation.[49] In case of LIF apparatus, operated @ 355 nm in a large area, surface oil slicks can be identified from the strong depression of water Raman scattering, due to the oil and CDOM overlapping contributions. Therefore a precise underlying fluorescence background at the water Raman scattering emission has to be performed. The film thickness d is obtained rationing lidar Raman signals from the polluted area (R_{in}) respect to a clean open sea waters (R_{out})[50] as:

$$d = \frac{1}{k_e + k_R} \ln\left(R_{in} / R_{out} \right) \tag{13}$$

where k_e and k_R, are the oil and water Raman extinction coefficients, both in the case of assignment to CDOM and crude oil.

As an example, a large oil spot (Fig. 16) was monitored inside a inner channel of the Venice lagoon, during the monitoring campaign performed by ELF systems installed on board of a small boat (10 m length, 3 m width and 1 m depth) in the overall lagoon and the surrounding sea-side.[33]

14. M. Losacco, V. Giannini, G. Lombardo, F. Pappalettera and Tedeschi, *Proceedings of the 22nd International Laser Radar Conference* (ILRC 2004), ESA Publications Division, 259, (2004).
15. C. Bellecci, S. Martellucci, M. Richetta, G. A. Dalu, P. Aversa, L. Casella, S. Federico and P. Gaudio, *Proceedings of SPIE* Volume 4070, 59, Vladimir I. Pustovoy, Vitaly I. Konov, Editors, (2000).
16. R. Barbini, F. Colao, R. Fantoni, A. Palucci, R. Habel, M. Cappai, A. Sollai, C. Bellecci, G. Artese, F. Dedonato, G. Frangella, M. Zupo, A. Giardini-Guidoni, A. Morone and M. Snels, *Il Nuovo Saggiatore*, Anno 7, no. 4, July - August 1991.
17. J. de Mello, *Lab Chip*, **3**, 29N (2003).
18. M. J. Castaldi and S. M. Senkan, *J. of the Air & Waste Management Association*, **48**, 77 (1998).
19. H. K. Lichtenthaler and U. Rinderle, *CRC Crit. Rev. Anal. Chem.*, 19, suppl. 1, S29-S85 (1988).
20. F.E. Hoge, R. N. Swift, *NASA Conf. Pub. "Chesapeake Bay Plume Study*, 349 (1981).
21. D. Diebel, T. Hengstermann, R. Reuter and R. Willkomm. AE Lodge (editor), J Wiley & Sons (Chichester), 165, pp. 127-142, (1989).
22. Z. Dai, L. K. Tseng and G.M. Faeth, *J. of Heat Transfer-Transactions of the Asme*. 117:918–26 (1995).
23. H. Neij, B. Johansson and M. Aldén. *Combust. Flame*, 99:449, (1994).
24. S. J. Sanders, R. F. Boivin, P. M. Bellan and R. A. Stern, *Phys. Plasmas*. 6, 4118(1999).
25. R. M. Measures, *Laser remote sensing*, Wiley-Interscience Publications, New York (1984).
26. D.V. Maslov, V. V. Fadeev and A. I. Lyashenko. *EARSeL-SIG-Workshop LIDAR*, 46 (1992).
27. G. Cecchi, L. Pantani, B. Breschi, D. Tirelli and G. Valmori, *EARSeL, Advances in Remote Sensing*, **1**, 72(1992).
28. J. Johansson, E. Wallinder, H. Edner and S. Svanberg, , *Proceedings of the 16th ILRC*, NASA CP 3158, 433-436, (1992).
29. R. Reuter, H. Wang, R. Willkomm, K.D. Loquay, T. Hengstermann and A. Braun, *EARSeL Advances in Remote Sensing*, **3**, 152 (1995).
30. F.E. Hoge, *Appl. Opt.*, **22**, 33, 3318 (1983).
31. S. Babichenko, S. Kaitala, A. Leeben Poryvkina and L. Seppala, *J. of Marine Systems* 23(1-3): 69-82 (1999).
32. R. Reuter, R. Willkomm, G. Krause and K. Ohm. *EARSeL Advances in Remote Sensing*, **3**, 15, (1995).
33. R. Barbini, R. Fantoni, F. Colao, A. Palucci and S. Ribezzo, *Int. J. Remote Sensing*, Vol. 20, no. 12, 2405 (1999).
34. V. Drozdowska, S. Babichenko and A. Lisin, *OCEANOLOGIA*, **44** (3), 339354 (2002).

35. S.Harsdorf, M. Janssen, R. Reuter, S. Tönebön, B. Wachowicz and R. Willkomm, *Measurement Science and Technology*, **10**, 1178 (1999).
36. R. Barbini, F. Colao, R. Fantoni, L. Fiorani, A. Palucci and S. Ribezzo, *Proceedings of the IATICE* 2002, 219 (2002).
37. R. M. Measures, *Laser remote sensing*, Wiley-Interscience Publications, New York (1984).
38. F. Barnaba, L. Fiorani, A. Palucci and P. Tarasov, *J. of Quantitative Spectroscopy & Radiative Transfer* 102, 11 (2006)
39. R. B. Slusher and V. E. Derr, *Appl. Opt.*, **14**, 2116 (1975).
40. M. Bristow, D. Nielsen, D. Bundy and F. Furtek, *Appl. Opt.*, **20**, 2889 (1981).
41. R. Barbini, F. Colao, G. d'Auria, A. Palucci and S. Ribezzo, *Proceedings of SPIE*, Vol. 3104, 167, (1997).
42. R. Barbini, F. Colao, R. Fantoni, A. Palucci and S. Ribezzo, *Int. J. Remote Sensing* **22**, 369 (2001).
43. R. Barbini, F. Colao, L. De Dominicis, R. Fantoni, L. Fiorani, A. Palucci and E. S. Artamonov, *Int. J. Remote Sensing*, **25**, 2095 (2004).
44. R. Barbini, F. Colao, R. Fantoni, L. Fiorani and A. Palucci, *J. of Optoelectronics and Advanced Materials*, **3**, 817 (2001).
45. R. Barbini, F. Colao, R. Fantoni, L. Fiorani and A. Palucci, *Int. J.of Remote Sensing* **24**, 3205 (2003).
46. R. Barbini, F. Colao, R. Fantoni, L. Fiorani and A. Palucci, E. S. Artamonov and M. Galli, *Antarctic Science*, **15**, 77 (2003).
47. R. Barbini, F. Colao, R. Fantoni, L. Fiorani, I.G. Okladnikov and A. Palucci, *J. of Optoelectronics and Advanced Materials*,**7**, 1091 (2005).
48. R. Barbini, S. Ceradini, F. Colao, R. Fantoni, G. M. Ferrari, A. Palucci, S. Sandrini, L. Tositti and O. Tubertini, *Int. J. Remote Sensing*, vol. 24, no. 1, 1 (2003).
49. R. Fantoni, R. Barbini, F. Colao, A. Palucci and S. Ribezzo, in *Excimer Lasers*, L. D. Laude Ed., NATO ASI series E-vol. 265, (Dordrecht: Kluwer Academic Publ.), pp. 289-305 (1994).
50. F. E. Hoge and J. S. Kincaid, *Applied Optics*, **19**, 1143 (1980),
51. J. Kasparian and J. P.Wolf, *Optic Express*, vol. 16, 1, 466 (2008).

LASER-BASED *IN SITU* GAS SENSORS FOR ENVIRONMENTAL MONITORING

Maurizio De Rosa*, Gianluca Gagliardi, Pasquale Maddaloni, Pietro Malara, Alessandra Rocco and Paolo De Natale

Istituto Nazionale di Ottica Applicata, CNR, Comprensorio "A. Olivetti", Via Campi Flegrei 34, 80078 Pozzuoli, Italy
**E-mail: maurizio.derosa@inoa.it*

Strong concern about anthropogenic effects on environmental and global climate issues motivated a large amount of studies and triggered development of sensitive techniques for monitoring relevant parameters. Laser spectroscopy offers powerful tools for fast and accurate measurements of very low concentrations of a large number of chemical species. We present an overview of the laser-based spectroscopic techniques and their application to in-situ measurements of gas species for environmental monitoring. We describe the basic principles of absorption spectroscopy and detection strategies. Major features of coherent radiation sources are described and several examples of laser-based sensors are given.

1. Introduction

One of the most relevant problems in environmental studies is the quantification of trace gases in the atmosphere, aimed at investigating phenomena such as stratospheric ozone depletion, acid rain formation and more generally global warming. The rising public concern towards these issues has lead, in recent years, to a strong interest in development of ultra-high sensitivity analytical techniques. Over the last few decades, growing attention has been devoted to infrared laser methodologies for molecular gas analysis in a large variety of environments and applications. Most efforts have focused on the possibility of detecting trace gases, such as NO, NO_2, HCl, CO, NH_3, C_2H_2, CH_4 as well as on quantitative chemical analysis of samples containing H_2O, CO_2, O_2 and others. A number of high sensitivity optical techniques, relying on long-path schemes and amplitude/frequency modulation, have been developed allowing detection of parts-per-billion (ppbv) concentrations.[1]

Many studies reported thorough investigations on the achievable signal-to-noise ratios and detection limits, with the important goal of defining the best experimental conditions to approach the shot-noise limited regime, pointing out the ultimate sensitivity limit for optical spectrometers. In this context, absorption spectroscopy based on semiconductor diode lasers has gained growing interest, especially for their low noise, high spectral quality and compatibility with telecommunications-grade optical fiber components. Using wavelength-division multiplexing techniques, integrated systems for simultaneous and remote multiple-species analysis can be implemented, thus offering potential improvements over current strategies in several field monitoring applications.

Moving to strong ro-vibrational bands, (beyond 2-μm wavelength), would allow the use of simpler detection methods, enabling for more precise and accurate measurements. Thanks to great advances in semiconductor materials and optoelectronic technologies, most of these features have been recently transferred to the mid-IR and part of the "thermal" IR window, where several atmospherically relevant molecules, like NO, NO_2, N_2O and CH_4, exhibit their strongest absorptions.[2] Indeed, novel coherent radiation sources have been developed, such as those based on difference-frequency generation or optical parametric oscillators in periodically-poled non-linear crystals. In this respect, diode lasers lend themselves to the realization of compact non-linear coherent-radiation systems able to directly access molecular fundamental vibrations. As a further development, Sb-based (between 2 and 3 μm) and quantum-cascade lasers (from 3.5 up to 24 μm) have become now commercially available, although development and demonstration of their actual performance are still in progress.

In daily use, a large majority of commercial spectrometers, which are widely adopted for practical applications, are still based on conventional incoherent sources that have benefited from decades of research and development activity, and have been validated in diverse experimental studies. Nevertheless, semiconductor diode lasers basically offer unique features to develop reliable and compact spectrometers, at a reasonable cost. Also, they possess several advantages in terms of tunability, spectral selectivity and power consumption. Use of such devices in combination with optical fibers offers an unique tool for development of real fiber-based nets, that can be employed in monitoring of large areas. Despite their great potential, most of such devices still need a deep engineering work to become really competitive and be suitable for long-term outdoor operation by non-expert end-users.

In the following, the authors will give an overview of state-of-the-art laser-absorption spectroscopy methods for gas sensing in the near-infrared and mid-infrared regions. Some examples will be described in details to point out benefits and critical issues.

2. Spectroscopic Detection and Concentration Measurements of Gases

Direct laser absorption spectroscopy used for quantitative measurements is based on Lambert-Beer law.[3] When a laser beam of frequency ν and intensity I_0 passes through a homogeneous gas sample of length L (expressed in cm), the transmitted intensity I is given by

$$I(\nu) = I_0 e^{-\sigma(\nu)\cdot N\cdot L} \tag{1}$$

where N (molecules per cm^3) represents the molecular concentration and $\sigma(\nu)$ is the absorption cross section (expressed in cm^2). If we denote with ν_0 the center frequency of an absorption line of the considered gas and with $g(\nu-\nu_0)$ the normalized lineshape function, the absorption cross section is conveniently expressed as

$$\sigma(\nu) = S(T)\cdot g(\nu-\nu_0) \tag{2}$$

where $S(T)$ (cm/molec) is the linestrength of the given molecular transition. This intensity factor is proportional to the lower-state population density and thus depends on temperature.

By inserting Eq. (2) into Eq. (1), in the limit of small absorption, we get

$$\frac{I_0 - I(\nu)}{I_0} \cong N \cdot L \cdot S(T) \cdot g(\nu - \nu_0) \tag{3}$$

Integration of the above formula leads to

$$IA \equiv \int_{-\infty}^{+\infty} \frac{I_0 - I(\nu)}{I_0} d\nu = N \cdot L \cdot S(T) \tag{4}$$

So, the measured integrated absorbance *IA*, namely the area under the recorded spectrum, yields the gas concentration *N*, provided that *S*(*T*) and *L* (or their product) are known.

The function $g(\nu - \nu_0)$ is basically related to the macroscopic parameters of the gas sample and in the most general case is described by a Voigt profile, that is a convolution of a Lorentzian and a Gaussian profile.[3] The former accounts for molecular collisions and is predominant at high pressures, while the latter, describing the effects of thermal motion, is only apparent at low pressures.

2.1. Laser Sources

Here, we restrict the attention to a few types of laser sources, based on semiconductor materials, which proved to be particularly well suited for field measurements, due to their properties and ease of operation. These sources cover an emission range from the visible to the mid-infrared depending on adopted materials and designs.

Semiconductor diode lasers (SDLs) are widely used as sources for gas spectroscopy. Laser emission from a diode laser is based on the recombination at a p-n junction of electron-hole pairs, which are generated by pumping current through the junction. Lead-salt diodes were the first SDLs ever lasing and emitted in the mid-infrared. Later, the emission range was extended to the near-infrared and visible windows with double-heterojunction lasers.[4,5]

Quantum cascade (QC) lasers are based on engineered structures made by a periodic series of different layers of semiconductor materials.[6]

This superlattice splits the allowed semiconductor band in a discrete number of electronic sub-bands and the laser emission depends on electronic transitions between different sub bands. Each electron contributes with more than one photon when tunneling between two adjacent periods. The relative position of the sub bands, and as a consequence the emission wavelength, is mainly determined by the geometry of the superstructure, rather than by the properties of the material itself. Therefore, a wider range of wavelengths (from ≈ 3 μm to ≈ 24 μm) can be generated by carefully tailoring the periodic superstructure.[7]

The resonant cavity is usually made by the cleaved facets of the semiconductor chip. The emitted wavelength can be continuously tuned over an interval of several nanometers by varying the injection current or the temperature of the junction. Emission is typically multimode, both for SDLs and QC lasers; however, single mode emission can be obtained, either by including the diode laser in an external cavity, or by introducing a grating in the gain structure (distributed feed-back), or introducing a selective reflective element as cavity mirror (distributed Bragg reflector).

Semiconductor laser sources need active stabilization of the chip temperature and a low-noise pump current. The operating range of temperature can be very different, going from cryogenic temperatures for lead-salt diode and cw QC lasers, to room temperature for lasers emitting in the visible and near infrared. Frequency scan of the emitted radiation can be accomplished by varying the injection current across the junction. Injection current can be changed very rapidly, allowing the implementation of very sensitive frequency modulation techniques, as described below.

2.2. Detection Techniques

There are two major strategies to improve the detection limit of a diode laser spectrometer. One is to improve the signal-to-noise (S/N) ratio, e.g. by using frequency modulation techniques; a second approach is to enhance the optical path of the radiation in the sample gas. These strategies can be used simultaneously.

2.2.1. Noise Reduction: Frequency Modulation Techniques

A primary source of noise in semiconductor lasers is amplitude noise. Its ultimate level is set by shot noise, due to quantum fluctuations of detected photons. However, lasers usually suffer from an excess noise with respect to the shot level. Excess noise is particularly relevant at low detection frequency, where it has a $1/f$ behavior, and reduces to shot level at detection frequencies greater than few MHz.

In order to move the detection frequency to such region, the optical frequency of the source is sinusoidally modulated and the absorption signal is coherently de-modulated by phase-sensitive electronics. In this way, noise contributions out of a narrow frequency band around the modulation frequency are suppressed.[8]

The laser field of a frequency modulated source emitting at frequency $\nu=\omega/2\pi$, can be described by

$$\begin{aligned} E(t) &= E_0 \exp[i \cdot (\omega t + \beta \sin \Omega t)] = \\ &= E_0 \exp(i\omega t) \sum_n J_n(\beta) \exp(in\Omega t) \end{aligned} \tag{5}$$

where Ω is the modulation frequency, β is the modulation index, $J_n(\beta)$ is the Bessel function of order n. In the frequency picture, (second line of Eq. (5)), modulation of the frequency adds a comb of sidebands around the carrier ω, equally spaced by the modulation frequency Ω and with amplitude $J_n(\beta)$. The detected signal contains the beat notes given by all the possible couples of probing frequencies. Mixing the signal with a properly phase-delayed reference signal having the same modulation frequency, and successive low-pass filtering will give a dc signal proportional to the probed absorption. FM techniques can operate with different modulation parameters Ω and β. A proper choice of the modulation frequency can lead to a reduction of the detected noise and to the improvement of the signal itself. The frequency modulation should be high enough to move detection in the shot noise limited region of the spectrum. Moreover, a sideband spacing comparable or greater than the typical width of the absorption line can enhance the resulting signal, contributing to the final S/N ratio. Even though practical limitations may prevent from meeting the optimal conditions, FM techniques can lead to a higher S/N ratio with respect to direct absorption. In fact, the first FM techniques (usually referred as Wavelength Modulation) used relatively low frequency for Ω, with respect to the width of the absorption feature, and large modulation index, so that many sidebands simultaneously probed the absorption. Techniques typically referred as FM[9] involve higher modulation frequency, comparable with the absorption width, and relatively small modulation index, so that only one pair of sidebands occurs. Another FM technique (Two-tone FM) uses two nearly equal modulation frequencies Ω_1 and Ω_2, both comparable to the absorption width Γ (0.1-1 GHz), whose difference, $\Delta\Omega=\Omega_1 - \Omega_2$, is smaller than Γ (1-10 MHz). Demodulation is then made at the difference frequency $\Delta\Omega$, which simplify the electronic design, while the absorption feature is probed by well separated sidebands, with optimal line contrast.[10]

FM can be easily implemented with diode lasers by simply modulating the injection current, provided that a corresponding fast detector can be used.

Fast photodetectors are available up to several GHz in the near IR, while in the mid IR detector frequency response is limited to some hundreds of MHz. In general, photodiode speed is limited by the diode capacitance which depends on the size of the diode junction and smaller active areas are required to increase response time. Moreover, high speed electronics requires a very careful design and can be very sensitive to environmental disturbances.

In a FM detection scheme, the final signal not only depends on the concentration of the probed molecular species, as for direct absorption, but also on the parameters of the detection electronics. So, for an accurate concentration determination, a FM-based spectrometer should be calibrated and a high degree of stability must be guaranteed for the electronic readout.

2.2.2. Signal Enhancement: Increasing the Optical Path

The simplest technique to increase the absorption path length is to use a multiple reflection cell (MRC) that is a device consisting of two or more facing mirrors separated by a distance d. The laser beam, entering the cell, undergoes N_r reflections, (depending only on the cell geometry), yielding an effective interaction length of $N_r d$. The main types of multiple reflection cells are the Herriott[11] and White type.[12] Typically, with cell lengths of several meters, absorption path lengths of a few hundred meters can be achieved. In particular, by combination of strong molecular transitions and MRCs it was possible to test fundamental principles of quantum mechanics at the 10^{-11} level.[13] The main drawback is the attenuation of the power throughput (1–2 % of the input) and the partial overlap of the reflected beams in the cell that may give rise to interference fringes that effectively limit the achievable signal-to-noise ratio.

The optical path enhancement can be realized much more efficiently by placing the sample in an optical resonator. In fact, the use of an optical cavity leads to an enhancement of the effective absorption path length of $2F/\pi$, where F is the cavity finesse, which ultimately depends on the mirror reflectivity. Considering the current state of the art for high reflectivity mirrors, a limit on available cavity finesse can be set in the range from 10^5 to 10^6. Therefore, the use of an optical resonator may result in effective absorption path lengths of several tens of kilometers.

An optical cavity of mirror separation L, reflectivity R, injected with intensity I_0 and containing a sample with per-pass absorption A, has a per-pass intensity loss $\Delta I = -I(1-R+A)$, while the per-pass optical transit time is $\Delta t = L/c$. If the laser is tuned into resonance with the cavity, when the number of passes is large, we can write the ratio between these two quantities as a differential equation

$$\frac{dI}{dt} = \frac{c}{L}\left[\frac{I_0 C_p T}{2} - L(1-R+A)\right] \tag{6}$$

The source term $I_0 C_p T/2$ represents the light effectively entering the cavity (C_p is a coupling parameter between 0 and 1 and T is the input mirror trasmittivity). The factor 1/2 takes into account that, while losses are from both cavity mirrors, light enters only through one. The general form of $I(t)$ is the sum of a steady state and a transient solution of Eq. (6)

$$I(t) = \frac{I_0 C_p T}{2L(1-R+A)}\left(1 - e^{-\frac{t}{\tau}}\right) \tag{7}$$

with

$$\tau = \frac{L}{c(1-R+A)} \tag{8}$$

From Eq. (7) it is easy to see that information on the intracavity-sample per-pass absorption A is encoded both in the transmitted steady state intensity $T \cdot I(t \to \infty)$ and in the decay rate τ of the transient solution. Recovering the absorption signal either from the steady-state transmitted intensity or from the cavity lifetime τ represents the basic difference between the two main families of cavity-based detection schemes, namely the Cavity-Enhanced Absorption Spectroscopy (CEAS) and the Cavity Ring-Down Spectroscopy (CRDS).

Cavity Enhanced Absorption Spectroscopy (CEAS) – In order to exploit the path length enhancement of an optical cavity, a possible approach relies on active frequency-locking between the laser source and the cavity resonance. In a typical CEAS setup, the radiation back-reflected from the input cavity mirror is monitored by a photodetector. Due to the FM-AM conversion, the detected signal level instantaneously depends on the overlap between the laser frequency and the cavity resonance lineshapes, and therefore can be used to generate an error signal that serves as input for a frequency-locking loop. The latter provides a suitable correction signal back to the laser or, alternatively, to a piezo transducer controlling the cavity mirror position. Finally, the transmitted intensity is used to retrieve information on the absorbing sample, according to Eq. (7). Minimum detectable absorption coefficients ranging from $10^{-8}cm^{-1}Hz^{-1/2}$ to $10^{-11}cm^{-1}Hz^{-1/2}$ have been demonstrated with CEAS-based spectrometers.[14-17]

Alternatively, integrated off-axis cavity-output spectroscopy (OA-ICOS) can be performed. The OA-ICOS scheme relies on a Herriott-like alignment of the optical resonator that allows the ray pattern inside the cavity to overlap only after a large number of round trips (depending only on the mirror radii and distance). This multipass-like alignment allows several TEM_{nm} cavity modes to be excited, resulting in an extremely dense resonant structure. Such dense ensemble of modes is then flattened by laser-frequency and cavity-length dithering, leading eventually to a quasi-continuous frequency response function of the resonator. Finally, after time integration of the output signal, the mode structure is suppressed and the cavity behaves effectively as always-resonant. With ICOS technique, a F/π factor is gained in the interaction pathlength, and the AM-FM noise is as much reduced as the cavity resonant structure is suppressed, without need for any active locking. Moreover, such a scheme is intrinsically insensitive to small vibrations and misalignments, and is therefore particularly suitable for outdoor applications. The main drawback of OA-ICOS is the extremely reduced output intensity. In fact, even though it has been successfully used for ultrasensitive spectroscopy and trace-gas detection at different wavelengths,[18-22] only recently OA-ICOS has been demonstrated in combination with low power nonlinear sources in the fingerprint spectral region.[23]

Cavity ring-down spectroscopy (CRDS) - Sensitivities comparable to CEAS and ICOS have been also obtained by the cavity-ring-down spectroscopy approach (also known as leak-out spectroscopy).[24-30] When the laser frequency is in resonance with a TEM_{00} cavity mode, the cavity injection is abruptly interrupted (for example by means of an acousto-optic modulator) and the decaying transmitted intensity is recorded by a fast photodetector. By a fitting procedure with Eq. (7), the decay rate τ is retrieved, and, as a consequence, the information about the sample absorption A (see Eq. (8)). Detection sensitivity for this technique ultimately relies on the cavity finesse and on the dynamic range of the signal processing electronics. In order to improve long-term reproducibility, our group combined CRDS with phase-locking of the laser source to an optical frequency comb synthesizer.[31]

Photo-acoustic spectroscopy (PAS) - A different approach for high sensitivity detection is represented by photoacoustic spectroscopy (PAS).[32] Excitation and non-radiative relaxation of a ro-vibrational molecular transition results in the absorbing sample to experience an abrupt temperature step. The consequent collisional mechanism generates a pressure wave that propagates within the sample, which carries information about the absorbing species concentration. In a typical PAS detection scheme, subsequent excitation and relaxation are induced in the sample by sweeping in and out of absorption the laser frequency, or by chopping the laser intensity. Typically, the sample is placed in a special absorption cell, namely an acoustic resonator, so that the generated wave matches an acoustic resonance.[33] The resonant pressure wave is then detected by a sensitive microphone embedded in the cell. Photoacoustic spectroscopy allows to achieve detection sensitivities down to sub-ppbv concentrations. Its main advantage is that the absorption signal increases with the incident laser power. Furthermore, since any absorbed wavelength is converted into an acoustic wave whose frequency depends on the chosen chopping period, the same microphone detector can be used to probe any spectral region, without changes in the responsivity. The main drawbacks of PAS technique is the selectivity, in many cases limited by the requirement of high sample pressures for an efficient collisional transfer. Moreover, the latter depends not only on the absorption strength, but also on the actual sample composition, for example temperature, humidity or presence of other non-absorbing species.

3. Near-Infrared Field-Deployed Sensors

In this section we present some examples of spectrometers based on diode lasers emitting below 2 μm which have been successfully employed in field applications. Among these, one of the most challenging is gas emission monitoring in volcanic areas. Indeed, concentration measurements in soil effluxes can provide useful information regarding volcanic dynamics. Particularly, combination with other surveillance data can provide valuable early warning of eruptions.

A first example of NIR spectrometer was used to monitor CO_2 and H_2O in volcanic fumaroles by a dual-wavelength approach. A beam with 1.57 and 1.39 μm diode-laser radiation was delivered to a 20 cm long cell placed upon the emitting fumarole by means of optical fibers. Such spectrometer enabled to perform measurements of water vapor concentration with a 3% accuracy, whereas it was quite inaccurate for carbon dioxide measurements due to the weakness of the available transitions.[34,35]

More recently, a new laser spectrometer with an emission wavelength of 2 μm, able to perform simultaneous measurements of CO_2 and H_2O emissions, was developed. Different detection configurations (concentration at different heights, diffusion as accumulation chamber, soil degassing) were used during field campaigns in Solfatara volcano and Eolian Islands, Italy.[36,37]

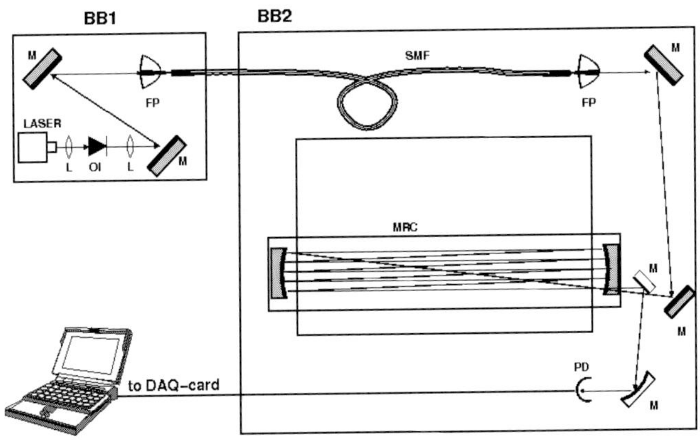

Figure 1. Experimental setup of the 2-micron diode laser spectrometer based on Herriott multiple reflection cell.[37]

The study of CO_2 concentration in atmosphere and its increase vs. anthropic activities has a major role in the study of global climate change. One of the main flux sources affecting the atmospheric CO_2 concentration is direct soil emission. They represent, on the other hand, a relevant uncertainty in the carbon cycle, as their magnitude and distribution have been so far poorly quantified. Typical techniques for measuring ground CO_2 fluxes can include micrometereological methods (as eddy covariance and gradient methods), closed circulation soil chambers, and open or flow-through chambers. In contrast with the latter approaches, use of tunable diode laser absorption spectroscopy allows real-time, fast, accurate, sensitive concentration measurements, without affecting the flux itself.

Using a tunable diode laser system, a gas analyzer was designed and

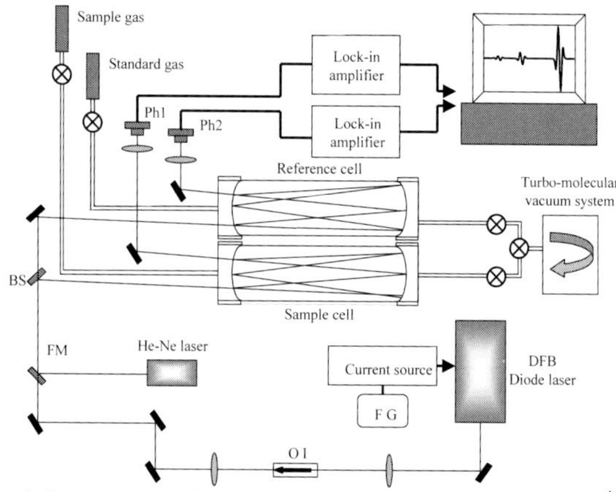

Figure 3. Laser set-up for isotope ratio measurements on CO_2.[46]

The apparatus is depicted in Fig. 3. A DFB diode laser is distributed among two multiple-reflection cells, kept at the same temperature, and absorption signals in pure carbon dioxide (50 Torr pressure) are recorded by lock-in phase-sensitive detection. Short-term (10 min) and long-term (100 min) reproducibility values of 0.16 ‰ and 0.3 ‰ (1σ), respectively, were observed by injecting exactly the same gas (same isotopic composition) in both the reference and the sample cell.
Another interesting example is for H_2O at 1.39 μm, for which a laboratory demonstration of a complete isotope analysis has been provided (see Ref. 47) and subsequently extended to volcanic field applications.[48] Analogously, Uehara et al. have reported on the precise determination of $^{13}CH_4/^{12}CH_4$ and $^{14}N^{15}N^{16}O/^{14}N^{14}N^{16}O$ using 1.7 μm and 2 μm wavelength DFB diode lasers.[49,50]

Over the last years, a tremendous progress occurred in the field of spectroscopic techniques based on high-finesse optical cavities.

Many efforts have been done by means of CRDS methods. A very efficient scheme was implemented by Paldus et al.,[51] who actively locked the laser to the resonator while measuring the absoprtion-related cavity decay. They achieved a detection limit in the absorption coefficient of $5\cdot10^{-9}$ cm^{-1} for ambient water vapor. Another interesting development has been reported (Ref. 52), using DFB diode lasers in the near-IR, where multiplexed CRDS has been demonstrated for simultaneous detection of multiple species with a sensitivity in the order of 10^{-9} $cm^{-1}/\sqrt{Hz}$.

Nowadays, portable CRDS-based spectrometers with conventional ring-down schemes have become commercial. Most of them are devoted to the quantification of air pollutants and, in few cases, the determination of isotope abundance ratios for atmospheric applications or breath analysis.

Extremely compact and stable apparatuses were recently devised by Romanini et al. that rely on optical-feedback CEAS.[53,54] Aircraft atmospheric water vapor detection and isotopic analysis were successfully demonstrated by such systems and also tested for detection of CO and CH_4 gas in volcanic effluxes.[55] The radiation source is a DFB diode laser emitting around 1312 nm and a V-shaped cavity is designed to optimize optical feedback, providing laser locking to the resonator.

As mentioned in previous section, most CEAS techniques are based on frequency stabilization via an electronic loop that acts externally on the laser. For example, in Ref. 17 the authors report on an optical resonator for detection and concentration measurement of H_2O in the wavelength region around 1393 nm. The work is performed by an extended-cavity DFB diode laser that is frequency locked to the cavity by the Pound-Drever-Hall method.[56] In this way, the laser transmitted beam brings information on inside cavity losses: if the laser is tuned across an absorption line its spectral features are reproduced at the cavity output as in conventional direct absorption spectroscopy. An example is reported in Fig. 4. A minimum detectable vapor pressure of 0.33 Pa was achieved in a pure water sample (see Fig. 4). Also, it has been demonstrated that frequency modulation techniques can be used in conjunction with cavity-enhanced detection schemes.

Despite a more complicated apparatus, it has been possible approaching the shot-noise-limited regime by means of the so-called NICEOHMS (Noise Immune Cavity Enhanced Optical-Heterodyne Molecular Spectroscopy) technique.[57] Particularly, by NICEOHMS, Ye et al. have obtained a sensitivity of $5 \cdot 10^{-13}$ in terms of minimum integrated absorption for a 1-s averaging time for detection of C_2H_2 at a 1.06-μm wavelength.[15] More recently, broad-bandwidth (from 1.5 to 1.7 μm), high-spectral-resolution analysis of human breath has been carried out by cavity-enhanced optical-frequency-comb-based spectroscopy with a minimum detectable absorption of $8 \cdot 10^{-10}$ cm^{-1} (Ref. 58). Such detection scheme could be applied as well in the 3-μm window where an optical frequency comb synthesizer (200-nm wide), based on difference frequency generation, has also been demonstrated.[59] However, these setups are not well-suited for field applications but mainly devoted to laboratory use.

In the area of resonator-based sensors, a novel approach has also emerged in the last years for development of miniaturized chemical probes, based on resonators fabricated in silica optical-fibers rings or micro-spheres for analysis in microfluidic devices and liquid samples. In this case, resonator losses are retrieved vs. evanescent interaction with chemicals in the surrounding medium either by ring-down or cavity-transmission measurements.[60,61]

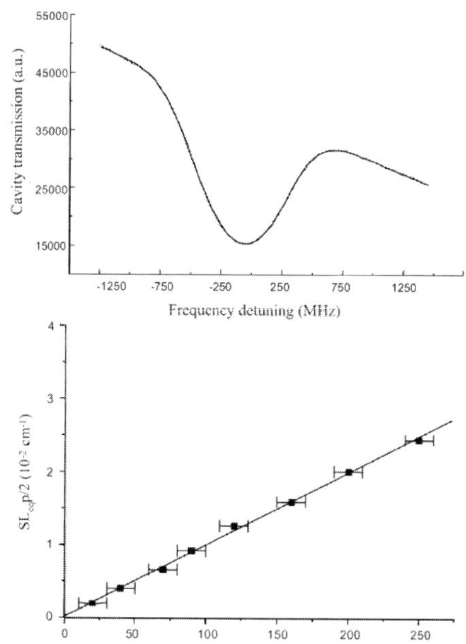

Figure 4. Water vapor absorption signal and absorbance measurements in the high-finesse cavity.[17]

4. Mid-infrared Sensors

Sensitive and selective gas detection can be much more efficiently carried out in the mid-infrared (MIR) spectral region, where strong and well-isolated rovibrational transitions occur for a large number of molecules. Common cw sources of coherent radiation include lasers (such as color center lasers, CO and CO_2 gas lasers, lead-salt diode lasers, quantum and interband cascade lasers) and devices based on nonlinear optical processes such as difference frequency generation and optical parametric oscillation.[62]

While providing an output power ranging from a few mW to several watts with a linewidth on the MHz order, color center lasers (2 – 3.5 μm) suffer from the need for liquid nitrogen cooling. Tunable cw optical parametric oscillators (OPOs), characterized by narrow linewidths (<MHz) and high-power levels up to 4.5-μm wavelength, are still too large and expensive for field deployable sensors. Similarly, use of CO (5–6 μm) and CO_2 (9–11 μm) lasers is prevented by their discrete line-tunability. So far, only lead-salt diode lasers, cascade lasers and difference-frequency generators (DFGs) have been successfully incorporated into portable spectrometers for *in situ* environmental monitoring.

4.1. Lead-Salt Diode Lasers

Historically, lead-salt diodes were the first working semiconductor lasers. They operate at cryogenic temperature, which complicates field operation. They also suffer from multi-mode emission, requiring dispersive optics to eliminate unwanted modes. Besides, mode emission can change dramatically when the diode undergoes thermal cycling.

Notwithstanding, lead salt diode have been used for several *in situ* measurements. A lead-salt diode spectrometer was mounted on an aircraft flying in the stratosphere for detection of species (N_2O and CO) of interest for atmospheric chemistry.[63] The laser was cooled by liquid nitrogen and direct absorption measurement was performed by continuously flowing the sampled gas in a multipass cell. The spectrometer operated unattended during the flight time (4 - 6 hours) averaging 1000 scanned spectra in 5 s, which is about the time of sample clearance. With the appearance of QC lasers, lead-salt diode lasers are foreseen to become less and less competitive for field operation.

4.2. Quantum and Interband Cascade Lasers

QC lasers exhibit high performance characteristics in terms of operating temperature, output power, and wavelength tunability. High power (10–100 mW) room temperature, cw DFB-QC lasers are now commercially available for specific wavelengths within the mid-IR region of 4.3–9.5 μm with a linewidth below 1 MHz and continuous frequency scanning by temperature tuning of ≈ 10 cm^{-1} (Ref. 64). Broad spectral coverage of up to 182 cm^{-1} (around 8.4 μm) with continuous mode-hop free tuning of ≈ 1.25 cm^{-1} has been recently reported using a bound-to-continuum QCL design in conjunction with an external cavity grating configuration.[65]

Their characteristics make possible the use of a variety of spectroscopic detection techniques such as long open path laser absorption spectroscopy, CRDS and ICOS, PAS, as well as evanescent field monitoring using fibers and waveguides.[66-69] Some examples in the field of environmental monitoring include: detection of traces of CO in propylene,[70] monitoring of atmospheric formaldehyde,[71] control of NH_3 concentration in bioreactors,[72] oxygen isotope ratio measurements in CO_2.[73]

Interband cascade (IC) lasers are based on interband transition (between the conduction and the valence bands) and emit between 2.7 to 5.5 μm. The applicability of IC lasers for trace-gas sensing has been demonstrated at cryogenic temperature. For example, the detection and quantification of H_2CO,[74] as well as aircraft and balloon *in situ* concentration measurements of CH_4 and HCl[75] were reported recently. Development of room-temperature cw operation of IC lasers is in progress[76] and a thermoelectrically-cooled cw IC laser operating at 264 K was demonstrated.[77]

4.3. DFG-based Sensors

The wavelength range between 2.5 and 3.5 μm is still hardly accessible to cascade lasers. Thanks to wide tunability, low noise and narrow linewidth, DFG sources proved to be the most valuable spectroscopic tool in this window.[78-80] Two lasers beams (pump and signal) at different frequencies are mixed in a non-linear optical crystal to generate coherent radiation at the difference frequency (idler).

Idler wavelength tuning is accomplished by tuning either the pump or the signal laser. For efficient frequency conversion, a critical requirement is that the interacting waves must stay in phase along their path through the nonlinear crystal (phase-matching condition). In fact, due to frequency dispersion in the crystal refractive index, a phase difference $\Delta\vec{k} \cdot L$ accumulates along the crystal length L, where $\Delta\vec{k} = \vec{k}_p - \vec{k}_s - \vec{k}_i$ ($k = 2\pi n / \lambda$). A phase shift of π is produced every coherence length ($L_c = \pi / \Delta k$), which leads to a reversal of the energy flow from the generated wave to the driving waves.

Maximum conversion efficiency is obtained when $\Delta\vec{k} = 0$. Such condition can be satisfied either by exploiting the birefringence of nonlinear crystals or by means of quasi phase matching (QPM).[81] In the latter scheme the sign of the optical nonlinearity of the crystal is modulated along the propagation direction so that the phase is periodically reset by π with a half-period equal to the coherence length.

The modulation period Λ should be appropriately selected so that the following QPM equation is satisfied

$$\frac{n(T,\lambda_i)}{\lambda_i} = \frac{n(T,\lambda_p)}{\lambda_p} - \frac{n(T,\lambda_s)}{\lambda_s} - \frac{1}{\Lambda} \qquad (10)$$

In contrast to birefringent phase matching, the QPM materials can be engineered to be phase-matchable at any wavelengths within the transparency range of the crystal by selecting the appropriate modulation period. Also, this method allows a free choice of polarization of the interacting waves and the use of the largest nonlinear susceptibility component. QPM interaction can be realized both in bulk crystal and in waveguide devices. The efficiency of bulk QPM is limited by a trade-off between tight focusing for high laser intensities and loose focusing for large interaction lengths and is proportional to the crystal length.

Vice versa, waveguide devices confine the interacting waves over a small cross sectional area along long paths yielding conversion efficiency proportional to L^2. For a review of phase-matchable materials as well as of techniques to realize waveguide QPM see Ref. 82.

In any phase-matching configuration the idler power increases with the ratio $d_{eff}^2 P_p P_s / \lambda_i^2$, where d_{eff} is the effective nonlinear coefficient and P_p (P_s) is the power of the pump (signal) laser.

The well-established grating-engineered QPM periodically-poled lithium niobate (PPLN) technology in combination with a large damage threshold and high thermal conductivity, as well as the commercial availability make periodically poled ferroelectric materials suitable for efficient DFG in the 3–5 μm region. The use of well developed telecommunications diode lasers, fiber amplifiers, optical fiber delivery systems and fiber couplers has led to the realization of compact and robust laser instruments based on DFG in PPLN for trace gas detection application.[83] CW DFG power scaling to the mW level has been recently achieved: 3.5-mW at 3.35 μm produced from a PPLN crystal pumped with high laser pump powers (P_p = 550 mW and Ps = 3.9 W),[84] and 15 mW DFG power at 3.52 μm generated in an efficient ridge waveguide PPLN crystal using P_p = 320 mW and P_s = 520 mW.[85]

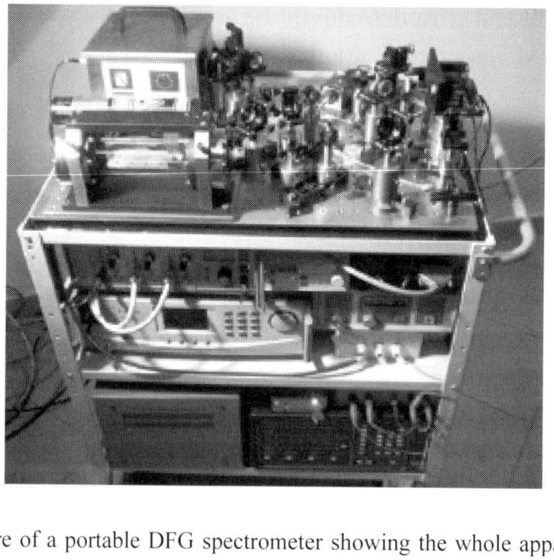

Figure 5. Picture of a portable DFG spectrometer showing the whole apparatus set on a trolley (55×70×75 cm). The laser sources, the wavemeter, and the control electronics are in the lower part, while a small breadboard on the top includes optical elements, fibers, the multi-pass cell, and the detector. The titanium box contains the PPLN crystal and the stages for its alignment. In this way, the breadboard can be used as a separate probe to operate even in challenging conditions without affecting the operation of the DFG pumping sources.[98]

The performance characteristics of DFG-based optical parametric sources have greatly advanced during the past decade. This new type source has become an attractive and useful mid-IR laser source for ultra-sensitive spectroscopic applications. Environmental and industrial applications include: measurements of volcanic gases,[86] vapor-phase benzene;[87] *in situ* and on-line monitoring of CO in an industrial glass furnace,[88] measurements of CO in the exhaust stream of a reactor,[89] tomographic imaging of CO in laminar flames[90] and time-resolved hydrocarbon fuel measurements;[91] isotopic ratio measurements of $^{13}CO_2/^{12}CO_2$ (see Ref. 92), $^{13}CH_4/^{12}CH_4$ (see Ref. 93).

The availability of relatively high DFG power scaling from some tens of μW to the mW level enables DFG-based laser instruments to achieve high sensitivity in conjunction with ultra-sensitive spectroscopic detection techniques based on: long optical pathlength spectroscopy (Herriott or White MRCs);[94,95] cavity ring-down and cavity enhanced spectroscopy;[96,23] a low noise approach, such as wavelength modulation spectroscopy[97] or two-tone frequency modulation spectroscopy.[98]

5. Conclusions

This chapter has given a survey of the current status of laser-based systems aiming at selective and sensitive detection of molecular species that are relevant for environmental monitoring applications.

We have elucidated the factors that are crucial for design and development of effective real-world sensors in different areas. In particular, we have shown how reliable infrared coherent sources, used in combination with high-sensitivity spectroscopic techniques as well as robust optical fiber devices, can provide fast and accurate gas concentration measurements even in extremely hostile environments.

Improving performances of such spectroscopic laser-based sensors, both in terms of ruggedness and detection sensitivity, will have in future a larger and larger impact on forecasting and managing environmental risks.

Acknowledgments

This work was partly performed in the frame of the activities of 'Analysis and Monitoring of the Environmental Risk'-AMRA s.c.a r.l.

References

1. P. Werle, *Spectrochim. Acta* **A 54**, 197 (1998).
2. Harvard Smithsonian Center for Astrophysics, *The Hitran Database 2004* (http://www.hitran.com).
3. W. Demtröder, *Laser Spectroscopy*, Springer-Verlag Berlin Heidelberg New York (2003).
4. C. E. Wieman and L. Hollberg, *Rev. Sci. Instrum.,* **62**, 1 (1991).
5. J. C. Camparo, *Contemp. Phys.* **26**, 443 (1985).
6. J. Faist, F. Capasso, D. L. Sivco, C. Sirtori, A. L. Hutchinson and A.Y. Cho, *Science* **264**, 553 (1994).
7. J. Faist, *Optics & Photonics News*, **17**, 32 (2006).
8. J. M Supplee, E. A. Whittaker and W. Lenth, *Appl. Opt.,* **33**, 6294 (1994).
9. G. C. Bjorklund, *Opt. Lett.*, **5**, 15 (1980).
10. D. E. Cooper and J. P. Watjen, *Opt. Lett.*, **11**, 606 (1986).
11. D. R. Herriott, H. Kogelnik and R. Kompfner, *Appl. Opt.*, **3**, 523 (1964).
12. J. U. White, *J. Opt. Soc. Am.*, **32**, 285 (1942).
13. D. Mazzotti, P. Cancio, G. Giusfredi, M. Inguscio and P. De Natale, *Phys. Rev. Lett.* **86**, 1919 (2001).
14. K. Nakagawa, T. Katsuda, A. S. Shelkovnikov, M. de Labachelerie and M. Ohtsu, *Opt. Commun.*, **107**, 369 (1994).
15. J. Ye, L.-S. Ma and J. L. Hall, *J. Opt. Soc. Am* **B 15**, 6 (1998).
16. R. Peeters, G. Berden, A. Apituley and G. Meijer, *Appl. Phys.* **B 71**, 231 (2000).
17. G. Gagliardi and L. Gianfrani, *Opt. Lasers Eng.*, **37**, 509 (2002).
18. J. B. Paul, L. Lapson and J. Anderson, *Appl. Opt.*, **40**, 4904 (2001).
19. V. L. Kasyutich, C.E. Canosa-Mas, C. Pfrang, S. Vaughan and R. P. Wayne, *Appl. Phys.* **B 75**, 755 (2002).
20. S. Williams, M. Gupta, T. Owano, D. S. Baer, A. O'Keefe, D. R. Yarkony and S. Matsika, *Opt. Lett.*, **29**, 1066 (2004).
21. Y. A. Bakhirkin, A. A. Kosterev, C. Roller, R. F. Curl and F. K. Tittel, *Appl. Opt.,* **43**, 2257 (2004).
22. M. L. Silva, D. M. Sonnenfroh, D. I. Rosen, M. G. Allenand and A. O'Keefe, *Appl. Phys.* **B 81**, 705 (2005).
23. P. Malara, P. Maddaloni, G. Gagliardi and P. De Natale, *Opt. Express*, **14**, 1304 (2006).
24. D. Romanini, A. A. Kachanov, N. Sadeghi, F. Stockel, *Chem. Phys. Lett.* **264**, 316 (1997).
25. G. Berden, R. Peeters and G. Meijer, *Int. Rev. Phys. Chem.*, **19**, 565 (2000).
26. S. Stry, P. Hering and M. Mürtz, *Appl. Phys.* **B 75**, 297 (2002).
27. G. von Basum, D. Halmer, P. Hering, M Mürtz, S. Schiller, F. Müller, A. Popp and F. Kühnemann, *Opt. Lett.*, **29**, 797 (2004).
28. J. T. Hodges and R. Ciurylo, *Rev. Sci. Instrum.,* **76**, 023112 (2005).

29. D. Halmer, G. von Basum, P. Hering and M. Mürtz, *Opt. Lett.*. **30**, 2314 (2005).
30. D. S. Baer, J. B. Paul, M. Gupta and A. O'Keefe, *Appl. Phys.* **B 75**, 261 (2002).
31. D. Mazzotti, P. Cancio, A. Castrillo, J. Galli, G. Giusfredi and P. De Natale, *J. Opt.* **A**: *Pure Appl. Opt.* **8**, S490 (2006).
32. F. J. M. Harren, J. Cotty, J. Oomens, S. L. Hekkert, in *Photo-acoustic spectroscopy in trace gas monitoring, in Encyclopedia of Analytical Chemistry*, R.A. Meyers ed. (Wiley, Chichester 2000), p 2203.
33. M. Nagele and M.W. Sigrist, *Appl. Phys.* **B 70**, 895 (2000).
34. L. Gianfrani, P. De Natale and G. De Natale, *Appl. Phys.* **B 70,** 467 (2000).
35. L. Gianfrani and P. De Natale, *Opt. & Phot. News.*, **11,** 44 (2000).
36. A. Rocco, G. De Natale, P. De Natale, G. Gagliardi and L. Gianfrani , *Appl. Phys.* **B 78**, 235 (2004).
37. M. De Rosa, G. Gagliardi, A. Rocco, R. Somma, P. De Natale and G. De Natale, *Geochem. Trans.*, **8:5** (2007).
38. L. Gianfrani, A. Rocco, G. Battipaglia, A. Castrillo, G. Gagliardi, A. Peressotti and M. F. Cotrufo, *Applied Spectroscopy,* **58**, 1051 (2004)
39. J. Balesdent, A. Mariotti and B. Guillet, *Soil Biol. Biochem.,* **19**, 25 (1987).
40. S.W. Leavitt, E. A. Paul, E. Pendall, P. J. Pinter and B. A. Kimball, *Nucl. Instr. Meth.* **B 123**, 451 (1997).
41. G. Gagliardi, R. Restieri, G. Casa and L. Gianfrani, *Opt. Lasers Eng.,* **37**, 131 (2002).
42. E. R. Th. Kerstel, R. van Trigt, N. Dam, J. Reuss and H. A. J. Meijer, *Anal. Chem.,* **71**, 5297 (1999).
43. P. Bergamaschi, M. Schupp and G. W. Harris, *Appl. Opt.,* **33**, 7704 (1994).
44. D.E. Murnick and B. J. Peer, *Science,* **263**, 945 (1994).
45. J. F. Becker, T. B. Sauke and M. Loewenstein, *Appl. Opt.*, **31**, 1921 (1992).
46. G. Gagliardi, A. Castrillo, R. Q. Iannone, E. R. Th. Kerstel and L. Gianfrani, *Appl. Phys.* **B 77**, 119 (2003).
47. L. Gianfrani, G. Gagliardi, M. van Burgel and E. R. Th. Kerstel, *Opt. Exp,* **11**, 1566 (2003).
48. A. Castrillo, G. Casa, M. van Burgel, D. Tedesco, L. Gianfrani, *Opt. Express* **12**, 6515 (2004).
49. K. Uehara, K. Yamamoto, T. Kikugawa and N. Yoshida, *Sens. and Act.* **B 74**, 173 (2001).
50. K. Uehara, K. Yamamoto, T. Kikugawa, S. Toyoda, K. Tsuji and N. Yoshida, *Sens. and Act.* **B 90**, 250 (2003).
51. B. A. Paldus, C. C. Harb, T. G. Spence, B. Wilke, J. Xie, J. S. Harris and R. N. Zare, *J. Appl. Phys.,* **83**, 3991 (1998).
52. G. Totschnig, D. S. Baer, J. Wang, F. Winter, H. Hofbauer and R. K. Hanson, *Appl. Opt.*, **39**, 2009 (2000).
53. E.R.T. Kerstel, R.Q. Iannone, M. Chenevier, S. Kassi, H.-J. Jost and D. Romanini, *Appl. Phys.* **B 85**, 397 (2006).
54. S. Kassi, M. Chenevier, L. Gianfrani, A. Salhi, Y. Rouillard, A. Ouvrard and D. Romanini, *Opt. Express,* **14**, 11442- (2006).
55. J. Morville, S. Kassi, M. Chenevier and D. Romanini, *Appl. Phys.* **B 80**, 1027 (2005).
56. R. W. P. Drever, J. L. Hall, F. V. Kowalski, J. Hough, G. M. Ford, A. J. Munley and H. Ward, *Appl. Phys.* **B 31**, 97 (1983).

57. J. Ye, L-S Ma and J. L. Hall, *J. Opt. Soc. Am.* **B 16**, 2255 (1999).
58. M. J. Thorpe, D. Balslev-Clausen, M. S. Kirchner and J. Ye, *Opt. Express,* **16**, 2387 (2008).
59. P. Maddaloni, P. Malara, G. Gagliardi and P. De Natale, *New J. of Physics,* **8**, 262 (2006).
60. R. S. Brown, I. Kozin, Z. Tong, R. D. Oleschuk and H.-P. Loock, *Chem. Phys.,* **117**, 10444 (2002).
61. G. Farca, S. I. Shopova and A.T. Rosenberger, *Opt. Express,* **15**, 17443 (2007).
62. P. De Natale, P. Cancio, D. Mazzotti, Infrared precision spectroscopy using femtosecond-laser-based optical frequency-comb synthesizers, in *Femtosecond Laser Spectroscopy*, P. Hannaford ed., (Springer, 2005), pp. 109-132.
63. M. Pantani, F. Castagnoli, F. D'Amato, M. De Rosa, P. Mazzinghi and P. Werle, *Infrared Phys. Technol.,* **46**, 109 (2004).
64. http://www.alpeslasers.com/.
65. G. Wysocki, R. F. Curl, F. K. Tittel, F. Capasso, L. Diehl, M. Troccoli, G. Höfler, R. Maulini and J. Faist, in *Conference on Lasers and Electro-Optics/Quantum Electronics and Laser Science Conference and Photonic Applications Systems Technologies 2007* Technical Digest, Optical Society of America, Washington, DC, 2007, CFB3.
66. G. Gagliardi, F. Tamassia, P. De Natale, C. Gmachl, F. Capasso, D. L. Sivco, J. N. Baillargeon, A.L. Hutchinson and A.Y. Cho, *Eur. Phys. J.* **D 19**, 327 (2002).
67. S. Borri, S. Bartalini, P. De Natale, M. Inguscio, C. Gmachl, F. Capasso, D. L. Sivco and A. Y. Cho, *Appl. Phys.* **B 85**, 223 (2006).
68. A. Castrillo, E. De Tommasi, L. Gianfrani, L. Sirigu and J. Faist, *Opt. Lett.,* **31**, 3040 (2006).
69. A. Kosterev, G. Wysocki, Y.A. Bakhirkin, S. So, R. Lewicki, M. P. Fraser, F. K. Tittel and R. F. Curl, *Appl. Phys* **B** (2007), DOI: 10.1007/s00340-007-2846-9.
70. A. Kosterev, Y.A. Bakhirkin, F.K. Tittel, S. Blaser, Y. Bonetti and L. Hvozdara, *Appl. Phys.* **B 78**, 673 (2004).
71. G. Wysocki, Y. A. Bakhirkin, S. So, F. K. Tittel, R. Q. Yang and M. P. Fraser, *Appl. Opt.,* **46**, 8202 (2007).
72. A. Kosterev, R. F. Curl, F. K. Tittel, R. Kohler, C. Gmachl, F. Capasso, D. L. Sivco and A.Y. Cho, *Appl. Opt.,* **41**, 573 (2002).
73. A. Castrillo, G. Casa and L. Gianfrani, *Opt. Lett.*, **32**, 3047 (2007).
74. J. H. Miller, Y. A. Bakhirkin, T. Ajtai, F. K. Tittel, C. J. Hill and R. Q. Yang, *Appl. Phys.* **B 85**, 391 (2006).
75. L. E. Christensen, C. R. Webster and R. Q. Yang, *Appl. Opt.,* **46**, 1132 (2007).
76. R.Q. Yang, C.J. Hill and B. H. Yang, *Appl. Phys. Lett.*, 87, 151109 (2007).
77. K. Mansour, Y. Qiu, C. J. Hill, A. Soibel and R. Q. Yang, *Electron. Lett.*, **42**, 1034 (2006).
78. F. K. Tittel, D. Richter and A. Fried: In Solid- State Mid-Infrared Laser Sources, Topics in Appl. Phys. 89, I.T. Sorokina, K.L. Vodopyanov (eds.) (Springer-Verlag, Berlin 2003), p. 445.
79. S. Borri, P. Cancio, P. De Natale, G. Giusfredi, D. Mazzotti and F. Tamassia, *Appl. Phys.* **B 76**, 473 (2003).
80. D. Mazzotti, P. De Natale, G. Giusfredi, C. Fort, J. A. Mitchell and L. Hollberg, *Opt. Lett.*, **25**, 350 (2000).
81. G. S. He and S. H. Liu, *Physics of nonlinear optics*, World Scientific (2000).

82. P. Günter (Ed.), *Nonlinear optical effects and materials*, Springer-Verlag Berlin Heidelberg (2000).
83. W. Chen, J. Cousin, E. Poullet, J. Burie, D. Boucher, X. Gao, M. W. Sigrist and F. K. Tittel, *C.R. Physique* (2007), DOI: 10.1016/j.crhy.2007.09.011.
84. P. Maddaloni, G. Gagliardi. P. Malara and P. De Natale, *Appl. Phys.* **B 80**, 141 (2005).
85. D. Richter, P. Weibring and A. Fried, *Opt. Express*, **15**, 564 (2007).
86. D. Richter, M. Erdelyi, R.F. Curl, F.K. Tittel, C. Oppenheimer, H. J. Duffell and M. Burton, *Opt. Lasers Eng*., **37**, 171 (2002).
87. W. Chen, F. Cazier, F. K. Tittel and D. Boucher, *Appl. Opt.,* **39**, 6238 (2000).
88. A. Khorsandi, U. Willer, L. Wondraczek and W. Schade, *Appl. Opt.,* **43**, 6481 (2004).
89. R. Barron-Jimenez, J. A. Caton, T. N. Anderson, R. P. Lucht, T. Walther, S. Roy, M. S. Brown and J. R. Gord, *Appl. Phys.* **B 85**, 185 (2006).
90. L. Wondraczek, A. Khorsandi, U. Willer, G. Heide, W. Schade and G. H. Frischat, *Combust. Flame,* **138**, 30 (2004).
91. A. Klingbeil, J. B. Jeffries and R. K. Hanson, *Proc. Combust. Inst.* **31**, 807 (2007).
92. M. Ederlyi, D. Richter and F. K. Tittel, *Appl. Phys.* **B 75**, 289 (2002).
93. M. E. Trudeau, P. Chen, G. A. Garcia, L.W. Hollberg and P. P. Tans, *Appl. Opt.,* **45**, 4136 (2006).
94. R. Bartlome, M. Baer and M. W. Sigrist, *Rev. Sci. Instrum.,* **78**, 0131101 (2007).
95. T. Yanagawa, O. Tadanaga, K. Magari, Y. Nishida, H. Miyazawa, M. Asobe and H. Suzuki, *Appl. Phys. Lett.,* **89**, 221115 (2006).
96. S. Stry, S. Thelen, J. Sacher, D. Halmer, P. Hering and M. Mürtz, *Appl. Phys.* **B 85**, 365 (2006).
97. H. Waechter and M. W. Sigrist, *Appl. Phys.* **B 87**, 539 (2007).
98. P. Maddaloni, P. Malara, G. Gagliardi and P. De Natale, *Appl. Phys.* **B 85**, 219 (2006).

LASER WELDING PROCESS MONITORING SYSTEMS: ADVANCED SIGNAL ANALYSIS FOR QUALITY ASSURANCE

Giuseppe D'Angelo*

FIAT Research Center – Manufacturing & Materials
Strada Torino 50, 10043 Orbassano, Italy
**E-mail: giuseppe.dangelo@crf.it*

Laser material processing today is widely used in industry. Especially laser welding became one of the key-technologies, e. g., for the automotive sector. This is due to the improvement and development of new laser sources and the increasing knowledge gained at countless scientific research projects. Nevertheless, it is still not possible to use the full potential of this technology. Therefore, the introduction and application of quality-assuring systems is required.

For a long time, the statement "the best sensor is no sensor" was often heard. Today, a change of paradigm can be observed. On the one hand, ISO 9000 and other by law enforced regulations have led to the understanding that quality monitoring is an essential tool in modern manufacturing and necessary in order to keep production results in deterministic boundaries. On the other hand, rising quality requirements not only set higher and higher requirements for the process technology but also demand quality-assurance measures which ensure the reliable recognition of process faults. As a result, there is a need for reliable online detection and correction of welding faults by means of an in-process monitoring.

The chapter describes an advanced signals analysis technique to extract information from signals detected, during the laser welding process, by optical sensors. The technique is based on the method of reassignment which was first applied to the spectrogram by Kodera, Gendrin and de Villedary[22,23] and later generalized to any bilinear time-frequency representation by Auger and Flandrin.[24]

Key to the method is a nonlinear convolution where the value of the convolution is not placed at the center of the convolution kernel but rather reassigned to the center of mass of the function within the kernel. The resulting reassigned representation yields significantly improved components localization. We compare the proposed time-frequency distributions by analyzing signals detected during the laser welding of tailored blanks, demonstrating the advantages of the reassigned representation, giving practical applicability to the proposed method.

1. Introduction

The application of laser welding technique in industry has gradually become a mature processing technology. Laser welding can be classified into pulse laser welding and continuous laser welding and the latter, in particular, can further be divided into heat conduction welding and deep penetration welding. With the uplift of laser output power, and also the development of high power laser apparatus in particular, the deep penetration technology of laser sees rapid development in Europe and overseas. The technology can now be applied to galvanized sheets, aluminum sheets, titanium sheets and ceramic materials as well, in addition to low-carbon steel. All in all, the wide use of laser welding technology in the automotive industry proves its maturity.

For instance, welding of car tops, high-speed welding of base plates, structural parts of cars (including car doors and bodies) and welding of transmission control gears have been widely applied to industrial production.

1.1. Pros and Cons of Laser Welding Technology

1.1.1. Pros

a - Laser welding technology provides, compared to traditional welding and fusion welding, a number of advantages:
i. Doubling processing accuracy: the high temperature area of laser welding seam will deform due to high heat, but because of the smaller seam width, the degree of deformation is very limited;
ii. Power and size of laser beam can be adjusted dynamically according to processing requirements;
iii. The solid laser can transmit laser at a place far from the operation site, making it no difficult task to spatially separate the energy source from processing equipment;
iv. The laser beam causes no wear and can work stably for a long time.
b - Statistics reveals that in developed industrial countries in Europe, and in the US, half to 70% of automotive parts and components are processed

with laser, indicating that laser welding has become standard technique in the auto industry. With laser, metal plates of different thicknesses and surface finishes can be welded and then stamped to produce car body panels of a structure with the most reasonable metal combination. And because laser introduces little deformation, there will not be the need for secondary processing. Laser welding has accelerated the replacement of stamped parts for their forged counterparts. The use of laser welding can also shorten lap widths, reduce the number of reinforcing parts, as well as decrease the volume of the structural parts of the car body.

Furthermore, laser welding ensures that welding spots are bonded on the molecular level, hence increases the rigidness and crash safety of car body as well as it effectively reduces noise inside cars.

1.1.2. Cons

a- Laser welding involves a complicated process and some defects, listed below, can occur:

i. Pores: diameter of normal pores should not exceed 1.0 mm;
ii. Micro-pore: pores less than 0.2 mm in diameter;
iii. Cavity: pores with diameter greater than 1.0 mm;
iv. Fusion-weld seam: no welding is present in the seam, which looks like laser fusion weld seam;
v. Poor connection of welding: welding is not connected to the sides of work pieces and the seam at the point of connection looks like "scattered wisps";
vi. Single-sided connection of welding: welding is connected to one side only;
vii. "Sausage" effect: the work pieces are not connected, with welding at the weld seam straightly extended or piled up;
viii. Irregular weld seam: weld seam is dented or raised;
ix. Scaled piling: the surface of weld seam is not smooth and looks very rough;
x. Problem at the front/end of welding seam: insufficient or excessive infill at the weld seam at the edge of the work piece, or un-melted welding residue is found on the orbit.

b- Laser welding is under the influence of a number of numerous factors which can affect the quality of laser welding in automotive applications. Some of them are reported as follow:
i. Dirty protective glass lens or aged arc lamp will decrease the power of laser;
ii. Position of the focal spot of laser incorrectly set;
iii. The feed rate is wrongly;
iv. The gap size between the welded parts exceeds the requirements of laser equipment.

2. Monitoring Devices

2.1. General Aspects

The function of the weld monitor is based on the collection of the photonic emission directly from the weld pool, and the conversion of these emissions to an electrical signal, by suitable sensors, which can be analyzed by computer software. Since the signals contain information about the beam-material interaction, welding defects can be detected during the process and recorded for each single work piece.

The detectable optical emissions which can be used as the process signals are:

i. The reflected laser, originated from the amount of the laser source radiation which is not absorbed by the material;

ii. Acoustic emissions, originated from the stress waves induced by changes in the internal structure of a work piece;

iii. Radiation emitted from the metal vapor and the molten pool.

The metal vapors and the molten pool emit continuous radiation which spectrum varies with different laser applications.

For instance, during Nd-YAG laser welding, the process radiation is in the visible and infrared range. For CO_2 laser keyhole welding, the plasma generated is known to emit light with a wavelength between 190 nm and beyond 400 nm, and the spatter emits light with a wavelength between 1000 nm and 1600 nm. In addition, the geometrical parameters of the keyhole and melt pool also contain useful information which can be used to inspect the welding quality.

The CO_2 wavelength can be directly measured by infrared detectors. The signals from plasma, metal vapors, weld spatter, heating, moltening and resolidifyng of material, can be detected by optical sensors such as photo-diodes or pyroelectric sensors.

Photo-diodes are able to detect signals from ultra-violet to near infrared. The availability of optical filters allows detecting different signals generated during the process. The advantages of this type of detectors are the high temporal resolution of the recorded signals and the low price compared to other devices like spectrographs or cameras.

Besides photo-diodes provide an high signal-to-noise ratio, the pyroelectric sensors allow detecting signals on a wide bandwidth. The main disadvantages of these sensors are their low mean power capability and the need of choppers.

A number of laser weld monitoring schemes have been developed, using detectors limited to spectral windows, and various optical schemes.

A lot of research activities has demonstrated that the information on the keyhole radiation depend on the angle of photodiode with respect to the laser beam direction. In case of a lateral position one or more detectors can be used in different angles to the surface. The main disadvantage of this set-up is the higher pollution sensitivity of the detector and the reduced handling capability of the welding optics.

To overcome this problem a coaxial sensor could be used. In this position the detector does not limit the moving capabilities of the welding optics and is even better protected against pollution.

Besides, in this position, the sensor could detect more information on the keyhole than on the surface plasma only.

Figure 1 reports a general lay-out of monitoring systems, used for CO_2 and Nd-YAG laser welding process, made by a) coaxial optical detectors coupled to optical elements, b) anti-aliasing low pass filter, c) PC computer with an analog to digital (A/D) converter board.

The computer and A/D board reads the signal levels at a certain frequency and analyses the data stream in real time by dedicated software.

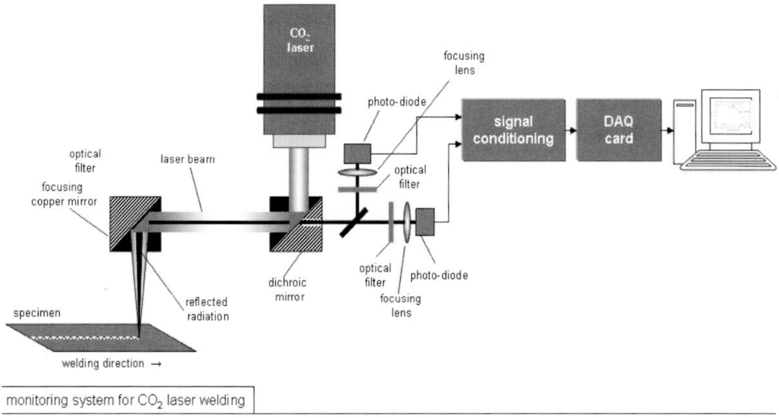

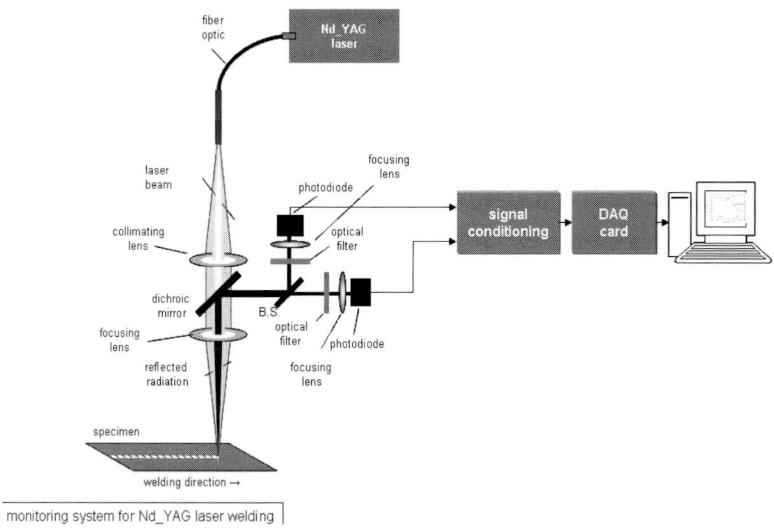

Figure 1. General layout of monitoring systems. The system for the CO_2 laser welding is based on a dichroic mirror able to reflect the laser wavelength (λ=10.6 μm) and transmit the reflected light (from UV to NIR). The system for Nd:YAG laser welding is based on optical elements transmitting the laser wavelength (λ=1.06 μm) and reflecting the reflected light (from UV to NIR). In both cases, the reflected light is divided by a beam-splitter, filtered by "ad-hoc" optical filter and finally detected by photodiodes.
The voltage signals are acquired by a multi-channel data acquisition card (DAQ) and elaborated by a specific software.

2.2. Why Should We Use the Anti-Aliasing Filter?

Let's start from the Nyquist theorem: the Nyquist theorem states that a signal must be sampled at least twice as fast as the bandwidth of the signal to accurately reconstruct the waveform, otherwise, the high-frequency content will alias at a frequency inside the spectrum of interest (pass band).

An alias is a false lower frequency component that appears in sampled data acquired at too low a sampling rate. To prevent aliasing in the pass band, the sampling frequency should be theoretically at least twice the maximum frequency of the sampled. But how do we ensure that this is definitely the case in practice? Even if we are sure that the signal being measured has an upper limit on its frequency, pickup from stray signals such as the power line frequency could contain frequencies higher than the Nyquist frequency. These frequencies may then alias into the appropriate frequency range and thus give you erroneous results.

To be completely sure that the frequency content of the input signal is limited, a low pass filter (a filter that passes low frequencies but attenuates the high frequencies) is added before the sampler and the ADC. This filter is called an anti-alias filter because by attenuating the higher frequencies (greater than Nyquist), it prevents the aliasing components from being sampled. Let's start, as example, from an ideal anti-alias filter. It passes all the appropriate input frequencies below a certain frequency value (let's say f_1) and cuts all the undesired frequencies above this frequency value. However, such a filter is not physically realizable. In practice, all the filters are characterized by the transition band, which contains a gradual attenuation of the input frequencies. Although we want to pass only signals with frequencies below a certain frequency value, those signals in the transition band could still cause aliasing. Therefore, in practice, the sampling frequency should be greater than two times the highest frequency in the transition band. So, this turns out to be more than two times the maximum input frequency (f_1). That is one reason why we may see that the sampling rate is more than twice the maximum input frequency.

2.3. Photodiodes

Photodiodes are semiconductor light sensors that generate a current or voltage when the p-n junction in the semiconductor is illuminated by light. The term photodiode usually refers to sensors used to detect the intensity of light. Photodiodes have no internal gain but can operate at much higher light levels than other light detectors. In contrast Avalanche Photodiodes (APD) do have internal gain. The materials used to realize the photodiodes are:

Silicon. It is now the most widely used material for photodiodes. The wavelength range is about 200 – 1100 nm with a peak near 850 nm, at which the spectral responsivity is up to 0.65 A/W and the quantum efficiency is close to 100%

Germanium. The wavelength range of these junction diodes extends further into NIR to about 2 μm. The responsivity wavelength (1.4 μm) is 0.66 A/W, which corresponds to a quantum efficiency of about 82%.

Other materials used are InGaAs, InAsCdTe, GaASP. Often these detectors are labeled according their structure: p-n, p-i-n. The terminology of these detectors has undergone several changes and it is ambiguous due to the ability of the junction detector to serve as photovoltaic or as photoconductive device. In photovoltaic mode no bias is applied, and under irradiation the photodiode generates a voltage of a certain polarity that may drive a current through an external circuit. In the photoconductive mode, an external bias of a polarity opposite to that of the unbiased mode is applied. Consequently, the current also flows in the direction opposite to that of the unbiased mode. The signal appears as voltage drop across the load resistor R_l. Following Palmer's suggestion (1980), photovoltaic mode corresponds to unbiased sensor, while photoconductive correspond to biased sensor.

Below the parameters characterizing the photodiodes sensors are reported:

▪ *Cut-off frequency* (f_c) - Measure used to evaluate the time response of high speed avalanche photodiodes and p-i-n photodiodes to a sine wave modulated light input. It is defined as the frequency at which the photodiode output decreases by 3 dB from the output at 100 kHz. The rise time t_r has a relation with the cut-off frequency f_c as follows: $t_r = 0.35/f_c$.

▪ *Dark Current (ID) and shunt resistance (R_{sh})* - The dark current is a small current which flows when a reverse voltage is applied to a photodiode even in dark state. This is a source of noise for applications in which a reverse voltage is applied to photodiodes (for example, as with p-i-n photodiodes). In contrast, for applications where no reverse voltage is applied, noise characteristics are figured out from the shunt resistance. This shunt resistance is the voltage-to-current ratio in the vicinity of 0V. The shunt resistance R_{sh} is defined as follows: R_{sh}=10 mV/I_0 (Ω), where I_0 is the dark current at V_R=10 mV.

▪ *Infrared sensitivity ratio* - Ratio of the output current (IR) measured with a light flux (2856 K, 100 I_x) passing through an R-70 (t=2.5 mm) infrared filter to the short circuit current (I_{sc}) measured without the filter. It is commonly expressed in percent:
Infrared sensitivity ratio= (I_r/I_{sc}) * 100 (%).

▪ *Noise equivalent power (NEP)* - The NEP is the amount of light equivalent to the noise level of a device. Stated differently, it is the light level required to obtain a signal-to-noise ratio of unity. Since the noise level is proportional to the square root of the frequency bandwidth, the NEP is measured at a band- width of 1Hz and thus expressed in units of $W/Hz^{1/2}$. NEP= [Noise current ($A/Hz^{1/2}$) / Photo sensitivity at λ_p (A/W)].

▪ *Quantum Efficiency (QE)* - The quantum efficiency is the number of electrons or holes that can be detected as a photocurrent divided by the number of the incident photons. This is commonly expressed in percent. The quantum efficiency and photo sensitivity have the following relationship at a given wavelength: QE= (S*1240)/λ*100(%), where *S* is the photo sensitivity in ANV at a given wavelength and λ is the wavelength in nm.

▪ *Rise time (t_r)* - This is the measure of the time response of a photodiode to a stepped light input, and it is defined as the time required for the output to change from 10% to 90% of the steady output level. The rise time depends on the incident light wavelength and load resistance.

3. Signal Analysis Techniques

The evaluation of laser welding quality can be carried out by analyzing the emissions which occur during the process. In particular it is possible to find out the relationship between emission characteristics and weld quality characteristics.[1-9] Since these techniques are indirect, they require accurate signal interpretation and processing to infer information about the actual condition of the weld: the more accurate signal analysis technique, the better weld quality characterization.

The purpose of the next pages is to show the application of reassigned time-frequency distribution, specifically the reassigned smoothed pseudo Wigner-Ville distribution (RSPWVD), for improving the analysis of the welding process.

3.1. Time – Frequency Representations (TFRs)

Most of the signals are time-domain signals in their raw format.

This representation is not always the best representation for most signal processing related applications. In many cases, the most distinguished information is hidden in the frequency content of the signal.

By performing a Fourier transform to obtain the signal spectrum, one can see how the energy of the signal is distributed in frequency. For stationary signals there is no need to go beyond the time or frequency domain. In presence of non stationary signals, it is necessary to create functions which represent the energy of the signals simultaneously in time and in frequency. These bi-dimensional functions, which indicate the time-varying frequency content of a signal, are referred to as a time-frequency distributions (TFDs). These functions have been developed for analyzing a wide variety of signals, including speech, acoustic signals, biological signals, radar and sonar signals.

The earliest and one of the most commonly used of these TFDs, is probably the spectrogram,[10, 11] defined as:

$$s_X(T,\omega) = \left| \int x(t)\, h^*(\tau - t)\, e^{-j\omega\tau} d\tau \right|^2 \qquad (1)$$

where *x* is the signal and *h* is a window function.

The spectrogram presents two main drawbacks: the first one is a consequence of the Heisenberg uncertainty principle, making it impossible to simultaneously have perfect resolution in both time and frequency. Given a specific spectrogram, the standard deviations for time and frequency of the window function, σ_t and σ_ω, respectively, are not independent of each other.

The Heisenberg uncertainty principle limits a spectrogram's time and frequency resolution by the following inequality: $\sigma_t^2\sigma_\omega^2 \geq 0.25$.

Note that the window type determines the time-frequency spread of a spectrogram. For example, the product of $\sigma_t^2\sigma_\omega^2$ is 0.2635 for a spectrogram calculated with a Hanning window.

A Gaussian window function satisfies the equality $\sigma_t^2\sigma_\omega^2 = 0.25$, but the current application aims to alter the shape of the time signal as little as possible while avoiding discontinuities across the boundaries of the windowed signal. The Hanning window is chosen as a compromise.

The time-frequency resolution of a spectrogram depends only on the window size and type and it is independent of frequency. A wide window gives better frequency resolution, but worsens the time resolution, whereas a narrow window improves time resolution but worsens frequency resolution.

The second drawback, with multi-components signals, is the presence of interference terms in regions of the time-frequency plane where the auto terms overlap. These interference terms will nearly be identical to zero, if the signal components are sufficiently distant.

In order to illustrate the interference structure, we consider a synthetic two-component signal composed of two parallel chirps.

Figure 2 shows the effect of the distance between the signals. From the figures, we can see that the interference terms are present where the signals are not sufficiently distant. Besides, the interference terms do not depend on the window length.

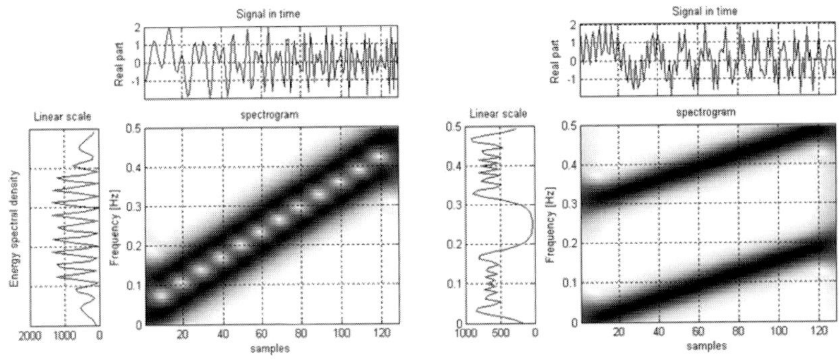

Figure 2. Spectrogram of two parallel chirps, (left) presence of interference terms, (right) signal with components sufficiently distant.

The trade-off of the spectrogram, which is controlled by the analysis window, has prompted the development of more advanced TFDs. Among them are the TFDs of Cohen's classes.[12] One of the most interesting is the Wigner-Ville distribution (WVD)[13,14] defined as:

$$W_x(t,\omega) = \int x\left(t+\frac{\tau}{2}\right)x^*\left(t-\frac{\tau}{2}\right)e^{-j\omega\tau}\,d\tau \tag{2}$$

The WVD is highly concentrated and can be interpreted as a short-time Fourier transform with the window matched to the signal. This distribution satisfies a large number of desirable mathematical properties. In particular, the WVD is always real valued, it preserves time and frequency shifts and satisfies the marginal properties.[14-16]

Figure 3a shows the Wigner-Ville distribution of a 128-points signal made up of a sinusoidal frequency modulation followed by a pure tone simultaneously with a chirp component and figure 3b shows the instantaneous frequency law of all the components, forming the time-frequency skeleton to which a time-frequency representation should be as near as possible.

The problem of the WVD is the so-called cross-term interference, which appears as frequencies that lie between the frequencies of any two

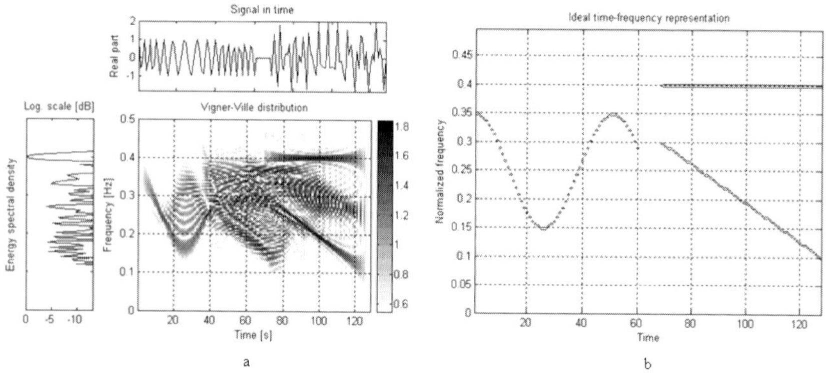

Figure 3. Wigner-Ville distribution (a), Instantaneous frequency law of the three signal components (b).

strong components. With respect to the spectrogram interferences, even if the signal components are distant, the interferences terms will never be nearly zero. The interference construction of the WVD can be summarized as follows: two points of the time-frequency plane interfere to create a contribution on a third point which is located at their geometrical midpoint. Besides, these interference terms oscillate perpendicularly to the line joining the two points interfering, with a frequency proportional to the distance between these two points. These interferences terms may overlap the auto-terms (signal components) and make difficult the WVD interpretation.

As the spectrogram suffers from the trade-off between time resolution and frequency resolution, the WVD suffers from the trade-off between the quantity of interferences and the number of good properties.

These interferences can often be reduced while preserving the time and frequency shift invariance property (and possibly other interesting theoretical properties) by a two-dimensional low-pass filtering of the WVD, leading to a time-frequency representation of the Cohen's class,[17,18] which can be written as:

$$\mathrm{TFR}_x(t,\omega) = \iint \Phi_{TF}(u,\Omega)\, \mathrm{WV}(x; t-u, \omega-\Omega)\, du\, d\Omega / 2\pi \qquad (3)$$

where Φ is the kernel of TFR. Of course, the WVD is the element of the Cohen's class with the function Φ=1. In case Φ is a smoothing function, Eq. (3) allows the interpretation of TFR_x as a smoothed version of the WVD.

Consequently, such a distribution will attenuate in a particular way the interferences of the WVD. Firstly, as a smoothing function, we can consider a short-time window *h*. In this case the smoothing function will be narrow in time and wide in frequency, leading to a good time resolution but bad frequency resolution, and vice-versa. This draw-back can be overcome by considering the separable smoothing function:

$$\Phi_{TF}(u,\Omega) = g(u)\,H(\Omega) \tag{4}$$

where H(Ω) is the Fourier transform of a smoothing window h(t), allowing a progressive and independent control, in both time and frequency, of the smoothing applied to the WVD.

The obtained distribution

$$SPW_x(t,\omega) = \int\int h(\tau)g(u)\,x\left(t-u+\frac{\tau}{2}\right)x^*\left(t-u-\frac{\tau}{2}\right)e^{-j\omega\tau}\,dt\,d\tau \tag{5}$$

is known as the smoothed-pseudo Wigner-Ville distribution (SPWVD). The previous compromise of the spectrogram between time and frequency- resolutions is now replaced by a compromise between the joint time-frequency resolution and the level of the interference terms: the more one smoothes in time and/or frequency, the poorer the resolution in time and/or frequency. Figure 4a shows the smoothed-pseudo Wigner-Ville distribution (SPWVD) of the same signal used for the Wigner-Ville distribution, while figure 4b shows the instantaneous frequency law of all components, forming the time-frequency skeleton to which a time-frequency representation should be as near as possible.

The smoothing action suppresses the cross-terms but it also produces a less accurate time-frequency localization of the signal components. Its shape and spread must therefore be properly determined so as to produce a suitable trade-off between good interference attenuation and good time-frequency concentration.[19-22]

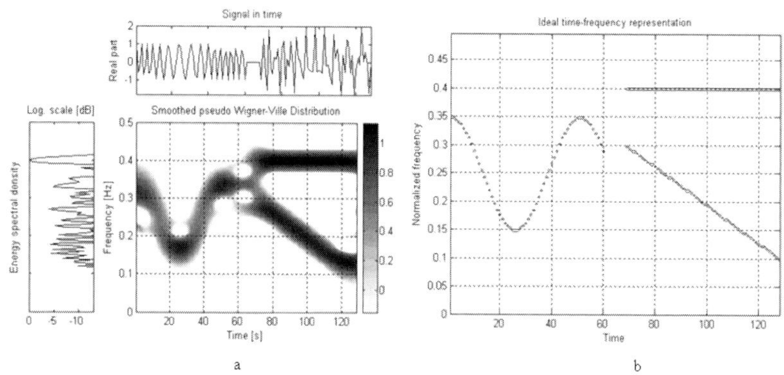

Figure 4. Smoothed Pseudo Wigner-Ville distribution (a), Instantaneous frequency law of the three signal components (b).

3.2. The Reassignment Method

In order to improve the readability of a signal representation, as a complement to the smoothing action, other processing can be used. The reassignment method improves the readability of a signal representation.

This method was first discovered by Kodera, Gendrin and de Villedary,[22,23] who used it only for the spectrogram. A new formulation of this method, leading to its practical use for a large family of TFR, was performed by Auger and Flandrin.[24] The aim of the reassignment method is to improve the sharpness of the localization of the signal components by reallocating the representation of the components in the time-frequency plane. In the reassignment method, "energy" is moved away from its original location, coordinates (t, ω), to a new location, the reassigned coordinates $(t^\wedge, \omega^\wedge)$ thus greatly reducing the "spread" of a spectrogram. The reassignment method improves the time-frequency resolution of any time-frequency shift invariant distribution of Cohen's class by concentrating its energy at a center of gravity. Auger and

Flandrin[24] showed that the reassigned coordinates $t^\wedge$ and $\omega^\wedge$,for any member of the Cohen's class, are:

$$t^\wedge(x;t,\omega)=t-\frac{\iint u\Phi_{TF}(u,\Omega)WV(x;t-u,\omega-\Omega)dud\Omega/2\pi}{\Phi_{TF}(u,\Omega)WV(x;t-u,\omega-\Omega)dud\Omega/2\pi} \tag{6}$$

$$\omega^\wedge(x;t,\omega)=\omega-\frac{\iint \Omega\Phi_{TF}(u,\Omega)WV(x;t-u,\omega-\Omega)dud\Omega/2\pi}{\Phi_{TF}(u,\Omega)WV(x;t-u,\omega-\Omega)dud\Omega/2\pi}$$

This reassignment leads to the construction of a modified version of this time-frequency representation, whose value at any point *(t', ω')* is therefore the sum of all the representation values moved to this point:

$$\mathrm{MTFR}(x;t',\omega')=\iint \mathrm{TFR}(x;t,\omega)\,\delta(t'-t^\wedge(x;t,\omega))\,\delta(\omega'-\omega^\wedge(x;t,\omega))\,dtd\omega 2\pi \tag{7}$$

where $\delta(t)$ denotes the Dirac impulse.

Figure 5 shows the reassigned Wigner-Ville distribution (SPWVD) of the same signal used for the Wigner-Ville distribution. Due to the fact that the aim of the reassignment method is to improve the sharpness of the localization of the signal components by reallocating its energy distribution in the time-frequency plane, when the representation value is zero at one point, it is useless to reassign it.

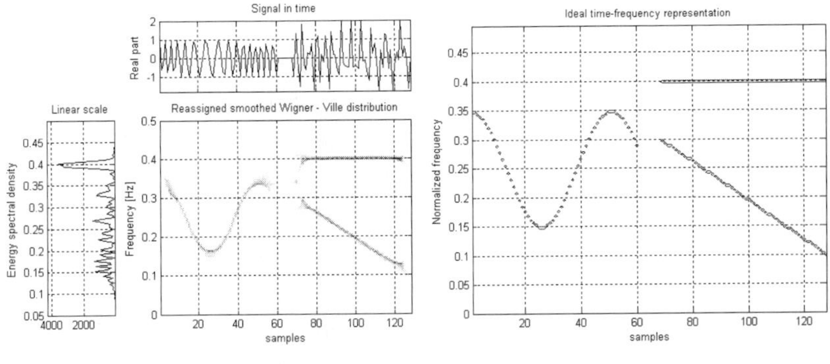

Figure 5. (a) Reassigned Wigner-Ville distribution; (b) Instantaneous frequency law of the three signal components.

4. Applications

The theory presented in the previous section has been used to monitor the laser welding processes. The aim of this paragraph is to demonstrate how the reassignment technique will improve the analysis of processing signals. Data acquisition was performed with a dual core based computer and a National Instruments multi-channel data acquisition board. The sampling frequency was set at 32 kHz using an anti-aliasing filter with a 12 kHz cut-off frequency.

Figure 6 displays two detected signals in time domain: left (or right) plot corresponds to a welded piece without defects referred to as a reference (or with defects). Figure 7 displays their spectrums.

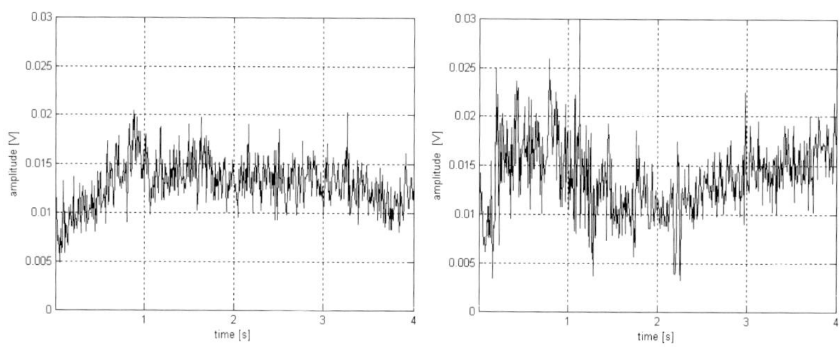

Figure 6. Detected signals: signal without defects (left), signal with defects (right).

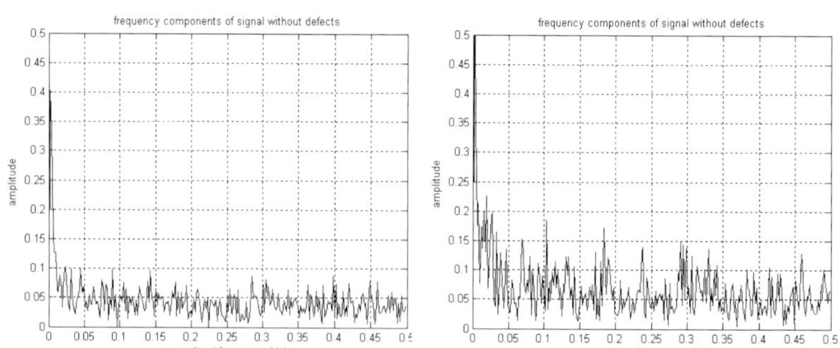

Figure 7. Signals spectrum: signal without defects (left), signal with defects (right).

From the analysis of spectrums, we can recognize that the signal with defects presents some frequency bands (bands of analysis) where the amplitude level is higher than the reference.

Without going beyond the Fourier analysis, we can only predict the presence of defects but we do not have any information about their number and their localization. In the following, we show how this information can be obtained by the reassignment method. Figures 8 and 9 show the plots of the reassigned distributions for the signal without and with defects respectively.

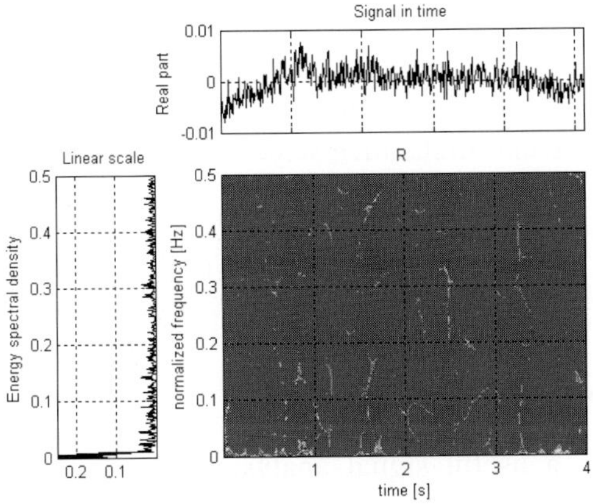

Figure 8. Reassigned distribution for signal without defects.

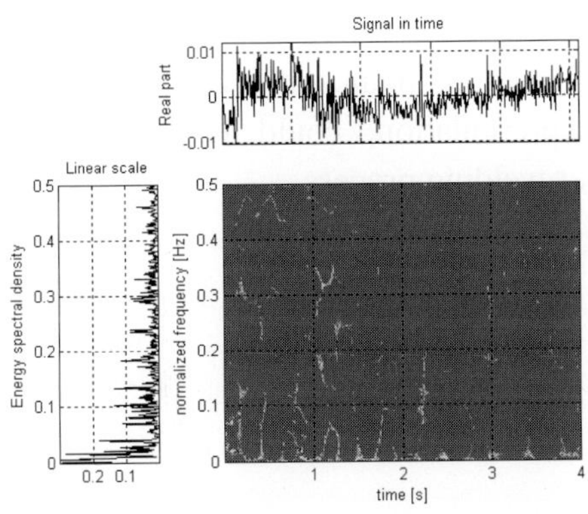

Figure 9. Reassigned distribution for signal with defects.

The reassigned distribution, as clearly shown, allow for a perfect localization of the signal terms.

The defects localization is executed by considering, firstly, the energy for both the reassigned representations in the bands of analysis by the formula:

$$E_i = \Sigma_j \text{ RSPWVD } (i, \omega_{inf} + j\ \Delta\omega) \tag{8}$$

where

$i=1..N$, $j=1..M$ with $M=(\omega_{sup}- \omega_{inf})/\Delta\omega$, $\Delta\omega=2\pi N/f_{acq}$

with N equal to the number of samples and ω_{inf} and ω_{sup} representing the lower and upper band limits, then selecting the samples where the energy of the signal with defects exceeds a threshold corresponding, for example, at the maximum energy level of the reference reassigned representation.

5. Conclusions

In this chapter, a useful signal analysis technique, for improving the analysis of laser welding process has been reported.

This method is based on the reassignment of the time-frequency distribution. It creates a modified version of a representation by moving the representation away from where they are computed.

The proposed method, thanks the advantages over the traditional time-frequency representation, could be practically applied for the monitoring of laser welding process.

References

1.H. B. Chen, L. Li, D. J. Brookfield, K. Williams and W. M. Steen, *Laser process monitoring with dual wavelength sensors,* Proceedings of ICALEO 91, San Jose, CA. Orlando, FL: Laser Institute of America, SPIE 1722, 113 (1991).

2.C. Bagger, I. Miyamoto, F. Olsen and H. Maruo, *On-line control of the CO2 laser welding process* in Beam Technology Conference Proceedings, Karlsruhe, Germany, March 13-14, DVS- Berichte 135, 1 (1991).

3.F. O. Olsen, H. Jørgensen, C. Bagger, T. Kristensen and O. Gregersen, *Recent investigations in sensoristics for adaptive control of laser cutting and welding* in Proc. LAMP2, Nagoka, Japan: High Temperature Society of Japan, p. 405 (1992).

4.D. U. Chang, *Monitoring laser weld quality in real time,* in Indust. Laser Rev. 15-16, November (1994).

5.D. Maischner, A. Drenker, B. Seidel, P. Abels and E. Beyer, *Process control during laser beam welding* in Proceedings of ICALEO 91, San Jose, CA. Orlando, FL: Laser Institute of America, SPIE 1722, 150 (1991).

6.G. D'Angelo, G. Pasquettaz and A. Terreno, *Laser process monitoring at FIAT group*in the Proceedings of EALA 07 - Bad Nauheim/Frankfurt, Germany, 30/31 January 2007.

7.C. Alippi et al., *Composite techniques for quality analysis in automotive laser welding*, CIMSA 2003 - Lugano, Switzerland, 29-31 July 2003.

8.G. D'Angelo, G. Pasquettaz and A. Terreno, Improving the analysis of laser welding process by the reassigned time-frequency representations in Proceedings of ICALEO 06, Scottsdale, AZ (2006).

9.P. G. Sanders, J. S. Keske, G. Kornecki and K. H. Leong, *Real-time Monitoring of Laser Beam Welding Using Infrared Weld Emissions*, Technology Development Division Argonne National Laboratory Argonne, IL 60439 USA.

10. D. Gabor , J. *Inst. Electron. Eng.*, vol. 93, no. 11, 429 (1946).

11. J. B. Allen and L. R. Rabiner, *A unified approach to short-time Fourier analysis and synthesis* in Proceedings of IEEE, vol. 65, 1558 (1977).

12. L. Cohen, *Time-frequency distributions. A review*, in Proceedings of IEEE, vol. 77, 941 (1989).

13. E. P. Wigner, *Phys. Rev.* vol. 40. 749 (1932).

14. J. Ville, *Câbles et Transmissions,* vol. 2A. 66 (in French) (1948).

15. T. A. C. M. Claasen and W. F. G. Mecklenbrauker, vol. 35. no. 3. 217, "Part 11: Discrete-time signals," vol. 35, no. 4/5, pp. 276300; "Part 111: Relations with other time- frequency signal (1980).

16. F. Hlawatsch and G. F. Boudreaux-Bartels, *IEEE Signal Processing Mag.,* 21 (1992).

17. L. Cohen, *J. Math Phys,* vol. 7, no. 5, 781(1966).

18. B. Escudié and J. Gréa, *c'. K. Acad. Sci.* vol. 283, 1049 (in French) (1976).

19. T. A. C. M. Claasen and W. F. G. Mecklenbrauker, The Wigner distribution, a tool for time-frequency analysis, Part I: Continuous-time signals, vol. 35. no. 3. p. 217, Part 11: Discrete-time signals, vol. 35, no. 4/5, p. 276300; Part 111: Relations with other time- frequency signal (1980).
20. F. Hlawatsch and G. F. Boudreaux-Bartels, *IEEE Signal Processing Mag.,* 21 (1992).
21. H. I. Choi and W. J. Williams, *IEEE Trans. Acoust., Speech, Signal Processing,* vol. 37, 862 (1989).
22. K. Kodera, C. de Villedary and R. Gendrin *Phys. Earth Planet. Interiors,* no. 12, 142 (1976).
23. K. Kodera, R. Gendrin and C. de Villedary, *IEEE Trans. Acoust., Speech, Signal Processing,* vol. ASSP-34, 6476 (1986).
24. F. Auger and P. Flandrin, *IEEE Transaction On Signal Processing*, vol.43, no. 5 May (1995).

APPLICATIONS OF OPTICAL SENSORS TO THE DETECTION OF LIGHT SCATTERED FROM GELLING SYSTEMS

Donatella Bulone, Mauro Manno, Pier Luigi San Biagio
and Vincenzo Martorana*

Istituto di Biofisica, CNR,
Via Ugo La Malfa 153, 90146 Palermo, Italy
**E-mail: vincenzo.martorana@pa.ibf.cnr.it*

Visible light, scattered within an angle of few degrees, (Small Angle Light Scattering, SALS) yields information on the spatial correlations and dynamical properties on the scale of the micrometers. In this way a quick and non-invasive characterization of a variety of samples is feasible. Lately the SALS instruments have been built around multi-element optical sensors (CCD, CMOS), allowing the simultaneous measurement of the complete structure factor even during fast kinetics. An assessment of some sensor matrices of different technology will be presented. The macromolecular assemblies produced by polysaccharides or proteins can be functional or dysfunctional, their properties being either desirable or detrimental. Anyhow, their morphology often depends, in a very delicate way, on the presence of cosolutes, on the thermal history, on the biopolymer concentration etc. We present some applications of low angle dynamic and static light scattering to the study of gelling systems (agarose, pectin, insulin).

1. Introduction

The sol-gel transition is a supramolecular self-organization process leading to the formation of a macroscopic, percolative structure starting from an initially homogeneous solution. A precise definition of a gel was given by Ferry many years ago[1] in terms of: "a substantially diluted system which exhibits no steady state flow". The currently widely accepted description of gel is that of a system where a macromolecular component, usually present in lower amount, is assembled in a network spanning through the entire sample's volume and swollen by a huge amount of solvent. Macromolecules may be associated in a solid-like, stress-bearing structure through covalent (chemical gels) or reversible (physical gels) interactions.

Physical gels based on biomolecules are stabilized by weak interactions that can be easily tuned by even small change of parameters such as pH, temperature or cosolvent addition. This opens the route towards obtaining supramolecular arrangements with tailored structural properties via exploitation of the competitive/cooperative effects induced by small changes of experimental parameters. Therefore, understanding the mechanism of sol-gel transition covers a high interest for technical applications. Gel matrices based on hydrogels are widely used as encapsulation and release systems in food, cosmetic, agricultural manifacturing. In pharmaceutical application, hydrogels are used for tissue engineering or drug delivery system. Hydrogels are also used for biosensors, artificial membranes and electrophoresis media.[2]

The long-standing theoretical interest in sol-gel transition has been recently renewed by issues raised by colloid physics. The central question is whether gel and colloidal glass can be described in the same conceptual frame.[3,4] Indeed, it has been shown[5] that the onset of gelation in a system of short-range attractive particles can be explained by an extension of Mode Coupling Theory,[6] originally developed for describing liquid-glass transition. In this scenario the gelation is the result of an arrested structural relaxation due to a trapping (or cage) effect similar to that operating in the glass transition.

Rheology is a natural technique for the investigation of gelling systems. It is based on the measurement of the response to a mechanical perturbation that, on the other hand, can alter the structure or kinetics of the sample.

Dynamic Light Scattering (DLS), instead, is a suitable technique for studying the gelation kinetics without mechanically perturbing the gelling system. Four characteristic signatures of gelation onset are observable using standard large angle DLS instrumentation. They are:

i) a change in the pattern of scattered intensity fluctuation, which reflects the onset of non ergodicity;
ii) the appearance of a power law behavior in the correlation function;
iii) a diverging increase of relaxation time and reduction of the amplitude of the correlation function.

The change of the dynamic properties of the system with time, during the gelation, may make unpractical to follow the process with standard DLS at large scattering angles. Indeed, when the relaxation time reaches the scale of seconds, the sampling over many correlation times requires a too long experiment's duration. Moreover the freezing of the scattering patterns makes the scattering volume seen by the detector not fully representative of the entire ensemble. In those cases where the gelation is anticipated or accompanied by a demixing of the solution, the measurement of static light scattering (SLS) can convey a number of information on the structural change of the gelling sample.

Usually the spatial concentration fluctuations accompanying the demixing, appear on the scale of micrometers, where standard large angle instrumentation is unable to provide good experimental results. These are the reasons that make the Small Angle Light Scattering (SALS) technique particularly interesting: it can be used to study the density fluctuations on the scale of micrometers and very slow dynamics can be investigated thanks to the concurrent nature of 2D detectors.

After a short introduction of the key concept of light scattering we will present some of the SALS implementation with a more detailed description of the instruments used in our laboratory. We also present a few applications of the technique to the study of aggregation/gelation of two polysaccharides (agarose and pectin) e of a protein (insulin).

2. Theory and Practice of Light Scattering

The region of the sample illuminated by the coherent radiation I_0 with wavelength, λ, will scatter light due to the spatial fluctuations of the dielectric constant, ε, with intensity, *I*, given by:

$$I(q) \propto I_o \int_V \langle \partial\varepsilon(0)\partial\varepsilon(\mathbf{r})\rangle e^{i\mathbf{q}\cdot\mathbf{r}}\, d\mathbf{r} \tag{1}$$

where q is scattering wave vector and V is the scattering volume. The magnitude of the wave vector is related to the scattering angle through $q=(4\pi n/\lambda)\sin(\theta/2)$ where n is the refraction index of the medium. If the scattering objects are identical interacting particles, the fluctuations of the dielectric constant are caused by fluctuations of local density and the latter can be decomposed in single particle and particle-particle contributions, so that,

$$I(q) = I_o \, Kc\, M \;\; P(q)\; S(q) \tag{2}$$

where K is a constant that depends on the contrast between the particles and the medium, c is the mass concentration and M is the molecular weight. The single particle contribution $P(q)$, the form factor, is given by

$$P(q) = \frac{\int d\mathbf{\Omega} \int \langle \rho(0)\, \rho(\mathbf{r}) \rangle e^{i\mathbf{q}\cdot\mathbf{r}}\, d\mathbf{r}}{4\pi \int \langle \rho(0) \rho(\mathbf{r}) \rangle d\mathbf{r}} \tag{3}$$

while the particle-particle structure factor S(q) can be expressed as:

$$S(q) = 1 + N/V \int_V (g(r) - 1) e^{i\mathbf{q}\cdot\mathbf{r}}\, d\mathbf{r} \tag{4}$$

where g(r) is the spherically averaged pair correlation function.

In the case of multiple species, Eq. 2 does not hold anymore, and the effects of the shape/size of the objects get mixed with the effects due to the structure of the solution. However, Eq. 2 is commonly used as a useful approximation in cases where detailed statistical/molecular models are not available.

Static light scattering is therefore able to supply information on the size of the objects and on their mutual interaction. The range of accessible wave vectors determines the length scale probed by the technique. Typical standard instruments measure scattering at different q by changing the scattering angle thanks to a rotatable detector (e.g. Brookhaven Instruments) or by optical fibers positioned at different angles (e.g. ALV), or by more unusual methods.[7]

Anyway the angles are typically limited to the range 160°-15° while the wave vectors varies in the range 3-30 μm^{-1} with a corresponding length scale ($2\pi/q$) in the range 2-0.2 μm. Note that these numbers are evaluated for the case of $\lambda = 0.633$ μm, so that it is possible to increase the wave vectors by changing the radiation source.

The range is however limited to a decade and this makes difficult a reliable study of molecular systems with mesoscopic spatial correlations.

Small angle light scattering instruments allow to widen the range of wave vectors down to 10^{-2} μm^{-1}, opening the way to a sort of Fourier optical microscopy with the added bonus of an implicit statistical averaging over the scattering volume (~ 30 mm^3).

A detailed view of a SALS instrument will be presented later to illustrate in particular the role of the optical sensor.

Another way to increase the dynamic range of a standard large angle light scattering instrument is to make use of the temporal fluctuations of the scattered electric field. These, in fact, contain information on the dynamics (diffusion, reptation, hopping, etc.) of the scatterers, and, thanks to fast detectors and electronics, they can be analyzed in the range from hundreds of nanoseconds to tens of seconds. The temporal fluctuations are measured through the real time computation of

$$g_I(q,\tau) = \langle I(q,t)\, I(q,t+\tau)\rangle_t \,/\, \langle I^2(t)\rangle_t = 1 + \beta\, g_E(\tau) \tag{5}$$

where the last equivalence is called Siegert relation, β depends on the number of coherence areas[8,9] seen by the detector, and

$$g_E(q,\tau) = \left\langle E(q,t)\, E^*(q,t+\tau)\right\rangle_t \,/\, \left\langle |E|^2(t)\right\rangle_t \tag{6}$$

is the autocorrelation of the scattered field. It can be shown[8] that:

$$g_E(q,t) = \int G(\mathbf{r},t)\, e^{-i\mathbf{q}\cdot\mathbf{r}} d\mathbf{r} \,/\, S(q) \tag{7}$$

where G(**r**,t) is the van Hove function, i.e. the conditional probability of finding the particle in **r** at time t if it was at the origin at time zero. In the simple case of (non interacting) Brownian particles the van Hove function is a 3D Gaussian function with a standard deviation which increases with time. Correspondingly, the autocorrelation is a simple exponential in time and a Gaussian in q-space: $g_E = \exp(-Dq^2t)$.

The diffusion coefficient D is related to the hydrodynamic radius R_h of a spherical particle through the Einstein relation $D = KT/6\pi\eta R_h$, where η is viscosity of the medium, k is the Boltzmann constant and T is the temperature. It has to be noted that the averages indicated with angular parentheses in Eqs. 5, 6 are performed by repeating the correlation measurements with different time origins.

When the dynamics of part of the system under study is arrested, as in the case of gelation, then the time average on a microscopic (scattering) volume does not yield the ensemble average. The latter can be recovered by averaging the autocorrelation functions measured at different spots of the sample or by using a smart correction to the normalization of the correlation function.[10] The trivial solution of increasing the size of the scattering volume does not work because it increases the number of coherence areas seen by the (unique) detector, thus degrading the amplitude of the field autocorrelation function (Eq. 5).

If part of the system cannot diffuse freely then, from Eq. 7, $g_E(q,t=\infty)=b\exp(-\langle \Delta r^2 \rangle q^2/6)$, where b weights the relative amount of intensity scattered by the bound objects and $\langle \Delta r^2 \rangle$ is the mean square displacement.

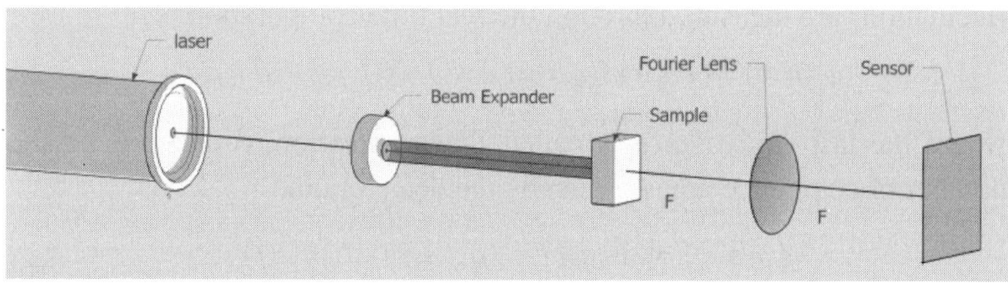

Figure 1. Schematic illustration a simple setup for SALS measurements. A laser beam, expanded to a few millimeters, illuminates the sample. The scattered and the transmitted radiation is collected by a lens at distance equal to its focal length. At the same distance is positioned the sensor, on which rings concentric with the optical axis correspond to a single value of the norm of the wave vector.

2.1. SALS Instruments

The measurement of the so called forward lobe of scattered light, with applications ranging from particle sizing[11,12] to the study of phase separations,[13-16] can be performed even with a very simple setup[17] as that shown in Fig. 1. Here, a filtered and expanded laser beam illuminates the sample, while a Fourier-transform lens is placed at its focal distance f from both the sample plane and the sensor plane. The role of the Fourier lens is that of concentrating the light scattered at angles $\vartheta \pm \Delta\theta$ into a corresponding annular region $f\theta \pm f\Delta\theta$ of the sensor plane, whatever the position of the scatterer within the sample volume.

The linear mapping between scattering angle and radial distance on the sensor plane holds for small angles where tg (ϑ) $\approx\vartheta$ and for an ideal, aberration-free lens. In this geometry the unscattered light beam is focused on the sensor plane at the optical axis and must be brought out of the optical path to i) avoid the saturation of the sensor and ii) to estimate the turbidity of the sample.

In fact, to get meaningful static and dynamic light scattering the probability of having multiply scattered photons must be kept very low, meaning that most of the energy (> 95%) is concentrated at scattering angle 0, while the rest is distributed at all the other angles. Thus a sensor *prepared* to measure the scattered photons can hardly measure the unscattered, focused light, whose reflected fraction would also contribute to the unwanted background (stray light). Solutions to this problem have been devised by piercing a hole in the sensor to let the transmitted beam out[18,19] or by placing a small mirror on the optical axis and using a second lens to relay the focal plane of the first lens onto the sensor plane.[20] The first solution requires an *ad hoc* sensor[11] or the use of optical fibers[15] and have a limited q- resolution, while the second solution allows the use of standard CCD or CMOS optical sensors, but imposes further geometric conditions to avoid the vignetting of the scattering pattern.

Other designs for SALS measurements can be found in the literature that requires a long focal lens between the incoming beam and the sample.[13,21] Whatever the design of the instrument, the stray light will represent a significant contribution to the measured sensor signal.

A separate measurement of the background has to be performed and subtracted from each subsequent measurement. Since the stray light pattern depends on the details of the optical surfaces and of their relative orientations, one should not move the optical cell after the background measurement. This has obvious adverse consequences on studies of long kinetics. On this side the heterodyne near field scattering (HNFS)[22,23] technique, which provides an absolute estimate of the scattered field at small angle using a simple setup,[23] looks very promising. We used this technique for a multi-spectroscopy study of the amyloid Aβ-protein fibrillization,[24] realizing that much care must be exercised in applying HNFS to solutions undergoing a progressive dynamical arrest.

We now illustrate in more detail the setup in Fig. 2 of the SALS instrument that we used for the applications shown later. The setup, following the suggestions of Ferri,[20] is based on a Fourier lens L that focuses the transmitted beam on a small mirror, acting also as a beam stop. The scattered field, as formed on the focal plane of lens L, is imaged with magnification ratio $M_2 = P_O / Q_O$ by the objective O onto the 2D sensor.

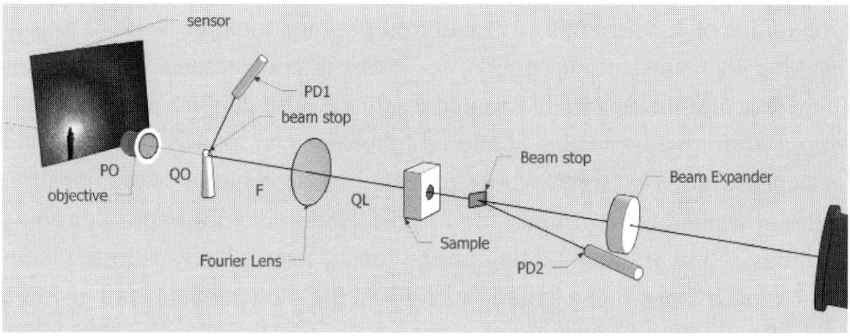

Figure 2. Schematic illustration of the SALS instrument built in our lab. With respect to the scheme in Fig. 1 a new lens (objective) has been added to relay the focal plane of the Fourier lens on the sensor, two photodiodes measure the incident (PD2) and the transmitted beam (PD1), and a beam stop with a mirror is placed at the focal length from the Fourier lens. The labels QL, QO, and PO indicates the distance between the sample and the Fourier lens, the distance between the objective and the beam stop and that between the objective and the sensor.

The theoretical accessible θ-range is given by $\lambda/(\pi w_0) \div s/(F_L M_2)$ where w_0 is the radius of the expanded beam and s is the size of the sensor.

The practical minimum accessible wave vector is actually limited by the size and imperfections of the beam stop, the size and pitch of the pixels and by the amount of stray light. The actual maximum wave vector is instead limited by the presence of aberration and vignetting.

Ferri suggests[20] to put the sample at a distance from the lens L such that an image of the sample is formed on the objective O. In this way all the light scattered from the illuminated sample and reaching the lens L will also reach the objective O, at the cost of moving the sample away from the Fourier lens. Using the law for thin lenses one can find the correct setup by solving the following equations:

$$\begin{cases} 1/Q_L + 1/\ (F_L + Q_O) = 1/F_L \\ 1/Q_O + 1/P_O = 1/F_O \end{cases} \tag{8}$$

The distance Q_L between the sample and the lens determines the maximum accepted scattered angle or, for a fixed maximum angle, the needed lens aperture. We have tested several setups chosen on the basis of Eqs. 8 using a Fourier lens D_L = 40 mm, F_L = 60 mm (achromatic doublet) and a Canon photographic objective with $f_0 = 0.95$, D_0 = 50 mm where D_L is the lens diameter, f_0 is the objective aperture, and capital F's indicate the focal lengths.

The setup was tested in our lab by using an opal glass, that should give I (q) $\propto$ cos $(\vartheta) \approx$ const for $\vartheta < 15°$, and a 250 μm pinhole that gives rise to a scattering pattern with concentric rings of decreasing intensity (Airy diffraction).

All the trials led to an disturbingly small range of wave vectors. Thus we decided to move the sample closer to the Fourier lens and to correct for the potential vignetting of lens O. By looking at the diffraction pattern of the pinhole we realized that a distortion was present for the outermost rings. The distortion was taken into account by introducing a third order polynomial in the mapping equation between pixel distance nd wave vector q. The polynomial coefficients were determined by nonlinear fitting to the pinhole data with the function $I(q) \propto (2J_1(x)/x)^2$, where $x = a_0 r(1 + a_1 r^3)$, J_1 is the Bessel function of the first order and r is the distance from optical axis, measured in number of pixels. No further correction was needed for the vignetting, most probably because of the lens aberration compensating effect.

A similar setup has been used[25] to perform *multi-speckle* dynamic light scattering a small angle.[26] The effect of speckle averaging, stray light, dark noise has been analyzed and an equation to recover the $g_E(q,\tau)$ from the raw autocorrelation has been proposed and tested. Ideally, each pixel of the sensor acts as the typical photomultiplier detector in the standard large angle DLS instrument, cutting down the time required to measure $g_E(q,\tau)$ by a factor proportional to the number of independent detectors. This is particularly important considering that at this small q-range the characteristic times go from seconds to hours (e.g. $\tau = 10^3$ s for a 5 μm Brownian sphere at $q = 0.1 \mu m^{-1}$).

Furthermore, by averaging the autocorrelation functions over many pixels one can recover the ensemble average even in the case of a system undergoing an ergodic to non-ergodic transition. Note also that in a single experiment the full intermediate dynamic structure factor can be measured. The technique, thus, appears appropriate for the study of gelling systems and some examples have appeared[27-30] and some will be presented below.

The role of the detector both for static and dynamic light scattering is crucial. The detector response must be linear, fast, with a high dynamic range and low dark noise. Further, the ideal detector is two-dimensional or one-dimensional and this excludes the use of a photomultiplier.

CCD cameras and photodiode arrays were used in the first SALS instruments. Nowadays CCD and CMOS sensors with varying degrees of sophistication (Peltier cooling, megapixel resolution, 16 bit ADC, etc.) are commonly used. Unfortunately these advanced features are expensive and can make the detector the major cost of a SALS instruments. We have found that it is possible to increase the dynamic range of a CCD camera with a 10 bit ADC, by taking consecutive frames a different exposure times. A simple algorithm will then choose the appropriate frame for each pixel so that the linear region of the detector is always used and the intensity will be scaled by the correspondent exposure time. In this way we are able to increase the dynamic range by two orders of magnitude at the cost of lengthening the acquisition process. However the typical kinetics and dynamics observed with SALS are most often on the time scale of seconds or slower. We have compared the PULNIX TM-765 CCD camera that we use in our SALS instruments with a CMOS, 10 bit, fire wire, monochrome camera from Pixelink (PL-A741) and with a the CMOS sensor of a digital reflex photographic camera (Canon 350D). The Pixelink camera offers a fast digital connection and is able of sustaining a high frame rate if a small *region of interest* is selected. This could be attractive when performing DLS in presence of quickly diffusing objects. We measured the dependence of the dark noise, measured as the camera digital output (DN), on temperature through a sensor attached to the CMOS and the scaling of the measured intensity with the integration time. The temperature dependence is strong and the readout noise amounts to $\sigma \approx 2.9\,\mathrm{DN}$ while the scaling with integration time is acceptable but for very low intensity. We found, in fact, that the dark level of the internal ADC of the PL-A741 camera is set by the firmware at the digital level zero thus distorting the measurement of the low level intensity signals. Furthermore the dark level was found to depend systematically on the analog gain of the sensor amplifier. These elements discourage the use of the PL-A741 for quantitative measurement of light intensity.

The CMOS sensor of the Canon 350 D is organized as a Bayer matrix[31] with red, green and blue filters on each of the 8 million pixels. The analog to digital conversion is performed with 12 bit resolution and the raw data are sent to a PC through a USB link. We measured the features of the red filtered subset of pixels, the most sensitive to our He-Ne laser radiation. The responsivity, defined as the digital output caused by unitary radiation energy, has been estimated[32] as 14 $DN/nJ/cm^2$, slightly larger than that stated for the PL-A741.

In this case the dark level was set to 256 DN and the linear behavior holds, within a few percents, in a range of almost three decades. The dark noise ($\sigma \approx 1.7\,\mathrm{DN}$) remains constant for exposure times up to 10 s and the dependence on temperature is weaker than that of the PL-A741.

Overall the Canon 350D features a surprisingly good optical sensor, and represents a practical evidence of the progress that CMOS sensor technology has accumulated in the last years. It is unfortunate that, from a practical point of view, the Canon camera, as it is, does not fit in the SALS instrument since it misses a *live view* feature that is particularly useful in the setup phase of the instrument and to diagnose the running experiments. We are thus considering the possibility of adapting it to the use in a SALS instrument.

3. Kinetics of Agarose Gel Formation

Agarose is a well characterized biostructural polysaccharide extracted from seaweed.[33] It provides a simple system for studying the self-assembly of a natural biopolymer and the influence of chemical and physical parameters on the gelation process and on the final gel structure. The phase diagram of water-agarose solutions has been explained[34] in terms of the mutual interaction of i) a demixing process, dominating in the low polymer concentration region, ii) a crosslinking between polysaccharidic molecules, and iii) a molecular conformational change.

Samples of agarose at 0.5% w/v were prepared[35] at 80 °C and quickly quenched at several temperatures, all below the spinodal temperature (T_{sp}=49.8°C). Similar experiments on more concentrated samples had shown,[34] using a very simple SALS instrument, the presence of a low angle ring for T > 40°C. The growth of the scattered intensity is shown in Fig. 3 for a 0.5%w/v agarose sample quenched at 38°C. A peaked structure factor is present since the beginning of the kinetics, with the peak slightly moving towards higher q, as shown in the upper inset.

The bottom inset, instead, shows that the intensity at the peak grows logarithmically with time. Both results are at odds with the expectations drawn from Cahn-Hilliard theory of spinodal demixing[36] of a linear regime with exponential growth of the intensity and fixed peak position, followed by a coarsening regime with the peak moving towards lower q. Experimental testing of the theory have been obtained with SALS instrumentation in the past with mixed results.[15,17,37,38]

The data on aqueous solutions of agarose at 2% actually show[34] a partial agreement with the Cahn-Hilliard theory, at least in the initial part of the kinetics, while the data shown in Fig. 3 speak more for a demixing in strong competition with a crosslinking/gelation process.

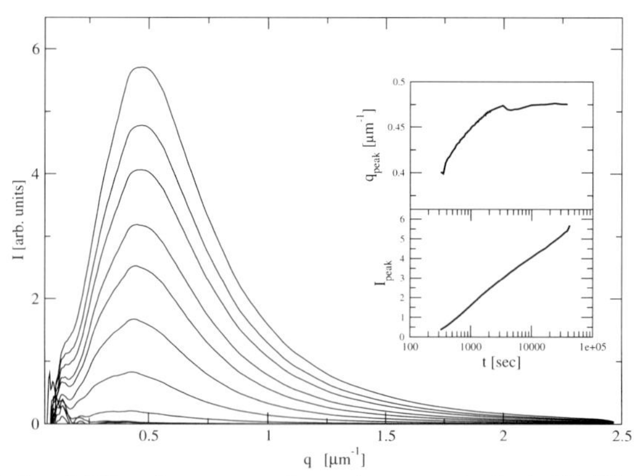

Figure 3. I(q) for 0.5% w/v agarose in water quenched at 38 °C. Higher curves correspond to increasing, logarithmically spaced, elapsed time. Upper inset: peak position vs. time. Bottom inset: peak amplitude vs. time. Note the logarithmic behavior of the amplitude. Redrawn from Ref. 35.

We thus modeled the experimental data with a fractal aggregation process[39] and a concurrent depletion process. The latter is thought as due to a slowed down diffusion caused by the ongoing crosslinking between agarose molecules.

By using this model we obtain good fit of the static scattered intensity both at small and large angle, as shown in Fig. 4 for the quenching temperatures 40 and 30 °C. Note that a scaling parameter had to be introduced to put together the results obtained from the two instrumentations, due to the absence of overlap in wave vector ranges.

By decreasing the quench temperature the infinite-time peak move at higher q (i.e. the higher density regions get smaller) and the kinetics gets faster. It is thus possible to set the wanted length scale of concentration fluctuations on the final agarose gel structure. An abrupt change in the behavior of scattering with quench depth was observed for quenching temperatures below 30 °C. In these cases the infinite-time peak position does not decrease anymore with the quenching depth as if the fingerprint of the initial demixing on the gel structure was overwhelmed by the much faster crosslinking process. Rheological measurements have shown that deeply quenched agarose gels are stronger but more fragile that those prepared at higher temperatures.[35]

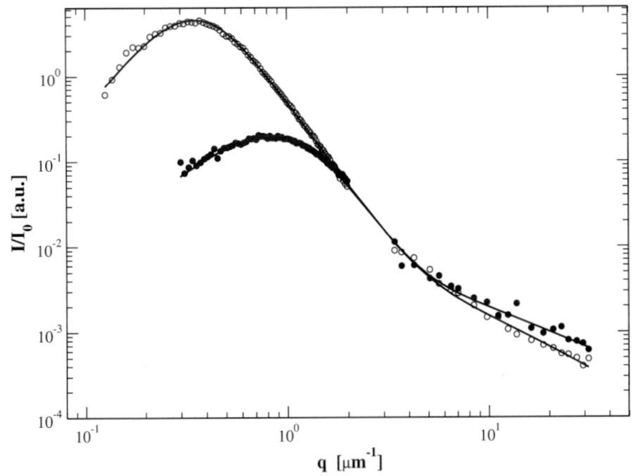

Figure 4. Composite of low- and high-angle scattered intensity for agarose solutions quenched at 40 °C (open circle) and 30 °C (full circle). Continuous lines are fits using the model discussed in the text.

By the way, the strong relationship between SALS measurement and rheological measurement has prompted the creation of scientific instruments that allow the concurrent measurement of (anisotropic) small angle scattering and of the elastic modulus.[40-42]

4. A DLS Time-Resolved Study of Pectin Gelation

As mentioned above, the multi-pixel structure of the typical SALS instruments, offers the possibility to concurrently (and separately) measure hundreds of individual occurrences of microscopic dynamical events. In other words it is possible to substitute the classical time averaged intensity correlation function of standard large angle instruments with an ensemble average, with the additional benefit of covering an entire range of wave vectors. This has prompted us to do a SALS time-resolved study of the dynamics of a gelling sample. A suitable system is represented by high methoxyl (HM) pectin in solution with a high sucrose content. This system is capable of producing transparent gels with different strength and kinetics, when brought at ambient temperature.[43]

Pectin is a natural polysaccharide, mainly constituted by a backbone of 1, 4-linked R-D-galactopyranosyluronic acid,[44] widely used as a gelling and thickening agent in the food industry. A concentration of a few g/L of pectin in an acidic solvent is sufficient to produce a strong gel, but, at high degree of esterification (> 50%), a significant amount of a cosolute is also needed (sucrose, ethanol, etc.).

The samples for the SALS experiments were prepared[43] with pectin at 0.2% w/w, pH 3.5 at temperature close to 100 °C and quenched at ambient temperature to start the gelation kinetics, whose duration ranged from few hours to many days, depending on the sucrose concentration ($58.5\% \div 56\% w/w$). During the experiments a number of speckle fields were acquired from the CCD sensor and digitized for the calculation of the correlation function.[45]

Since each correlation function is supposed to refer to an equilibrium situation, during a kinetics one must select a number of time windows that can be reasonably considered as in equilibrium. The correlation functions are then calculated within each of these time windows. An alternative interesting approach for the study of time-resolved time correlation functions has been proposed in the field of diffusing wave spectroscopy.[46] We show a few of these *windowed* autocorrelation functions in Fig. 5, measured during the gelation kinetics of a sample with 56% w/w sucrose concentration. None of the curves shown in Fig. 5 can be fitted to a simple exponential. They are more similar to Gaussian function (ballistic-like motion) and some of them present oscillations and negative values, probably caused by the mixing of scattered and unscattered (stray light) radiation at the sensor (heterodyne DLS[8]).

At the beginning the correlation functions go to zero in few minutes, then time increases up to tens of minutes. A substantial change is observed between 70 and 74 hours from the initial quench.

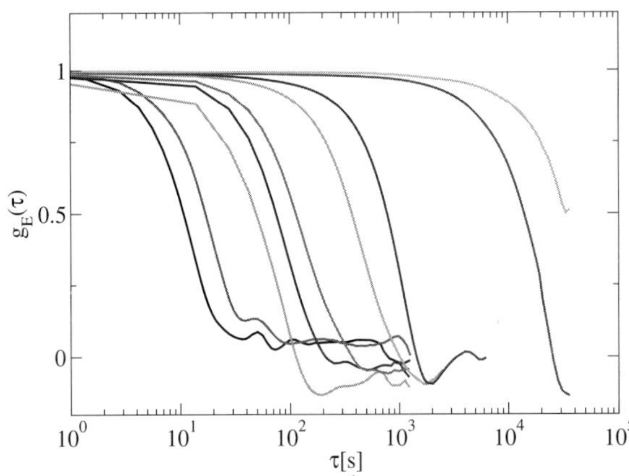

Figure 5. Autocorrelation functions (q=1.1 μm^{-1}) for successive time windows during the gelation of a sample of HM pectin 0.2% w/w, sucrose 56% w/w. The time windows were centered at (curves from left to right) 1, 2, 10, 24, 48, 56, 70, 74, 80 hours after the quench at ambient temperature. Note the big jump between the dark maroon (70 h) and the violet (80h) curves.

We fitted the data in Fig. 5 with the empirical law $g(t) = \exp(-(t/\tau)^{\mu})$ and plotted the characteristic time τ as function of the elapsed time in Fig. 6. Note the steepness of the growth after 40 hours from the quench at ambient temperature. This growth of the characteristic time of the dynamics coincide with the growth of the viscoelastic signal measured in a parallel rheological experiment,[45] thus demonstrating that we are able to detect the creation a macroscopic elastic network by DLS SALS measurements.

5. Kinetics of Insulin Fibrillization and Gelation

Insulin is a known hormone, important for sugar blood regulation. Its aggregation and stability are major concerns for the treatment of diabetic patients, and for the improvement of pharmaceutical formulations.[47]

In recent works, we have studied the mechanisms of human insulin aggregation at high temperatures and acidic pH.[48-50] In such conditions, insulin solutions are initially made of small size oligomers.[48]

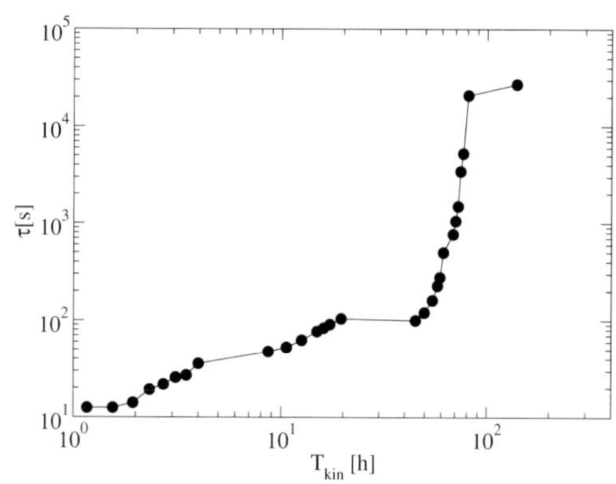

Figure 6. Evolution with elapsed time of the characteristic time of autocorrelation functions ($q = 1.1\ \mu m^{-1}$) as in Fig. 5. A slowing down of the dynamics starts since the beginning of the kinetics, but a striking arrest comes only after ca. 40 hours.

After this lag-phase, insulin forms a complex hierarchy of supramolecular assemblies from single amyloid-like filaments[51] to macroscopic floccules of fibrils.[52] These large size objects may either precipitate or remain interconnected into a gel structure.[50]

The kinetic of insulin aggregation has been studied by the combined usage of small angle and large angle light scattering, which allowed us to observe the growth of structures on many length scales, from hundreds of nanometers to hundreds of microns. The structure functions S(q) after the initial lag-phase are displayed for one typical case in Fig. 7.

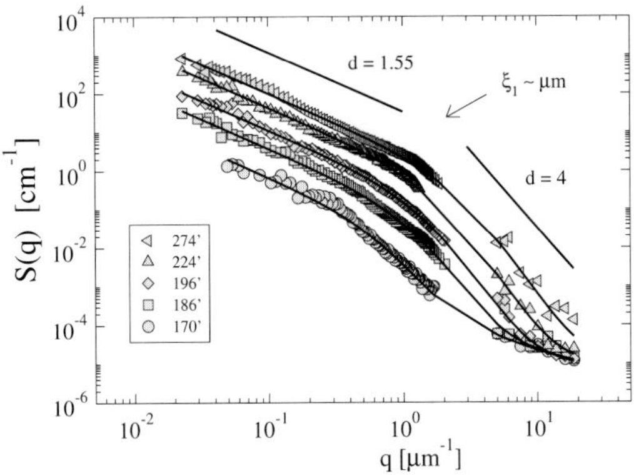

Figure 7. Scattering structure function of a 800 μM insulin solution incubated at 60 °C at selected times, as reported in the legend (redrawn from Ref. 50).

They exhibit a power law dependence in the range from 10^{-2} to 10^{-1} μm^{-1}: $S(q) \approx q^{-d}$. The exponent d, which is about 1.5 to 1.6, is related to the packing of molecular aggregates. At a scattering vector q of the order of a few micrometers, the slope of the structure functions change and reaches a value of 4, typical of compact objects.

These features of the structure functions hint the existence of huge loosely-packed clusters with a characteristic size of the order of hundreds of microns or higher. Such clusters can be formed either by fiber elements with a characteristic length of about 100 nm or by compact bundles of fibers with a characteristic size of microns.[50]

This analysis is confirmed by optical microscopy images recorded at the end of the aggregation (Fig. 8, right) and, on a smaller length scale, by AFM images (Fig. 8, left).

18. F. Ferri, M. Giglio, E. Paganini and U. Perini, *Europhys. Lett.*, 599 (1988).
19. M. Carpineti, F. Ferri, M. Giglio, E. Paganini and U. Perini, *Phys. Rev.* **A**, 7347 (1990).
20. F. Ferri, *Rev. Sci. Instrum.*, 2265 (1997).
21. F. Ferri, M. Greco, G. Arcovito, F.A. Bassi, M. De Spirito, E. Paganini and M. Rocco, *Phys. Rev.* **E**, 031401 (2001).
22. D. Brogioli, A. Vailati and M. Giglio, *Europhys. Lett.*, 220 (2003).
23. F. Ferri, D. Magatti, D. Pescini, M.A.C. Potenza and M. Giglio, *Phys. Rev.* **E**, 041405-1 (2004).
24. R. Carrotta, J. Barthes, A. Longo, V. Martorana, M. Manno, G. Portale and P. L. San Biagio, *Eur. Biophys. J.*, 701 (2007).
25. L. Cipelletti and D. A. Weitz, *Rev. Sci. Instrum.*, 3214 (1999).
26. A. P. Y. Wong and P. Wiltzius, *Rev. Sci. Instrum.*, 2547 (1993).
27. L. Cipelletti, S. Manley, R. C. Ball and D.A. Weitz, *Phys. Rev. Lett.*, 2275 (2000).
28. L. Ramos and L. Cipelletti, *Phys. Rev. Lett*, 245503 (2001).
29. L. Cipelletti, L. Ramos, S. Manley, E. Pitard, D. A. Weitz, E.E. Pashovski and M. Johansson, *Farday Discuss.*, 237 (2002).
30. A. E. Bailey et al., *Phys. Rev. Lett.*, 205701 (2007).
31. T. Kreis, *Handbook of Holographic Interferometry: Optical and Digital Methods* (Wiley, Weinheim, 2005).
32. A. Acquaviva, Thesis, University of Palermo, Electronic Engineering Dept. (2007).
33. C. Araki and K. Arai, *Bull. Chem. Soc. Jpn.*, 1452 (1967).
34. M. Manno, A. Emanuele, V. Martorana, D. Bulone, P. L. San Biagio, M. B. Vittorelli and M. U. Palma, *Phys. Rev.* **E**, 2222 (1999).
35. D. Bulone, D. Giacomazza, V. Martorana, J. Newman and P. L. San Biagio, *Phys. Rev.* **E**, 041401 (2004).
36. J.W. Cahn, *J. Chem. Phys.*, 93 (1965).
37. F. Mallamace, N. Micali, S. Trusso and S. H. Chen, *Phys. Rev.* **E**, 5818 (1995)
38. F. Mallamace and N. Micali, in *Light Scattering, Principles and development*, W. Brown ed., (Clarendon Press, Oxford, 1996).
39. T. Nicolai, D. Durand and J. C. Gimel in *Light Scattering, Principles and development*, W. Brown ed., (Clarendon Press, Oxford, 1996).
40. J. Läuger and W. Gronski, *Rheol. Acta*, 70 (1995).
41. C. Chou and P. Wong, *Macromolecules*, 7331 (2003).
42. C. Chou and P. Wong, *Macromolecules*, 5596 (2004).
43. D. Bulone, V. Martorana, C. Xiao and P.L. San Biagio, *Macromolecules*, 8147 (2002).
44. R. Noto, V. Martorana, D. Bulone and P. L. San Biagio, *Biomacromolecules*, 2555 (2005).
45. G. Napoli, Thesis, University of Palermo, Physics Dept. (2005).
46. H. Bissig, S. Romer, L. Cipelletti, V. Trappe and P. Schurtenberger, *Phys. ChemComm*, 21 (2003).

47. J. Brange, L. Andersen, E. D. Laursen, G. Meyn and E. Rasmussen, *J. Pharm. Sci.*, 517 (1997).
48. A. Podestà, G. Tiana, P. Milani and M. Manno, *Biophys. J.* 589, (2006).
49. M. Manno, E.F. Craparo, V. Martorana, D. Bulone and P. L. San Biagio, *Biophys. J.*, 4585 (2006).
50. M. Manno, E. F. Craparo, A. Podestà, D. Bulone, R. Carrotta, V. Martorana, G. Tiana and P. L. San Biagio *J. Mol. Biol.*, **366**, 258 (2007).
51. J. L. Jimenez, E. J. Nettleton, M. Bouchard, C.V. Robinson, C.M. Dobson and H. R. Saibil, *Protein. Sci.*, 1960 (2000).
52. M. R. H. Krebs, C. E. MacPhee, A. F. Miller, I. E. Dunlop, C. M. Dobson and A. M. Donald, *Proc. Natl. Acad. Sci. USA.*, 14420 (2004)
53. R. Jansen, W. Dzwolak and R. Winter, *Biophys. J.*, 1344 (2004).
54. L. Nielsen, et al., *Biochemistry*, 6036 (2001).

CONTACTLESS CHARACTERIZATION FOR ELECTRONIC APPLICATIONS

Lucio Rossi,[a] Giovanni Breglio,[a*] Andrea Irace[a] and Antonello Cutolo[b]

[a]Dipartimento di Ingegneria Elettronica e delle Telecomunicazioni, Facoltà di Ingegneria, Università degli Studi di Napoli "Federico II

[b]Università del Sannio, Dipartimento di Ingegneria Palazzo Bosco Lucarelli, Corso Garibaldi 107, 82100 Benevento, Italy

[*]E-mail: breglio@unina.it

Modern day electronic applications are increasingly demanding ultrahigh-speed electronic and optoelectronic devices as well as integrated circuits (ICs) that operate at frequencies over 10 GHz for a wide range of applications such as communications network systems and instruments. Because of this, a thorough design and a reliable production of these devices require a novel approach to contactless characterization. This indeed is a wide topic, and it is hard to be considered in full detail in a single chapter.

We have selected only some peculiar contactless techniques favoring those which can be easily employed for circuits and materials characterization; we include: scanning electron, photoexcitation and force microscope, electro-optical sampling techniques charge sensing probes and SNOM. As we will explain, such techniques are applied to signal survey, detection and measurement of microcracks, temperature, lifetime, surface recombination velocity, and diffusivity.

1. Introduction

Over the last years, we have witnessed the rapid growth of Si-based semiconductor electronics and GaAs or InP-based optoelectronics in the market. In research laboratories, their operation frequencies or switching speeds have already outpaced the ability of conventional instruments. For instance, the cutoff frequency of such devices has entered the submillimeter-wavelength (> 300 GHz) region.

The technological improvements in integrated electronic circuits and optoelectronic devices, in the last two decades[1-6] have determined a demand for new characterization techniques with wide bandwidth[7-27] and very high spatial resolution features as well as based on a contactless approach to reduce interference and loads on the device under test (DUT) during operation conditions.

Among the contactless techniques, a large variety of physical effects have been exploited: the scanning electron microscope (SEM) and the scanning photoexcitation microscope (SPM) are based on the spectral analysis of a secondary electron beam whereas the scanning tunnel force microscope (SFM) is based on the atomic force interaction. Since the early 1980s, ultrafast laser and optoelectronics technologies have produced a variety of novel measurement techniques whose potential bandwidth approaches the terahertz regime.

Optical contactless techniques provide all the frequency components of the Fourier transform and can excite or trigger devices with a light beam.[29-32] This property, which is largely exploited in optical switching of high power electrical commutation,[33-35] may provide additional data about the device, which is not otherwise accessible. Optical testing can be used to measure both linear and nonlinear behavior with equal ease.

In addition, contactless techniques are largely employed for an accurate characterization of semiconductor material impurity content, electron-hole recombination lifetime and surface recombination velocity too.

Because most of the techniques described are based on the use of optical beams, we report the requirements of the optical sources in Table 3. The precision and the goodness of the optical contactless techniques are strictly related to the quality of the optical beam. In the case of c.w. laser beams a single transverse mode laser is useful, while, for pulsed operation, it becomes important to use pulses characterized by a good stability.

Technique	Sensitivity	Invasivity	Spatial Resolution	Bandwidth	Related Problems
SEM	≈ 1 mV[a,e,f]	Charges in the oxide layer[e]	≈ 1 μm[e]	> 2 GHz[f]	Vacuum Chamber
Photo emission	≈ 1 mV[a]	High electric fields	≈ 2 μm	> 20 GHz[d]	Vacuum Chamber
Plasma optic Effect	≈ 2 mV[a,b]	None	≈ 2 μm	d	Use of interferometric apparatus
Indirect E.O.	≈ 10 mV[a]	None	≈ 10 μm	d	Use of a probe crystal
Direct E.O.	≈ 50 mV	None	≈ 2 μm	d	None

Table 1. Comparison between some selected techniques (see Secs. II and III for more details) for the contactless measurements of the electric field distributions inside the electronic devices. The bandwidth refers to the case where the technique is used for a sampling analysis of the under test signal. SEM = Scanning electron microscopy, EO = Electro-optic.

a. The shot noise can limit the sensitivity of the optical techniques when the photodiode current becomes smaller than 100 μA and the sensitivity of a SEM when the electron current becomes smaller than 1 nA.
b. The exact sensitivity can be a function of the characteristics of the under test device.
c. The exact value of the bandwidth is a function of the time duration of the laser pulses. As a rule of thumb we can say that the bandwidth is roughly equal to the inverse of the laser pulse length.
d. The bandwidth is limited by the transit time of the electrons.
e. In order to reduce the induced charges inside the oxide layer, the electron beam energy should be smaller than 3KeV. With these low energies the spatial resolution can be improved with a related worsening of the sensitivity.
f. It is expected a noticeable improvement of the available bandwidth up to values higher than 70 GHz.

2. Non-Optical Techniques

The contactless techniques for electronics applications typically involve many different physical phenomena such as: ion and electron beam interactions, atomic forces and X-rays. In this section, we discuss some non-optical techniques used in the contactless characterization of electronic devices and circuits. Although capable of very accurate results, often not available by other methods, these techniques are often very bulky and complex and hence costly. Recently, the spatial resolution of ion beam techniques has become smaller than 20 nm by exploiting focused ion beam stations based on the use of a liquid Ga+ source.

Typically sensitivity is a decreasing function of the spatial resolution, since high sensitivity requires large excitation beam diameters. Some parameters (e.g. the energy) of the emitted entities depend on the characteristics of the target. Then, the distribution of the unknown quantities can be mapped on the surface of the device and frequently as a function of the depth. Differences between the various techniques include sensitivity, impurity detection, spatial resolution, invasiveness, speed, imaging capability, and cost. Electron beam techniques can be classified as:

- Techniques based on the analysis of an emitted electron beam which include: Auger electron spectroscopy, Cathodoluminescence, Electron microprobe and ultraviolet photo electron spectroscopy;[36-44]
- Techniques based on the analysis of the reflected electron beam which include: Scanning Electron Microscopy, Low Energy Beam Diffraction, High Energy Beam Diffraction and Surface Voltage Contrast methods;[45-48]
- Techniques based on the analysis of the absorbed electron beam which include: Electron Beam Induced Current method and Thermal Wave Imaging;
- Techniques based on the analysis of the transmitted electron beam that includes: Transmission Electron Microscopy and Electron Energy Loss Spectroscopy.

Scanning electron microscopy is one of the most largely used for electronic applications.

The electron beam technique can be easily used for the characterization of test chips, devices, very large scale integration structures and surface analysis.

beam current was realized.[49-53] In addition, the SEM presents a very good linearity. The signal detected with the SEM can be linearized by means of a preamplifier/feedback circuit: changing the analyzer grid voltage it is possible to maintain the maximum detected signal strength.

2.2. Scanning Photoexcitation Microscope

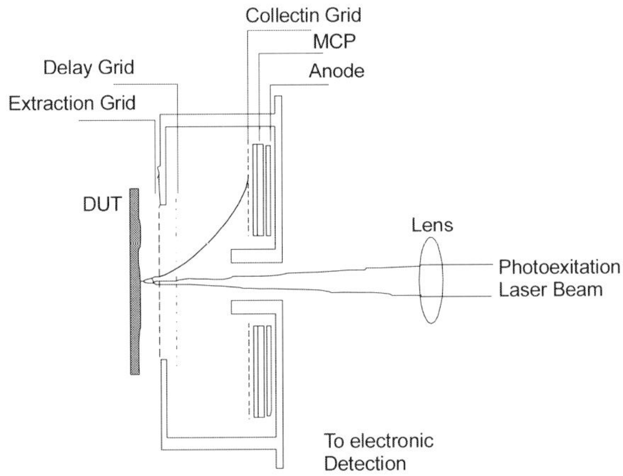

Figure 2. The vacuum chamber of a SPM with the laser beam and the different grids used to control and detect the photo-emitted electrons.

We include the Scanning Photoexcitation Microscope (SPM) (see Fig. 2) among the electron beam techniques for its strong similarity with the SEM. The only difference is that the secondary electron beam is generated by the interaction, in vacuum, of a high intensity laser beam with the target under test.[54-56] As the voltage on the circuit under test is increased, the energy of the secondary electrons drops, and the analyzer grid prevents these low energy electrons from reaching the detectors in a way similar to the SEM. Using optical pulses rather than electron pulses the bandwidth and sensitivity can be improved. Practically, the sensitivity is limited by the amplitude fluctuation of the laser pulses. Experimental results suggest that the photoemission probing system has a sensitivity of about 1 mV/√Hz.

There are two invasive features of the SPM. First, in order to achieve a bandwidth of the order of a few GHz, very high extraction fields, higher than 50 kV/cm, must be applied between the IC and the extraction grid. These fields can influence the behavior of the test IC. Hence, we can increase the bandwidth at the expenses of a perturbation of the circuit. Second, the photons that do not fall on the metal lines can be absorbed by the semiconductor. Since very high peak optical pulses are used, these photocurrents can strongly modify the values of the current signals being measured. The spatial resolution is related to the diffraction-limited spot diameter. With conventional lenses we can achieve a spot-size diameter of about 0.5 μm. Because high peak power and short laser pulses are possible, the bandwidth is only limited by how fast the secondary electrons leave the IC surface. In order to increase this speed, we can apply a high extraction field ($> 5 \cdot 10^4$ V/cm), which may perturb the test circuit, but temporal resolutions up to 10ps ($\approx$ 40 GHz) have been achieved.[53] Very large linearity ranges can be achieved using an equivalent feedback circuit. The detection system is similar to that of the SEM. However, the setups built on this technique are very critical to calibrate.

2.3. Scanning Force Microscope

The principle of sensing the electrical force to measure the potential distribution above an IC can be understood referring to Fig. 3.[57,58] The deflection of the conducting Scanning Force Microscope (SFM) tip, which is at fixed height h above the sample, is caused by the potential difference V between tip voltage V_S (oscillating at resonance) and sample voltage V_P. The local electric field E_{loc} above the conducting sample interacts with the SFM-tip through the Coulomb force F. This force is attractive or repulsive depending on the sign of the voltage V.

Several simple methods to calculate the potential distribution exist.[55]

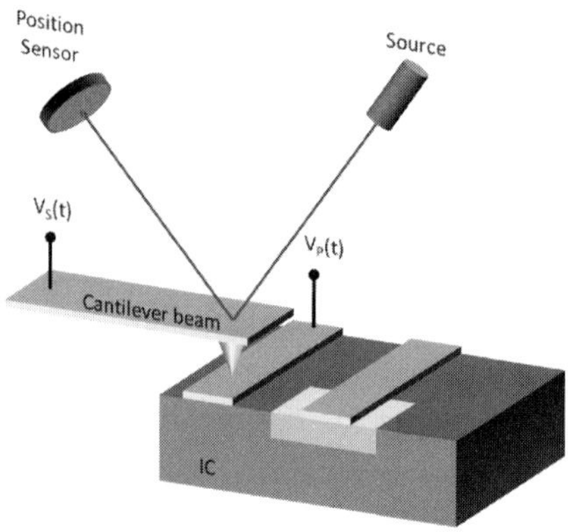

Figure 3. Schematic view of the SFM.

Modeling the geometry as a two plate capacitor with the SFM-tip as one plate and the sample as the other the effective Coulomb force is given by:

$$F(h) = \frac{1}{2} \cdot \varepsilon_o \cdot \frac{A}{h^2} \cdot V^2 \tag{2}$$

where ε is the permittivity of vacuum, A is the effective area of the plate capacitor and h is the effective tip-sample distance, the effective force F is caused by the tip-sample voltage and the tip position. Because F varies as V^2 the two oscillating voltages V_S and V_P mix and the difference frequency, which can be considered a down-conversion of the sample voltage V_P is detected. The SFM is very attractive because it features a large acquisition bandwidth together with a good spatial resolution. Bandwidths up to 20 GHz have been obtained and characterization of devices at nanometer-scale has been shown.

3. Optical Techniques

Optical measurement techniques are gaining an increasing popularity as they are non-contacting, which is a noticeable advantage since contacting the sample can be detrimental for its performance and even potentially dangerous.

Optical measurements[10-26] can be categorized in three broad categories:

- photometric measurements: the amplitude of reflected or transmitted light is measured;
- interference measurements: the phase of reflected or transmitted light is measured;
- polarization measurements: the state of the polarization of the reflected light is detected.

With the recent advent of practical solid-state laser technologies including semiconductor lasers, the optical technique has become the laboratory standard for characterizing modern devices and circuits.

Optical technologies allow measurements up to THz frequencies. A barrier to industrial employment of optical testing is the need of sub-picosecond laser sources.

Most of the optical tests for high-frequency circuits depend on a system that splits a sub-picosecond pulse from a laser into two parts. One part, the pump pulse, strikes a target while the second part, the probe pulse, is reflected through a time delay loop before reaching its target.
In optoelectronic sampling, the pump pulse generates short-lifetime charge carriers in a gap between transmission lines, known as an Auston switch.[24] The material in which the carriers are generated is the substrate itself or is integrated in a small probe bridging two contact points in the circuit. Switching speeds of less than 20 ps have been measured correlating the transmission factors of two gates in. Integrated switches and contact probes are equally accurate.

The development of lasers sources able to achieve ultra short optical pulses has increased the interest in optical testing of high-speed devices and circuits. Recently, techniques based on laser pulses have been used to investigate the propagation of electrical signals in waveguides and devices mainly via the Electro Optic (EO) technique. At the same time, other optical characterization techniques have been developed (i.e., the Plasma Optic effect sampling and the DC-induced second harmonic generation effect sampling).

	50 fs – 500 fs	500 fs – 5 ps	5 ps – 50 ps
Mid 1980s	Dye laser • CPM • Synch-pumped [600 – 800 nm]	Flash-lamp-pumped YAG laser + pulse compressor [1.06 um/532 nm:SHG]	Laser diode • gain-switched • mode-locked • chirp-compensated [800 nm/1.3 μm/1.5 μm]
Recent years	Ti:Sapphire laser [700 –1000 nm]	Diode-pumped YLF laser + pulse compressor [1.05 um] Laser diode + soliton-compression [1.55 μm]	Diode-pumped YLF laser [1.05 um] Fiber laser [1.55 μm]

Table 2. Evolution of laser sources for optical sampling applications.

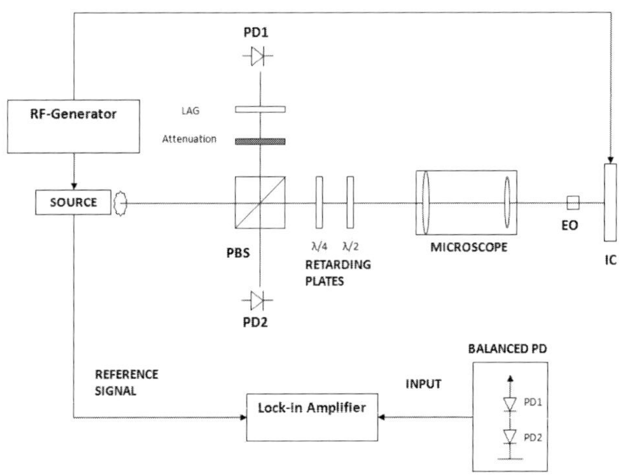

Figure 4. Block Diagram of an E-O Sampling System.

Non-invasive optical techniques, which can take advantage on the electro-optic effect in a semiconductor, are able to map the electric fields in planar III-V semiconductor circuits. Another advantage of the optical techniques is evident in the measurement of nonlinear responses.

Electronic tests work in the frequency domain and must sweep the frequency. Optical techniques work in the time domain and offer all frequency components of the Fourier transform of data from one pulse. A third advantage is the possibility of exciting or triggering devices with a light beam. This property may give additional data on the device that are not otherwise accessible.

3.1. Electro-Optic Sampling Technique

One of the most attractive optical techniques for measuring and testing electronic devices and ICs is electro-optic sampling (EOS).[90,91] This is because of the simplicity of both its principle and setup and because it offers the highest performances with respect to temporal resolution, sensitivity and invasiveness. In addition, EOS's capability to measure electrical signals internal to ICs is of great advantage in diagnosing the failure of ultrahigh-speed ICs, since circuit-design models and methodologies for handling frequency and layout-dependent parasitic have not yet been established.

This technique exploits the linear electro-optic (Pockels) effect according to which the polarization plane of an optical beam is rotated when it passes through a non-center-symmetric crystal subjected to the action of an electric field. If a linearly polarized laser beam passes through an electro-optical crystal subjected to the action of a properly oriented electric field, the intensity (I) of the transmitted laser beam analyzed after passing through a polarization filter can be written in the form:

$$I = I_0\left[1-\sin^2\left(\frac{\pi V}{V_\pi}\right)\right] \cong I_0\pi\left(\frac{V}{V_\pi}\right) \tag{3}$$

where V is the voltage applied to the electro-optical crystal associated to the electric field and V_π is a parameter, usually referred to as half-λ voltage, which depends on both the crystal properties and the beam wavelength. Materials that exhibit electro-optic effects are usually called electro-optic materials. Both silicon and germanium do not exhibit an electro-optic effect while compound semiconductors (e.g. gallium arsenide and indium phosphide) exhibit a fairly large electro-optic effect. Later on we describe how the EOS can be exploited for any semiconductor independently of its electro-optical properties.

EO sampling measurements can be performed either directly on the metallization of the DUT or, with some loss in sensitivity, through a passivation layer of typically 1 μm thickness. If a real time voltage transient reconstruction is required, a different approach is needed. The DUT is driven electrically via a periodic electrical signal applied to the

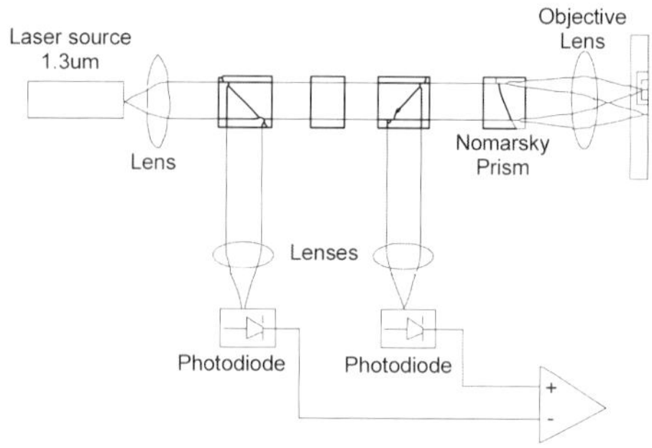

Figure 6. Optical configuration of a charge sensing probe technique.

Many experimental results indicate the sensitivity of the order of few μV/√Hz.[62] The invasiveness of this technique can be strongly reduced by using a laser wavelength that hamper the photogeneration effect into the test device. As in the direct electro-optic technique, the minimum spatial resolution of the plasma optic technique is due to the diffraction limits of the laser probe. It cannot be reduced to less than a few microns. The intrinsic time constant of the plasma optic effect is of the order of a few tens of femtoseoconds. Accordingly, the bandwidth of the plasma optic technique is dependent on the used laser source. With a c.w. laser source the bandwidth is limited by the desired signal to noise ratio. With a pulsed laser source the bandwidth is related to the laser pulse width.

The relation between the free carriers concentration and the retardation phase of the probe beam shows a very good linearity.

3.3. Other Optical Techniques

Several other optical-crystal interactions can be exploited to realize a contactless optical probing. We remember among them the electrically induced second harmonic generation, the Kerr effect and the Franz-Keldysh effect.

The electrically induced second harmonic generation effect is common to a large number of crystals. For particular polarization conditions of the incident laser beam the incident and the refracted beam on/from the surface of a crystal can interact and generate a third beam at double frequency. The generated second harmonic beam is dependent on any electric field applied to the crystal, via the third order susceptibilities as:[66,67]

$$\mathrm{PNL} = \chi^{(3)}(\omega,\omega,0,2\omega)E_\omega E_\omega E_{DC} \tag{4}$$

where E_ω is the electric field of the incident laser beam, E_{DC} is the biasing electric field applied to the crystal and $\chi^{(3)}$ is the third order susceptibility of the non-linear material.

This technique is not very invasive and its spatial resolution is diffraction limited to about 2-3 μm. The bandwidth of this technique is a function of the laser pulse width as the interaction between the probe beam and the test device is limited by the semiconductor resonance in the THz regime.

The optical Kerr effect exploits the third-order optical susceptibility of the test device as well. In this case the third-order optical susceptibility is at the fundamental frequency of the probe laser beam and the nonlinear polarizability is given by:[66,67]

$$\mathrm{PNL} = \chi^{(3)}(\omega,0,0,\omega)E_\omega |E_{DC}|^2 \tag{5}$$

This effect produces a polarization dependent phase shift in the probe laser beam, which can be detected by using a system similar to that used by Kolner and Bloom[68] for the electro-optical sampling in GaAs integrated circuits.

Franz and Keldysh theoretically predicted that the semiconductor bandgap could be shifted by means of an applied electric field. This field creates a cluster of tunneling states close to the conduction and valence bands that determines the change of the absorption characteristics near the edge of the bandgap. Wendland and Chester[69] experimentally proved that, at λ=1.06 μm, the differential absorption is a square function of the electric field. By using the Kramers-Kroning relation, it is possible to relate the differential electro-absorption to the differential electro-refraction:

$$\Delta n(\omega) \propto \frac{c}{\pi}\int_0^\infty \frac{\Delta\alpha(\omega')d\omega'}{\omega'^2-\omega^2} \tag{6}$$

that shows the relation between the differential electro-refraction and the electric field similar to that between the differential electro-absorption and the electric field ($\Delta n = aE^2$, where a, for the silicon, is a constant equal to $1{,}3\times10\text{-}15\ cm^2/V^2$). This interaction, as the Kerr effect, is not isotropic. Hence, it is possible to use a detection system similar to the direct electro optical sampling described in a previous section.

3.4. SNOM

The Scanning Near-field Optical Microscopy, which was first proposed by Binning and Rohrer at the IBM Research Center in Zurich in 1983,[84,86] is a microscope technique based on photon tunneling as opposed to the electron tunneling effect of STM (Scanning Tunneling Microscopy).

The main advantage of such a technique can be identified in the possibility of overcoming the diffraction-limit common to all the usual optical systems if the measurement is carried out in the near-field. This non-radiating component of the EM field emitted or reflected by a surface is strictly related to the sub-wavelength features of the surface itself but since it is non-radiating (the smaller the details the bigger the attenuation) the sensor must be placed very close to the DUT at a distance smaller than $\lambda/2$.

This is not usually a problem since piezoelectric actuators capable of movements in the nanometer scale are commercially available, but nonetheless, because of the typical small S/N ratio which is intrinsic to the measurement, the near field condition which grants the sub-wavelength resolution can be easily maintained only over small areas.

It is interesting to note that the possibility of having a resolution smaller than the wavelength of the radiation used can be explained by means of the Heisenberg uncertainty principle applied to the position (or equivalently the superficial details dimension) and the vector of propagation:

$$\Delta x \cdot \Delta k > 1 \tag{7}$$

Well if the vector k is real as it happens in the far-field region it can be showed that Δx cannot be smaller than:

$$\Delta x \geq \frac{\lambda}{2n\sin(\theta)} \tag{8}$$

This actually is the Rayleigh principle. But if we consider the possibility of *k* to be complex in one direction as in the case of evanescent fields, the uncertainty in the propagation vector can be far bigger so that Δx can be reduced.

Anyway, since the technique is not diffraction-limited, it is possible to use a light source in a broad spectrum, but typically visible light is preferred.

The applications of the SNOM range from local spectroscopy if a broadband or tunable source is employed, to nanometer data storage applications as well as lithography of nano-devices.

Because of its distinct features, namely the fact that the probe has to be put close to the DUT surface, SNOM is intrinsically scanning and this, together with the fact that because of the usually low S/N ratio signals involved a lock-in amplifier has to be used, is the reason why it is considered somewhat slow.

3.5. All Optical Testing with ad-hoc Structures

This approach has to be considered limited to the digital circuit testing framework since it relies on the use of some ad-hoc structures like integrated LED or Photodiodes which have to be inserted in strategic positions inside the circuit.

LEDs can be turned on and emit light depending on the voltage of the node they are bonded to, and this permits to measure the logic status of the node, while photodiodes can be used as optical signals transducers to provide a stimulus to the DUT.

The simplicity of the approach is nonetheless counterbalanced by the need of sacrificing a large chip area to realize such structures, so that, even if this technique is usually numbered among the contactless ones, it cannot be stated that it has no impact on the behavior of the electronic system.

Besides, even if the theoretical bandwidth of the system could reach the GHz regime, being it limited by the LEDs switching speed only, the low efficiency of these integrated sources, especially when realized on silicon, is the cause of the bottleneck in the readout system, since avalanche photodiodes or photon multipliers, which are intrinsically slow, have to be used.

Recent works[87,88] have estimated a real time acquisition bandwidth of around 100 kHz.

4. Measurement of Recombination Lifetime and Surface Recombination Velocity

In the characterization of semiconductor materials, among the others, two very important parameters are the bulk recombination lifetime (τ_B) and the surface recombination velocity (SRV).[70-73, 92-96] Their knowledge allows the optimization of semiconductor devices design and a direct control of the semiconductor material manufacturing process. These parameters depend on the semiconductor growth technique, doping, superficial finishing and, for any semiconductor, on the free carrier density injected in the material under measurement conditions. It is interesting to notice that contactless measurements of lifetime and surface recombination velocity are important for two reasons. The first one is that they can be easily performed at any stage of all the processes required by semiconductor technology without the need of particular test structures. In addition, the deep level spectroscopy can be noticeably improved if lifetime measurements as a function of temperature are available. As they are typically based on the use of electromagnetic radiation, contactless measurements can be performed at any temperature of the sample either high, in a furnace, or low, in a cryostat.[74]

In the previous sections we have described some contactless techniques based on different physical phenomena and we have shown how they can be employed in a large variety of practical cases.

In this section we will discuss, in some detail, the use of the previously described methods for the simultaneous measurement of the electron hole recombination lifetime and of the surface recombination velocity in semiconductor materials.

Technique	Measurable quantities	Further information
Microwave reflectivity measurements	τ_{eff}	Possibility of mapping large areas
Photoacoustic techniques	τ_{eff} τ_b S	Computer code required for the extraction of the results
Infrared absorption Measurements	τ_{eff} τ_b S	—
Infrared interferometry	τ_{eff} τ_b S	Optical finishing of both sample surfaces. Relative errors smaller than 10% in many practical cases

Table 3. Comparison between different contactless techniques for the measurement of the electron hole pairs recombination lifetime in semiconductors. τ_{eff} is the effective recombination lifetime which is a function of both the bulk recombination lifetime (τ_b) and the surface recombination velocity (S).

4.1. Typical Experimental Configurations

Theoretical considerations suggest that we must be able to measure the excess carrier density and the injection level with a very sensitive technique for an accurate monitoring of the excess carrier decay process.

In the low injection regime, due to the linearity of the problem, the system can be investigated by analyzing either its pulse responsivity or its harmonic responsivity. From a theoretical point of view the two approaches are exactly equivalent. on this line of argument, the contactless techniques for the measurement of τ_B and S can be divided in two main classes: the first one[75,76] is based on the detection of the transient evolution of the excess carrier density injected into the sample by means of a suitable laser pulse. Transient methods are attractive because they give a direct impression of the velocity of the recombination process. The second one[77-80] is a steady state technique based on the measurement of the frequency response of the semiconductor when excited by a modulated laser beam. Harmonic techniques require a simpler experimental apparatus but the final result comes only from a complex numerical analysis. In both cases the recombination process can be analyzed either by a laser beam or by a microwave beam and the physical mechanism which produces the modulation of the probe beam is the free carrier scattering. The main problems concerning these techniques are the knowledge of the injection level and separation between surface and bulk contribution to the recombination process.

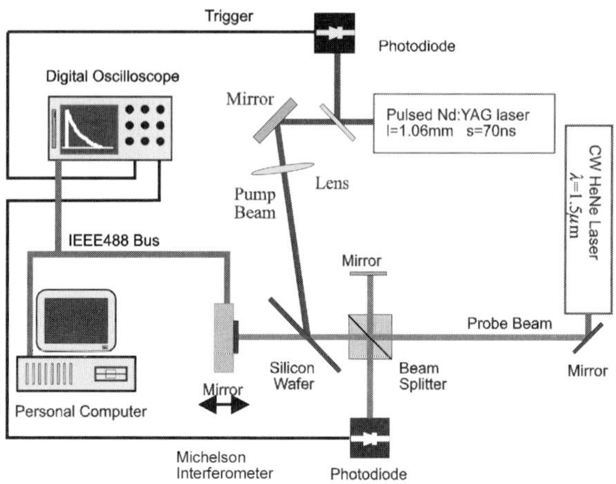

Figure 7. Experimental set-up for the optical characterization of recombination lifetime in silicon.

Separation between bulk and surface contributions in transient methods is often achieved by the so called "dual slope" technique.[77-80]

This technique is based on the identification of a change of slope in the excess carrier decay curve. In fact for a given Silicon wafer the decay curve may exhibit an evident difference between the initial slope immediately after the laser pulse and the asymptotic slope toward the end of the measured transient, when the SRV becomes high enough. Unfortunately the correct identification of the above slopes is very difficult because of the unavoidable measurement noise and on non-perfect linearity of measurement system.[81] Furthermore this technique does not take advantage of the information contained in the entire decay curve. The extraction of bulk and surface contributions from experimental results has been performed by a numerical algorithm based on a comparison between the Luke and Cheng model and experimental results. The procedure has been shown to be very effective and robust to the noise. In addition, it is possible to measure the transient carrier evolution on the same sample under different surface conditions. From an experimental point of view the contactless techniques used for the measurement of the electron hole recombination lifetime share the common feature of using a laser beam in order to generate the excess electron-hole pairs. The main difference between the contactless techniques for lifetime measurements is in the apparatus used for monitoring the responsivity of the semiconductor.

4.2. Microwave Method

In this technique, a pulsed laser beam is exploited for the generation of the excess electron hole pairs. Due to the plasma optic effect the dielectric constant of the semiconductor is a function of the free carrier concentration. Accordingly, the reflection coefficient of an electromagnetic wave incident on the under test semiconductor is a function of the free carrier concentration as well. On this line of argument, by monitoring the time evolution of the reflection coefficient it is possible to monitor a signal proportional to the free carrier concentration from which we can extrapolate both the lifetime and the surface recombination velocity. In spite of its relative simplicity, the main limit of this technique is its sensitivity and the impossibility to perform accurate measurements of the injection level which are instrumental for measuring the above mentioned parameters as a function of the free carrier concentration.

4.3. Optical Methods

After the excess electron hole pairs have been optically generated, the recombination is analyzed by monitoring the variations of either the optical absorption coefficient or the refractive index due to the plasma optic effect. The pump optical beam can be either pulsed or harmonically modulated. In the first case, the lifetime and the surface recombination velocity can be extrapolated by observing the decay curve. In the second case, where the required instrumentation is less complex than that required for the previous case, the extrapolation of the lifetime and the surface recombination velocity from experimental results requires a much more complex numerical algorithm. This circumstance can generate higher errors if the algorithm is not implemented with enough care. The advantage of the pulsed optical pumping has been recognized also in view of the possibility of directly measuring the injection level which permits accurate measurement of both lifetime and surface recombination velocity as a function of the free carrier concentration.

4.4. Other Methods

We explicitly note that, when the pump beam is periodically modulated, the measurement of lifetime and surface recombination velocity can be performed also by taking advantage on the photo-acoustic techniques

In addition, scanning electron microscopy can be used for lifetime measurements as well. Although capable of a quite good resolution, this technique requires a very involved instrumentation which cannot be used for on line or temperature dependent measurements.

4.5. Diffusivity Measurements

All the previously described techniques share the common feature of assuming the diffusivity (D) as known. In most of practical cases this is a fairly well verified assumption also in view of the fact that its typical variations do not cover a wide range a values. However, in this section we want to describe a contactless technique which allows the direct measurement of this parameter. The basic apparatus is a pump-probe configuration equal to that described in the previous section. The basic idea requires monitoring the decay of the excess electron hole pairs excited by laser beams under different focusing conditions. On this line of argument, we, first, observe that, by solving the linearized continuity equation, we can write the expression of the concentration of the excess free carriers (Δ(t)) in the form:

$$\Delta(t) = \frac{F(t)}{w_p{}^2 + 2D \cdot t} \tag{9}$$

where $F(t)$ is a function which depends only on the temporal structure of the pump laser beam without depending on its spatial characteristics, w_p is the waist of the pump laser beam and t is the time computed from the moment when the pump laser beam hits the sample under test. It is obvious that any error arising from the determination of t strictly depends on the length of the pump pulse. For this reason, pump pulses much shorter than the characteristic times of the phenomena are required. If we perform the same decay measurement with two pump laser beams characterized by different spots (let us say w_{p1} and w_{p2}) on the semiconductor surface, the diffusivity can be easily calculated by the relation:

$$\frac{\Delta_1(t)}{\Delta_2(t)} = \frac{w_{p1}^2 + 2D \cdot t}{w_{p2}^2 + 2D \cdot t} \tag{10}$$

where $\Delta_1(t)$ and $\Delta_2(t)$ have been derived from the experimental results. The validity of the previous result rests on the hypotheses that the waist of the probe laser beam is much smaller than that of the pump laser beam (at least a factor 10) and that the pump laser beam is a pure TEM_{00} Gaussian mode. While the first can be easily accomplished, the second one requires a very high quality laser. Although lasers, with an almost pure TEM_{00} Gaussian transverse mode are available on the market the previous results can be extended, if necessary, to the case of probe lasers characterized by an arbitrary number of transverse modes.

5. Thermal Characterization of Power Semiconductor Devices

The complete characterization of power electron devices can require the exact knowledge of the temperature distribution on the test device, from which the possible presence of hot spots can be detected. In this section, we describe the basic principles of infrared radiometry, which is instrumental to solve the aforementioned problem.

Infrared radiometry is a contactless characterization technique for temperature measurements where no external pump source is required.[82,83] This is a very important property because it allows to perform accurate on line and real time measurements without affecting at all the typical working conditions of the under test device. In addition, the use of appropriate microscopic equipment can provide a very high spatial resolution which is essential for applications to microelectronics.

This is mainly due to the fact that typical dimensions of electronic devices are continuously decreasing while their power dissipation tends to increase. This problem is of a particular relevance in integrated power (smart-power) devices which typically dissipate very high powers. Irregular current distributions produce a not uniform temperature distribution. The basic principle of infrared radiometry consists in the measurement of the heat irradiated by the device under test. This can be done by taking advantage on a large variety of infrared detectors or cameras. As an example, Indium-Antimonide sensors are active in the 2-5 μm range.

Radiometric measurements are based on the detection of self emitted infrared radiation. In order to get an accurate temperature mapping, the knowledge of the sample emissivity is required. Unfortunately, a typical electron device is thermally inhomogeneous and therefore the emissivity characterization can be particularly challenging. To cope with this problem the use of a uniform temperature controlled frame allows an accurate emissivity computation.

One possibility to measure the infrared radiation is to use a single IR sensor in a scanning microscope system to map the temperature on the whole surface of an electronic device, but since infrared CCD Cameras are lately available not only for military applications, their price is getting lower and can be accessible to industrial and academic research and development as well. The first advantage in using a camera-based approach is that all the thermogram is available in a single shot. Nonetheless this means that a considerable amount of time is needed for the complete readout of the CCD so that the frames per second (FPS) and hence the *real time* sampling frequency is usually limited to a maximum of some tenths of Hz.

It is important to stress the concept that we are referring to a *real time* FPS, because if the transient to be measured can be made periodic (as in non-destructive testing), it is possible to use an *equivalent time* stroboscopic technique to acquire temperature signals limited by the pixel integration time only, which usually is in the order of some μs.[89]

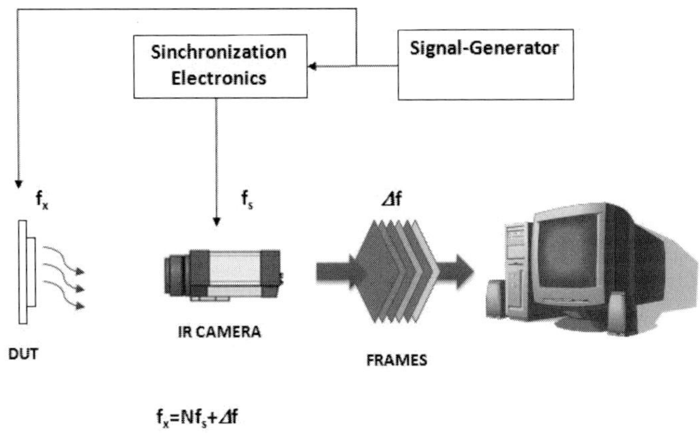

Figure 8. Layout of the high frequency IR camera based contactless temperature measurement system with an *equivalent time* approach.

This approach which is identical to that explained in the EO sampling context is based on the measurements of low frequencies replicas of the thermogram generated by the mixing of the sampling frequency and the signal repletion rate if they are not in an integer ratio between each other. Even the temperature sensitivity, which for modern day thermocameras is in the range of 10 mK, can be greatly improved if the signal is made periodic and a Lock-in acquisition technique is employed.[89]

References

1. L. Treitinger and M. Miura-Mattausch, in *Ultra fast silicon bipolar techn.*, Springer (1989).
2. G. Morou and D. Bloom, in *Picosecond electronics and optoelectron.*, Springer (1988).
3. B. Hallbach, in *High speed electronics*, Springer Verlag, (Berlin, 1992).
4. F. Capasso, in *Physics of quantum electron devices*, Springer Verlag, (Berlin, 1992).
5. G. Morou and D. Bloom, in *Picosecond electronics and optoelectron. II*, Springer Verlag (Berlin, 1991).
6. T. K. Gustaffson and P. W. Smith in *Photonic switching*, Springer Verlag (Berlin, 1993).

7. D. Schorder, in *Semiconductor material and device characterization*, J. Wiley (1990).
8. J G. Grimes, Photonics Spectra, 23, 101 (1994)
9. M. Levenson, in *Introduction to non linear spectroscopy*, Academic Press (N.Y.1982).
10. T. Yajima, K. Yoshihara and C. B. Harris, in *Ultrafast phenomena VI*, Springer Verlag (Berlin, 1993).
11. K. Izuka, in *Engineering Optics*, Springer Verlag (Berlin, 1988).
12. R. M. Azzam and N. M. Bashara, in *Ellipsometry and polarized light*, North-Holland (N.Y.1987).
13. W. R. Runyan, in *Semicond. Meas. and Instr.*, McGraw-Hill (N.Y.1975).
14. J. Attal, in *Advanced study for non destructive evaluation of semiconductors*, Plenum Press (N.Y. 1979).
15. Proceedings of the 24th ESSDERC Conf. Edinburgh Sept. 11th-15th 1994.
16. ESSDERC 95.
17. V. Privitera, W. Vandervost and V. Raineri, *J. Electrochem. Soc.,* **140**, 262 (1993).
18. C. P. Wu, E. C. Douglas and C. W. Muller, *J. Electrochem. Soc.,* **126**, 1982 (1979).
19. S. C. Choo, *Solid State Electron.*, **35**, 269 (1992).
20. J. Lagowsky, *Semiconduct. Science Technol.*, **7**, 185 (1992).
21. P. Heremans, *IEEE Trans. Electron Dev.,* **ED36**, 1318 (1989).
22. M. Schultz, *J. Appl. Phys.* **74**, 326 (1993).
23. M. H. Tsai, *Appl. Phys.Lett.*, **61**, 1691 (1992)
24. D. H. Auston, *Appl. Phys.Lett* ., **26**, 101 (1975).
25. C. W. Oatley and T. E. Everhart, *J. Electron.* **2**, 568 (1956).
26. T. E. Everhart, O. Wells, R.K. Matta, *J. Electrochem. Soc.* **111**, 929 (1964).
27. C.W. Oatley, *J. Phys.* **E2**, 742 (1968).
28. A. Gopinath, *J. Phys.* **E10**, 911 (1977).
29. H. Feuerbaum, *Electron Beam Testing: Methods and Applications*, Scanning 5, 14 (1983) FACM Inc.
30. C. T. Sah, *IEEE Trans. Nucl. Sci.* **NS-23**, 1563 (1976).
31. F. Catalano, *Scanning Electron Microscopy I*, pp.521-528 (1976).
32. L. J. Balk, H. P. Feuerbaum, E. Kubalek and E. Menzel, *Scanning Electron Microscopy J.*, 615 (1976)
33. G. Breglio, R. Casavola, A. Cutolo and P. Spirito, *IEEE Trans. on Power Electron.*, **11**, 6, 755 (1996).
34. R. Simmons, in *Optical control of microwave devices*, Artech House (London, 1993).
35. K. J. Lau, *Appl. Phys. Lett.*, **52**, 2214 (1988).
36. K. F. J. Heinrich and D. E. Newbury, "Electron Probe Microanalysis," in *Metals Handbook*, (R. E. Whan, coord.), Am. Soc. Metals (Metals Park OH, 1986).
37. C. E. Fiori and D. E. Newbury, *Scanning Electron Microsc.*, 1, 401 (1978).

38. R. B. Marcus and T. T. Sheng, in *Electron Microscopy of Silicon VLSI Circuits and Structures*, J. Wiley, (New York, 1983).
39. C. H. Spence, in *Experimental High-Resolution Electron Microscopy*, Oxford Univ. Press (Oxford, 1988).
40. D. Cherns, *High-Resolution Transmission Electron Microscopy of Surface and Interfaces*, in Analytical Techniques for Thin Film Analysis (K. N. Tu and R. Rosenberg, eds.), Academic Press (San Diego, 1988).
41. V. E. Cosslett and R. Barer, in *Advances in Opt. and Electron Microscopy*, Academic Press, (San Diego, 1988).
42. J. I. Goldstein, D. E. Newbury, P. Echlin, D. C. Joy, C. Fiori, and E. Lifshin, in *Scanning Electron Microscopy and X-Ray Microanalysis*, Plenum Press (New York, 1984).
43. K. Joardar, C. O. Jung, S. Wang, D. K. Schroder, S. J. Krause and G. H. Schwuttke, IEEE Trans. Electron Dev. ED35, 911, (1988).
44. B. G. Yacobi and D. B. Holt, *J. Appl. Phys.* **59**, 1 (1986).
45. E. Mentzel, *Microelectron. Engin.*, **16**, 3 (1992).
46. V. Kazmiruk, S. Kudriasev, V. Mordisev, *Microelectron. Engin.*, **16**, 69 (1992).
47. P. Garino, M. Battu, *Microelectron. Engin.*, **16**, 111 (1992).
48. J. Kolzer, H. Richter, *Microelectron. Engin.*, **16**, 251 (1992).
49. P. May, J-M. Halbout, G. Chiu, *Laser Pulsed E-beam System for High Speed IC Testing*, Picosecond.
50. E-beam Tester Data Sheet, Lintech Instruments Inc., Suite 1500, 655 Madison Ave, NewYork, NY 10021.
51. A. Gopinath and C. Sanger, *J. Phys. E: Sci. Instrum.* 4, 334 (1971).
52. A. Blacha, R. Clauberg, H. K. Seitz, H. Beha, *IEEE Trans. Electron Dev.* **ED-33**,1859 (1986).
53. J. Bokor, A. M. Johnson, R.H. Storz, W.M. Simpson, *Appl. Phys. Lett.* **49**, 226 (1986).
54. R. B. Marcus, A. M. Weiner, J. H. Abeles and P. S. D. Lin, *Appl. Phys. Lett.* **49**, 357 (1986).
55. D. Sarid, in *Scanning Force Microscopy*, Oxford University Press, N.Y. (1991).
56. C. H. Lee, Ed. Picosecond Optoelectronic Devices, Academic Press, New York, (1984).
57. T. Trenker, P. De Wolfe, ESSDERC 95, 477 (1995).
58. W. Vandervost, *Nucl. Instrum. Meth., B*, **96**, 123 (1995).
59. J. Allam, K. Ogawa, J. White, N. B. Baynes, J. R. A. Cleaver, I. Ohbu, T. Tanoue, and T. Mishima, OSA Proc. on Ultrafast Electronics and Optoelectronics, 14 (1993).
60. J. Kim, S. Williamson, J. Nees, S.-I. Wakana, and J. Whitaker, *Appl. Phys. Lett.* **62**, 2268 (1993).
61. Inserisci NagathsumaT. Pfeifer, H.-M. Heiliger, E. Stein von Kamienski, H. G. Roskos and H. Kurz, *J. Opt. Soc. Am. B* (1994).

62. D. Jager, G. David and W. von Wendorff, Proceedings of the 4th EOBT, (1993).
63. G. David, S. Redlich, W. Mertin, R. M. Bertenburg, S. Coblowski, F. J. Tegude, E. Kubalek and D. Jager, Proceedings of the 23rd EuMC, (1993).
64. F. Taenzler and E. Kubalek, *Microelectron. Engin.*, **16**, 325 (1992).
65. C. A. Eldering, S. T. Kowel and P. F. Brinkel, *Appl. Opt.*, **29**, 1149 (1990).
66. A. Yariv, in *Quantum Electronics*, J. Wiley and Sons (N.Y.1975).
67. Y. Shen, in *The principles of nonlinear optics*, J. Wiley and Sons (N.Y.1989).
68. B. Kolner, D. Bloom, *IEEE J.Quant. Electron.*, QE22, 134 (1986).
69. P. Wendland, M. Chester, *Phys.Rev.*, **140**, 456 (1965).
70. E. Suzuki, *IEEE Trans. Electron Dev.*, **ED36**, 1150 (1989).
71. D. K. Schroder, *IEEE Trans. Electron Dev.*, **ED31**, 462 (1984).
72. S. Daliento, N. Rinaldi, A. Sanseverino and P. Spirito, *IEEE Trans. Electron Dev.* **ED42**, 2 (1995).
73. P. Spirito, G. Cocorullo, *IEEE Trans. Electron Dev.*, **ED35**, 2546 (1987).
74. A. Sanseverino and P. Spirito, *Solid State Electron.* 37, 1429 (1994).
75. U. Bhatacharya, *IEEE Microw. and Guided Waves Lett.*, 5, 50 (1995).
76. C. Hu, *Solid State Electron.*, **21**, 965 (1978).
77. M. Schöfthaler, U. Rau, G. Langguth, M. Hirsch, R. Brendel, J. H. Werner, 12th *European Photovoltaic Solar Energy Conference*, 534, Amsterdam (1994).
78. F. Shanii, F. P. Giles, R.J. Schwartz and J. L. Gray, *S. S. Electron.*, **35**, 311 (1992).
79. T. Otaredian, *Solid State Electronics*, **36**, 153 (1993).
80. K. L. Luke and L. Cheng, *J. App. Phys*, **61**, 2282 (1987).
81. M. Schöfthaler and R. Brendel, *J. Appl. Phys*, **77**, 3162 (1995)
82. G. Breglio, F. Frisina, A. Magrì and P. Spirito, ISPSD'99, (1999).
83. G. Breglio, N. Rinaldi and P. Spirito, *Microelectronic*, **31**, 9, 753 (2000).
84. G. Binning and H. Rohrer, *Helv. Phys. Acta*, **55** 726-35 (1982).
85. U. Dürig, D. W. Pohl and H. Rohrer *J. Appl. Phys.* **59** 3318-27 (1986).
86. D. Courjon and C. Bainier, *Rep. Prog. Phys.* **57**, 989-1028 (1994).
87. S. Sayil, D. V. Kerns and S. E. Kerns *IEEE Trans. on Instr. and Meas.* 54/5 (2005).
88. J. J. Brown, J. T. Gardner and S. R. Forrest, *IEEE J. Quant. Electron.* **29** 715-26 (1993).
89. O. Breitenstein, M. Langenkamp, in *Lock-in Thermography*, Springer (2003).
90. A. Cutolo, G. Breglio, *IEEE Transaction on Instr. and Meas.*, IM43, 7-13 (1994).
91. G. Breglio, A. Cutolo, L. Zeni, F. Corsi, D. De Venuto, G. Portacci, *Optics Comm.*, **111**, 276 (1994).
92. R. Bernini, A. Cutolo, A. Irace, P. Spirito, L. Zeni, *Alta Frequenza*, **7**, 72 (1995).
93. R. Bernini, A. Cutolo, A. Irace, P. Spirito, L. Zeni, *S. S. Electron.*, 39, 1165 (1996).
94. A. Cutolo, A. Irace, P. Spirito and L. Zeni, *App. Phys. Lett.*, **71**, 1691 (1997).
95. A. Cutolo, S. Daliento, A. Sanseverino, G. F. Vitale, L. Zeni, *Solid State Electron.*, **42**, 1035 (1998).
96. A. Irace, L. Sirleto, G.F. Vitale, A. Cutolo, L. Zeni, J. Horzel and J. Szlufcik, *Solid State Electron.*, **43** (12), 2235 (1999).

INDEX